TRENDS IN EUROPEAN FOREST TREE PHYSIOLOGY RESEARCH

Tree Physiology

VOLUME 2

Series Editor:
Professor Heinz Rennenberg, *University of Freiburg, Germany*

Aims and scope

The series **"Tree Physiology"** is aimed to cover recent advances in all aspects of the physiology of woody plants, i.e.: environmental physiology including plant-atmosphere, plant-pedosphere/hydrosphere, and organismic interactions; whole tree physiology including storage and mobilization as well as maturation and senescence; woody plant tissue culture and molecular physiology for micropropagation, transformation of chemicals, preservation of gen-pools, membrane transport, transformation of tree species, etc. In addition, technological advances in tree physiology will be covered together with the achievements obtained with these techniques.
Although the main emphasis will be on basic research, applied questions will also be addressed. It is the scope of the series to provide advanced students in forestry and plant biology as well as researchers working in the various fields of tree physiology and woody plant physiology. Since also applied questions will be addressed, part of the series will also be of interest for environmental and applied biologists.

TRENDS IN EUROPEAN FOREST TREE PHYSIOLOGY RESEARCH

Cost Action E6: EUROSILVA

Edited by

SATU HUTTUNEN
Botany Division, Department of Biology, University of Oulu, Oulu, Finland

HANNELE HEIKKILÄ
Thule Institute, University of Oulu, Oulu, Finland

JÜRG BUCHER
Swiss Federal Institute for Forest, Snow and Landscape Research, Zürich, Switzerland

BJÖRN SUNDBERG
Department of Forest Genetics and Plant Physiology,
Swedish University of Agricultural Sciences, Umeå, Sweden

PAUL JARVIS
Institute of Ecology and Resource Management,
University of Edinburgh, Edinburgh, United Kingdom

and

RAINER MATYSSEK
Technical University of Münich, Münich, Germany

KLUWER ACADEMIC PUBLISHERS
DORDRECHT / BOSTON / LONDON

A C.I.P. Catalogue record for this book is available from the Library of Congress.

ISBN 1-4020-0023-5

Published by Kluwer Academic Publishers,
P.O. Box 17, 3300 AA Dordrecht, The Netherlands.

Sold and distributed in North, Central and South America
by Kluwer Academic Publishers,
101 Philip Drive, Norwell, MA 02061, U.S.A.

In all other countries, sold and distributed
by Kluwer Academic Publishers,
P.O. Box 322, 3300 AH Dordrecht, The Netherlands.

Printed on acid-free paper

Printed in the Netherlands.

Contents

Preface

The increasing concern for the serious problems of forest decline that occurred in the Northern Hemisphere in the late 1970's and early 1980's led to an emphasis on the necessity of promoting and setting up investigations into the basic physiological mechanisms of forest trees. Since then, the concern about rapid changes has decreased along with the increase of monitored data on European forests health status. But tree physiology has faced new questions about changing climate and increasing atmospheric carbon dioxide concentrations. Advances in plant molecular biology and forest genetics have opened up new avenues in the research on forest tree physiology. At the same, time it has become evident that molecular and genetic tools give only a basis for further research on tree structure and function, which needs basic tree physiology again.

On the other hand, the problems of forest decline in Europe are not over. They are no longer discussed daily in the media, but stress is an everyday phenomenon experienced by European forest trees. For instance, in southern Europe and mountainous regions, drought stress and many other abiotic or biotic factors are stressors and cause problems to forests with many important social and protective functions. Stress physiology is a branch of everyday physiology in traditional forestry. How to grow a forest with maximal carbon-binding functions and optimal wood quality and rich in biodiversity. In the northern parts of Europe, discussions on forest ecosystem biodiversity have suggested a need to have more intensive tree growth in some areas in order to ensure wildlife, recreation and biodiversity in other forest areas.

Problems related to the development and ageing of forest trees are discussed widely and the role of tropospheric ozone in the forest injuries in Europe remains obscure. The European forest health monitoring programme has so far not included tropospheric ozone questions.

This book presents some of the research topics addressed by COST Action E 6 EUROSILVA Forest Tree Physiology Research since 1995. The numbert of publications somehow related to the action was over 800 by the end of the year 1997, and preliminary calculations seem to reveal additional 1000 publications on European tree physiology. Our knowledge of this field has thus been improved. The name EUROSILVA was launched in Paris in February 1986. Now, fifteen years later, we have had a real PanEuropean network on tree physiology.

I would like to thank all contributors for their valuable work, all anonymous reviewers of the articles and all working group chairmen and editors for helping me.

Oulu, March 2001

Satu Huttunen
Professor
Chair

SECTION I

Basic Tree Physiology

CAMBIAL CELL BIOLOGY COMES OF AGE

NIGEL CHAFFEY
IACR–Long Ashton Research Station, Department of Agricultural Sciences, University of Bristol, Long Ashton, Bristol BS41 9AF, UK

Key words: cambium, models, poplar, secondary vascular system, trees, wood formation

Abstract: Although trees, and their wood in particular, are one of our most important resources, study of their cell biology is difficult and progress has generally been slow. However, over the last decade, that situation has changed with the advent of new technologies. This overview considers some of what is now possible regarding the cell biology of the tree, particularly of the cambium and wood formation, and identifies several 'drivers' that should encourage further research into tree biology. A major driver is the realisation that lack of fundamental knowledge of wood formation is holding back progress in the attempts of molecular biologists to engineer trees with 'better' lignin. Another is the knowledge that the model herb, *Arabidopsis*, can undergo substantial secondary thickening. A third is the demonstration that trees are amenable to the techniques of modern cell biological investigation, with focus on poplar, the model hardwood tree. Some of the 'new cambial cell biology' is briefly discussed. Adoption of *Arabidopsis* as an honorary tree, its exploitation, in tandem with poplar, and development of new methods of *in vivo* examination of cambial cell biology hold out the promise of a bright future for tree research. However, it is stressed that trees must remain the model *species* for studying wood formation with model *systems* in a supporting role.

1. INTRODUCTION

It should be possible to gauge the importance of a research area by the number of conferences it commands. Sadly, that is not always the case. Take, for example, the vascular cambium of trees. To my knowledge the only 'conference' devoted to this topic in the last 40 years was that held in March 1999 (see Savidge et al., 2000). In the same period, however, the cambium has been reviewed many times (e.g. Morel, 1960; Catesson, 1964, 1974, 1980, 1981, 1984, 1990, 1994; Philipson and Ward, 1965; Reinders-Gouwentak,

S. Huttunen et al. (eds.),
Trends in European Forest Tree Physiology Research, Cost Action E6: EUROSILVA, 3-18

1965; Phillipson et al., 1971; Srivastava, 1973; Paliwal and Sajwan, 1975; Jacquiot, 1984; Iqbal and Ghosh, 1985, Catesson and Lachaud, 1993; Iqbal, 1994, 1995; Larson, 1994; Lachaud et al., 1999). The large number of reviews suggests that the tree cambium is important, the small number of conferences implies that its importance is not sufficiently appreciated.

Is the tree cambium really important? Yes! As the meristem that generates secondary phloem ('inner bark') and secondary xylem ('wood'), it has an almost unique place in the history of Mankind, and importance to our future well-being. The marvellous utility of tree products, particularly wood, has long been associated with turning points in human development, e.g. as fire-wood to provide light, help keep us warm, and cook our food; as timber to build our ships and enable the great voyages of discovery of the 15th and 16th centuries; as a source of drugs to remove some of our suffering and cure our ills, and as wood-pulp to enable thoughts, beliefs, and ideas to be recorded on paper and widely disseminated.

Increasingly, we are concerned about the future of our planet and environmental issues such as global warming. Here, too, trees are important. Trees have a significant influence over climate, and act as tremendous sinks for one of the major greenhouse gases, carbon dioxide, which is locked away as carbon compounds, primarily in the wood. The energy thus stored in the wood can subsequently be used to generate electricity; the carbon dioxide released during the combustion process can be used to promote growth of more trees, which can later be harvested for generation of more energy.

However, it is past and current concerns of the wood-pulp/paper industry that have contributed most to creating the present climate where more widespread work on cambial cell biology can at last be contemplated. The major focus of those interested in paper production has been the lignin, which is such an integral part of wood cells, but whose removal is required for production of high-quality white paper. However, this removal is expensive and generally uses chemicals that are harmful to the environment. Thus, production of wood with reduced lignin (so there is less to remove), or with altered lignin (so that it is easier to remove), has been a highly prized goal, and has proved particularly amenable to a molecular-biological approach. The importance of this subject can readily be appreciated by the plethora of lignin reviews that have appeared over the last five years (e.g. Boudet et al., 1995; Whetten and Sederoff, 1995; Boerjan et al., 1996; Boudet and Grima-Pettenati, 1996; Campbell and Sederoff, 1996; Douglas, 1996; Boudet et al., 1997; Ros Barceló, 1997; Baucher et al., 1998; Boudet, 1998; Whetten et al., 1998; Grima-Pettenati and Goffner, 1999; Lewis, 1999; Sederoff et al., 1999; Christensen et al., 2000; Halpin et al., 2000).

2. THE KNOWLEDGE GAP

"Widespread application of this technology [genetic engineering of wood] awaits two developments: better methods of gene transfer, and a fundamental understanding of the developmental process of wood formation." (Whetten and Sederoff, 1991).

In the decade since those words were written, and since the first report of simultaneous transformation and regeneration of a forest tree species – poplar – by Filatti et al. (1987), gene transfer to trees has become almost commonplace (e.g. Tuominen et al., 1995; chapters in Klopfenstein et al., 1997). The results to date are encouraging, and molecular approaches to tree improvement are currently the focus of much research effort (e.g. Jung and Ni, 1998; Sederoff, 1999).

So, it seems that we have the better methods of gene transfer, what of the "fundamental understanding of wood formation"? The answer to this question is not so encouraging; until a very few years ago, and with only a handful of exceptions, that topic had been much neglected. Whether this neglect has been due to the difficulties – real or perceived – of working with trees (e.g. Chaffey, 1997), or the concentration of research effort on understanding the primary plant body of such model organisms as *Arabidopsis* (e.g. Somerville, 2000), is not clear. However, I believe that the scene is now set for a renaissance in the field of cambial cell biology, which will begin to provide that fundamental understanding. Why am I so optimistic? What is driving current interest in cambium?

3. DRIVERS

Several factors have recently come together to provide the right conditions for a dramatic increase in cambial cell biological work.

3.1 We need wood

Several years ago, at the beginning of the computer revolution, we were promised the 'paperless office'. Patently that has not arrived. Indeed we seem to use more paper now than before. And, generally, use of wood and wood-derived products shows no sign of decreasing. Rather, it is predicted to increase in the near future (e.g. Haygreen and Bowyer, 1996). That need has generated a requirement for more trees. With increasing pressure on land-use, this has to be met in part by more efficient use of existing tree plantations, in part by producing 'better' trees. As pressures also mount to harvest younger trees, it is essential to understand the cell biology of wood formation in order to find ways to manipulate the type of wood that is produced.

3.2 GM backlash

There is a clear economic impetus to provide trees that have altered lignin for the paper industry, and other properties that are better-suited to modern processes/needs. However, the approach that has been adopted to date, genetic engineering (genetic modification, GM), has a serious image problem, and is suffering as a result of the general backlash against the whole issue of GM crops (James et al., 1998; Poppy, 2000), which has given rise to widespread public disquiet and private misgivings.

That, together with concerns about such technical issues as 'gene silencing' (Gallie,

1998) and variation in the pattern of transgene expression (Ellis et al., 1996) should encourage attempts to identify more acceptable solutions to the problem of tree improvement. And given the tremendous variety that exists within the great range of tree types, it is likely that 'environmentally sound' alternatives already exist within the tree gene-pool.

A telling comment in the lignin review by Halpin et al. (2000) is the boost given to GM work by the natural variability in lignin in trees. However, that realisation can equally be used to justify attempts to modify trees by more traditional tree-breeding practice. In that context, it is interesting to note a recent report demonstrating that cellulose content increased and lignin decreased in a transgenic poplar tree in which the gene 4-coumarate:coenzyme A ligase (4CL) had been down-regulated (Hu et al., 1999). Interestingly, natural variation in the cellulose:lignin ratio had previously been reported in shortleaf pine by Burkart and Watterston (1968). Although these differences were found between trees growing on sites of different qualities, they were considered to be more a reflection of the differing ratio of earlywood to latewood between the trees. Whether the differences recorded by Hu et al. (1999) are related to such differences in the anatomy of the woods between wild type and transgenic trees remains to be demonstrated.

3.3 Modelling in vogue

Although the widely used *Zinnia* mesophyll cell system (McCann et al., 2000) has several shortcomings as a xylogenetic model (Chaffey, 1999a), it will – and should – continue to be used for developing hypotheses concerning xylem formation (Chaffey, 1999a). However, with the recent elevation of members of the genus *Populus* to the status 'model trees' (Stettler et al., 1996; Klopfenstein et al., 1997; Leplé et al., 2000; Tzfira et al., 2000), less reliance will need to be placed on such models in future.

Amongst the advantages of poplar are the following (see Chaffey, 1999a, 2001b for further details):

1 It has a relatively small genome.
2 A large number of expressed sequence tags (ESTs) exist (Sterky et al., 1998), opening the way for an increase in molecular-genetical study of tree biology (see 3 and 4).
3 It is susceptible to the techniques of molecular-genetical investigation, e.g. gene expression studies (Hawkins et al., 1997), characterisation of regulatory genes (Hertzberg and Olsson, 1998), and *in situ* hybridisation (Regan et al., 1999a).
4 It is readily transformable (e.g. Filatti et al., 1987; Tuominen et al., 1995; chapters in Klopfenstein et al., 1997) permitting insertion of genes for their further molecular-genetical study.
5 Its anatomy and physiology is well known (e.g. Sauter and van Cleve, 1990).
6 It is suitable for immunocytochemical investigation (e.g. reserve proteins - Sauter and van Cleve, 1990; enzymes of lignification – Šamaj et al., 1998; enzymes of cell wall polysaccharide modification – Micheli et al., 2000; cytoskeletal proteins – Chaffey and Barlow, 2000).

7 Its cell wall chemistry/biology is well studied (e.g. Simson and Timell, 1978; Viëtor et al., 1995; Ermel et al., 2000; Micheli et al., 2000).
8 It is amenable to hormone analysis (e.g. Tuominen et al., 1997).
9 It is fast-growing (e.g. Isebrands and Bensend, 1972, particularly such inter-specific hybrids as the triploid *Populus tremula* x *P. tremuloides* – Einspahr, 1984).
10 It is commercially important, especially as a biomass fuel (Douglas, 1989; Smith et al., 1997).

Although there remain good economic reasons why work is carried out on other trees, e.g., eucalyptus (Bossinger and Leitch, 2000), walnut (Label et al., 2000), and *Robinia* (Magel, 2000), concentrating on *Populus* will allow the work to be more focused, and generate the 'critical mass' of interest that has made study of *Arabidopsis* so successful (e.g. Somerville, 2000). However, it must be stressed that *Populus* is an example of only one type of tree, a non-storied, diffuse-porous, northern temperate hardwood. It is necessary to study several species to appreciate the full range of tree types, e.g. softwood; tropical; and ring-porous species.

3.4 A Multiplicity of models

Most of the cell biology that has been undertaken to date on trees has relied for the large part on use of fixed material. That approach is necessitated by the difficulties of working with the intact tree, but I am mindful that, "In order to understand the full story, it will almost certainly be necessary to improve techniques for *in vivo* examination of xylem formation..." (Barnett, 1981). Exploitation of the explant culture technique (Leitch and Savidge, 2000) will enable the cell biology of wood formation to be studied *in* semi-*vivo*. This will permit use of some of the new techniques of non-invasive imaging (e.g. Fricker and Oparka, 1999), and, more importantly, allow experimental manipulation of xylogenesis.

Another system that has largely been overlooked is that of the graft union. As Barnett and Asante (2000) point out, the development of cambium from callus here is analogous to the development of interfascicular cambium from parenchyma in stems. In particular, they show that the differentiation of callus to cambium requires both a physical (pressure) and a biochemical stimulus. This provides an obvious parallel to cambial development in the tree, which has also been linked to a positional flow of chemical signals (see section 4.3 below) and pressure exerted by the cortex and bark. The graft union thus has virtues as a 'natural' model system for studying the formation of cambium and the maintenance of its meristematic activity once formed. There are also parallels between the *Zinnia* system and the 'graft union'. The first 48 hours of *Zinnia* mesophyll cell culture are seen as a conditioning period, which increases the competence of the cells to respond to the hormones that are required for trans-differentiation to tracheary elements (McCann et al., 2000). This conditioning period is thought to be triggered as a wound response to cell isolation; wounding is a prominent feature of the grafting process.

3.5 Troublesome trees

Without doubt, from the cell biological point of view, trees are rather difficult organisms to work with (e.g. Chaffey, 1997). The general problems of studying trees can be summed up in the words of Hawkins et al. (1997), "a standard method of preparing sections for determining reporter gene expression patterns, commonly used for herbaceous species, is unsuitable for woody plants". However, a new book, which contains detailed protocols for anatomical, biochemical, and molecular-genetical techniques that have been specifically developed for the study of trees (Chaffey, 2001a), demonstrates that trees *are* amenable to study by the techniques of modern cell biology. As these techniques become better known, and more widely practised, we will generate Whetten and Sederoff's "fundamental understanding of the developmental process of wood formation".

3.6 *Arabidopsis* a model tree?

It has been known for many years that *Arabidopsis* undergoes *some* secondary thickening (e.g. Dolan et al., 1993), but its true extent has only recently been appreciated (Regan et al., 1999b). Utility of this process as a model system for understanding xylogenesis in trees is demonstrated by work using mutants of the ethylene biosynthesis-perception pathway, which has identified a role for this hormone in wood formation in *Arabidopsis* (Regan et al., 1999b). The effect of ethylene on tree xylogenesis is now being examined in hybrid aspen. Although a little uncomfortable for the purists among us, this use of *Arabidopsis*, in tandem with hybrid aspen, probably represents the way forward for the immediate future of wood research (Chaffey, 1999b). The attractions of working with *Arabidopsis* are undeniable (see below) and will permit hypothesis-*development*, while focussed hypothesis-*testing* can then be performed in the tree.

Three examples indicate the benefits that *Arabidopsis* can bring to tree work. Caño-Delgado et al. (2000) review recent work on xylem development mutants of *Arabidopsis* root. Amongst the mutants they consider are: those in which lignification occurs in cells that are not normally lignified (*eli* – ectopic lignification); others which have an increased number of xylem strands (*mux* – multiple xylem); and a group which display differences in the timing of xylem development (*tpx* – timing of protoxylem). Another interesting series of mutations relates to those with altered xylem function, screened for using their ability to transport – or not! - a membrane-impermeant fluorescent probe, pyranine. Further characterisation of all these mutants is likely to illuminate the process of xylogenesis in trees. And we must not forget that the ability to generate mutants is in addition to the existing, but largely untapped, variation within natural populations of *Arabidopsis* (Alonso-Blanco and Kornneef, 2000).

As is widely recognised, changes to the cell walls are amongst the most obvious, and economically important, events during xylogenesis. Further insights into wood formation within the tree, and controls thereover, can be envisaged by extending the work of Turner and co-workers, who have characterised several irregular xylem (*irx*) mutants of *Arabidopsis*. The collapsed mature xylem cells of such plants display a many-fold reduction in cellulose, but a lignin content that is apparently unaffected (Turner and Somerville, 1997), and should prove useful for study of cell wall synthesis during xylogenesis. Additionally, *irx3* encodes a component of cellulose synthase that appears to be specifically required for synthesis of cellulose in *secondary* cell walls (Taylor et al., 1999).

Baima et al. (2000) have used *Arabidopsis* to investigate homeobox genes (genes coding for transcription factors, which are involved in 'switching on' other genes that control particular pathways of development). One such, ATHB-2, is of particular interest because elevated levels of this gene interfere with auxin response pathways, and appear to affect secondary thickening by acting as a negative regulator of gene expression (Steindler et al., 1999).

3.7 Old fuel for the new millennium

Biomass energy, particularly that from trees, offers an environmentally-friendly, and renewable, alternative to fossil fuels. In light of present concerns about global warming and CO_2 levels, it also has the added bonus of being 'politically correct' (hence, fundable?). The likely increase in the demand for such fuels increases the need for basic information on the cell biology of wood formation, particularly in relation to sites of deposition of lignin, which has a calorific value almost twice that of cellulose (Chaffey, 2000).

3.8 Cerebral cambium

Let us also not forget one of the most important driving forces behind any scientific enquiry, the thirst for knowledge and understanding. Bailey (1952) identified six stages to xylogenesis. Combine that with the numerous different wood and cambial cell types. Add a seasonal component to cambial activity and wood-cell production. Mix in the fact that there are two differently oriented sub-systems – axial and radial. And, extend all this into the phloem tissue, and you have a very complicated system indeed. Few greater challenges exist in plant biology than to unravel the mysteries of the secondary vascular system!

4. SOME RECENT ADVANCES IN CAMBIAL CELL BIOLOGY

Competence and commitment (determination) are considerations at the heart of modern cell biology. They are also at the very centre of wood biology, since it is the daughter cells arising from cambial initials which subsequently go on to produce the wide variety of wood cells, whose unique characteristics and associations define wood properties. Knowledge of what determines which cell pathway a given daughter will follow is essential to any attempt to 'design' better wood. Such questions can be studied at many levels, for example: anatomical, biochemical, and molecular-genetical.

4.1 Anatomical

In many respects, anatomy is the 'Grand Old Man' of cambial biology, dating back at least as far as Grew (1682). However, despite its antiquity, new discoveries are still being made (e.g. it is only recently that the essential similarity of root and shoot cambia has been demonstrated at the ultrastructural level – Chaffey et al., 1997, 1998). Nevertheless, the basics have been known for a long time, and have been well covered in previous cambial reviews (cited above).

4.2 At the interface

The advent of technological advances, particularly immunolocalisation, is permitting new understanding of cambial cell biology (Chaffey, 1999a) at the interface between the anatomical and biochemical levels.

For example, examination of the cytoskeleton of cambial cells and their differentiating derivatives (Chaffey, 2000; Funada et al., 2000) is providing valuable insights into the details of the process of wood – and phloem (Chaffey et al., 2000) – formation. The involvement of both major cytoskeletal components, microtubules and microfilaments, in development of wood-cell form and wall-elaboration (Chaffey et al., 1999, 2000b) has led to the suggestion that their manipulation might facilitate development of 'designer' wood, which is better-suited to a range of uses (Chaffey, 2000). Such an idea indicates the sort of questions we can now ask in the light of application of 'new technologies' to study of cambial cell biology, and is in keeping with the view of the plant cell wall as a target for biotechnology (Chapple and Carpita, 1998).

In the context of cambial biology, it is noteworthy that the emphasis is primarily on wood formation. Whilst the importance of wood cannot be denied, it is equally important to study the phloem, since, without this cambial-derived tissue there would be very little wood to study. This is because the phloem represents the major source of photosynthate that is required to maintain cambial activity, whether as respirable substrate for energy production or as a source of precursors for biosynthesis. Unfortunately, this is such a neglected area of study that a recent review was forced to conclude that, "...we know very little about the structural and physiological environment of photosynthate transfer in the cambial zone." (Van Bel and Ehlers, 2000). In particular, they highlighted gaps in knowledge relating to the symplasmic organization of the zone from sieve elements to cambium, and information about the nature, frequencies and gating of plasmodesmata, which determine the fluxes of materials and messages towards the phloem. Recent identification of cytoskeletal arrays within long-lived ray cells in trees (Chaffey and Barlow, 2001), and their postulated role in intracellular transport may contribute to an understanding of the symplasmic route involved in photosynthate transfer here. Additionally, localisation of myosin at the pit fields of ray cells in angiosperm trees (Chaffey, 2001c; Chaffey and Barlow, 2001) suggests a way in which the plasmodesmata at those sites can be gated, thereby influencing cell-cell exchange of solutes, and development of symplasmic domains within the secondary vascular system (Lachaud et al., 1999).

4.3 Biochemical

Some of the greatest advances in cambial cell biology in recent times have come from biochemical studies of this system, particularly when applied in combination with other techniques (e.g. Catesson et al., 1995). Lack of space precludes a fuller account of this aspect of study, however, one recent advance, which offers insight into the hormonal control of secondary vascular differentiation in trees, is considered below.

Progress in understanding the role of hormones in xylogenesis in trees is summarised by Sundberg et al. (2000). Those workers present evidence that the amount of indole-acetic acid (auxin, IAA), supplied by polar transport, influences the rate of xylem production through changes in the concentration and distribution pattern of IAA across the cambial tissues (from phloem to xylem). They propose that this control operates in two ways: primarily by affecting the radial population of dividing cells/size of the cambial zone, via changes in radial width of the IAA distribution, and secondarily, and when the supply is large, by affecting the rate of cell cycling via changes in IAA concentration in cells of the cambial zone. In this way, IAA is envisaged as providing positional information. Such insight is of relevance to an appreciation of the formation of biomass, since the radial width of these zones is critical for determining the time each cambial derivative spends in each phase of differentiation/development, and hence impacts directly on tree productivity.

However, they also point out that supplies of additional IAA, as in appropriately altered transgenic poplar, which result in modification of other aspects of cambial activity (such as changes in wood-cell size and cell-type ratios – Tuominen et al , 1995), suggests that IAA has additional roles.

The success of this high resolution analysis of IAA augers well for similar work investigating the distribution of other hormones that are involved in aspects of cambial growth or secondary vascular differentiation, such as gibberellins, cytokinins and ethylene (e.g. Aloni et al., 2000). Armed with that knowledge, we should be in a much better position to appreciate the role that these substances play in the overall xylogenetic process.

4.4 Molecular-genetical

If we regard anatomy and biochemistry as the past and present, respectively, of cambial cell biology, 'molecular approaches' are very much the future. And exciting times are ahead, particularly with use of double and triple mutants in which the activity of 2 or 3 lignin biosynthetic genes is reduced (Halpin et al., 2000). Another area where progress is being made is in understanding the 'master genes' that control pathways of differentiation, such as xylogenesis (e.g. Hertzberg and Olsson, 1998). All developmental processes depend for their success upon the regulated and co-ordinated expression of many genes. In turn, those genes are 'controlled' at the gene-transcription level, which is mediated via transcription factors. Newman and Campbell (2000) recently presented evidence from loblolly pine that MYB proteins could act as transcription factors regulating some aspects of xylogenesis. Rather satisfyingly, they have identified orthologues of their pine genes in *Arabidopsis*, paving the way for use of both systems in tandem to understand this aspect of cambial activity.

5. THE SHOW MUST GO ON!

Although the performance has long been advertised, the audience is waiting, technical problems have been overcome, the players are assembled, and the scene is set for a dramatic retelling of 'the Story of Wood', one major element is missing: the Producer. Regardless of the fact that cambial cell biology can now be explored at all levels of enquiry, without the requisite funding very little can happen. It is to be hoped that the topicality of such environmental concerns over global warming and exploitation of biomass energy will generate the political will that will ensure that sufficient funds can be directed to this essential area of research. In that regard, the 'biomass initiative' of the USA (Anon, 1999) is to be welcomed.

ACKNOWLEDGEMENTS

I am grateful to all those colleagues who kindly sent me preprints of their work, particularly those whose contributions could not be included due to lack of space, and to Anne-Marie Catesson for introducing me to the idea of symplasmic domains within the secondary vascular system. I thank Prof. Huttunen for inviting me to share some of my thoughts with a wider audience. IACR receives grant-aided support from the Biotechnology and Biological Sciences Research Council of the UK.

REFERENCES

Anon, 1999. President Clinton announces biomass initiative. *Agric & E Intl* **51**: 132-134.

Aloni R, Feigenbaum P, Kalev N and Rozovsky S (2000) Hormonal control of vascular differentiation in plants: the physiological basis of cambium ontogeny and xylem evolution. In: Savidge, R. Barnett J, Napier R (eds) Cell and molecular biology of wood formation. Oxford: Bios Scientific Publishers, pp. 223-236.

Alonso-Blanco C and Koornneef M (2000) Naturally occurring variation in *Arabidopsis*: an underexploited resource for plant genetics. *Trends Plant Sci* **5**: 22-29.

Bailey, IW (1952) Biological processes in the formation of wood. Science 115: 255-259.

Baima S, Tomassi M, Matteucci A, Altamura MM, Ruberti I and Morelli G (2000) Role of the ATHB-8 gene in xylem formation. In: Savidge, R. Barnett J, Napier R (eds) Cell and molecular biology of wood formation. Oxford, Bios Scientific Publishers, pp. 445-455.

Barnett JR (1981) Current research into tracheary element formation. Comment Plant Sci 1: 161-174.

Barnett JR and Asante AK (2000) The formation of cambium from callus in grafts of woody species. In: Savidge, R. Barnett J, Napier R (eds) Cell and molecular biology of wood formation. Oxford, Bios Scientific Publishers, pp. 155-167.

Baucher M, Monties B, Montagu M and Boerjan W (1998) Biosynthesis and genetic engineering of lignin. *Crit Rev Plant Sci* **17**: 125-197.

Boerjan W, Meyermans H, Chen C, Leplé J-C, Christensen JH, van Doorsselaere J, Baucher M, Petit-Conil , Chabbert B, Tollier M-T, Mopnties B, Pilate G, Cornu D, Inzé D, Jouanin L and van

Montagu M (1996) Genetic engineering of lignin biosynthesis in poplar. In: Ahuja MR, Boerjan W, Neale DB (eds) Somatic cell genetics and molecular genetics of trees. Dordrecht, Kluwer Academic Publishers, pp. 81-88.

Bossinger G and Leitch MA (2000) Isolation of cambium-specific genes from *Eucalyptus globulus* Labill. In: Savidge, R. Barnett J, Napier R (eds) Cell and molecular biology of wood formation. Oxford, Bios Scientific Publishers, pp. 203-207 and the molecular control of cambial activity. *J Exp Bot* **50** (Supplement): 23.

Boudet A-M (1998) A new view of lignification. *Trends Plant Sci* **3**: 67-71.

Boudet A-M, Goffner D and Grima-Pettenati J (1997) Lignins and lignification: recent biochemical and biotechnological advances. *CR Acad Sci, Paris* **319**: 317-331.

Boudet A-M and Grima-Pettenati J (1996) Lignin genetic engineering. *Mol Breed* **2**: 25-39.

Boudet AM, Lapierre C and Grima-Pettenati J (1995) Biochemistry and molecular biology of lignification. *New Phytol* **129**: 203-236.

Burkart LF and Watterston KG (1968) Effect of environment on ratio of cellulose to lignin in shortleaf pine. *For Prod J* **18**: 25-28.

Campbell MM and Sederoff RR (1996) Variation in lignin content and composition. *Plant Physiol* **11**: 3-13.

Caño-Delgado A, Metzlaff K and Bevan M (2000) Genetic analysis of xylem development in the primary root of *Arabidopsis thaliana*. In Savidge, R. Barnett J, Napier R (eds) Cell and molecular biology of wood formation. Oxford, Bios Scientific Publishers, pp. 471-482.

Catesson A-M (1964) Origine, fonctionnement et variations cytologiques saisonnières du cambium de l'*Acer pseudoplatanus* L. (Acéracées). *Ann Sci Nat, bot (12e, sér)* **5**: 229-498.

Catesson A-M (1974) Cambial cells. In: Robards AW (ed.) Dynamic Aspects of Plant Ultrastructure. London, McGraw-Hill Book Co., pp. 358-390.

Catesson A-M (1980) The vascular cambium. In: Little CHA (ed.) Control of shoot growth in trees. IUFRO Workshop Proc, Maritime Forest Research Centre, Fredericton, New Brunswick, Canada, pp. 12-40.

Catesson A-M (1981) Le cycle saisonier des cellules cambiales chez quelques feuillus. *Bull Soc Bot Fr* **128**: 43-51.

Catesson A-M (1984) La dynamique cambiale. *Ann Sci Natur Bot Paris 13e Sér* **6**: 23-43.

Catesson A-M (1990) Cambial cytology and biochemistry. In: Iqbal M (ed.) The vascular cambium. Taunton, UK, Research Studies Press Ltd, pp. 63-112.

Catesson A-M (1994) Cambial ultrastructure and biochemistry: changes in relation to vascular tissue differentiation and the seasonal cycle. *Int J Plant Sci* **155**: 251-261.

Catesson A-M, Bonnemain JL, Eschrich W and Magel E (1995) The cambium and its derivative tissues: biochemical changes in relation to cell differentiation and seasonal activity. In: Sandermann H, Jr, Bonnet-Masimbert M, (eds) EUROSILVA: contribution to forest tree physiology. Paris, INRA, pp. 57-77.

Catesson A-M and Lachaud S (1993) Le cambium, structure, fonctionnement et contrôle de l'activité saisonière. *Acta Bot Gall* **140**: 337-350.

Chaffey NJ (1997) Big *is* beautiful: science (and art) in wood microscopy. Micro Today November: 12-14.

Chaffey NJ (1999a) Cambium: old challenges new opportunities. *Trees* **13**: 138-151.

Chaffey NJ (1999b) Wood formation in forest trees: from *Arabidopsis* to *Zinnia*. *Trends Plant Sci* **4**:

203-204.

Chaffey NJ (2000) Cytoskeleton, cell walls and cambium: new insights into secondary xylem differentiation. In: Savidge, R. Barnett J, Napier R (eds) Cell and molecular biology of wood formation. Oxford, Bios Scientific Publishers, pp. 31-42.

Chaffey NJ (ed.) (2001a) Wood formation in trees: and molecular cell biology techniques. Harwood Academic Publishers (in press).

Chaffey NJ (2001b) An introduction to the problems of working with the tree secondary vascular system. In: Chaffey NJ (ed.) Wood formation in trees: and molecular cell biology techniques. Harwood Academic Publishers (in press).

Chaffey NJ (2001c) Secondary growth of roots: a cell biological perspective. In: Waisel Y, Eshel A and Kafkafi U (eds) Plant roots: the hidden half, 3rd edn. Marcel Dekker (in press).

Chaffey NJ and Barlow PW (2000) Actin in the secondary vascular system of woody plants. In: Staiger C, Baluška F, Volkmann D and Barlow PW (eds) Actin: A dynamic framework for multiple plant cell functions. Dordrecht, Kluwer Academic Publishers, pp. 587-600.

Chaffey NJ and Barlow PW (2001) The cytoskeleton facilitates a three-dimensional symplasmic continuum throughout the long-lived ray and axial pasenchyma cells q trees. *Planta* (in press).

Chaffey NJ, Barlow PW and Barnett JR (1998) A seasonal cycle of cell wall structure is accompanied by a cyclical rearrangement of cortical microtubules in fusiform cambial cells within taproots of *Aesculus hippocastanum* L. (Hippocastanaceae). *New Phytol* **139**: 623-635.

Chaffey NJ, Barlow PW and Barnett JR (2000a) Structure-function relationships during secondary phloem development in *Aesculus hippocastanum*: microtubules and cell walls. *Tree Physiol* **20**: 777-786.

Chaffey NJ, Barlow PW and Barnett JR (2000b) A cytoskeletal basis for wood formation in angiosperm trees: the involvement of microfilaments. *Planta* **210**: 890-896.

Chaffey NJ, Barnett JR and Barlow PW (1997) Endomembranes, cytoskeleton and cell walls: aspects of the ultrastructure of the vascular cambium of taproots of *Aesculus hippocastanum* L. (Hippocastanaceae). *Int J Plant Sci* **158**: 97-109.

Chaffey NJ, Barnett JR, and Barlow PW (1999) A cytoskeletal basis for wood formation in angiosperm trees: the involvement of cortical microtubules. Planta 208:19-30.

Chapple C and Carpita N (1998) Plant cell walls as targets for biotechnology. *Curr Opin Plant Biol* **1**: 179-185.

Christensen JH, Baucher M, O'Connell A, van Montagu M and Boerjan W (2000) Control of lignin biosynthesis. In: Jain SM and Minocha SC (eds) Molecular biology of woody plants, Volume 1. Dordrecht, Kluwer Academic Publishers, pp. 227-267.

Dolan L, Janmaat K, Willemsen V, Linstead P, Poethig S, Roberts K and Scheres B (1993) Cellular organisation of the *Arabidopsis thaliana* root. *Development* **119**: 71-84.

Douglas CJ (1996) Phenylpropanoid metabolism and lignin biosynthesis. *Trends Plant Sci* **1**:171-178.

Douglas GC (1989) Poplar (*Populus* spp.). In: Bajaj YPS (ed.) Biotechnology in agriculture and forestry 5, Trees II. Berlin, Heidelberg, New York, Springer, pp. 300-323.

Einspahr DW (1984) Production and utilization of triploid hybrid aspen. *Iowa State J Res* **58**: 401-409.

Ellis DD, Rintamaki-Strait J, Francis K, Kleiner K, Raffa K and McCown B (1996) Transgene expression in spruce and poplar: from the lab to the field. In: Ahuja MR, Boerjan W, Neale DB (eds) Somatic cell genetics and molecular genetics of trees. Dordrecht, Kluwer Academic Publishers, pp. 159-163.

Ermel FF, Follet-Gueye M-L, Cibert C, Vian B, Morvan C, Catesson A-M and Goldberg R (2000) Differential localization of arabinan and galactan side chains of rhamnogalacturonan 1 in cambial derivatives. *Planta* **210**: 732-740.

Filatti JJ, Sellmer J, McCown B, Haissig B and Comai (1987) *Agrobacterium* mediated transformation and regeneration of *Populus*. *Mol Gen Genet* **206**: 192-199.

Fricker MD and Oparka KJ (1999) Imaging techniques in plant transport: meeting review. *J Exp Bot* **50**: 1089-1100.

Funada R, Furusawa O, Shibagaki M, Miura H, Miura T, Abe H and Ohtani J (2000) The role of cytoskeleton in secondary xylem differentiation in conifers. In: Savidge, R. Barnett J, Napier R (eds) Cell and molecular biology of wood formation. Oxford, Bios Scientific Publishers, pp 255-264.

Gallie DR (1998) Controlling gene expression in transgenics. *Curr Opin Plant Biol* **1**: 166-172.

Grew N (1682) The anatomy of plants, with an idea of a philosophical history of plants and several other lectures read before the Royal Society. New York, London, Johnson Reprint Corporation.

Grima-Pettenati J and Goffner D (1999) Lignin genetic engineering revisited. *Plant Sci* **145**: 51-65.

Guglielmino N, Liberman M, Jauneau A, Vian B, Catesson AM and Goldberg R (1997) Pectin immunolocalisation and calcium visualization in differentiating derivatives from poplar cambium. *Protoplasma* **199**:151-160.

Halpin C, Abbott J and Barakate A (2000) Investigating lignin biosynthesis using transgenic mutant plants. In: Savidge, R. Barnett J, Napier R (eds) Cell and molecular biology of wood formation. Oxford, Bios Scientific Publishers, pp. 425-436.

Hawkins S, Šamaj J, Lauvergeat V, Boudet A and Grima-Pettenati J (1997) Cinnamyl alcohol dehydrogenase (CAD): identification of important new sites of promoter activity in transgenic poplar. *Plant Physiol* **113**: 321-325.

Haygreen JG and Bowyer JL (1996) Forest products and wood science, 3rd edn. Ames, Iowa State University Press.

Hertzberg M and Olsson O (1998) Molecular characterisation of a novel plant homeobox gene expressed in the maturing xylem zone of *Populus tremula* x *tremuloides*. *Plant J* **16**: 285-295.

Hu W-J, Harding SA, Lung J, Popko JL, Ralph J, Stokke DD, Tsai C-J and Chiang VL (1999) Repression of lignin biosynthesis promotes cellulose accumulation and growth in transgenic trees. *Nature Biotech* **17**: 808-812.

Iqbal M (1994) Structural and operational specializations of the vascular cambium of seed plants. In: Iqbal M (ed.) Growth patterns in vascular plants. Portland, USA, Dioscorides Press, pp. 211-271.

Iqbal M (1995) Structure and behaviour of vascular cambium and the mechanism and control of cambial growth. In: Iqbal M (ed.) The cambial derivatives. Berlin, Gebrüder Borntraeger, pp. 1-67.

Iqbal M and Ghosh AKM (1985) Cell events of radial growth with special reference to cambium of tropical trees. In: Malik CP (ed.) Widening horizons of plant science. New Delhi, Cosmo Publications, pp. 219-252.

Isebrands JG and Bensend DW (1972) Incidence and structure of gelatinous fibers within rapid-growing eastern cottonwood. *Wood and Fiber* **4**: 61-71.

Jacquiot C (1984) Sur la physiologie du cambium des arbres forestières. *Rev For Francaise* **2**:113-121.

James RR, Difazio SP, Brunner AM and Strauss SH (1998) Environmental effects of genetically engineered woody biomass crops. *Biomass Bioenergy* **14**: 403-414.

Jung H-JG and Ni W (1998) Lignification of plant cell walls: impact of genetic manipulation. Proc

Natl Acad, Sci, USA **95**: 12742-12743 (& **96**: 795 (1999).

Klopfenstein NB, Chun YW, Kim M-S and Ahuja MR (eds), Dillon MC, Carman RC and Eskew LG (tech. eds) (1997) Micropropagation, genetic engineering, and molecular biology of *Populus*. Gen Tech Rep RM-GTR-297. Fort Collins, CO, US Dept of Agriculture, Forest Service, Rocky Mountain Forest and Range Experimental Station.

Label P, Beritognolo I, Burtin P, Dehon L, Couée I, Breton C, Charpentier J-P and Jay-Allemand C (2000) Cambial activity and xylem differentiation in walnut (*Juglans regia* L.). In: Savidge, R. Barnett J, Napier R (eds) Cell and molecular biology of wood formation. Oxford, Bios Scientific Publishers, pp. 209-221.

Lachaud S, Catesson A-M and Bonnemain J-L (1999) Structure and functions of the vascular cambium. *CR Acad Sci, Paris, Sér III Sci de la vie* **322**: 633-650.

Larson P (1994) The vascular cambium: development and structure. Berlin, Springer.

Leitch MA and Savidge RA (2000) Cell, tissue and organ culture for the study of cambial activity and wood formation – a resurgence of interest in an old technique. In: Savidge, R. Barnett J, Napier R (eds) Cell and molecular biology of wood formation. Oxford, Bios Scientific Publishers, pp. 493-512.

Leplé J-C, Pilate G and Jouanin L (2000) Transgenic poplar trees (*Populus* species). In: Bajaj YPS (ed) Biotechnology in agriculture and forestry 44: Transgenic trees. Berlin, Heidelberg, New York, Springer, pp. 221-244.

Lewis NG (1999) A 20th century roller coaster ride: a short account of lignification. *Curr Opin Plant Biol* **2**: 153-162.

McCann MC, Domingo C, Stacey NJ, Milioni D and Roberts K (2000) Tracheary element formation in an *in vitro* system. In: Savidge, R. Barnett J, Napier R (eds) Cell and molecular biology of wood formation. Oxford, Bios Scientific Publishers, pp. 457-470.

Magel E (2000) Biochemistry and physiology of heartwood formation. In: Savidge, R. Barnett J, Napier R (eds) Cell and molecular biology of wood formation. Oxford, Bios Scientific Publishers, pp. 363-376.

Micheli F, Bordenave M and Richard L (2000) Pectin methyl esterases: possible markers for cambial derivative differentiation. In: Savidge, R. Barnett J, Napier R (eds) Cell and molecular biology of wood formation. Oxford, Bios Scientific Publishers, pp. 295-304.

Morel G (1960) Physiologie du cambium. *Bull Soc Bot Fr Mémoires*, **1960**: 50-61.

Newman LJ and Campbell MM (2000) MYB proteins and xylem differentiation. In: Savidge, R. Barnett J, Napier R (eds) Cell and molecular biology of wood formation. Oxford, Bios Scientific Publishers, pp. 437-444.

Osakabe Y, Nanto H, Kawei S, Kondo Y, Fujii T, Takabe K, Katayama Y and Morohoshi N (1996) Immunocytochemical localization of phenylalanine ammonia-lyase in tissues of *Populus kitakamiensis*. *Planta* **200**: 13-19.

Paliwal GS and Sajwan VS (1975) The vascular cambium. *Botanica (silver jubilee volume)* **25**: 104-114.

Philipson WR and Ward JM (1965) The ontogeny of the vascular cambium in the stem of seed plants. *Biol Rev* **40**: 534-579.

Phillipson WR, Ward JM and Butterfied BG (1971) The vascular cambium: its development and activity. London, Chapman and Hall Ltd.

Poppy G (2000) GM crops: environmental risks and non-target effects. *Trends Plant Sci* **5**: 4-6.

Regan S, Bourquin V, Tuominen H and Sundberg B (1999a) Accurate and high resolution in situ hybridization analysis of gene expression in secondary stem tissues. *Plant J* **19**: 363-369.

Regan S, Chaffey NJ and Sundberg B (1999b) Exploring cambial growth with *Arabidopsis* and *Populus*. *J Exp Bot* **50** *(Supplement)*: 33.

Reinders-Gouwentak CA (1965) Physiology of the cambium and other secondary meristems of the shoot. In: Ruhland W (ed.) Encyclopedia of plant physiology XV (1). Berlin, Springer, pp. 1077-1105.

Ros Barceló A (1997) Lignification in plant cell walls. *Int Rev Cytol* **176**: 87-132.

Šamaj J, Hawkins S, Lauvergeat V, Grima-Pettenati J. and Boudet A. (1998) Immunolocalisation of cinnamyl alcohol dehydrogenase 2 (CAD 2) indicates a good correlation with cell-specific activity of CAD 2 promoter in transgenic poplar shoots. *Planta* **204**: 437-443.

Sauter JJ and van Cleve B (1990) Biochemical, immunochemical, and ultrastructural studies of protein storage in poplar (*Populus x canadensis* 'robusta') wood. *Planta* **183**: 92-100.

Savidge R, Barnett J and Napier R (eds) (2000) Cell and molecular biology of wood formation. Oxford, Bios Scientific Publishers, pp. 71-83.

Sederoff R (1999) Building better trees with antisense. *Nature Biotech* **17**: 750-751.

Sederoff RR, MacKay JJ, Ralph J and Hatfield RD (1999) Unexpected variation in lignin. Curr Opin *Plant Biol* **2**: 145-152.

Simson BW and Timell TE (1978) Polysaccharides in cambial tissues of *Populus tremuloides* and *Tilia americana*. I. Isolation, fractionation, and chemical composition of the cambial tissues. *Cellulose Chem Technol* **12**: 39-50.

Smith NO, Maclean I, Miller FA and Carruthers SP (1997) Crops for industry and energy in Europe. Luxembourg, European Commission Directorate General XII E-2 Agro-industrial Research Unit.

Somerville C (2000) The twentieth century trajectory of plant biology. *Cell* **100**: 13-25.

Srivastava LM (1973) Cambial activity in trees. *Arnoldia* **33**: 46-66.

Steindler C, Matteucci A, Sessa G, Weimar T, Ohgishi M, Aoyama T, Morelli G and Rubertis I (1999) Shade avoidance responses are mediated by the ATHB-2 HD-Zip protein, a negative regulator of gene expression. *Development* **126**: 4235-4245.

Sterky *et al.* (1998) Gene discovery in the wood-forming tissues of poplar: analysis of 5,692 expressed sequence tags. *Proc Natl Acad Sci, USA* **95**: 13330-13335.

Stettler RF, Bradshaw HD, Jr, Heilman PE and Hinckley TM (eds) (1996) Biology of *Populus* and its implications for management and conservation. Ottawa, NRC Research Press, National Research Council of Canada.

Sundberg B, Uggla C and Tuominen H (2000) Cambial growth and auxin gradients. In: Savidge R, Barnett J, Napier R (eds) Cell and molecular biology of wood formation. Oxford, Bios Scientific Publishers, pp. 169-188.

Taylor NG, Scheible W-R, Cutler S, Somerville CR and Turner SR (1999) The *irregular xylem3* locus of *Arabidopsis* encodes a cellulose synthase required for secondary cell wall synthesis. *Plant Cell* **11**: 769-779.

Tuominen, H. Puech L, Fink S and Sundberg B (1997) A radial concentration gradient of indolacetic acid is related to secondary xylem development in *Populus*. *Plant Physiol* **115**:577-585.

Tuominen H, Sitbon F, Jacobsson C, Sandberg G, Olsson O and Sundberg B (1995) Altered growth and wood characteristics in transgenic hybrid aspen expressing *Agrobacterium tumefaciens* T-DNA indoleacetic acid-biosynthetic genes. *Plant Physiol* **109**: 1179-1189.

Turner SR and Somerville CR (1997) Collapsed xylem phenotype of *Arabidopsis* identifies mutants deficient in cellulose deposition in the secondary cell wall. *Plant Cell* **9**: 689-701.

Tzfira T, Wang W and Altman A (2000) Genetic transformation of *Populus* toward improving plant performance and drought tolerance. In: Jain SM and Minocha SC (eds) Molecular biology of woody plants, Volume 2. Dordrecht, Kluwer Academic Publishers, pp. 135-160.

van Bel AJE and Ehlers K (2000) Symplasmic organization of the transport phloem and the implications for photosynthate transfer to the cambium. In: Savidge, R. Barnett J, Napier R (eds) Cell and molecular biology of wood formation. Oxford, Bios Scientific Publishers, pp. 85-99.

Viëtor RJ, Renard CMGC, Goldberg R and Catesson A-M (1995) Cell-wall polysaccharides in growing poplar bark tissue. *Int J Biol Macromol* **17**: 341-344.

Whetten RW, MacKay JJ and Sederoff RR (1998) Recent advances in understanding lignin biosynthesis. *Annu Rev Plant Physiol Plant Molec Biol* **49**: 585-609.

Whetten R and Sederoff R (1991) Genetic engineering of wood. For Ecol Manag 43: 301-316.

Whetten R and Sederoff R (1995) Lignin biosynthesis. *Plant Cell* **7**: 1001-1013.

PHYSIOLOGY OF CAMBIAL GROWTH, STORAGE OF RESERVES AND HEARTWOOD FORMATION

MAGEL ELISABETH
Physiologische Ökologie der Pflanzen, Institut für Botanik, Universität Tübingen, Auf der Morgenstelle 1, D-72076 Tübingen, Germany
Tel.: ++49 7071 2976941, Fax: ++49 7071 295635, E-mail: elisabeth.magel@uni-tuebingen.de

Key words: Cambial growth, conifer, deciduous trees, fats, heartwood, nonstructural carbohydrates, sapwood, transition zone, triacylglycerids, vegetative storage protein

Abstract: Wood constitutes a renewable bio-product of huge ecological and economical value. Within axes of trees, wood formation starts during cambial growth and inwards produced cells which differentiate and mature during the first growing season forming functional units. In a standing tree, living, mature wood parts serve three functions: water and mineral transport, storage of food reserves and mechanical stability. Wood is thus composed of different cell types which enable the different functional purposes and which show a highly ordered arrangement. The dominant reserve substances starch, triacylglycerids, and storage proteins, are accumulated in living parenchyma cells during favorable periods and consumed at times of demand. However, they behave differently with respect to the deposition period and the pool sizes within the cells, but show similar behavior with respect to mobilization in spring. In most tree species, the final step in the life cycle of living xylem cells, is a genetically determined, programmed cell death which is characterized by the activation of hydrolytic enzymes, gene expression and *de novo* protein synthesis. The activation of metabolic pathways leads to the formation of phenolic heartwood extractives, which are responsible for the biological, chemical and physical features of heartwood.

S. Huttunen et al. (eds.),
Trends in European Forest Tree Physiology Research, Cost Action E6: EUROSILVA, 19-32

1. INTRODUCTION

Coal, oil, and wood are the most important organic energy resources on earth. In contrast to coal and oil, wood is a renewable bio-product. Nowadays, more then 30 % of the earth's landscape (surface) is covered by woody plants, bushes, shrubs and trees. Wood constitutes up to 40 % of the total phytomass. Most of it is found (sequestered) in the subtropical and tropical forest. Based on its chemical an physical features, wood is of large economic importance and is used for a wide range of indoor and outdoor construction purposes.

From the botanical point of view, wood is a secondary plant tissue which is produced from a vascular cambium. This meristematic tissue forms cells in centripetal direction which differentiate into xylem elements. In centrifugal direction, secondary phloem elements (= inner, functional and outer, dead bark) are formed (Fig. 1). Thus, wood constitutes the bulk tissue of woody axes in gymnosperms and dicotyledonous angiosperms. Based on the very thick, highly lignified cell wall of xylem elements, wood is generally regarded as a very hard and rigid tissue. In perennial woody plants, secondary xylem serves three functions: mechanical stability, long distance water and mineral transport, and storage of reserve substances. Thus it is composed of different tissues, enabling the different functional purposes. In softwoods (gymnosperm trees), mechanical stability and water conductance is achieved by tracheids, constituting up to 95% of total volume percentage (Higuchi 1998). In hardwoods (dicotyledonous angiosperms), however, the cells are more differentiated and vessels and tracheids serve exclusively for water conductance, while fibers establish mechanical stability (Fig. 1 inset). Living parenchyma cells which show axial (only in hardwoods) and radial (in softwoods and hardwoods) orientation store reserve substances. In the latter case they are arranged in a highly coordinated way and form functional units, the rays. For proper functioning, all parenchyma cells are connected. In contrast to tracheids, vessels and fibers which die soon after their development has been completed, xylem parenchyma cells stay alive for a considerable time.

From the macroscopic and physiological point of view, wood is not a homogenous tissue. On cross sections of most tree stems, a sequence of different layers can be observed (Fig. 1). Taking away the outer bark (a very effective protection shield against biotic and abiotic threats) and the inner bark (the assimilate conducting tissue), the bulk-tissue of the stem, the wood, remains. Within the wood, a pale-colored outer zone – the sapwood – and a dark-colored inner core – the heartwood – can be distinguished. Between both, the sapwood heartwood transition zone is located, a tissue of prominent physiological features.

By definition (International Associations of Wood Anatomists, IAWA), sapwood is "living, physiological active wood of pale color" or "the portion of the wood that, in the living tree, contains living cells and reserve materials, e.g. starch". In contrast, heartwood is "non-living, usually darker colored wood, which is surrounded by sapwood", or, heartwood constitutes "the inner layers of the wood, which, in the growing tree, have ceased to contain living cells, and in which the reserve materials (e.g. starch) have been removed or converted into heartwood substances". The transition zone which in some instances is also referred to as "intermediate wood" (Anonymous 1957) is "a narrow, pale-

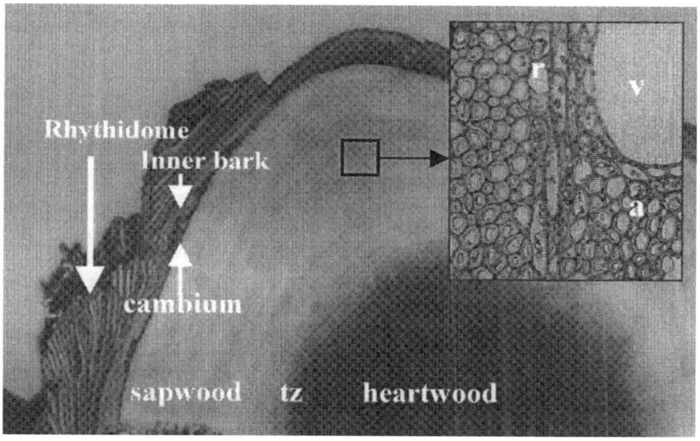

Figure 1. Cross section through the stem of a heartwood forming broad-leaved tree species. A sequence of different tissues is seen from the outer to inner stem parts: the dead rhythidome, the living inner bark (secondary phloem), the vascular cambium, the living, pale-colored sapwood, the sapwood heartwood transition zone (tz) and the dead, dark-colored heartwood. In the inset a vessel (v), ray parenchyma cells exhibiting starch grains (r), and axial tissue (a; fibers and axial parenchyma) are presented.

colored zone surrounding some heartwoods and injured regions, often containing living cells, usually devoid of starch, often impermeable to liquids, with a moisture content lower than the sapwood and sometimes also than the heartwood" (Hillis 1987). The physiological functions of wood such as water transport and food storage are restricted to the sapwood. In a living tree, the area of the sapwood before it is transformed into heartwood varies considerably between families, genera, and species, and ranges from 12 % (e.g. *Larix occidentalis*) up to 60 % in *Pinus ponderosa* (Hillis 1987). Moreover, in some species the sapwood area (or number of sapwood growth rings) is dependent on the age of the tree and of environmental conditions.

Tree species which do not normally form colored heartwood contain within the inner core wood of lower moisture content compared to sapwood and very small amounts of colored substances. This part is ripewood" (Büsgen and Münch 1929).

Moreover, there are tree species which do not from colored or uncolored heartwood at all. These sapwood trees from e.g. *Acer pseudoplatanus* or *Betula alba*, can grow up to considerable size, and contain starch and living cells up to the center of the trunk (for more sapwood tree species see Hillis 1987).

In this paper I briefly summarize the physiology of wood. I will start with cambial differentiation and describe the fate and physiological function of mature wood in storage. Finally, the last step in the life cycle of living wood cells, the formation of heartwood is addressed. The present paper illustrates the quite complex reactions involved in the secondary differentiation processes of woody axes. A more thorough understanding of these complex processes was enabled by the expertise and interdisciplinary cooperation of European scientists and turned Europe into a leading region of research in wood physiology.

2. CAMBIAL DIFFERENTIATION

The meristematic tissue, the vascular cambium, resumes activity in spring and remains active in forming tissues up to early fall. Since long, it is well accepted that phytohormones are important regulators of cambial differentiation (Roberts et al. 1988). Besides gibberellic acids which seem to be involved in the determination of the phloem/xylem ratio and formation of specific wood characteristics such as e.g. reaction wood (Waisel et al. 1966), polar transport of auxin is important for the establishment and maintenance of the vascular system (e.g. Aloni 1987). Moreover, it is proven that IAA plays an important role in defining the boundaries of the individual tissues and thus acts as a competent morphogen for their formation, both in softwoods (Uggla et al. 1996) and hardwoods (Tuominen et al. 1997). The biochemical and molecular basis of the functions of the different phytohormones which are present in the cambium and its derivatives are largely unknown. In addition to IAA, cytokinins, gibberellins, brassinosteroids and ethylene are found in a variety of cambial tissues (for review see Little and Savidge 1987, Higuchi 1998). Exogenously applied, cytokinins and gibberellins showed interactive effects with auxin such as altered mitotic activity of the cambial initials or fiber elongation. Ethylene on the other hand is assumed to impact on the lignification pattern and thus xylogenesis. In addition, minerals such as Ca^{2+}-ions are important regulators of xylogenesis. As no functional xylem is formed in the absence of phloem, it is suggested that an up to now unknown tracheid differentiation factor is needed (Savidge and Barnett 1993).

Metabolic and growth processes within the cambial tissues are highly dependent on carbon skeletons. Like in other herbaceous plants, carbon is predominantly transported in trees as sucrose (Ziegler 1975). In deciduous trees, storage components in the youngest sapwood are important carbon-suppliers. In conifers, however, net-assimilating one-year-old needles export photoassimilates which are imported into the meristematic tissue via the assimilate long distance transport paths, the phloem. In the cambial tissue of *Pinus*, sucrose and monosaccharide distribution are inverse, with highest sucrose contents in the phloem and a sharp decrease towards the cambial initials and the inwards-directed derivatives. In the latter a transient increase in monosaccharides is obvious (Uggla, Magel, Sundberg, unpublished data). In order to make carbon and energy transported as, and sequestered in, sucrose available, sucrose cleavage dominates over synthesis in the cambial differentiation zone (Sung et al. 1993, Uggla, Magel, Sundberg, unpublished data). The energetically favorable cleavage of sucrose by sucrose synthase (SuSy) dominates by far and the dominance of SuSy in the current increment in spring can be related to the high demand of cell wall material of this rapidly growing sink tissue. With maturation of the xylem tissue and reduced cambial activity toward autumn there is also a decline in SuSy activity, enzyme specific protein and transcript pools (Fig. 2; Hauch and Magel 1998). Moreover, sucrose cleaving enzyme activities show a tissue specific distribution, with a

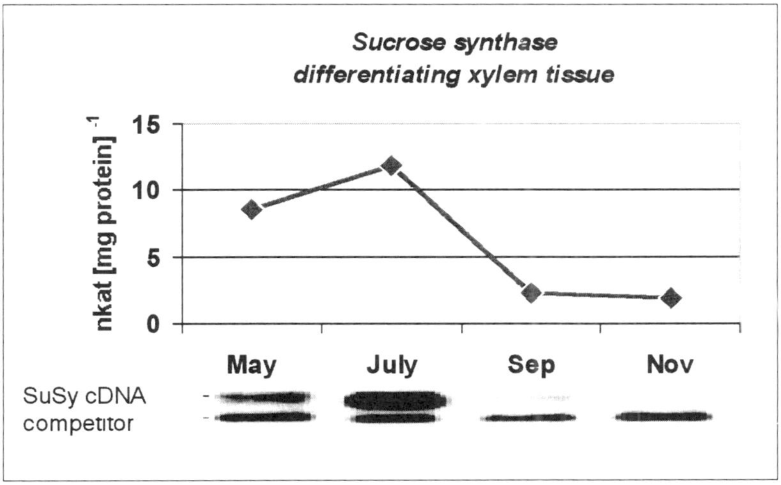

Figure 2. Catalytic activities and enzyme specific transcript of sucrose synthase (SuSy; determined by competitive RT-PCR) during xylogenesis of the differentiating inwards directed cambial derivatives during the growing season from May up to November.

relative preponderance of invertases in the cambial initials and expanding tissues (Uggla, Magel, Sundberg, unpublished data). In contrast, SuSy peaks in the phloem and in the xylem-oriented derivatives which undergo cell wall formation and is highest during vigorous cell wall lignification (Sung et al. 1993, Uggla, Magel, Sundberg, unpublished data). It is likely to speculate that SuSy is not only part of the cellulose synthase complex and thus involved in cellulose synthesis but also an important factor enabling vigorous lignification (Hauch and Magel 1998, Uggla, Magel, Sundberg, unpublished data). This shift in biochemical patterns towards the formation of cell wall phenols including the lignin-network is exhibited even in very early stages of differentiation of the xylem-oriented cambial descendents as shown in black locust (Fig. 3, Magel et al 1996). Quantitative histochemical analyses revealed a preponderance of the NADP(H)-system which coincided with increased catalytic activities of a NAD-kinase in the inwards-directed cambial derivatives. In addition, the extractable activities of the key enzymes of the oxidative pentose phosphate pathway, glucose-6-phosphate dehydrogenase and 6-phosphogluconate dehydrogenase, dominated by far in these cells. Together with ratios of NADPH:NADP of approximately 1, this is indicative for increased rates of reductive biosynthesis such as the formation of phenols and lignin monomers (Magel et al. 1996).

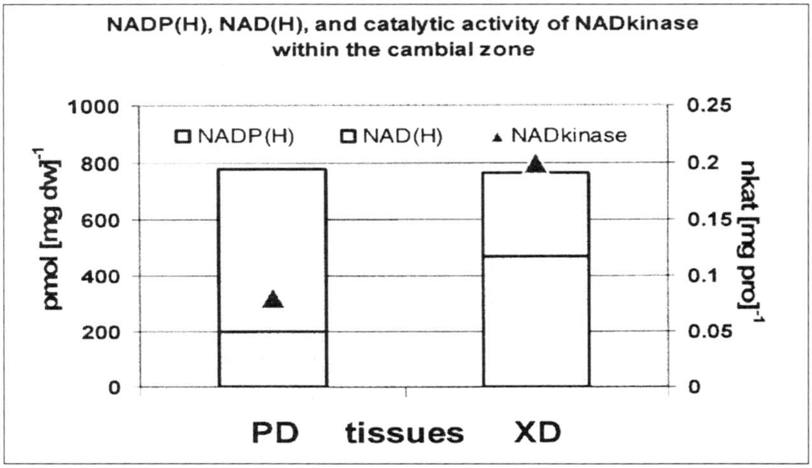

Figure 3. Pools of pyridine nucleotides, and catalytic activities of NAD-kinase in the phloem (PD) and xylem (XD) oriented cambial descendants within the cambial zone of *Robinia pseudoacacia* L. in May.

3. STORAGE OF RESERVES IN MATURE SAPWOOD TISSUES

In angiosperm and gymnosperm trees, bark and living wood parts of the trunk are the dominating storage compartments. These storage tissues are filled during favorable periods and cleared at times of demand such as leaf growth, formation of new roots and twigs, and cambial growth of trunks and roots in early spring and during summer. Concomitant with the biochemical changes of accumulation and depletion of storage compounds, ultrastructural alterations of the living xylem parenchyma cells are manifested during the seasons (for detailed information refer to Sauter and Witt 1997). Carbon is stored as nonstructural carbohydrates (starch, fructosans, soluble sugars) and fat (triacylglycerol) (Höll 1997, 2000; Sauter and Witt 1997). In addition, specific storage proteins are accumulated seasonally. According to the predominant carbon storage compounds, starch or fat, trees are classified, as "starch trees" like most ring-porous angiosperms and some conifers like species of *Abies* and *Picea* or "fat trees" like most diffuse-porous angiosperms and species of *Pinus* (Fischer 1891, Sinnott 1918).

Carbohydrates and enzymes of starch and sucrose turnover. In conifers and broad-leaved trees, starch and the soluble carbohydrates sucrose, glucose, fructose, and in some species (e.g. *Acer, Araucaria, Populus*), maltose, are the dominating carbohydrates. Considerable amounts of sucrose and its galactosides accumulate during the cold season (for a review see Magel et al. 2000). In addition, cyclitols, isocyclic polyols, which are regarded as stress metabolites, are found in all tree tissues, including the sapwood (Popp et al. 1997).

All sapwood trees show a more or less even distribution of starch and soluble sugars within the wood. In heartwood trees, however, the pool of nonstructural carbohydrates decreases steadily with increasing age of the tissue. Both, in conifers and deciduous trees, this decline follows different patterns in different species. It is more pronounced in *Robinia pseudoacacia* and *Juglans nigra*, and less pronounced in a *Juglans* hybrid (*J. major* x *J. regia*) and *Pinus sylvestris* (Magel et al. 2000). Only trace amounts are present in the sapwood heartwood transition zone and the heartwood proper is –almost– free of food reserves. In the heartwood, small amounts of sugars like arabinose (*Pinus sylvestris*) or glucose (*Juglans nigra*) seem to be released from the cell wall (Saranpää and Höll 1989) or glucosidic precursors of heartwood phenolics (Magel unpublished), respectively, during the active process of heartwood formation. In heartwood trees, starch extracted from younger sapwood tissues exhibits a higher degree of branching ($\alpha(1\rightarrow5)$ glycosyl chains of amylopectin) compared with starch found in older sapwood tissues close the heartwood boundary (Magel et al. 1997). The increasing linearization of the starch molecule results from the activity of a debranching enzyme. However, the physiological meaning neither of the enzyme nor of the highly unbranched starch in older sapwood tissues is hitherto known.

In trunks of *Robinia*, a reduced capacity for starch synthesis (e.g. catalytic activity of ADP-glucose pyrophosphorylase) together with enhanced capacities of phosphorolytic and amylolytic starch degradation (e.g. exoamylases and endoamylases) account for the decline in starch with ageing of the sapwood tissues. It is obvious, that in the younger wood tissues, starch breakdown is controlled by starch phosphorylases and exoamylases whereas in the inner tissues, endoamylases dominate (Magel et al. 1997).

Like in herbaceous plants, sucrose turnover in living wood tissues is catalyzed by the activity of the sucrose-metabolizing enzymes, sucrose-phosphate synthase (SPS), sucrose synthase (SuSy), and, dependent on the species, by acid or neutral invertases. Across a stem-transect, the distribution of sucrose-synthesizing and sucrose-cleaving activities is inverse: SPS is preferably located in the storage part of the sapwood, the mature middle and inner wood, whereas the sucrose-cleaving activities dominate in the outermost and the innermost living wood tissues (Hauch and Magel 1998).

In sapwood trees as well as in heartwood trees, in conifers like in deciduous trees, starch and sucrose constitute the dominating nonstructural carbohydrates throughout the year (Höll 1997, 2000; Magel 2000, Magel et al. 1997, 2000; Sauter and Witt 1997, and references therein). In winter, soluble carbohydrates such as glucose, fructose, sucrose, and its galactosides accumulate at the expense of starch and act as effective cryoprotectants (Guy et al. 1992). During this starch-sucrose interconversion, the pool sizes of starch are adjusted to the environmental conditions by the activity of starch synthesizing (AGPase, starch synthase) and hydrolyzing (e.g. α-amylases) enzyme activities, with maltose as a regulating agent (for more detail see Sauter and Witt 1997). Seasonal fluctuations of starch, however, differ in conifers and deciduous tree species. In conifers, the winter-minimum in starch is followed by conspicuous starch formation, both on the expense of soluble carbohydrates and from current photoassimilates (Fig. 4, Höll 1997, 2000). As the latter provide the dominant carbon-source for newly forming tissues by the cambium, starch pools are more or less constant up to the end of May/June. Thereafter, at the end of shoot elongation and needle expansion, starch contents within the sapwood decrease up to the beginning of dormancy due to a balance between degradation and synthesis of reserves (Höll 1997). In deciduous trees, however, starch is resynthesized in early spring. This starch pool is depleted during bud break and reinitiation of cambial activity and sustains cambial growth processes in spring and early summer. A continuous deposition of starch in the sapwood starts immediately after leaves are fully developed and lasts until the yellowing of leaves in fall (Fig. 4, Sauter and Witt 1997, Höll 2000, Magel et al. 2000). In all sapwood tissues under investigation, the observed starch-sugar and sugar-starch interconversion are enabled by the close interlinkage of enzymes of the starch and sucrose (hexose) turnover.

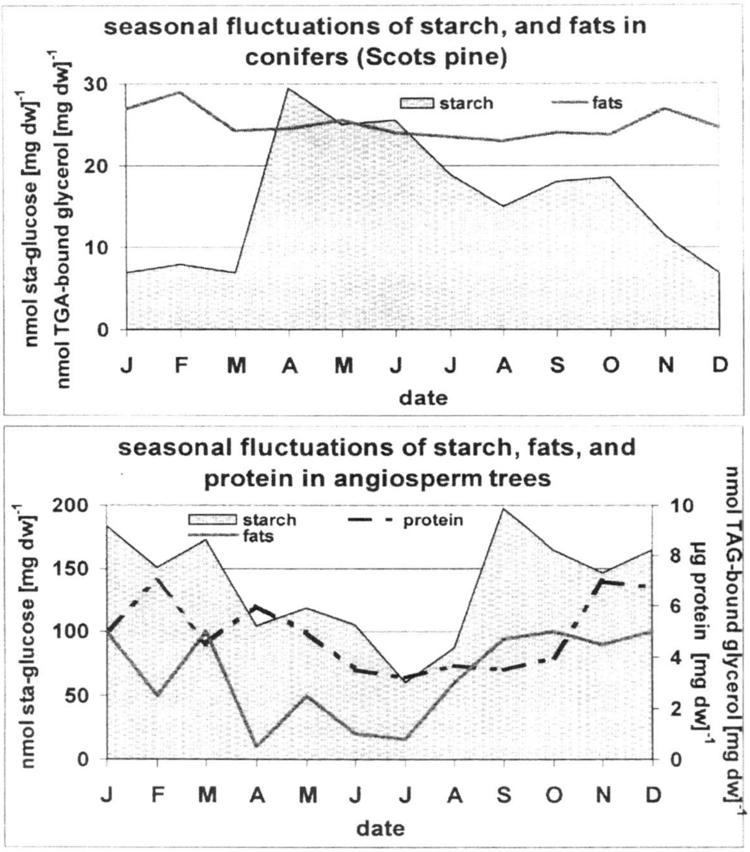

Figure 4. Seasonal fluctuations of the dominating reserve substances starch, triacylglycerols (TAG, or fats; expressed as TAG-bound glycerol), and protein in the sapwood of a conifer, Scots pine, *Pinus sylvestris* (upper panel; after Höll 1997) and deciduous trees (lower panel, after Sauter and Witt 1997, Magel unpublished data).

Air pollution affects the radial distribution and the seasonal fluctuations of nonstructural carbohydrates in wood. In declining Norway spruce, pools of nonstructural carbohydrates are elevated in the bark, whereas starch contents in the sapwood are decreased when compared with healthy trees. Altered storage processes are a consequence of changed carbon allocation between bark and wood tissues and thus of reduced radial transport capacities of ray cells. Imbalanced ratios of soluble carbohydrates and starch point towards either impaired starch synthesis or higher sensitivity of the woody tissues towards climatic changes (Magel et al. 1995a). Elevated CO_2 concentrations, however, lead to species specific changes in the total nonstructural carbohydrate pool (see references in Magel et al. 2000).

Lipids, fats and lipase activity. In contrast to carbohydrate reserves, investigations on the composition and the annual fluctuations of lipids in wood are scarce and contradictory. In the following, a brief summary on lipids is given, addressing exclusively aspects of storage lipids. For more detailed information on radial and seasonal distribution of all lipid fractions, refer to reviews given by Höll (1997, 2000).

The neutral lipid fraction of woody tissues consists of steryl esters, triacylglycerols (TAG), free fatty acids (FFA), diacylglycerols (DAG), free sterols and phospholipids (Höll 1997). Only TAG are regarded as true fats and thus can contribute to the food storage fraction of living sapwood areas and are thus particularly considered. The fatty acid moieties of TAG vary in their chain length from 14 to 24 carbon atoms and have up to three double bonds. E.g. in pine sapwood, oleic and linoleic acid are found almost exclusively (up to 90%) in the TAG accumulated. In so-called "fat-trees", TAG are the dominant food reserves. TAG are also present in the sapwood of "starch-trees". From the mass and energetic point of view, however, they constitute only a minor fraction (less than 1%). Both, in heartwood forming softwood and hardwood trees, pools of TAG decrease steadily towards the heartwood boundary and are still present in traces within the dead, colored wood tissue. Moreover, an altered double bound index in the fatty acid moieties of the TAG fraction indicates qualitative alterations. At least in conifers (genera *Pinus*), the TAG degradation coincides inversely with an increase of free fatty acids (Magel et al. 1997). In stemwood of *Robinia*, the catalytic activities of a triacylglycerol acylhydrolase (lipase) and of four phospholipases increase towards the heartwood.

Seasonal fluctuations of the lipid fraction are reported for two softwood (Scots pine, a predominantly fat storing tree, and Norway spruce, a predominantly starch storing tree; Höll 1994, 1997, 2000) and two hardwood species (linden tree, a fat-tree, Höll 2000; and poplar, a starch-tree, Sauter and Witt 1997). In general, in sapwood specimen of softwoods, TAG pools are not significantly altered from March up to August/September. Thus, this is a further proof that internal food reserves of the trunkwood are not required for cambial growth (see above). In contrast to softwoods, dramatic changes in the amounts and nature of storage lipids are found in the sapwood of linden tree from the dormant to the growing period. Long before budbreak and leaf-unfolding, fats undergo conspicuous changes, and the bulk of them are hydrolyzed. It is likely, that cleavage products are translocated to sites of carbon demand e.g. the roots, or cambial region. In the latter tissue, high lipase activity and the accumulation of sugars indicate that storage lipids are converted into sugars during cambial activity and wood formation (Höll 2000). In sapwood tissues of poplar trees, a similar seasonal trend of fats is found. However, as poplar stores predominantly starch in the wood tissues, total amounts are much lower. In both hardwood species, fats start to accumulate within the trunk, when leaves are developed and pools remain more or less constant up to the following spring (Höll 2000, Sauter and Witt 1997).

Protein. Besides nonstructural carbohydrates and storage lipids, protein constitutes the third group of storage components in living trunkwood. Typical storage proteins have been discovered for the first time in fruit trees in the seventies (Tromp and Ovaa 1973) and some years later in forest trees (for references see Sauter and Witt 1997). In the meantime, specific storage proteins are identified in a variety of gymnosperm and angiosperm trees. During times of their accumulation these vegetative storage proteins can be found in specific protein bodies. By biochemical and molecular approaches, these storage proteins are described as glycoproteins with molecular masses ranging from 32 up to 38 kDa. It is assumed that only a few typical vegetative storage proteins have evolved in individual tree species and that these polypeptides are closely related between different species, genera or even families.

Storage protein pools in the sapwood exhibit a seasonal pattern clearly different from that of starch or lipids. The rapid and prominent decrease in spring, parallels the mobilization of starch and fats during the budbreak and leaf-unfolding. It seems likely, that glutamine is being predominantly formed, released into and transported in the xylem. Protein pools in the sapwood are low during summer and resequestration of protein starts with the beginning of yellowing of leaves (Sauter and Witt 1997). In contrast to other storage protein syntheses which are controlled by the photoperiod, protein synthesis in wood tissues is initiated by increased levels of nitrogenous components.

4. MOLECULAR AND BIOCHEMICAL BASIS OF HEARTWOOD FORMATION

Heartwood formation in the innermost living wood tissues of a standing tree is a kind of programmed cell death and represents the final step in the life cycle of living xylem and ray parenchyma cells. In the transition zone between sapwood and heartwood, cells die genetically determined and phenolic heartwood extractives are accumulated. They determine the biological, chemical and physical properties of wood and thus ascribe for its ecological and economical value. Despite the importance of heartwood formation, only little was known up to now about the origin of these substances. Based on very recent findings, at least two different types of heartwood formation have to be distinguished.

• Type 1 or *Robinia*-Type: This type of heartwood formation is characterized by a *de novo*, *in situ* synthesis of phenolic heartwood extractives within the sapwood heartwood transition zone. The sapwood is almost free of any kind of phenols; neither glycosylated nor deglycosylated precursors are present.

• Type 2 or *Juglans*-Type: With ageing of the sapwood tissue, glycosylated precursors of the heartwood phenolics accumulate. Characteristic heartwood extractives are formed within the sapwood heartwood transition zone, both by transformation of the precursors (oxidation, hydrolysis) and by *in situ* biosynthesis (Magel 2000 and references therein).

In the following, a brief summary of the rather well characterized Type 1 or *Robinia*-

Type of heartwood formation is given. For detailed information refer to Magel (2000) and references therein.

Cell death in the sapwood heartwood transition zone starts in July and lasts until mid winter (November/December). Physiologically, it starts with the hydrolytic cleavage of storage substances, starch and fats, enabled by enhanced activities of endoamylases and triacylglycerol acylhydrolases (lipases). In addition, sucrose is imported from outer source tissues into this dying sink tissue and is primarily cleaved by sucrose synthase with limited contribution of the activity of a neutral invertase. As catalytic activities, enzyme specific protein pools and SuSy-specific transcripts are highly correlated, regulation of this enzyme is by gene expression (Fig. 5). In the beginning, products of sucrose cleavage are predominantly used for glycolytic and respiratory energy production and to a minor extent, for the synthesis of extractives. Later on, respiratory metabolic pathways are blocked and carbon-skeletons are increasingly used for the *in situ* synthesis of heartwood phenolics via the general phenyl propanoid metabolism. Concomitantly, a transient activation (phenylalanine ammonia lyase, PAL) and expression (chalcone synthase, CHS) of key enzymes of the biosynthesis of secondary substances is correlated with the accumulation of heartwood extractives which reach highest contents in the heartwood tissues. At least in hardwoods, the accumulation of phenolic compounds seem to involve also a mechanism of an enzymatically initiated but chemically driven copolymerization between the phenolic derivatives and the preexisting cell-wall macromolecular components including lignins and lignin-polysaccharide complexes (Magel et al. 1995b). This mechanism explains the occurrence of a relative increase in lignin content of heartwood which, however, is dependent on the biological conditions. Thus, at least in angiosperm trees, a *de novo*

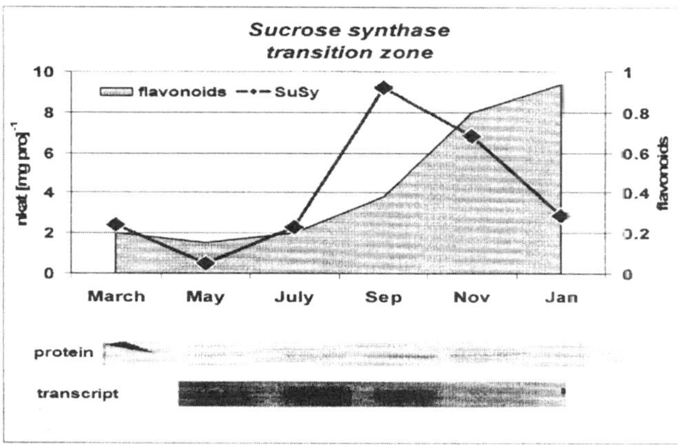

Figure 5. Catalytic activities, enzyme specific protein and transcript of sucrose synthase (SuSy) and accumulation of phenolic heartwood extractives (flavonoids) during heartwood formation in the sapwood heartwood transition zone of *Robinia pseudoacacia* L from March up to January of the following year.

synthesis of lignin during heartwood formation can be excluded. The term "secondary lignification" coined by Hergert (1977) to describe the increase in lignin content during heartwood formation, should therefore be changed into "*pseudo lignification*".

Our results point out that a concerted action of sucrose cleavage and metabolism via the sucrose synthase pathway and neutral invertase is needed to meet the high demand for energy and substrates for the *in situ* biosynthesis of heartwood extractives. This indicates that the sucrose synthase pathway is not only characteristic for metabolic sinks associated with polysaccharide biosynthesis as reported for herbaceous plants but also with the biosynthesis of phenolic compounds, heartwood extractives and lignin during heartwood formation and cambial xylem differentiation. Moreover, the type of programmed cell death that occurs during heartwood formation shares some features with e.g. the type of programmed cell death that mature leaves undergo during senescence: Both processes require gene expression, *de novo* protein synthesis and activation of hydrolytic enzymes.

For the second type of heartwood formation, the *Juglans*-Type, up to now, only few data are available. From the few publications available it seems likely that this type of heartwood occurs in genera such as *Juglans* (Burtin et al. 1998), *Pinus*, *Pseudotsuga* (Dellus et al. 1997) and *Platycarya* (Tanaka et al. 1998). Whether parts of the above given pathways can also be applied to this type of heartwood formation has still to be shown.

Owing to the presence of secondary substances, one of the most important features of heartwood is its natural durability. Searching for biotechnological approaches to alter wood properties, target genes are those which are differentially expressed during heartwood formation. RAPD-PCR techniques can be chosen as a tool not only for scanning the plant genome for polymorphism in DNA fingerprinting of species and hybrids, but also for detecting nonspecific changes in DNA as a result of major metabolic changes. Using this approach, we were able to show polymorphism not only between bark and wood, but also between differently aged sapwood tissues of one individual tree. By analyzing and sequencing the detected polymorphic RAPD-PCR DNA bands, more information will be available (DeFilippis and Magel 1998). Employing the method of RT-PCR and using degenerated primers for sucrose synthase (see above, marker enzyme for sink tissues and for heartwood formation), at least one gene specifically expressed during heartwood formation was identified (Fig. 5; Hauch and Magel 1998).

ACKNOWLEDGEMENTS

As far as our own results are presented we acknowledge financial support by COST E6, DAAD, EC, EUROSILVA and Deutsche Forschungsgemeinschaft. I am indebted to Prof. Dr. R. Hampp for critical reading of the manuscript.

REFERENCES

Aloni R (1987) Differentiation of vascular tissues. *Ann Rev Plant Physiol* **38**: 179-204.

Anonymous (1957) International glossary of terms used in wood anatomy: Prepared by the Int. Assoc. of Wood Anatomists. *Trop Woods* **107**: 1-36.

Büsgen M, Münch E (1929) The structure and life of forest trees, 3rd revised edn by E Münch (ed), Chapman Hall London

Burtin P, Jay-Allemand C, Charpentier JP, Janin G (1998) Natural wood colouring process in *Juglans* sp. (*J. nigra, J. regia* and hybrid *J. nigra 23 x J. regia*) depends on native phenolic compounds accumulated in the transition zone between sapwood and heartwood. *Trees* **12**: 258-264.

Dellus V, Mila I, Scalbert A, Menard C, Michon V, Herve du Penhoat C (1997) Douglas-fir polyphenols and heartwood formation. *Phytochem* **45**: 1573-1578.

De Filippis L, Magel E (1998) Differences in genomic DNA extracted from bark and from wood of different zones in Robinia trees using RAPD-PCR. *Trees* **12**: 377-384.

Fischer A (1891) Beiträge zur Physiologie der Holzgewächse. *Jahrb Wiss Bot* **22**: 73-160.

Guy CL, Huber JL, Huber SC (1992) Sucrose-phosphate synthase and sucrose accumulation at low temperature. *Plant Physiol* **100**: 502-508.

Hauch S, Magel EA (1998) Extractable activities and protein content of sucrose phosphate synthase, sucrose synthase and neutral invertase in trunk tissues of *Robinia pseudoacacia* L. are related to cambial wood production and heartwood formation. *Planta* **207**: 266-274.

Hergert HL (1977) Secondary lignification in conifer trees. Amer Soc Symp Series, Vol. 48, "Cellulose Chemistry and Technology": 227-243.

Higuchi T (1998) Biochemistry and molecular biology of wood. In: Springer Series in Wood Science, TE Timell (ed). Springer, Berlin, München.

Hillis WE (1987) Heartwood and tree exudates. Springer, Berlin, München.

Höll W (1994) Zur Physiologie verholzter Achsen. *Naturwissenschaften* **81**: 250-259.

Höll W (1997) Storage and mobilization of carbohydrates and lipids. In: Trees - Contribution to modern tree physiology, Rennenberg H, Eschrich W, Ziegler H (eds), SFB Academic Publ., The Hague, 197-211.

Höll W (2000) Distribution, fluctuation and metabolism of food reserves in the wood of trees. In: Cell and molecular biology of wood formation, Savidge R, Barnett J, Napier R (eds), BIOS, Oxford, 347-362.

Little CHA, Savidge RA (1987) The role of plant growth regulators in forest trees cambial growth. *Plant Growth Reg* **6**: 137-169.

Magel E (2000) Biochemistry and physiology of heartwood formation. In: Cell and molecular biology of wood formation, Savidge R, Barnett J, Napier R (eds), BIOS, Oxford, 363-376.

Magel E.A., Bleuel H., Hampp R. (1995a) Verteilung von Stärke und löslichen Kohlenhydraten in Stämmen unterschiedlich geschädigter Fichten (*Picea abies* L. Karst.) am Standort Schöllkopf. In: Waldschäden im Schwarzwald, Bittlingmaier L, Reinhardt W, Siefermann-Harms D (eds), Ecomed, Landsberg, 194-203.

Magel EA, Monties B, Drouet A, Jay-Allemand A, Ziegler H (1995b) Heartwood formation: biosynthesis of heartwood extractives and "secondary" lignification. In: EUROSILVA - Contribution to forest tree physiology. Sandermann H, Bonnet-Masimbert M (eds.), INRA edition, Paris, 35-56.

Magel EA, Bleuel H, Hampp R, Ziegler H (1996) Pyridine nucleotide levels and activities of dehydrogenases in cambial derivatives of *Robinia pseudoacacia* L. *Trees* **10**: 325-330.

Magel E, Einig W, Hampp R (2000) Carbohydrates in Trees. In: Carbohydrate reserve in plants - Synthesis and regulation, Gupta AK, Kaur N (eds), Elsevier Amsterdam, 317-336.

Magel EA, Hillinger C, Höll W, Ziegler H (1997) Biochemistry and physiology of heartwood formation: Role of reserve substances. In: Trees - Contribution to modern tree physiology, Rennenberg H, Eschrich W, Ziegler H (eds), SFB Academic Publ., The Hague, 477-506.

Popp M, Lied W, Bierbaum U, Gross M, Große-Schulte T, Hams S, Oldenettel J, Schüler S, Wiese J (1997) Cyclitols – stable osmotica in trees. In: Trees - Contribution to modern tree physiology, Rennenberg H, Eschrich W, Ziegler H (eds), SFB Academic Publisher, The Hague, 257-270.

Roberts LW, Gahan PB, Aloni R (1988) Vascular differentiation and plant growth regulators. Springer, Berlin Heidelberg New York, 154 pp.

Saranpää P, Höll W (1989) Soluble carbohydrates of *Pinus sylvestris* L. sapwood and heartwood. *Trees* **3**: 138-143.

Sauter JJ, Witt W (1997) Structure and function of rays: storage, mobilization, transport. In: Trees - Contribution to modern tree physiology, Rennenberg H, Eschrich W, Ziegler H (eds), SFB Academic Publisher, The Hague, 177-195.

Savidge RA, Barnett JR (1993) Protoplasmic changes in cambial cells induced by a tracheid-differentiation factor from pine needles. J Exp Bot 44: 395-405.

Sinnott EW (1918) Factors determining character and distribution of food reserves in woody plants. *Bot Gaz* **66**: 162-175.

Sung SS, Kormanik PP, Black CC (1993) Vascular cambial sucrose metabolism and growth in loblolly pine (*Pinus taeda* L.) in relation to transplanting stress. *Tree Physiol* **12**: 243-258.

Tanaka T, Jiang ZH, Kouno I (1998) Distribution of ellagic acid derivatives and a diarylheptanoid in wood of *Platycarya strobilacea*. *Phytochem* **47**: 851-854.

Tromp J, Ovaa IC (1973) Spring mobilization of protein nitrogen in apple bark. *Physiol Plant* **29**: 1-5.

Tuominen H, Puech L, Fink S, Sundberg B (1997) A radial concentration gradient of indoleacetic acid is related to secondary xylem development in *Populus*. *Plant Physiol* **115**: 577-585.

Uggla C, Moritz T, Sandberg G, Sundberg B (1996) Auxin as a positional signal in pattern formation in plants. Proc Natl Acad Sci USA 93: 9282-9286.

Waisel Y, Noah I, Fahn A (1966) Cambial activity in *Eucalyptus camaldulensis* Dehn II. The production of phloem and xylem elements. *New Phytol.* **65**: 319-324.

Ziegler H (1975) Phloem transport. Nature of transported substances. In: Encyclopedia of Plant Physiol, new series Vol 1, Transport in plants I, Zimmermann MH, Milburn JA (eds), Springer, Berlin, 59-100.

INSIGHTS INTO BUD DEVELOPMENT AND DORMANCY IN POPLAR

A. ROHDE AND W. BOERJAN
Vakgroep Moleculaire Genetica, Departement Plantengenetica, Vlaams Interuniversitair Instituut voor Biotechnologie, Universiteit Gent, K.L. Ledeganckstraat 35, B-9000 Gent, Belgium

Dormancy is a survival strategy that enables plants or plant parts to grow synchronized with seasons and to endure periods that are unfavorable for growth. In addition, dormancy can be imposed upon plant organs to control the architecture of the whole plant in ways that are advantageous for growth and survival. Because bud dormancy is utilized for both adaptive and morphogenetic reasons, trees can adapt to a wide range of growth conditions while developing their characteristic form. Here, we will refer to dormancy as the temporary absence of visible growth of any plant structure containing a meristem (Lang, 1987).

Depending on species, organ, and ontogeny, bud dormancy may include cessation of shoot extension, formation of the bud, formation of embryonic leaves inside the bud, arrest of meristematic activity, entry into dormancy, increase in drought resistance, leaf fall, and acquisition of frost resistance. Once dormancy has been imposed, no apparent changes in bud morphology can be observed. Nonetheless, metabolic changes are continuous, because when chilling temperature is perceived dormancy is released gradually. Finally, growth resumes when the environmental conditions are favorable.

Dormancy is best described and regarded as a developmental process (as much as active growth is). The type, intensity, and duration of dormancy in a given organ is determined by its previous ontogeny and the dormancy-triggering process. The notion of dormancy as a developmental process (and not a state) implies that it consists of many interrelated subprocesses. The systematic dissection of the dormancy process into subprocesses may, in the end, give us a basis to distinguish mechanisms that are active during the different periods of this developmental process.

Here, we will focus on the morphological events that occur at the onset of dormancy in photoperiodically-determined species. We will give a short survey on organogenesis during active growth and introduce the molecular network that controls meristem function

S. Huttunen et al. (eds.),
Trends in European Forest Tree Physiology Research, Cost Action E6: EUROSILVA, 33-52

in *Arabidopsis thaliana*. Then, we will describe how organogenesis is arrested during dormancy induction, with particular emphasis on the development of the bud and its physiological relationships with other plant parts. Finally, a glimpse of the molecular framework underlying bud formation in poplar will be presented. We will mainly emphasize on poplar, as it has become a model tree for molecular and genetic studies notwithstanding the relatively weak dormancy in buds and seeds. Poplars exhibit an indeterminate growth when young; growth cessation occurs in response to short days and trees shed leaves at the time of bud set (Nitsch, 1957; Critchfield, 1960). Photoperiodic ecotypes, *i.e.*, subgroups within the genus with a defined critical day length according to the latitude of their natural origin, have been recognized in *Populus tremula* already in the 1940s (Sylven, 1940). For full coverage of the physiology of bud dormancy we refer to the comprehensive reviews of Samish (1954), Vegis (1964), Saure (1985), Borchert (1991), and Crabbé (1994). The still scarce molecular basis of bud dormancy has been reviewed more recently by Lang (1996), Rowland and Arora (1997) and Rohde *et al.* (2000a).

1. ORGANOGENESIS AT THE SHOOT APEX

Organogenesis takes places at the shoot apical meristem (SAM) and has a direct impact on the number and type of plant organs. Leaf-like organs are the most apparent products of a vegetative meristem, but stem tissues and axillary buds are as well derived from it. In the apical meristem itself, a set of pluripotent cells is maintained in the central zone, from which cells in the peripheral and rib zones originate (Figure 1). These cells are recruited into organogenesis and differentiate the organs that descend from the apical meristem.

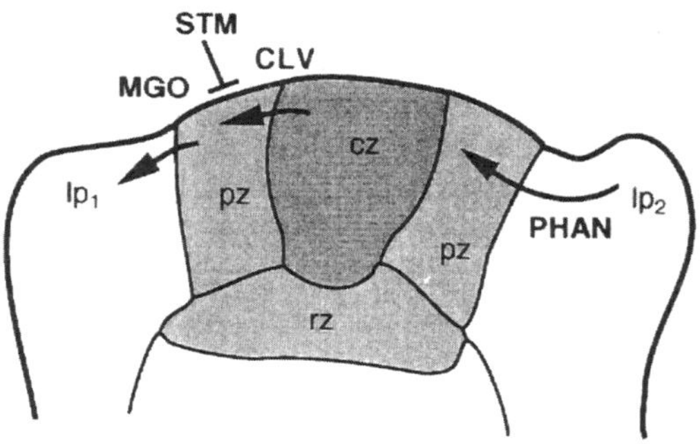

Figure 1. Genes involved in the regulation of shoot meristem activity in *Arabidopsis* and *Antirrhinum* (modified from Lenhard and Laux, 1999). CLAVATA1 and CLAVATA3 (CLV) enhance the transition of cells from the central zone (cz) to the peripheral zone (pz) of the apical meristem, a step that is counteracted by SHOOTMERISTEMLESS (STM). MGO (MGOUN) promotes the formation of leaf primordia from cells in the peripheral zone. PHANTASTICA (PHAN) is necessary for maintenance of shoot meristem activity. lp_1, last initiated leaf primordium; lp_2, penultimately initiated leaf primordium; rz, rib zone.

Besides this cell proliferation and organogenesis at the meristem, growth consists of elongation, which occurs immediately below the meristem. In many trees that exhibit determinate (fixed) growth, the timing of the formation of an organ and its elongation are separated by a period of dormancy. To facilitate such a growth interruption, perennial plants form a perennating organ, *e.g.*, a bud. A bud is a short axis bearing a densely packed series of leaf primordia that is produced by the SAM. In active buds, the axis elongates with progressive development of the primordia, whereas dormant buds do not elongate and are surrounded by protective bud scales.

1.1 Organogenesis during active growth

Through a combination of mutant analyses and *in situ* hybridization, the molecular network of meristem function has become better understood in model plants (Figure 1; Barton, 1998). The *Arabidopsis SHOOTMERISTEMLESS (STM)* gene is required for the formation of the apical meristem in the embryo, and its expression is later on restricted to those cells that form the SAM. The downregulation of *STM* mRNA is one of the earliest signs for leaf cell specification at the SAM. *STM* expression is not only extinguished from the primordium cells, but also in the cells beneath the future primordium. These cells will form the leaf trace, *i.e.*, establish the connection with the vascular system (Long and Barton, 2000). The presence of a new leaf trace that develops acropetally towards the new leaf primordium is apparent earlier than the primordium itself. The *PINHEAD* gene becomes expressed in this trace below the position where the next primordium will be initiated (Lynn *et al.*, 1999). During recruitment of meristematic cells into organogenesis in *Arabidopsis*, the CLAVATA1 and CLAVATA3 (CLV) proteins are thought to enhance the transition rate of cells from the central to the peripheral zone, a step that is counteracted by the STM protein (Laufs *et al.*, 1998a; Lenhard and Laux, 1999). On the other hand, the MGOUN protein partitions cells from the peripheral zone into organ primordia (Laufs *et al.*, 1998b). In the very young organ primordium, *AINTEGUMENTA* is expressed, being reciprocal in its expression to *STM* (Elliott *et al.*, 1996). The outgrowth and elaboration of lateral organs further involves the gene products of *PHANTASTICA* in *Antirrhinum* and *PHABULOSA* in *Arabidopsis* (McConnell and Barton, 1998; Waites *et al.*, 1998).

It is still not known when the distinct leaf domains, such as the lamina and the stipules, become specified during the formation of a leaf primordium. In maize, a model for cell fate acquisition has been put forward to explain the later differentiation steps (Freeling, 1992). Cells are predicted to go through competency stages, in which they can respond to time- and position-dependent signals. The fate is finally determined by the stage in which a particular cell receives the signal to differentiate, having received a discrete amount of the time- and position-dependent information.

In annual plants, organogenesis can be arrested when growth-limiting conditions are imposed (quiescence). However, unlike perennial plants, they are unable to temporarily redirect organogenesis at the apical meristem to create a perennating structure, such as a bud. Only perennial plants dispose of developmental strategies to modify leaf development so that bud scales are made to encase a winter bud.

1.2 Organogenesis during natural dormancy induction

Leaf expansion and internode elongation gradually cease during dormancy induction (Figure 2). In fact, cessation of elongation growth is the first visible event during dormancy induction and is primarily based on the perception of day length in many deciduous trees (Garner and Allard, 1923; Nitsch, 1957). The longest photoperiod that induces growth cessation, the critical photoperiod, is inherited as a quantitative character (Eriksson *et al.*, 1978; Junttila, 1982; Riemenschneider and McMahon, 1993). Tree species with wide geographic distributions often consist of populations, or photoperiodic ecotypes, that are adapted to different climatic and photoperiodic regimes. The molecular basis of photoperiodic ecotypes is poorly understood. However, phytochromes are involved in the perception of the critical photoperiod in poplar (Olsen *et al.*, 1997; Howe *et al.*, 1998). Moreover, changes in the concentration of active gibberellins occur during short-day-induced growth cessation prior to any visible reduction of growth (Junttila, 1990; Moritz, 1995; Olsen *et al.*, 1995).

Most leaves, which have been present as primordia at the perception of the critical day length, still expand beyond a (though ever decreasing) critical size and their laminae mature (organs 5-0; Figure 2A; Goffinet and Larson, 1981). Their laminae abscise in the normal way after bud set, although the stipules may persist and sometimes enlarge to form bud scales (organs 1 and 0; Figure 2B). Starting from the first primordium that was initiated after the perception of the critical day length (organ –1; Figures 2A and 2C), however, the laminae of the leaf primordia senesce and abscise prematurely, and the stipular domain of the leaf primordia enlarges (abortive bud-scale leaf, absl; Figure 2C).

The stipular origin of the bud scales can be inferred from their anatomy: they lack the typical leaf venation, the midrib, and the petiole. In *Populus deltoides* plants, two to five such abortive bud-scale leaves are developed under natural conditions, while only one abortive bud-scale leaf develops under controlled short-day conditions (Curtis and Lersten, 1974; Goffinet and Larson, 1981). Embryonic leaves immediately distal to the abortive bud-scale leaves do not abort their laminae (Figure 2C). Nevertheless, the first embryonic leaves may be incipient or potential bud-scale leaves because the development of their midrib is suppressed and the lamina appears broad and short (ibsl; Figure 2C; Goffinet and Larson, 1982a).

Inside the bud, organogenesis proceeds to form embryonic leaves and leaf primordia (el and lp; Figure 2C). The former develop a recognizable lamina with some vascularization, whereas the latter do not. Each embryonic leaf has two stipules, the length of which can exceed that of the leaf itself (Richards and Larson, 1981). In *Populus trichocarpa*, the embryonic leaves and leaf primordia give rise to two kinds of morphologically distinct leaves, the early and late leaves, respectively (Critchfield, 1960). However, *Populus deltoides* does not exhibit such a heterophyllous leaf expansion from the older or younger leaf primordia inside the bud (Curtis and Lersten, 1974). After a characteristic number of leaves has been initiated in the bud, the activity of the apical meristem is arrested.

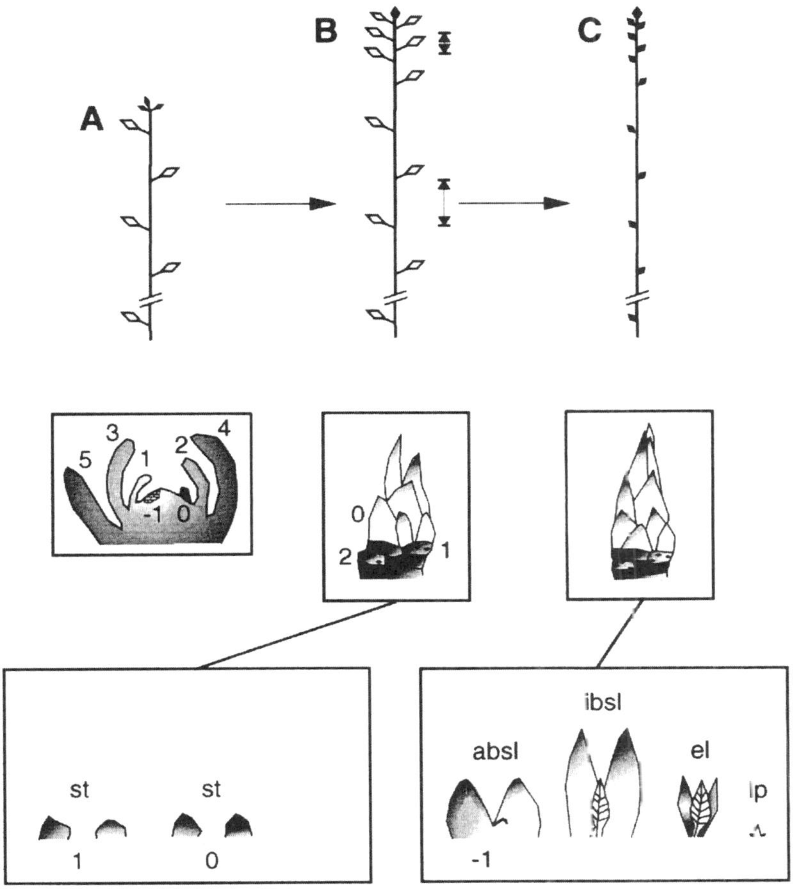

Figure 2. Schematic representation of organogenesis during dormancy induction in *Populus*. Dormancy induction is illustrated by three characteristic stages in the process: at the perception of the critical day length (A), after cessation of internode elongation and leaf expansion (B), and after completion of bud set and shedding of the leaves (C). Considerable time may elapse between these stages (see text). **(A)** At the perception of the critical day length, the apex of the plant contains already committed primordia that will develop into leaves. Only the primordia that have been initiated after the perception of the critical day length will change morphogenetic fate (organ –1 and younger ones). **(B)** Committed primordia grow out into leaves while the internode length between the leaves gradually decreases (compare the arrows next to the internodes). The laminae of these leaves abscise in the normal way after bud set (leaving a leaf scar), although the stipules (st) may persist to form bud scales (organ 1 and 0). The bud becomes apparent. **(C)** After completed bud set, the following organs are found inside the bud (in successive order from outside to inside): abortive bud-scale leaves (absl), incipient bud-scale leaves (ibsl), embryonic leaves (el) and leaf primordia (lp). These organs are derived from the first primordium that was initiated after the perception of the critical day length (organ –1) and all younger ones.

A winter bud of *Populus trichocarpa* typically contains six to nine leaf primordia (Critchfield, 1960). The formation of the dormant bud is not a rapid process. Most morphological changes are elaborated weeks after the onset of the supposed inductive short-day signal. In Belgium (50° Northern latitude), the critical day length for *Populus trichocarpa* is perceived in mid-August, but the apical bud formation is not finished before approximately two months later. Whether a particular primordium will develop into a bud scale, a foliage leaf, or a leaf primordium is not predetermined. The morphogenetic controls are rather derived from physiological influences and are constantly adjusted with the environment and the vigor of the plant. The first abortive bud-scale leaf has been suggested to originate from the first primordium that has been initiated after the critical day length was perceived (Goffinet and Larson, 1981). Its differentiation, however, is not elaborated before it attains its final position at the bottom of the developing bud. Thus, a morphogenetic fate is progressively imprinted during dormancy induction. The time elapsing during the dormancy induction may facilitate the plasticity of the process, might allow controls to be incorporated, and even signals to be re-checked, especially under a very variable environment. For example, poplar clones may repeatedly flush after the first autumnal bud set, when they are grown at higher northern latitudes than at their latitude of origin, because the specific critical day length is not reached (Ilstedt, 1996). However, after a certain point, bud set is irreversible. Rinne and van der Schoot (1998) demonstrated that in *Betula pubescens* the ability to resume growth is severely reduced after 10 short days. From the cessation of growth until the establishment of endodormancy, a considerable time may again pass by.

1.3 Organogenesis during bud set under controlled short-day conditions

Bud-set conditions can be mimicked by the application of short days. In *Populus deltoides*, complete dormancy, in which plants returned to long days and favorable growth conditions cannot spontaneously resume growth, can be achieved by a 7-week-exposure to short days (Goffinet and Larson, 1981). However, all plants set bud by the fourth week in short days (Goffinet and Larson, 1982a). Under conditions of 8 h/day photoperiod, bud set is achieved within 5 weeks in aspen (*Populus tremula x P. alba*). During the first three weeks under short days, all primordia that had been committed before the onset of short days will grow out, though internode elongation between them ceases gradually and becomes fully arrested after 3-4 weeks. During this time, the number of primordia at the shoot apex (or inside the bud) decreases until about the fourth week (Goffinet and Larson, 1981). When the leaf initiation rate slows down while the number of source leaves is kept constant, transiently more sugar will be available in the apical meristem. After the fourth week, the first bud-scale leaves become visible and the number of primordia increases again to reach the typical number of organs in the bud for a given species. As under natural conditions, the abortive bud-scale leaf is derived from the first primordium initiated after the onset of short days (Figure 2C; organ –1 in Figure 3A). The last leaf that had been committed before the onset of short days, is the last to mature, often not to its full size and to subtend the bud (organ 0 in Figure 3A; Goffinet and Larson, 1981). Similarly to what happens under natural conditions, the next primordia are elaborated along a morphogenetic gradient: as abortive and incipient bud-scale leaves, embryonic leaves, and leaf primordia (Figure 2C; Goffinet and Larson, 1982a). In addition to these morphogenetic processes, some other aspects of apical bud formation cannot be accomplished under short-day conditions. *Populus tremula x P. alba* plants, for instance, do not shed leaves and do not cold-acclimate under such controlled conditions.

Concluding the possible developmental paths of leaf primordia under conditions of growth arrest, we find out that leaf primordia or specific domains of the primordium can acquire two different fates: either a leaf-like or a stipule/scale-like. With respect to the leaf-like organs, the following dormancy period separates the organogenesis of embryonic leaves and their elongation in the next growing season. Bud scales as well as stipules that are often already quite elongated in the bud will not develop much more after the dormant period. They rather served the persistence of the other leaf organs. Thus, dormancy can be regarded as an interruption of development for some, but not all, primordia.

1.4 Organogenesis in apical and axillary buds

Generally, two types of buds have to be distinguished: the apical bud and the axillary buds (Figure 3). The SAM of the apical bud is formed during embryogenesis, whereas the axillary buds descend from the SAM. The apical bud is thought to ensure the correct temporal/seasonal growth, *i.e.*, its growth and dormancy is largely directed by the environment. In contrast, the growth of axillary buds is limited by correlative inhibition to balance growth with the available resources (Crabbé and Barnola, 1996). The limits for growth of the axillary buds are dictated by the growth potential of the apical bud. The growth of a seedling is characterized by absolute apical dominance; the axillary buds produced under such conditions remain inferior. In the secondary stem region (in *Populus deltoides* characterized by a change in phyllotaxis from 2/5 to 3/8 and later 5/13), axillary buds develop continuously and may even grow out into sylleptic branches.

When dormancy induction occurs, the axillary buds at successive nodes are in different developmental stages. A vigorously growing *Populus deltoides* plant contains approximately 15 primordia beyond the presently elongating leaf, and axillary buds originate normally in the 5th primordium below the apex (Larson and Pizzolato, 1977). The axillary buds are in a developmental gradient all the time: axillary buds at "youngest" positions will only have initiated bud scales, whereas those in more mature positions will have produced bud scales and foliage leaves. Environmental restrictions, signaleds by short days, for example, become superimposed onto these ontogenetic differences. As a consequence, axillary buds of plants exposed for 6 weeks to short days have fewer primordia than those at similar positions of plants grown in long days (Richards and Larson, 1981).

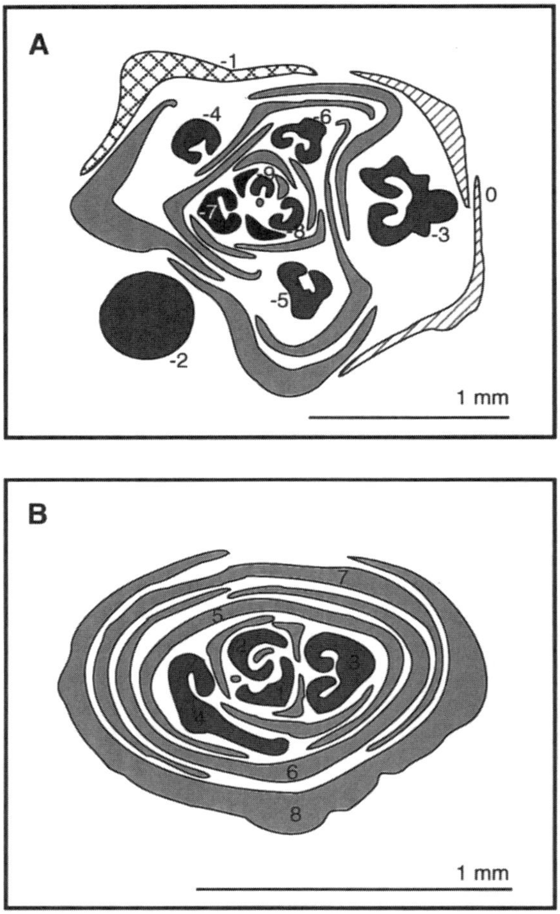

Figure 3. The radial structure of a dormant apical and a dormant axillary bud.

Buds of *Populus tremula x P. alba* plants had been exposed for 6 weeks to short days (8 h/day photoperiod) under controlled conditions before they were subjected to microscopy. **(A)** Apical bud, exhibiting clear helical phyllotaxis. Organ 0 (hatched, corresponding to organ 0 in Figure 2B) is the last leaf, of which the lamina has matured and abscised. The stipules, however, persist to form bud scales. Organ –1 (cross-hatched, corresponding to organ -1 in Figure 2C) is the abortive bud-scale leaf, forming a single bud scale. All younger primordia (organs –2 to –9) developed into embryonic leaves, each consisting of a leaf and two stipules (as highlighted for organ –7). **(B)** Axillary bud, exhibiting decussate phyllotaxis for the bud scales and helical phyllotaxis for the embryonic leaves. Typically, four bud scales (organs 8 to 5) encase four to five embryonic leaves, each comprising a leaf and two stipules (organs 4 to 1). Organs of stipular origin are light gray, whereas leaf-like organs are dark gray (A and B).

Bud-scale formation differs between apical and axillary buds (Figure 3). As explained above, the bud scales of apical buds are modified stipules (Figure 3A). On the contrary, the bud scales of axillary buds arise directly from the primordium (called true bud scales or cataphylls). Bud-scale formation in the apical bud involves increasingly earlier lamina abortion (Figures 2B and 2C), but this is not the case for axillary buds (Richards and Larson, 1981). As a consequence, axillary buds differ from apical buds with respect to their connection with the vascular cylinder and their phyllotaxis (Figures 3A and 3B). The scales of the apical bud are connected with the procambial strands in a 5/13 phyllotaxis, as is the case for leaves. In contrast, the first two scale traces of the axillary bud lead to a decussate phyllotaxis of the cataphylls, and higher orders of phyllotaxis are established in the embryonic leaves (Figure 3B; Pizzolato and Larson, 1977). Generally, only the foliage leaves are arranged in a helical phyllotaxis. Axillary buds thus recapitulate the phyllotactic history of the apical bud they are descending from.

Curiously enough, a two-dimensional gel analysis on protein dynamics during bud set in poplar (*Populus deltoides*) did not reveal any changes at the protein level that corresponded to these clear-cut morphological differences between apical and axillary buds (Jeknic´and Chen, 1999). Changes in proteins were similar in apical and axillary buds, while being distinct from leaves or bark.

2. ORGANS IN THE BUD FORM A CONTINUOUS SYSTEM

During the induction of dormancy, internode elongation ceases before the formation of new leaf primordia and finally meristem activity stop (Figure 2), implying that cell divisions are not synchronized and that the cell cycle is regulated differently in these distinct parts of the apex. Most likely, short-distance signaling between the meristem proper, the bud scales and leaf primordia, the peripheral and rib meristem zone, as well as cell-to-cell communications are central components in the regulation of dormancy induction. The integration of this short-distance signaling probably takes place in the subapical meristem and, therefore, its activity and cell-to-cell communication through plasmodesmata will need to be considered more explicitly in the future.

For instance, plasmodesmata connect plant cells symplasmically and allow the controlled exchange of critical molecules. The apical meristem is subdivided into symplasmic fields. These morphogenetic fields bring cells together, involve boundary interactions between different fields, and shield distal cells from organogenesis (Rinne and van der Schoot, 1998). In *Populus deltoides*, cells of dormant apical buds have either fewer plasmodesmata or non-functional plasmodesmata (Jian *et al.*, 1997). Rinne and van der Schoot (1998) have shown that short-day-induced growth cessation in *Betula pubescens* seedlings is correlated with a breaks down of symplasmic fields into symplasmically isolated cells. This example illustrates that a bud has to be broken down into its organs and domains in order to understand its complex development.

2.1 Relations of the apical meristem with the previous primordium/primordia

The descendants from the apical meristem, *i.e.*, the leaves, influence the continued activity of the apical meristem, as shown most convincingly in model plants. In *Antirrhinum*, the *phantastica* mutation that disrupts the correct dorsoventral organization of leaves (abaxial tissue also present on the adaxial leaf side) prevents younger leaf primordia from outgrowth, indicating that the activity of the meristem is arrested (Figure 1; Waites *et al.*, 1998). The *phabulosa* mutation in *Arabidopsis* has an opposite effect on both dorsoventrality and meristem function: the adaxial tissues are also present on the abaxial side of the leaf and ectopic meristems form at the lower side of the petiole base (McConnell and Barton, 1998). In both cases, signals from the adaxial side of the leaf primordium are required for continued meristem function and primordium initiation. This clearly shows that the organs generated by the SAM determine the fate and position of later descendants from the same SAM.

While these recent findings in model plants illustrate that each primordium receives signals from the preceding primordium, the details of primordium initiation that occur in a forming bud cannot be explained yet. In most species, a precise number of bud-scale leaves is made and then the bud switches to the formation of embryonic leaves. What the signals and controls are that provoke such an abrupt change remains to be investigated. Classical experiments with *Aesculus hippocastanum* have already demonstrated that a primordium "knows" its predecessor (Dostál, 1952). Defoliation of the whole shoot that is setting an apical bud led to the insertion of foliage leaves in between the bud-scale leaves. When, on the contrary, the bud scales of a young bud were removed, bud scales were inserted between the foliage leaf primordia. Dostál (1952) argued that the type of leaf primordium is controlled by definite correlations: scales induce the initiation of embryonic foliage leaves, whereas leaves are required for the formation of scales. Whatever the basis for this phenomenon is in molecular terms, it emphasizes once more that organs immediately below the apex influence the organogenesis at the SAM.

2.2 Sink-source relationship within the bud

Within the bud, the different organs differ for virtually all characteristics, including such important physiological parameters as growth capacity, hormone concentrations, and hormone sensitivities. The functions of the different organs may change with seasons as well as their sink/source characteristics.

In *Tilia cordata* buds, interesting changes in the sink/source relations between the inner bud scales (of stipular origin in *Tilia*) and leaf primordia have been hypothesized from observations during bud break (Himanen, 1999). During re-activation after winter dormancy, the inner scales become green and, therefore, most probably also photosynthetically active before the young leaves. Soon after the leaves have become green and photosynthetically active, the scales become pale and are abscised during later stages of bud outgrowth. In *Tilia cordata* buds, the inner scales grow/elongate only for the short period until the leaf is self-sustaining. During this time, they might provide the yet white leaf primordia with sugars – as nutritional but also as signaling components – to finally re-activate their growth.

2.3 Interrelations of the bud with subapical tissues and vascular bundles

To become a sink, a new leaf primordium must be connected with the vascular system by a leaf trace. Continued vascularization of such a leaf trace is required to transform a sink leaf into a source leaf. In many plants, the procambial leaf traces develop acropetally to conform precisely with phyllotaxis, *i.e.*, to meet the site of a new leaf primordium.

In *Populus deltoides*, the traces for the axillary bud arise from the central trace of the axillant leaf, a few plastochrons after the initiation of the leaf primordium (Larson and

Pizzolato, 1977). This procambial system provides the backbone for further vascularization: the protophloem differentiates acropetally, whereas the protoxylem differentiates basipetally from the leaf basal meristem (Larson *et al.*, 1980).

The changes in leaf morphology during dormancy induction are paralleled by changes in the vascular cylinder just below the bud. The primary vascular cylinder of a poplar stem consists of an often "open" ring of leaf traces of all above initiated leaves and leaf primordia. Secondary xylem is formed when the maturation of the leaves they serve is progressing (Goffinet and Larson, 1982b). A vascular transition zone can be observed, where the vascularization changes from primary to secondary xylem characteristics (Larson, 1980). During growth, both leaf maturation and the vascular transition zone move forwards in acropetal direction. When a plant is induced to dormancy, *de novo* leaf formation decreases to finally cease. As a consequence, the number of leaves to mature is finite and the vascular transition reaches closer to the apex than in long-day plants (Goffinet and Larson, 1982b). Moreover, the leaf traces serving leaf primordia within the bud are made of primary xylem. Thus, the vascular cylinder beneath a dormant bud becomes fragmented into sectors with secondary and primary xylem characteristics that have served the leaves of this season and the new leaf primordia, respectively. During bud burst, these sectors are presumed to be differentially reactivated, with the primary xylem responding first. Those sectors that had served the abscised leaves of the previous year can be re-activated only, if a leaf is emerging. Moreover, in new and re-activated (after dormancy period) leaf traces, xylem vessel formation correlates with leaf expansion, whereas fiber differentiation is associated with leaf maturation (Goffinet and Larson, 1982b).

In conclusion, phyllotaxis is realized through both the SAM that generates new leaf primordia and the procambium that establishes their connection with the vascular bundle. Furthermore, the procambial system of *Populus deltoides* progresses from a 1/2 phyllotaxis at the cotyledon stage, through a 2/5 phyllotaxis in the young seedling to higher orders of 3/8 and finally 5/13 (Larson, 1980). The greater the vigor of a plant, the greater the demand for leaf production is, which can be best realized at higher phyllotactic orders. Moreover, the more vigorously a plant is growing, the greater will be the distance between the vascular transition zone and the apex.

It is disputed of what nature the relationship is between the apical meristem and the secondary cambial meristem. The "connecting" tissue between these two meristem (the subapical region) can be expected to govern at least some of the components that adjust the two meristems to each other. The idea that the function of the cambium depends on the activity and signals derived from the primary meristems (SAM and axillary meristems in the leaf axil) is somewhat too simple. Barnola and Crabbé (1993) argue that especially the temporary storage of reserves in the cambium could be important in the regulation of growth and dormancy of the primary meristems.

There are additional indications that the corpus-derived tissues, *i.e.*, among others the procambial strands moving acropetally, govern important growth controls over the primary meristems. The *CLV1* gene, which is involved in delimiting the SAM, is expressed in the corpus (L3 layer), while it exerts the growth control in the whole SAM (Figure 1). This expression pattern suggests that the SAM has to adjust to the growth rate of the corpus (Clark *et al.*, 1997). In periclinal chimeras of *Nicotiana tabacum* and *Nicotiana glauca*, both clearly differing in their potential to generate axillary buds, the genotype of the L3 layer influences the number and position of the axillary meristems to a greater extent than does that of the L1 and L2 layers; yet, the genotype of the L3 layer does not determine the development of these meristems alone (Tian and Marcotrigiano, 1994).

3. UNRAVELING MOLECULAR AND GENETIC COMPONENTS OF BUD SET IN POPLAR

3.1 Can any conclusions be drawn from similar systems?

Having had this discourse in developmental complexity of bud development, the question can be raised as to whether any parallels can be found of bud dormancy with other dormant systems and conclusions drawn from them.

From an evolutionary point of view, the bud was the first dormant organ that developed, when branching was first used to adjust growth of several meristems in a single plant (Crabbé, 1994). The dormant seed, as the more prevalent dormant organ nowadays, appeared later when the first conifers moved into relatively dry, previously uninhabited areas. This emphasizes that seed dormancy developed as an adaptive strategy secondary to the morphogenetic strategy of bud dormancy. As dormancy was co-opted in other organs, such as seeds and bulbs, at the same time it became a strategy to cope with more extreme environmental limits, as opposed to the initially highly correlative purposes of bud dormancy (Crabbé, 1994). Although dormancy in different organs has different reasons and results, it is still not excluded that certain mechanisms were re-used, especially at the cellular and molecular levels.

Along this line, seed and bud dormancy can be judged to have numerous features in common. The analogy in the developmental programs of seed and bud dormancy was first put forward in 1956 by Wareing on physiological grounds (Wareing, 1956). The rationale for assuming that there is a common basis for both processes include: (i) a similar chilling requirement for a given genotype in some species, (ii) similar influences of the plant hormones abscisic acid (ABA) and gibberellins on dormancy induction and dormancy release, (iii) similar changes in water availability in the cells together with the acquisition of desiccation tolerance, and (iv) a similar accumulation of reserve proteins and lipids. In seeds, the inhibition of germination can be caused by the dormant embryo itself or by the covering structures (endosperm, testa, pericarp; Bewley, 1997). In buds, the meristem is covered by bud scales, which, at least in part, participate in establishing and maintaining dormancy (Myking, 1997). Although bud and seed dormancy may involve common metabolic and physiological processes, not only their purposes are different, but also their integration into the plant life cycle. Seed dormancy occurs when the mother plant is reproducing. The dispersal of the seed is facilitated by a temporary arrest of growth inside the seed. In contrast, bud dormancy is integrated into the morphogenesis of one and the same individual, it takes place simultaneously with active growth of other plant parts, and it is recurrent throughout the life-span of a tree. Nevertheless, the analogy approach, if treated cautiously and evaluated rigorously, holds promise to unravel components of processes for which by virtue nothing is known.

3.2 Seed and bud dormancy

As the photoperiod is perceived by (most probably) the leaves, it has been postulated in early days that the dormancy-controlling factors are exported from leaves to effectuate their growth-inhibiting function in the buds (Hemberg, 1949). Despite very limited data, since its isolation ABA was suggested to be the growth-inhibiting substance in dormant buds (originally called "dormin"). Indeed, ABA has often been found in dormant tissues and has growth-arresting effects when applied exogenously (Lavender and Silim, 1987). However, the expected peaks in endogenous ABA prior to dormancy induction were only sometimes, but never consistently, shown throughout several species and/or conditions. Barros and Neill (1986), who investigated the response to ABA in lateral *Salix viminalis* buds, found that the induction and breaking of dormancy and the response to ABA did not correlate with the endogenous ABA levels in the buds. However, sensitivity to exogenously provided ABA coincided reproducibly with the periods prior to bud set (Barros and Neill, 1986).

To better understand the importance of ABA in the establishment of bud dormancy, we have studied genes that had been shown previously to be important in ABA signaling during seed dormancy in *Arabidopsis*. The maintenance of mid and late embryogenesis has been correlated strongly with the presence of ABA (Bewley, 1997). ABA is part of the signal that initiates the regulatory pathways that promote maturation and acquisition of desiccation tolerance and that also prevents precocious germination (McCarty, 1995). The abscisic-acid-insensitive genes *ABI1* and *ABI3* play a crucial role during seed dormancy and encode central components of ABA signal transduction (for reviews, see Bonetta and McCourt, 1998; Leung and Giraudat, 1998). The *Arabidopsis abi3* mutant has a reduced seed dormancy and the orthologous maize *vp1* mutant is viviparous (Koornneef *et al.*, 1984; McCarty *et al.*, 1989). The main function of ABI3 in annual plants, such as *Arabidopsis*, is in seeds (Parcy *et al.*, 1994). However, several experiments have shown that ABI3 plays a role outside the seed as well (Rohde *et al.*, 2000b). The fact that ABI3 functions in vegetative tissues of *Arabidopsis*, particularly during growth-arresting conditions and quiescence (Rohde *et al.*, 1999), suggested that ABI3 could also be involved in the regulation of bud dormancy in trees. Given the fact that these genes possibly link ABA and dormancy, it was challenging to study this relationship in buds under conditions of "true" dormancy.

3.3 ABI3 is involved in seed and bud dormancy

To show that ABI3 plays a role in bud dormancy, the expression of *ABI3* during bud development in poplar was studied. By rtPCR, *ABI3*-homologous transcripts were detected in autumn buds, at approximately the time of vegetative growth arrest. At the same time, the expression of *PtABI1b*, another component of the ABA signaling pathway, peaked (A. Rohde, unpublished results). Neither transcript was detected during later stages of dormancy. Together, this evidence argues for a role for *ABI3* and *ABI1* in the creation of seed and bud dormancy.

The *ABI3*-homologous gene and cDNA, *PtABI3* (*Populus trichocarpa ABI3*), were isolated from *Populus trichocarpa* (Rohde *et al.*, 1998). To understand the function of PtABI3 in dormancy induction, *Populus tremula* x *P. alba* trees were transformed with sense and antisense constructs of the *PtABI3* cDNA as well as with a chimeric *PtABI3::GUS* fusion and analyzed under controlled short-day conditions (Rohde *et al.*, in preparation).

PtABI3 was found to be expressed towards the end of the bud formation process, from approximately 28 days after the onset of short days. This "late" expression implies that PtABI3 is not actively involved in the perception of the short-day signal, as is, for example, the phytochrome PHYA (Olsen *et al.*, 1997). Overexpression of *PHYA* from *Avena fatua* in *Populus tremula* x *P. tremuloides* resulted in changes in the critical day length. Of a total of 22 independently *35S::PHYA*-transformed lines, 16 lines did not cease their growth under a 12 h/day photoperiod, as was the case for untransformed wild-type plants. Thus, a strong *PHYA* overexpression can apparently prevent the cessation of internode elongation and the initiation of bud scales, the two main morphogenetic processes associated with the formation of the apical bud.

Cessation of internode elongation as well as the formation of bud scales happened well before the time that *PtABI3::GUS* expression was first detected. Therefore, PtABI3 functions after or at the end of the leaf differentiation program. The expression occurs exclusively in organs and cells, where growth processes need to be fixed and/or stopped: the young leaf primordia, the subapical meristem, and the procambial strands. Growth in these tissues has to be retarded, arrested, and acclimation to be initiated to prepare the plant for the expected environmental changes signaled by short days. The suggestion that PtABI3 might function in the leaf primordia, procambial strands, and apical cells to fix a state of differentiation is supported by the phenotypes of transgenic *Populus tremula x P. alba* down-regulating or overexpressing *PtABI3*. Constitutive overexpression of *PtABI3* led to plants with normal embryonic leaves but with very small bud scales, leaving the apex exposed in an "open" bud. The growth of bud scales was suppressed to varying degrees, suggesting that ectopic overexpression of *PtABI3* in these organs has rendered bud scales sensitive to a factor to which they are normally not sensitive. This factor must be responsible for the retardation of growth and differentiation in the young embryonic leaves, in which *PtABI3* is normally expressed. In antisense and co-suppressed lines, bud scale growth is more pronounced. The embryonic leaves inside the bud tend to be smaller than in wild type. These observations together with the expression of *PtABI3::GUS* in leaves suggest that wild-type PtABI3 is also required for the relative growth rate of the embryonic leaves, before being involved in their growth arrest. At first sight it seems contradictory to the suggestion that PtABI3 renders cells sensitive to a factor that retards growth. Perhaps PtABI3 may need to be regarded, however, as a factor that allows a short cellular maturation that finally results in a complete arrest. When leaf growth cannot be finalized because of the absence of PtABI3 and is thus arrested prematurely, increased bud-scale growth might represent a compensatory response.

Further evidence for a role of PtABI3 in cellular maturation comes from the observation that *PtABI3::GUS* and *PtABI3* expression are occasionally seen in apical buds under long-day conditions, but these are not correlated with phenotypic alterations in *Populus tremula* x *P. alba* that down-regulate or overexpress *PtABI3*. This observation suggests that under long-day conditions an additional factor that is present under short days is missing. Future investigations will need to clarify whether ABA may be one of these growth-retarding factors. Alternatively, the signals that induce *PtABI3* expression might belong to the developmental factors that report a certain cellular developmental state or a discrepancy between the cellular developmental state and the environment. Such a discrepancy can, for example, be solved by rendering sensitive to ABA those cells that are normally determined to grow rapidly. If this scenario were true, PtABI3 would occupy a central place in the network assuring the complex morphogenesis during the apical bud formation described above. PtABI3 would act downstream of the molecular network that regulates initiation of leaf primordia at the SAM, as established in plants such as *Arabidopsis*. Moreover, ABI3 could possibly interact with the components of leaf determination and differentiation, when growth limitations are signaled from the environment.

3.4 Quantitative trait loci for bud set and candidate gene mapping

Poplar trees from Northern locations initiate bud set earlier than do trees originating from Southern locations, when cultivated in a common environment (Pauley and Perry, 1954). When a clone is displaced from its latitude of natural occurrence towards the North, *i.e.*, into a regime with longer days, the period of active growth will be prolonged. When a clone is displaced towards the South, into a regime with shorter days, the period of active growth will be shortened. This knowledge can be applied to tree breeding programs. In poplar, for instance, knowledge of the critical photoperiod of various clones enables the tree breeder to create systematically inter- and intraspecific hybrids with a length of the growing season adapted to a given latitude (Steenackers, 1996).

The search for quantitative trait loci (QTLs) becomes a feasible approach to identify genes involved in a trait of interest such as bud set. Autumnal bud set will not be determined by a single gene, but by several or many genes. A QTL is a locus or region of the chromosome that has a significant effect on a quantitatively inherited trait (Tanksley, 1993). Genetic markers in a particular region of the genome are identified that associate with a particular phenotypic elaboration of the trait in a population. However, it remains difficult to break a QTL down to the gene level, at least in outbreeding tree species where the creation and analysis of appropriate populations is tedious.

The genetic control of bud phenology in poplar was studied by mapping QTLs affecting the timing of autumn bud set and spring bud flush (Frewen *et al.*, 2000). In order to maximize segregation of dormancy-related traits in the F_2 generation, a female *Populus trichocarpa* parent from 48° Northern latitude was crossed with a male *P. deltoides* parent from 31° Northern latitude. Among others, bud set was measured in the F_2 generation in a replicated clonal field trial. Using a linkage map constructed of AFLP and microsatellite markers, three QTLs that controlled bud set were detected. These QTLs each accounted for 6-12% of the phenotypic variation observed in the date of bud set. The difference in bud-set date amounts to 9 to 16 days between F_2 individuals that were homozygous for either *Populus trichocarpa* or *Populus deltoides* at the QTLs. Additionally, five candidate genes believed to be involved in the perception of photoperiod (*PHYB1*, *PHYB2*) or the transduction of ABA signals (*PtABI1b*, *PtABI1d*, and *PtABI3*) were placed on the same map. *PHYB2* and *PtABI1b* were found to coincide with a QTL affecting bud set, suggesting that allelic differences in these genes may be responsible for ecotypic variation in the timing of bud set in poplar. The allelic differences at the *PtABI1b* locus need to be checked in the appropriate, natural populations in order to confirm the association of the bud set QTL with the map position of *PtABI1b*.

4. CONCLUSIONS

The successful completion of bud set is an important event prior to the establishment of bud dormancy and cold acclimation. Moreover, proper bud set assures that the growth potential for the next spring is preserved. In this respect, it will be most interesting to evaluate the transgenic plants with altered *PtABI3* expression and an inappropriate bud morphology for their properties in cold acclimation and survival.

Altogether, molecular information has been provided on genes that play potentially important roles in bud development and dormancy. Based on the analysis of PtABI3 and PtABI1, the picture emerges that at least the signaling pathway involving ABI3 and ABI1 is active during both bud and seed dormancy. As PtABI1 and PtABI3 appear to be important regulatory proteins, their further characterization will give us key access to the induction of bud dormancy. The integration of molecular components, such as the ABA signaling proteins, into the classical morphogenetic description of dormancy entry illustrates that such an approach provides new insight into its regulation. More players in the underlying molecular network will be unraveled in the future through the identification of new candidate genes by exploring previous physiological knowledge and the advanced understanding of meristem function in *Arabidopsis thaliana*. Alternative approaches to compare gene expression in different stages of bud development and bud dormancy, such as cDNA-AFLP, EST sequencing, and DNA microarrays, are new powerful methods that will provide a more detailed picture of bud development and dormancy.

REFERENCES

Barnola, P., and Crabbé, J. (1993). L'activité cambiale, composante active ou passive dans les réactions de croissance de l'arbre? *Acta Bot. Gallica* **140,** 403-412.

Barros, R.S., and Neill, S.J. (1986). Periodicity of response to abscisic acid in lateral buds of willow (*Salix viminalis* L.). *Planta* **168,** 530-535.

Barton, M.K. (1998). Cell type specification and self renewal in the vegetative shoot apical meristem.

Curr. Opin. Plant Biol. **1,** 37-42.

Bewley, J.D. (1997). Seed germination and dormancy. *Plant Cell* **9,** 1055-1066.

Bonetta, D., and McCourt, P. (1998). Genetic analysis of ABA signal transduction pathways. *Trends Plant Sci.* **3,** 231-235.

Borchert, R. (1991). Growth periodicity and dormancy. In *Physiology of Trees*, A.S. Raghavendra (Ed.). New York, John Wiley & Sons, pp. 221-245.

Clark, S.E., Williams, R.W., and Meyerowitz, E.M. (1997). The *CLAVATA1* gene encodes a putative receptor kinase that controls shoot and floral meristem size in Arabidopsis. *Cell* **89,** 575-585.

Crabbe, J.J. (1994). Dormancy. In *Encyclopedia of Agricultural Science*, Vol. 1, C.J. Arntzen, and E.M. Ritter (Eds.). San Diego, Academic Press, pp. 597-611.

Crabbé, J., and Barnola, P. (1996). A new conceptual approach to bud dormancy in woody plants. In *Plant Dormancy: Physiology, Biochemistry and Molecular Biology*, G.A. Lang (Ed.). Wallingford, CAB International, pp. 83-113.

Critchfield, W.B. (1960). Leaf dimorphism in Populus trichocarpa. *Am. J. Bot.* **47,** 699-711.

Curtis, J.D., and Lersten, N.R. (1974). Morphology, seasonal variation, and function of resin glands on buds and leaves of *Populus deltoides* (Salicaceae). *Am. J. Bot.* **61,** 835-845.

Dostál, R. (1952). Experimentální morfogenese pupen jírovce (*Aesculus hippocastanum* L.). *Acta Acad. Sci. Natural. Moravo-Silesiacae* **24,** 109-146 (Engl. summ.).

Elliott, R.C., Betzner, A.S., Huttner, E., Oakes, M.P., Tucker, W.Q.J., Gerentes, D., Perez, P., and Smyth, D.R. (1996). *AINTEGUMENTA*, and *APETALA2*-like gene of Arabidopsis with pleiotropic roles in ovule development and floral organ growth. *Plant Cell* **8,** 155-168.

Eriksson, I., Dormling, I., Erikson, G., and von Wettstein, D. (1978). The inheritance of bud-set and bud-flushing in *Picea abies* (L.) Karst. *Theor. Appl. Genet.* **52,** 3-19.

Freeling, M. (1992). A conceptual framework for maize leaf development. *Dev. Biol.* **153,** 44-58.

Frewen, B.E., Chen, T.H.H., Howe, G.T., Davis, J., Rohde, A., Boerjan, W., and Bradshaw, H.D. Jr (2000). Quantitative trait loci and candidate gene mapping of bud set and bud flush in Populus. *Genetics* **154,** 837-845.

Garner, W.W., and Allard, H.A. (1923). Further studies in photoperiodism, the response of the plant to relative length of day and night. *J. Agr. Res.* **28,** 871-920.

Goffinet, M.C., and Larson, P.R. (1981). Structural changes in Populus deltoides terminal buds and in the vascular transition zone of the stems during dormancy induction. *Am. J. Bot.* **68,** 118-129.

Goffinet, M.C., and Larson, P.R. (1982a). Lamina abortion in terminal bud-scale leaves of Populus deltoides during dormancy induction. *Bot. Gaz.* **143,** 331-340.

Goffinet, M.C., and Larson, P.R. (1982b). Xylary union between the new shoot and old stem during terminal bud break in Populus deltoides. *Am. J. Bot.* **69,** 432-446.

Hemberg, T. (1949). Growth-inhibiting substances in terminal buds of Fraxinus. *Physiol. Plant.* **2,** 37-44.

Himanen, K. (1999). Meristem reactivation during bud burst in *Tilia cordata*. M.Sc. Thesis, Department of Biosciences, University of Helsinki.

Howe, G.T., Bucciaglia, P.A., Hackett, W.P., Furnier, G.R., Cordonnier-Pratt, M.-M., and Gardner, G. (1998). Evidence that the phytochrome gene family in black cottonwood has one *PHYA* locus and two *PHYB* loci but lacks members of the *PHYC/F* and *PHYE* subfamilies. *Mol. Biol. Evol.* **15,** 160-175.

Ilstedt, B. (1996). Genetics and performance of Belgian poplar clones tested in Sweden. *Forest*

Genet. **3,** 183-195.

Jeknic´, Z., and Chen, T.H.H. (1999). Changes in protein profiles of poplar tissues during the induction of bud dormancy by short-day photoperiods. *Plant Cell Physiol.* **40,** 25-35.

Jian, L.-c., Li, P.H., Sun, L.-h., and Chen, T.H.H. (1997). Alterations in ultrastructure and subcellular localization of Ca^{2+} in poplar apical bud cells during the induction of dormancy. *J. Exp. Bot.* **48,** 1195-1207.

Junttila, O. (1982). Cessation of apical growth in latitudinal ecotypes and ecotype crosses of *Salix pentandra. J. Exp. Bot.* **33,** 1021-1029.

Junttila, O. (1990). Gibberellins and the regulation of shoot elongation in woody plants. In *Gibberellins*, N. Takahashi, B.O. Phinney, and J. MacMillan (Eds.). Berlin, Springer Verlag, pp. 199-210.

Koornneef, M., Reuling, G., and Karssen, C.M. (1984). The isolation and characterization of abscisic acid-insensitive mutants of *Arabidopsis thaliana. Physiol. Plant.* **61,** 377-383.

Lang, G.A. (1987). Dormancy: a new universal terminology. *HortScience* **22,** 817-820.

Lang, G.A. (1996). *Plant Dormancy: Physiology, Biochmistry and Molecular Biology*. Wallingford, CAB International, 386 p.

Larson, P.R. (1980). Interrelations between phyllotaxis, leaf development and the primary-secondary vascular transition in *Populus deltoides. Ann. Bot.* **46,** 757-769.

Larson, P.R., and Pizzolato, T.D. (1977). Axillary bud development in Populus deltoides. I. Origin and early ontogeny. *Am. J. Bot.* **64,** 835-848.

Larson, P.R., Isebrand, J.G., and Dickson, R.E. (1980). Sink to source transition of *Populus* leaves. *Ber. Deutsch. Bot. Ges.* **93,** 79-90.

Laufs, P., Grandjean, O., Jonak, C., Kiêu, K., and Traas, J. (1998a). Cellular parameters of the shoot apical meristem in Arabidopsis. *Plant Cell* **10,** 1375-1389.

Laufs, P., Dockx, J., Kronenberger, J., and Traas, J. (1998b). *MGOUN1* and *MGOUN2*: two genes required for primordium initiation at the shoot apical and floral meristems in *Arabidopsis thaliana. Development* **125,** 1253-1260.

Lavender, D.P., and Silim, S.N. (1987). The role of plant growth regulators in dormancy in forest trees. *Plant Growth Regul.* **6,** 171-191.

Lenhard, M., and Laux, T. (1999). Shoot meristem formation and maintenance. *Curr. Opin. Plant Biol.* **2,** 44-50.

Leung, J., and Giraudat, J. (1998). Abscisic acid signal transduction. *Annu. Rev. Plant Physiol. Plant Mol. Biol.* **49,** 199-222.

Long, J., and Barton, M.K. (2000). Initiation of axillary and floral meristems in *Arabidopsis. Dev. Biol.* **218,** 341-353.

Lynn, K., Fernandez, A., Aida, M., Sedbrook, J., Tasaka, M., Masson, P., and Barton, M.K. (1999). The *PINHEAD/ZWILLE* gene acts pleiotropically in *Arabidopsis* development and has overlapping functions with the *ARGONAUTE1* gene. *Development* **126,** 469-481.

McCarty, D.R. (1995). Genetic control and integration of maturation and germination pathways in seed development. *Annu. Rev. Plant Physiol. Plant Mol. Biol.* **46,** 71-93.

McCarty, D.R., Carson, C.B., Stinard, P.S., and Robertson, D.S. (1989). Molecular analysis of *viviparous-1*: an abscisic acid-insensitive mutant of maize. *Plant Cell* **1,** 523-532.

McConnell, J.R., and Barton, M.K. (1998). Leaf polarity and meristem formation in *Arabidopsis. Development* **125,** 2935-2942.

Moritz, T. (1995). Biological activity, identification and quantification of gibberellins in seedlings of Norway spruce (*Picea abies*) grown under different photoperiods. *Physiol. Plant.* **95,** 67-72.

Myking, T. (1997). Effects of constant and fluctuating temperature on time of budburst in *Betula pubescens* and its relation to bud respiration. *Trees* **12,** 107-112.

Nitsch, J.P. (1957). Photoperiodism in woody plants. *Proc. Am. Soc. Hort. Sci.* **70,** 526-544.

Olsen, J.E., Junttila, O., and Moritz, T. (1995). A localised decrease of GA_1 in shoot tips of *Salix pentandra* seedlings precedes cessation of shoot elongation under short photoperiod. *Physiol. Plant.* **95,** 627-632.

Olsen, J.E., Junttila, O., Nilsen, J., Eriksson, M.E., Martinussen, I., Olsson, O., Sandberg, G., and Moritz, T. (1997). Ectopic expression of oat phytochrome A in hybrid aspen changes critical daylength for growth and prevents cold acclimatization. *Plant J.* **12,** 1339-1350.

Parcy, F., Valon, C., Raynal, M., Gaubier-Comella, P., Delseny, M., and Giraudat, J. (1994). Regulation of gene expression programs during Arabidopsis seed development: roles of the *ABI3* locus and of endogenous abscisic acid. *Plant Cell* **6,** 1567-1582.

Pauley, S.S., and Perry, T.O. (1954). Ecotypic variation of the photoperiodic response in Populus. *J. Arnold Arboretum* **35,** 167-188.

Pizzolato, T.D., and Larson, P.R. (1977). Axillary bud development in Populus deltoides. II. Late oxygen and vascularization. *Am. J. Bot.* **64,** 849-860.

Richards, J.H., and Larson, P.R. (1981). Morphology and development of *Populus deltoides* branches in different environments. *Bot. Gaz.* **142,** 382-393.

Riemenschneider, D.E., and McMahon, B.G. (1993). Genetic variation among Lake States balsam poplar populations is associated with geographic origin. *Forest Sci.* **39,** 130-136.

Rinne, P.L.H., and van der Schoot, C. (1998). Symplasmic fields in the tunica of the shoot apical meristem coordinate morphogenetic events. *Development* **125,** 1477-1485.

Rohde, A., Ardiles-Diaz, W., Van Montagu, M., and Boerjan, W. (1998). Isolation and expression analysis of an abscisic acid-insensitive (*ABI3*) homologue from *Populus trichocarpa*. *J. Exp. Bot.* **49,** 1059-1060.

Rohde, A., Van Montagu, M., and Boerjan, W. (1999). The *ABSCISIC ACID-INSENSITIVE 3* (*ABI3*) gene is expressed during vegetative quiescence processes in *Arabidopsis*. *Plant Cell Environ.* **22,** 261-270.

Rohde, A., Howe, G.T., Olsen, J.E., Moritz, T., Van Montagu, M., Junttila, O., and Boerjan, W. (2000a). Molecular aspects of bud dormancy in trees. In *Molecular Biology of Woody Plants, Volume 1*, (Forestry Sciences, Vol. 64), S.M. Jain, and S.C. Minocha (Eds.). Dordrecht, Kluwer Academic Publishers, pp. 89-134.

Rohde, A., Kurup, S., and Holdsworth, M. (2000b). ABI3 emerges from the seed. *Trends Plant Sci.* **5,** 418-419.

Rowland, L.J., and Arora, R. (1997). Proteins related to endodormancy (rest) in woody perennials. *Plant Science* **126,** 119-144.

Samish, R.M. (1954). Dormancy in woody plants. *Annu. Rev. Plant Physiol.* **5,** 183-204.

Saure, M.C. (1985). Dormancy release in deciduous fruit trees. *Hort. Rev.* **7,** 239-300.

Steenackers, V. (1996). Photopériodisme et amélioration des peupliers. 20ième Session de la Commission Internationale du Peuplier, Budapest (Hongrie), 1-4 Octobre, 1996.

Sylven, N. (1940). Lang- och kortdagstyper av de svenska skogstraden. Long day and short day types of Swedish forest trees. *Svensk Papperstidn.* **43,** 317-324; 332-342; 350-354.

Tanksley, S.D. (1993). Mapping polygenes. *Annu. Rev. Genet.* **27,** 205-233.

Tian, H.-C., and Marcotrigiano, M. (1994). Cell-layer interactions influence the number and position of lateral shoot meristems in *Nicotiana*. *Dev. Biol.* **162,** 579-589.

Vegis, A. (1964). Dormancy in higher plants. *Annu. Rev. Plant Physiol.* **15,** 185-224.

Waites, R., Selvadurai, H.R.N., Oliver, I.R., and Hudson, A. (1998). The *PHANTASTICA* gene encodes a MYB transcription factor involved in growth and dorsoventrality of lateral organs in *Antirrhinum*. *Cell* **93,** 779-789.

Wareing, P.F. (1956). Photoperiodism in woody plants. *Annu. Rev. Plant Physiol.* **7,** 191-214.

PHYSIOLOGY OF TREE ROOT/FUNGUS SYMBIOSIS

RÜDIGER HAMPP AND UWE NEHLS
Physiologische Ökologie der Pflanzen, Universität Tübingen, Auf der Morgenstelle 1, D-72076 Tübingen, Germany

Key words: Amino acid transport, *Amanita muscaria*, carbon allocation, C-N interaction, ectomycorrhiza, gene expression, *Picea abies*, sugar transport, sugar sensing, transgenic aspen

Abstract: Roots of trees of boreal, temperate and some tropical forests form symbiotic structures with 5000-6000 species of fungi belonging to the asco- and basidiomycetes. These structures are called ectomycorrhizas. The high diversity of fungal partners allows for improved foraging and mobilisation of various forms of nitrogen and phosphorus from organic soil layers. In this review current knowledge on the physiology, biochemistry and the molecular biology of this symbiosis is summarized. Special emphasis is given to the role of host derived carbohydrates on the regulation of the symbiotic interaction.

1. INTRODUCTION

The term 'mycorrhiza' describes the symbiotic association of plant roots with fungal hyphae. Major forms are arbuscular mycorrhiza (AM) and ectomycorrhiza (ECM). As ECM is the dominating symbiotic structure in boreal forest ecosystem, AM will only be treated marginally in this review. ECM is established with fine roots of, e.g., trees of the families Betulaceae, Pinaceae, Fagaceae, and Salicaceae (Read, 1991). The fungal partners belong to the basidiomycetes and ascomycetes. ECM is characterized by fungal structures of different function. Typically, hyphae form a mantle of varying thickness around the fine roots. From there, hyphae or more specialized hyphal aggregates (rhizomorphs) radiate into the substrate in order to exploit nutrients and water. Mantle hyphae also extend into

S. Huttunen et al. (eds.),
Trends in European Forest Tree Physiology Research, Cost Action E6: EUROSILVA, 53-62

the apoplast of the root cortex. Here, they form highly branched networks which establish a large surface area for solute exchange. This structure is called the Hartig net (Fig. 1). The development of the ECM structures is a complex process and depends on a series of external and internal factors of which the supply of photoassimilates by the host plant is most important. In the following we will summarize current knowledge about the plant-fungus-interaction in ECM on the basis of carbon and nitrogen availability. For overviews on the current state of knowledge about the physiology of mycorrhiza in general, see Smith and Read (1997) and Hampp et al. (2000a).

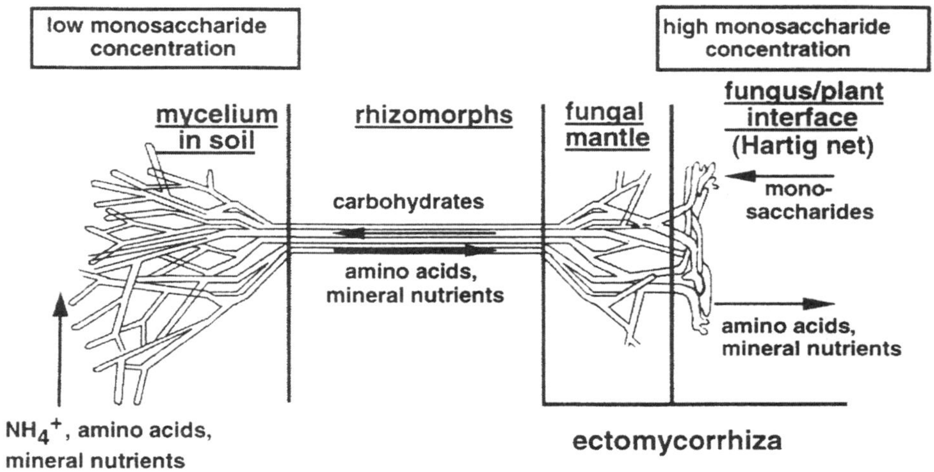

Figure 1. Functional diversity of ectomycorrhiza structures. Soil mycelia exploit nutrients, rhizomorphs have transport function, the fungal mantle covers the fine roots of the host, and the Hartig net with finger-like extensions of hyphae in the root cortex cell walls constitutes the area of solute exchange between the partners of symbiosis.

2. THE ROLE OF CARBON IN PLANT/FUNGUS INTERACTION

The production of photoassimilates is the most fundamental activity of plants. Formation, storage, transport and consumption of carbohydrates are dynamic processes which are closely linked to the physiological situation of the plant. Obviously, plants are able to monitor and to respond to changing sugar levels, thereby integrating external environmental conditions. There is a lot of evidence that sugar levels affect plant and fungal functions via control of gene expression. Recent data indicate that sensing and signalling by sugars occurs already in the lower mM concentration range of the respective molecules. Targets are sugar binding proteins (enzymes, transporters; see e.g. Sheen et al., 1999).

This obviously also holds for the situation in ECM where the availability of hexoses can affect fungal gene expression. One of the first examples studied is sugar transport. Using a molecular approach, a monosaccharide uptake system was identified in *A. muscaria* (*AmMST1*; Nehls et al., 1998).

The function of the AmMST1 protein as an active monosaccharide transporter was confirmed by heterologous expression of the full length cDNA in a yeast mutant, lacking a functional endogenous monosaccharide uptake system (Wiese et al., 2000). *A. muscaria* hyphae grown at glucose concentrations below 5 mM expressed the *AmMST1* gene at a basal level, while monosaccharide concentrations above this threshold caused a four-fold increase of the amount of the *AmMST1* transcript. An increase of *AmMST1* expression, similar to that found in fungal mycelia when cultivated at elevated monosaccharide concentrations, was also observed in symbiosis of *Amanita muscaria* with both the gymnosperm *Picea abies* and the angiosperm *Populus tremula x tremuloides*. It can thus be assumed, that the increase of *AmMST1* expression in mycorrhizas is also sugar-regulated and depends on the *in vivo* concentration of monosaccharides at the fungus/plant interface. At the threshold value of 5 mM glucose, the uptake capacity of the *Amanita* monosaccharide transporter was nearly saturated. Obviously, the fungus can sense saturation of monosaccharide transport and in due course enhances *AmMST1* expression.

The increase of *AmMST1* expression was shown to be a slow process. A transition from constitutive to maximal content of *AmMST1* transcripts occured between 18 and 24 h of fungal culture. We assume that this is possibly due to the fact that mycelia growing in the soil are exposed to low concentrations of carbohydrates (Wainwright, 1993). Higher monosaccharide concentrations were found only locally and for limited times. In contrast to hyphae growing in the soil, those hyphae at the symbiotic interface are exposed to a continuous supply of plant-derived carbohydrates. In order to adapt the flux of carbohydrates into the fungus, monosaccharides must be quickly taken up and either metabolized or converted into compounds for intermediate or long term storage (e.g. trehalose, mannitol, glycogen). It can thus be speculated that both the extended lag phase for enhanced AmMST1 expression, and its threshold response to elevated monosaccharide concentrations are adaptations of the ectomycorrhizal fungus to the conditions found at the symbiotic interface.

In contrast to the fungal partner, monosaccharide transport in spruce root cells was not increased upon mycorrhization (Nehls et al., 2000). Expression of the respective gene appeared even decreased under conditions of symbiosis. Similar results were obtained for birch (Wright et al., 2000). This implies that regulation of hexose flow is by the fungal partner, not by the host. This is especially of interest as hexoses rather than sucrose are involved in metabolic regulation, including gene expression (see above and Sheen et al., 1999).

AmMST1 is not the only gene responding to glucose availability. Differential screening of cDNA libraries from sterile cultures of *A. muscaria* and of mycorrhizas formed between *A. muscaria* and *Picea abies* delivered a range of clones which were up-regulated (as shown for *AmMST1*), but also down-regulated by a continuous supply of hexoses above the 5 mM threshold. One of these clones could be identified as phenylalanine ammonia lyase (*AmPAL*; Nehls et al., 1999a). The transcript of this gene was abundant in hyphae grown at low external glucose concentrations but exhibited a 30-fold decrease in amount in hyphae cultured at glucose concentrations above 2 mM. In contrast to *AmMST1*, this gene is possibly regulated in a hexokinase-dependent manner (see Nehls et al., 1999a). PAL is a key enzyme of secondary metabolism and thus of the production of phenolic compounds. In plants, phenolics are widely used in protection against pathogenic invasion. ECM forming fungi have also been reported to use phenolic compounds for the protection of themselves but also of their host plant against bacterial or fungal attacks (Garbaye, 1991). We hypothesize that free living mycelia (i.e. low glucose supply) produce more of these protective compounds than those involved in symbiotic structures. This assumption is supported from experiments, where the hyphal mantle was physically separated from Hartig net structures. Down-regulation of *AmPAL* expression was only observed in Hartig net hyphae where hexose concentrations are expected to be high and where protection is not needed (Nehls et al., unpublished).

3. THE FUNCTION OF NITROGEN IN PLANT/FUNGUS INTERACTION

Due to low rates of nitrogen mineralization, supply of inorganic nitrogen is limiting in many forest ecosystems (Read, 1991). In addition, nitrification is usually slow and the poorly mobile ammonium ion predominates as inorganic nitrogen source (Keeney, 1980). Association with ectomycorrhizal fungi can improve nitrogen acquisition by (1) facilitating access to organic nitrogen sources (Read et al., 1989) and (2) by increasing uptake of ammonium (Rygiewicz et al., 1984) via extensive growth of soil mycelia and circumvention of ammonium depletion zones. Therefore, mycorrhiza formation is an important strategy for improving the development and competitive ability of forest trees (Bledsoe and Zasoski, 1983; Ingestad et al., 1986).

The ability of ectomycorrhizal fungi for uptake of inorganic nitrogen and transport of nitrogen-containing solutes to their host plant is well established (Melin and Nilsson, 1952; Plassard et al., 1986; Finlay et al., 1988; for reviews see France and Reid, 1983; Chalot and Brun, 1998).

In recent studies it was possible to clone an amino acid transporter in *A. muscaria.* Heterologous expression in a yeast mutant indicated a preference for basic and aromatic amino acids (Nehls et al. 1999b). The k_m values for transport were between 22 (histidine) and 100 μM (proline). According to its properties, we assume that the function of this general amino acid permease is primarily in the uptake of amino acids from the soil rather than in the export toward the host. From uptake studies with hyphae from *Paxillus involutus* biochemical evidence for such a general amino acid transporter was obtained (Chalot et al., 1996).

Fungal enzymes involved in nitrogen metabolism have been characterized and their genes were cloned. Obviously there is a co-regulation of fungal and plant activities, with different fungal preferences for alternative pathways (Martin and Botton, 1993; Botton and Dell, 1994; Lorillou et al., 1996; Marmeisse et al., 1998).

A surplus of N availability in the soil can affect the interaction of both partners of symbiosis. In fertilizer experiments, carried out both in natural conifer forests and in pot experiments, a reduction of the mycorrhizal development at elevated levels of N has been reported (Arnebrandt and Söderström, 1992; Wallander and Nylund, 1991; Wiklund et al., 1995). This is most probably a consequence of decreased photoassimilate allocation to the root if photoassimilates are primarily used for amino acid synthesis in the leaf cells. The regulatory steps involved in switching from sucrose (gluconeogenesis) to amino acid synthesis (glycolysis) have been described in detail elsewhere (Wingler et al., 1994; Champigny, 1995; Wallenda et al., 1996; Hampp and Schaeffer, 1999).

There is, however, also evidence that the fungal partner can store a surplus of assimilated nitrogen in its vacuoles. This is balanced with the fungal glycogen content when rates of photoassimilation by the host plants are increased under elevated CO_2 (Turnau et al. 2000).

4. REGULATION OF THE DEGREE OF MYCORRHIZATION BY HEXOSE AVAILABILITY: AN INTEGRATED VIEW

The studies of gene expression described above clearly show that host carbohydrates have a profound effect on the fungus in that hexose availability at the plant-fungus-interface causes changes in levels of fungus-specific transcripts.

It is still an open question how hexose signals are sensed. As glucose and/or glucose-6 phosphate act as signal molecules, hexokinase is discussed as a possible sensor protein (Fig. 2), but for yeast there is also evidence against this assumption. Halford et al. (1999) suggested that increasing pool sizes of AMP rather than hexokinase could trigger gene expression after glucose application. In *A. muscaria* we could, however, not detect a significant alteration of pools sizes of the different adenine nucleotides (unpublished data).

In higher plants hexokinase-dependent and independent pathways have been identified (Sheen et al., 1999). Evidence for the former is given by the finding that substrates of this enzyme such as mannose and 2-deoxyglucose can mimic glucose-mediated effects (Graham et al., 1994; Jang and Sheen, 1997). For *A. muscaria*, our data show that a regulation in a hexokinase-dependent manner could hold for the expression of PAL (this is in contrast to PAL from higher plants, see Sheen et al., 1999), but not for the fungal monosaccharide transporter. In the latter case, the transporter by itself could be the sensor (Fig. 2). Support for this assumption comes from experiments with yeast, where two hexose transporter-like sensors, SNF3 and RGT2, have been identified which are involved in a glucose-dependent regulation of gene expression (Johnston, 1999).

In addition, hexose supply has a profound effect on fungal metabolism. We have evidence for the involvement of a cascade of phosphorylation/dephosphorylation processes which could start with increasing the pool size of cAMP (Fig. 2). At least for *Amanita muscaria* we could show that levels of this second messenger are affected by sugar supply (Hoffmann et al., 1997), and that a cAMP-regulated protein kinase A is present and alters its extractable activity upon changes in glucose supply (unpublished results). In yeast, one of the definite targets of cAMP-dependent activity of protein kinase A is trehalase (Dickinson, 1999). *A. muscaria* shows activities of both acid (Wisser et al., 2000) and neutral trehalase (Wisser et al., unpublished), and the product of trehalase activity, trehalose, is an important intermediate storage carbohydrate in this fungus. As trehalose is obviously not exchanged with host tissue, conversion of host-derived hexoses into this fungal disaccharide could be of importance for creating a sink for host photoassimilates and thus carbon flow toward the fungal partner.

If hexoses have indeed an impact on fungal gene expression and metabolism then it can be assumed that any environmental interference with the ability of the host plant to supply photoassimilates to the symbiotic fungus will also affect the degree of mycorrhization.

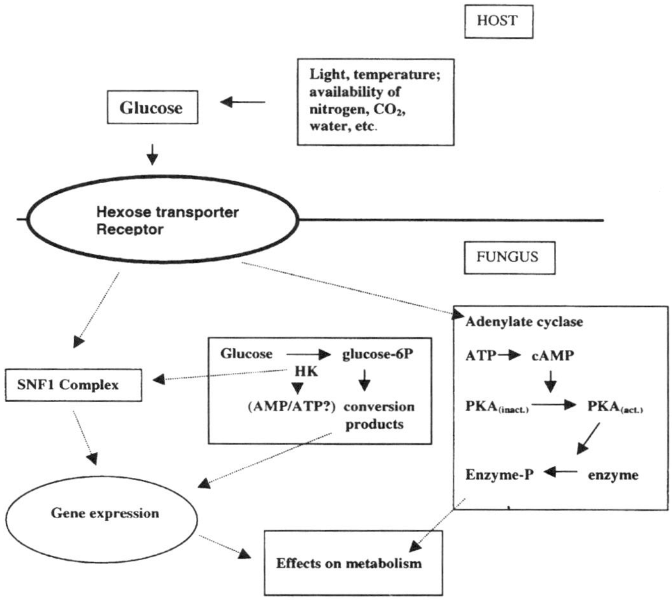

Figure 2. Possible ways of glucose sensing and signal transduction in *Amanita muscaria*. The whole scenario depends on the availability of hexoses, which can be affected in many ways. The scheme is mainly derived from data obtained for yeast. Cyclic AMP (cAMP) and cAMP-dependent proteinkinase A (PKA) have been identified in *A. muscaria*. HK: hexokinase; SNF: sucrose non fermenting. Arrows with broken lines indicate regulatory steps.

Increased supply of CO_2 can, for example, increase the source properties of leaves and will thus improve provision of photoassimilates to the mycorrhizal sink, with the consequences of increased mycorrhiza formation and, finally, better water and nutrient exploitation (Norby et al., 1987; Ineichen et al., 1995; Lewis and Strain, 1996; Loewe et al., 2000; but see Kytöviita et al., 1999).

Reduced supply of photoassimilates by the host plant can have the opposite effect. In experiments with beech clones exposed to drought stress under field conditions, the degree and the type of mycorrhiza were affected (Hampp et al., 2000b). Under such conditions, rates of photosynthesis are decreased due to increased stomatal resistance. In due course, storage pools (starch) are depleted, and less carbohydrates are available for the fungal partner. Those mycorrhizal fungi which are able to deal with such a situation convert available host carbon into compatible solutes, mainly sugar alcohols (mannitol, arabitol) in order to increase their water potential (Hampp et al. 2000b).

Similarly, increased nitrogen supply can reduce mycorrhiza formation under limiting carbon supply. As discussed above, under these conditions photoassimilates will preferentially be directed to mesophyll cell glycolysis and in due course support synthesis of amino acids, rather than sucrose formation and carbon export to sink organs. We assume that in consequence the availability of hexoses in the Hartig net will be rather limited and, by decreasing below the threshold concentration, hexose-regulated expression of fungal genes will be halted. In consequence, the symbiotic interaction should disintegrate and this is clearly shown by the decrease of fungal markers (ergosterol, mannitol, trehalose) in pot experiments with supplemented nitrogen (Wallenda et al., 1996).

In summary, we conclude that the availability of hexoses at the plant fungus interface plays an important role in the establishment and perpetuation of ECM. In consequence we suggest that any environmental factor which impairs carbon allocation to the symbiotic interface will also affect the degree of mycorrhization and, thereby, the viability of the host tree.

ACKNOWLEDGEMENT

We are indebted to Margret Ecke, Elke Klenk, and Andrea Bock for excellent technical assistance. As far as own work is concerned, we gratefully acknowledge financial support by the Deutsche Forschungsgemeinschaft and the county of Baden-Württemberg (PEF).

REFERENCES

Arnebrandt K and Söderström B (1992) Effects of different fertilizer treatments on ectomycorrhizal colonisation potential in two Scots pine forests in Sweden. *Forest Ecology and Management* **53**: 77-89.

Bledsoe CS and Zasoski R J (1983) Effects of ammonium and nitrate on growth and nitrogen uptake by mycorrhizal Douglas-fir seedlings, *Plant and Soil* **71**: 445-54.

Botton B and Dell B (1994) Expression of glutamate dehydrogenase and aspartate aminotransferase in eucalypt ectomycorrhizas. *New Phytol* **126**: 249-257.

Chalot M and Brun A (1998) Physiology of organic nitrogen acquisition by ectomycorrhizal fungi and ectomycorrhizas. *FEMS Microbiol Rev* **22**: 21-44.

Chalot M, Brun A, Botton B and Söderström B (1996) Kinetics, energetics and specificity of the general amino acid transporter from the ectomycorrhizal fungus Paxillus involutus. *Microbiol* **142**: 1749-1756.

Champigny M-L (1995) Integration of photosynthetic carbon and nitrogen metabolism in higher plants. *Photosynthesis Res* **46**: 117-127.

Dickinson JR (1999) Nitrogen metabolism. In: The metabolism and molecular physiology of Saccharomyces cerevisiae (Dickinson JR, Schweizer M, eds.) pp. 57-77, Taylor & Francis, UK

Finlay RD, Ek H, Odham G and Söderström B (1988) Mycelial uptake, translocation and assimilation of nitrogen from ^{15}N-labelled ammonium by *Pinus sylvestris* plants infected with four different ectomycorrhizal fungi, *New Phytol* **110**: 59-66.

France RC and Reid CPP (1983) Interactions of nitrogen and carbon in the physiology of ectomycorrhizae. *Can J Bot* **61**: 964-984.

Garbaye J (1991) Biological interactions in the mycorrhizosphere. *Experientia* **47**: 370-375.

Graham IA, Denby KJ and Leaver CJ (1994) Carbon catabolite repression regulates glyoxylate cycle gene expression in cucumber. *Plant Cell* **9**: 1825-1841.

Halford NG, Purcell PC and Hardie DG (1999) Is hexokinase really a sugar sensor in plants? *Trends Plant Sci.* **4**: 117-120.

Hampp R, Nehls U and Wallenda T (2000a) Physiology of mycorrhiza. Progress in Botany, Vol. 61, pp. 223-254.

Hampp R, Shi L, Guttenberger M, Nehls U (2000b) Mykorrhizierung und Stresstoleranz von Ökotypen der Buche (*Fagus sylvatica*) L.): "Conventwaldprojekt". FZKA/PEF-Berichte http://bwplus.fzk.de/pef/diskpef00/hampp/hampp.htm.

Hampp R, Schaeffer C (1999) Mycorrhiza – Carbohydrate and energy metabolism. In: Mycorrhiza: Structure, function, molecular biology and biotechnology, Varma A and Hock B (eds). pp 273- 303.

Hoffmann E, Wallenda T, Schaeffer C and Hampp R (1997) Cyclic AMP, a possible regulator of glycolysis in the ectomycorrhizal fungus Amanita muscaria. *New Phytol* **137**: 351-356.

Ineichen K, Wiemken V and Wiemken A (1995) Shoots, roots and ectomycorrhiza formation of pine seedlings at elevated atmospheric carbon dioxide. *Plant Cell Environm* **18**: 703-707.

Ingestad T, Arveby AS and Kähr M (1986) The influence of ectomycorrhiza on nitrogen nutrition and growth of *Pinus sylvestris* seedlings. Physiol. *Plant* **68**: 575-582.

Jang J-C, Sheen J (1997) Sugar sensing in higher plants. *Science* **2**: 208-214.

Johnston M (1999) Feasting fasting and fermenting. *Trends Genet* **15**: 29-33.

Keeney DR (1980) Prediction of soil nitrogen availability in forest ecosystems: a literature review. *Forest Sci* **26**: 159-171.

Kytöviita M-M, Pelloux J, Fontaine V, Botton B and Dizengremel P (1999) Elevated CO_2 does not ameliorate effects of ozone on carbon allocation in *Pinus halepensis* and *Betula pendula* in symbiosis with *Paxillus involutus*. *Physiol Plant* **106**: 370-377.

Lewis JD and Strain BR (1996) The role of mycorrhizas in the response of *Pinus taeda* seedlings to elevated CO_2. *New Phytol* **133**: 431-443.

Loewe A, Einig W, Shi L, Dizengremel P and Hampp R (2000) Mycorrhiza formation and elevated CO_2 both increase the capacity for sucrose synthesis in source leaves of spruce and aspen. *New Phytol* **145**: 565-574.

Lorillou S, Botton B and Martin F (1996) Nitrogen source regulates the biosynthesis of NADP-glutamate dehydrogenase in the ectomycorrhizal basidiomycete Laccaria bicolor. *New Phytol* **132**: 289-296.

Marmeisse R, Jargeat P, Wagner F, Gay G and Debaud JC (1998) Isolation and characterization of nitrate reductase deficient mutants of the ectomycorrhizal fungus Hebeloma cylindrosporum. *New Phytol* **140**: 311-318.

Martin F and Botton B (1993) Nitrogen metabolism of ectomycorrhizal fungi and ectomycorrhiza. *Adv Plant Path* **9**: 83-102.

Melin E and Nilsson H (1952) Transport of labelled nitrogen from an ammonium source to pine seedlings through mycorrhizal mycelium. *Svensk Botanisk Tidskrift* **46**: 281-285.

Nehls U, Ecke M and Hampp R (1999a) Sugar and nitrogen-dependent regulation of an *Amanita muscaria* phenylalanine ammonium lyase gene. *J Bacteriol* **181**: 1931-1933.

Nehls U, Kleber R, Wiese J and Hampp R (1999b) Isolation and characterization of a general amino acid permease from the ectomycorrhizal fungus *Amanita muscaria. New Phytol* **144**: 343-349.

Nehls U, Wiese A and Hampp R (2000) Exchange of carbohydrates between symbionts in ectomycorrhiza. In: The Mycota, Vol. IX, Hock B (ed.). Springer-Verlag, in press.

Nehls U, Wiese A, Guttenberger M and Hampp R (1998) Carbon allocation in ectomycorrhiza: identification and expression analysis of an *A. muscaria* monosaccharide transporter. *Molec Plant Microb Int* **11**:167-176.

Norby RJ, O'Neill EG, Hood, WG and Luxmore RBJ (1987) Carbon allocation, root exudation and mycorrhizal colonization of *Pinus echinata* seedlings grown under CO_2 enrichment. *Tree Physiol* **3**: 203-210.

Plassard C, Scheromm P and Llamas H (1986) Nitrate assimilation by maritime pine and ectomycorrhizal fungi in pure culture. In: Gianinazzi-Pearson V and Gianinazzi S (eds). Physiological and genetical aspects of mycorrhizae. pp 383-388. INRA, Paris, France.

Read DJ (1991) *Mycorrhizas in ecosystems Experientia* **47**: 376-391.

Read DJ, Leake JR and Langdale AR (1989) The nitrogen nutrition of mycorrhizal fungi and their host plants. In: Boddy L, Marchant R and Read DJ (eds) Nitrogen, phosphorus and sulphur utilization by fungi. pp. 181-204. Cambridge University Press, Cambridge.

Rygiewicz PT, Bledsoe CS and Zasoski RJ (1984) Effects of ectomycorrhizae and solution pH on [^{15}N]ammonium uptake by coniferous seedlings. *Canadian J Forest Research* **14**: 885-892.

Sheen J, Zhou Li, Jang J-C (1999) Sugars as signalling molecules. *Current Opinion in Plant Biology* **2**: 410-418.

Smith SE and Read DJ (eds) (1997) Mycorrhizal Symbiosis, 2nd Edition. Academic Press, San Diego.

Turnau K, Berger A, Loewe A, Einig W, Hampp R, Chalot M, Dizengremel P and Kottke I (2000) CO_2 concentration and nitrogen input affect the C and N storage pools in *Amanita muscaria / Picea abies* mycorrhizas. Tree Physiol, in press.

Wainwright M (1993) Oligotrophic growth of fungi - stress or natural state. In: Jennings DJ (ed) Stress tolerance of fungi. pp 127-144, Marcel Dekker, New York.

Wallander H and Nylund JE (1991) Effects of excess nitrogen on carbohydrate concentration and mycorrhizal development of *Pinus sylvestris* L. seedlings. *New Phytol* **119**: 405-411.

Wallenda T, Schaeffer C, Einig W, Wingler A, Hampp R, Seith B, George E and Marschner H (1996) Effects of varied soil nitrogen supply on Norway spruce (Picea abies [L.] Karst.) *Plant Soil* **186**: 361-369.

Wiese J, Kleber R, Hampp R and Nehls U (2000) Functional characterization of the *Amanita muscaria* monosaccharide transporter AmMst1. *Plant Biol* **2**:278-282.

Wiklund K, Nilsson L-O and Jacobsson S (1995) Effect of irrigation, fertilization and artificial drought on basidioma production in a Norway spruce stand. *Can J Bot* **73**: 200-208.

Wingler A, Einig W, Schaeffer C, Wallenda T, Hampp R, Wallander H and Nylund J-E (1994) Influence of different nutrient regimes on the regulation of carbon metabolism in Norway spruce (*Picea abies* [L.] Karst.) seedlings. *New Phytol* **128**: 323-330.

Wisser G, Guttenberger M, Hampp R and Nehls U (2000) Identification and characterization of an extracellular acid trehalase from the ectomycorrhizal fungus Amanita muscaria. *New*

Phytol, **146**: 169-175.

Wright DP, Scholes JD, Read DJ and Rolfe SA (2000) Changes in carbon allocation and expression of carbon transporter genes in Betula pendula Roth. colonized by the ectomycorrhizal fungus Paxillus involutus (Batsch) Fr. *Plant, Cell and Environment* **23**: 39-49.

SECTION II

STRESS PHYSIOLOGY

UNDERSTANDING THE ROLE OF OZONE STRESS IN ALTERING BELOWGROUND PROCESSES

CHRISTIAN P. ANDERSEN
U.S. Environmental Protection Agency, National Health and Environmental Effects Research Laboratory, Western Ecology Division, 200 SW 35th Street, Corvallis, OR 97333, USA

Key words: ozone, ponderosa pine, roots, carbon allocation, respiration

1. INTRODUCTION

Forested ecosystems are comprised of tremendous biological diversity and functional complexity both above and belowground. Soil ecosystems are known to contain thousands of species, with many more that have not yet been identified. Soil heterotrophic organisms depend on green plants for their energy, while autotrophs depend on soil heterotrophs for processing and cycling nutrients that are needed for continued growth. Together, heterotrophs and autotrophs form a complex web of interacting entities, the balance of which is always changing as ecosystems change in response to stress.

Our relative lack of understanding about ecosystem processes in general and soil ecosystems in specific makes it difficult to understand how O_3 may be affecting forested ecosystems. Ecosystem fluxes of carbon and nitrogen are highly regulated by coupled processes occurring above and belowground. Microbial interactions belowground have been shown to contribute to plant biodiversity and productivity (van der Heijden, *et al.*, 1998). Belowground processes also contribute substantially to fluxes of carbon into and out of ecosystems (Law *et al.*, 1999). A better understanding of mechanistic processes occurring in plant roots and subsequent changes in food web organisms is a necessary foundation for understanding how a stress such as O_3 affects ecosystem structure and function.

Food webs are highly integrated systems that have a regular trophic structure that

S. Huttunen et al. (eds.),
Trends in European Forest Tree Physiology Research, Cost Action E6: EUROSILVA, 65-79

typically consist of three or four transfers within the food chain (Pimm and Lawton, 1977; Pimm, 1982; Moore *et al.*, 1993). Species diversity can be high, particularly in soil food webs, but trophic structure is thought to be constrained. Food web organisms process energy at different rates and respond to disturbance differently. Moore *et al.* (1993) examined 7 food webs at four different sites to examine the importance of species in ecosystem structure and function. The results showed that predicting the impact of a stress on a system can be difficult since a species may control a disproportionate fraction of a systems biomass relative to its' own biomass. Moore *et al.* (1993) demonstrated this concept experimentally and through model simulations by showing that in some systems productivity was related to the levels of mycorrhizal fungi present in the soils after the stress. Naeem *et al.* (1994) showed experimentally that removing soil faunal groups, i.e., reducing trophic level diversity, could result in a change in plant productivity in controlled mesocosms.

Biological processing of carbon compounds by soil organisms is a fundamental property of a functional ecosystem. Soil organisms decompose organic materials, sequester nutrients, and fix nutrients such as nitrogen from the atmosphere. Processing of soil carbon residues leads to the formation of soil organic matter, which influences physical and chemical properties of any given soil (Brady, 1974). Therefore, the soil food web is not only responsible for recycling and retention of nutrients needed by plants, but it is responsible for establishing soil physical and chemical characteristics such as aggregate structure, porosity, water holding capacity, and cation exchange capacity. Any change in soil food web structure or function due to oxidant stress could alter the fundamental properties of ecosystems.

2. OZONE STRESS AND CHANGES BELOWGROUND

Ozone may alter belowground processes through several different pathways, however all are considered indirect since the site of O_3 action is in the foliage, with very little penetration of O_3 into soil due to scavenging of O_3 molecules (Turner *et al.* 1973; Blum and Tingey 1977). The working hypothesis is that a change in carbon flux to soils from plants in response to O_3 will alter food web activity, thereby altering ecosystem properties. An idealized ecosystem emphasizing belowground properties is illustrated in Figure 1.

A host of physiological changes occur in plant foliage when plants are exposed to O_3 which will not be discussed here (e.g., Weber *et al.*, 1994; Ozone Criteria Document, US EPA, 1996). Belowground, studies have shown that O_3 may affect the soil ecosystem through one or more of the following: 1) changes in litter quality and quantity, 2) decreased carbon allocation to roots; 3) altered root exudation and soil CO_2 flux; and 4) decreased root growth, and possibly increased root mortality.

2.1 Litter Quantity and Quality

Ozone has been found to increase rates of foliar senescence, and in some cases, nitrogen remobilization from foliage into the plant is not complete at the time of foliage abscission (Stow *et al.*, 1992; Matyssek *et al.*, 1993; Patterson and Rundel, 1995). Any change in litter quality, quantity, or the timing of senescence resulting from O_3 exposure will have an impact on populations of soil arthropod shredders, and these changes would be propagated through other trophic levels in the food chain (Figure 1).

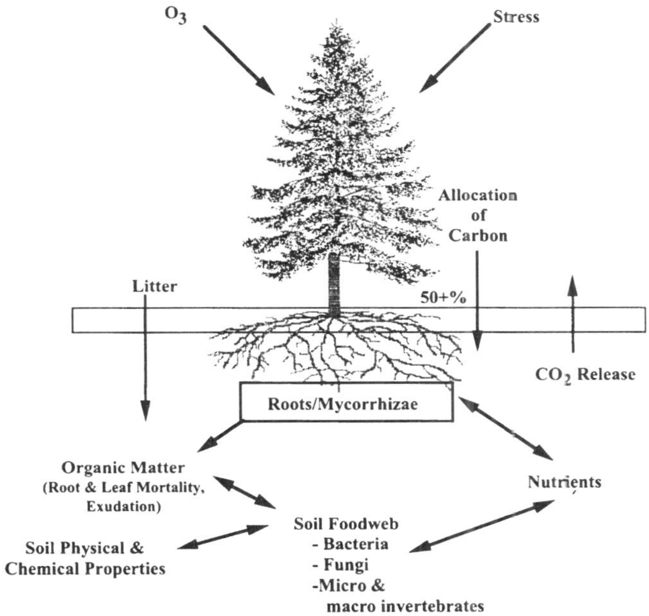

Figure 1. Conceptual diagram of carbon flow in a tree-soil system, including transfer between biotic and abiotic components belowground. See text for details.

Fenn and Dunn (1989) found that litter of ponderosa pine (*Pinus ponderosa* Laws.) and Jeffrey pine (*Pinus jeffreyi* Grev. & Balf.) decomposed more rapidly in plots receiving the highest pollutant loads across a pollution gradient in Southern California. There was evidence that litter in the highest pollutant areas originated from younger foliage, based on the high N and relatively low Ca contents of foliage from these sites. Higher N content of litter in the high pollution plots did not appear to be due to reduced translocation in of N from senescing foliage, but rather due to increased site fertility or atmospheric deposition of nitrogen on foliage and litter (Fenn 1991).

In short term, controlled studies with cottonwood (*Populus deltoides* (Bar)), senesced foliage was found to decompose more slowly, suggesting an opposite response than was observed in ponderosa and Jeffrey pines (Findlay and Jones, 1990). Exposure to O_3 for 5 hours (200 $\mu L\ m^{-3}$) caused both early senescence and a change in the chemical composition of leaves that decreased subsequent rates of decomposition (Findlay and Jones, 1990; Findlay *et al.*, 1996). Foliage of plants exposed to O_3 was found to contain higher N concentrations, but also higher concentrations of phenolic materials which were thought to be responsible for the decreased decomposition rates observed.

Nitrogen lost via early senescence of foliage is not available for internal recycling, and therefore represents a significant cost to the plant. Preliminary results with aspen suggest that O_3 treated plants retranslocate less N from senescing foliage than controls, and therefore additional root uptake of nitrogen is necessary to maintain the same N balance (J. Fletcher, personal communication). In addition, senescence earlier in the growing season may affect decomposition rates simply due to edaphic conditions present at the time of leaf drop. Additional research is needed on the short and long-term effects of O_3 on litter quality due to the potential to alter soil properties, which may lead to changes in ecosystem structure and function.

2.2 Altered Carbon Allocation

Ozone may alter soil processes through decreased carbon allocation to roots and thus root-associated soil organisms. Research emphasis has been on quantifying how O_3 alters the energy flow from plant shoots to roots and mycorrhizae. The results illustrate how O_3 may alter soil processes by altering carbon flux to roots and subsequently soils. While many of these studies have been done in mesocosms and not yet verified in natural ecosystems, they provide the mechanistic foundation to understand ways in which O_3 may be altering ecosystem structure and function.

Studies have shown that O_3 decreases allocation to roots, leading to decreased root growth and increased shoot/root ratio (Manning *et al.*, 1971, McLaughlin and McConathy 1983, McCool and Menge, 1983, Cooley and Manning, 1987, Gorissen and van Veen, 1988, Spence *et al.*, 1990, Gorissen *et al.*, 1994, Andersen and Rygiewicz, 1995; Rennenberg *et al.*, 1996). The threshold level of O_3 necessary for response varies by species. Short-term pulse-chase experiments using ^{14}C showed that mycorrhizal roots received more carbon than nonmycorrhizal roots in the absence of O_3 (i.e., increase root sink strength) (Andersen and Rygiewicz, 1991; Rygiewicz and Andersen, 1994). Exposure to O_3 for 25 days (39.3 ppm hr total accumulated exposure) decreased the maximum rate of carbon transport to roots in both mycorrhizal and nonmycorrhizal plants, and there was no evidence that mycorrhizal plants were less susceptible to O_3 stress as hypothesized (Andersen and Rygiewicz, 1995). Extramatrical hyphae of O_3 exposed seedlings received only 60% of the carbon that was allocated to hyphae of control plants (Table 1).

Decreased carbon allocation belowground has been observed to reduce root carbohydrates and root growth in several species, including ponderosa pine (Tingey *et al.*, 1976, Cooley and Manning, 1987, Hogsett *et al.*, 1985, Andersen *et al*, 1991; Gorissen *et al.*, 1994; Andersen *et al.*, 1997). In at least one study, there is evidence that O_3 reduces root system size in the natural forests (Grulke *et al.*, 1998). Grulke *et al.* (1998) found decreased coarse, medium and fine root biomass with increased pollutant load across a gradient in southern California (Figure 2). Although the effects of O_3 could not be separated

Table 1. Biomass-weighted retention of ^{14}C in needles, coarse root, fine root, and fungal components of mycorrhizal ponderosa pine exposed to ozone. Values were obtained by dividing percent ^{14}C allocated to each tissue component by that component dry weight. Total ozone exposue in ozone treatment was 39.3 ppm hr (calculated by summing the hourly mean concentration, 24 hrs per day). Data from Andersen and Rygiewicz (1995).

	Needles	Course Root	Fine Root	Fungus
Control	39.5	20.9	25.1	18.7
± SE	± 4.0	± 2.8	±2.8	± 2.0
O_3	39.8	17.2	24.2	11.2
± SE	±6.7	± 1.5	±7.9	± 2.3
P Value	0.886	0.221	0.836	0.031

completely from other known stresses across the pollutant gradient, there was evidence that O_3 was a key factor in the patterns observed. The field results suggesting cumulative effects of O_3 on roots are consistent with controlled experiments that revealed growth and carbohydrate concentrations of new roots are lower in seedlings exposed to O_3 during the previous year, providing evidence of carry-over effects of O_3 on roots (Andersen *et al.*, 1991, 1997).

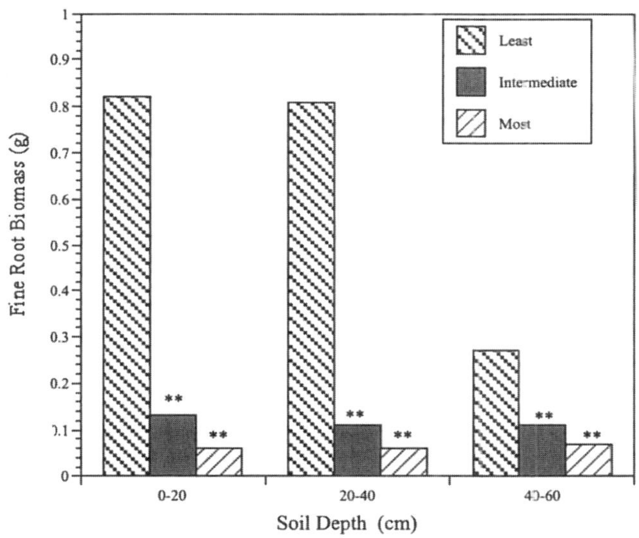

Figure 2. Fine root biomass (g) from 20 X 20 cm pits at 3 depths in July across an air pollution gradient in S. California. Least polluted site = Camp Osceola, intermediate site = Strawberry Peak, and most polluted site = Camp Paivika. **Denotes statistical significance at p<0.05. Data from Grulke et al. (1998).

Natural stresses alter carbon allocation to roots of plants, and the co-occurance of ozone with these natural stresses may interact in unpredictable ways. Temple *et al.* (1992) found that drought-stressed ponderosa pine seedlings had less ozone injury than well watered trees. Similarly, Lippert, *et al.* (1996) found that effects of ozone on photosynthetic parameters of Norway spruce (*Picea abies* (L.) Karst.) only occurred with adequate water and nutrients. In contrast, birch (*Betula pendula* Roth) response to O_3 was greatest at low N supply (Pääkkönen and Holopainen, 1995). Soil CO_2 flux increased in when pine were exposed to ozone, and the response was greatest at low nutrient supply (Andersen and Scagel, 1997). Changes in allocation due to ozone are likely to alter plant response to other stresses; clearly, additional research is necessary on multiple stress interactions.

2.3 Root Respiration, Exudation and Soil CO_2 Flux

Smaller root systems containing less soluble carbohydrates, coupled with less carbon allocation to mycorrhizal hyphae, would be expected to reduce carbon flow to soil organisms. Since microbial growth in soil is limited by carbon availability (Zak, *et al.*, 1994), the combined effects of ozone on roots and mycorrhizae would be expected to decrease soil CO_2 flux.

Edwards (1991) observed lower root respiration in loblolly pine (*Pinus taeda* L.) seedlings exposed to O_3 for one season but not after two exposure seasons. In studies with ponderosa pine seedlings, an opposite response was observed such that O_3 exposure resulted in increased CO_2 flux from soil (Andersen and Scagel, 1997; and Scagel and Andersen, 1997) (Figure 3, panel b). In nutrient amended potting soil, lacking natural food web organisms, there was no apparent increase in soil CO_2 flux over the same time interval (Figure 3, panel a).

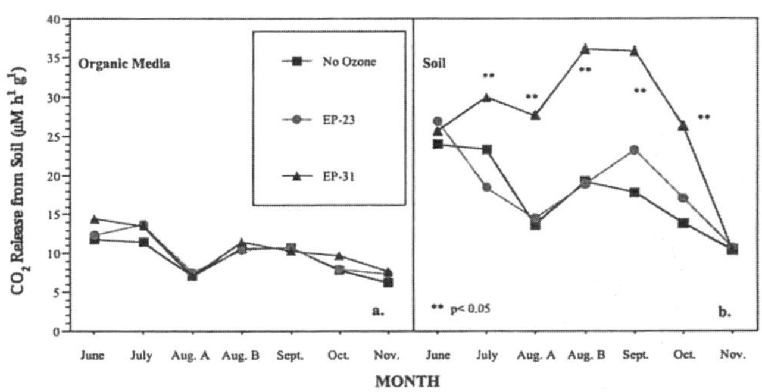

Figure 3. Soil CO_2 flux (µmol CO_2 g-1) from ponderosa pine exposed to ozone. Panel a) is from pots where organic media was employed. Panel b) is from same sized pots containing soil from a ponderosa pine forest. Total ozone exposure in the EP-23 treatment was 213.1 ppm hr and for the EP-31 treatment 239.5 ppm hr over the 19 wk exposure period (calculated by summing the hourly mean concentration, 24 hrs per day). ** Denotes statistical significance at the p<0.05 level. Date from Scagel and Andersen (1997).

The extent to which O_3 alters root versus microbial respiration through altered carbon availability remains to be determined. Kelting *et al.* (1995) found that O_3 exposure to mature red oak trees increased fine root respiration through out the growing season, while total soil CO_2 flux was increased by O_3 in May but decreased in July and September. In other studies, increased CO_2 release from soil was associated with an increase in O_2 uptake in O_3 treated plants, but not in equal proportions (Andersen and Scagel, 1997). The results showed that O_3 reduced RQ (Respiratory Quotient= CO_2/O_2 ratio), suggesting that O_3 may have a direct affect on pine root metabolism (Andersen and Scagel, 1997). Decreased RQ may reflect a change in substrate utilization to more highly reduced carbon forms by roots of plants exposed to O_3. However, preliminary studies on individual roots of ponderosa pine exposed to O_3 have not revealed large changes in respiration, suggesting that O_3 may affect soil organisms more than root metabolism *per se* (not shown).

In the short-term, it is possible that decreased allocation of carbon to roots causes an increase in membrane leakiness, which initially may stimulate rhizosphere microflora by increasing carbon flux to the rhizosphere (McCrady and Andersen, 2000). Short-term studies using wheat (*Triticum aestivum* L. 'ANZA') showed that O_3 increased soluble root exudation, which may stimulate microbial populations in the rhizosphere (McCrady and Andersen, 2000; Table 2). Shaffer (1988) found increased bacteria and fungi in the rhizosphere of O_3 exposed plants. Reddy *et al.* (1991) found that low levels of O_3 plus simulated acid rain at pH 5.3 and 3.3 stimulated soil arylsufatase activity; at higher O_3 concentrations, activity was reduced. Scagel and Andersen (1997) found increased bacterial and fungal populations in soils of seedlings exposed to O_3.

Increased soluble root exudation may facilitate mycorrhizal colonization by providing substrates in the rhizosphere to support the colonization process. Reich *et al.* (1985) and Stroo *et al.* (1988) both found increased mycorrhizal colonization of plants exposed to low levels of O_3. Over time, however, reduced carbon availability to the fungus would be expected to impair hyphal proliferation, and may alter host-species compatibility, perhaps favoring less carbon demanding fungal species over others.

Others have found O_3 decreased root exudation and mycorrhizal colonization due to O_3. McCool and Menge (1983) found decreased amino acid concentrations in mycorrhizal root exudates of tomato exposed to O_3. Simmons and Kelly (1989), Adams and O'Neill (1991), and Edwards and Kelly (1992) all found a trend towards decreased number of short roots of loblolly pine colonized by mycorrhizal morphotypes with increased O_3. Coupled with reports of increased colonization in response to O_3 (above), these results appear to raise more questions than answers, however they illustrate the potential importance of altered carbon availability in support of rhizosphere organisms, including mycorrhizal fungi. Whether or not a change in mycorrhizal diversity or an increase in activity of rhizosphere organisms results in a change at higher levels of biological organization is not yet known due to our relatively poor understanding of the structural arrangement and functional redundancy of soil food web organisms.

	^{14}C Activity	^{14}C Act./Root DW
O_3	0.227	1.72
± SE	± 0.005	± 0.03
Control	0.208	1.33
± SE	± 0.005	± 0.03
P Value	0.039	0.0005

Table 2. ^{14}C Activity recovered in the root eluate (fixed root exudate plus $^{14}CO_2$), normalized to the total ^{14}C recovered in the root plus elate (Kbq), and normalized to root dry weight (kBq g^{-1}). Total ozone exposure in ozone treatment was approximately 39 ppm hr (calculated by summing the hourly mean concentration, 24 hrs per day). Data from McCrady and Andersen (2000).

2.4 Root Mortality and Turnover

A third way O_3 may increase carbon flux to soil and CO_2 flux from soil is through increased root mortality and turnover. Ozone may decrease root longevity by reducing root starch (Marshall and Waring, 1985), a process that over the short-term would increase flux of carbon to food web organisms and stimulate heterotrophic respiration. Kelting *et al.* (1995) used a compartment -flow model to estimate fine root production and turnover in seedlings and mature trees of red oak (*Quercus rubra* L.). Mature red oak showed decreased fine root production and turnover in 2X O_3 compared to ambient O_3, while seedlings showed decreased fine-root production and no change in turnover compared to the ambient O_3 treatment. They also observed a decrease in belowground carbon use efficiency which was hypothesized to be due to increase root respiration costs in the 2X O_3 treatment. The study revealed the potential for O_3 to alter belowground carbon allocation in mature red oak, and points to the need for additional studies with other species.

3. OZONE AND CO_2 FLUX FROM ECOSYSTEMS

If O_3 alters litter inputs, root exudation, and rates of root turnover, then it may affect CO_2 flux from ecosystems and ultimately ecosystem carbon budgets. Law *et al.* (1999) estimated that 76% of ecosystem respiration from an old growth ponderosa pine ecosystem (*Pinus ponderosa* Laws.) originated from root and microbial respiration of soils. The degree of the response to O_3 would be dependent on the frequency and magnitude of exposure. Figure 4 shows hypothesized respiratory flux of CO_2 from soils of ponderosa pine exposed to O_3. Under 'typical' conditions, CO_2 flux approaches a historical 'steady state', which is a function of temperature, moisture, soil type, species composition, etc. of trees occupying the site. During years when O_3 is high, CO_2 release from soil is hypothesized to increase above this historical average, due to accelerated leaf senescence, altered root metabolism, increased exudation and perhaps increased root turnover. Heterotrophic soil CO_2 flux increases as newly released carbon is metabolized by soil microflora. One consequence of increased carbon flux to soil might be increased microbial sequestration of N and other nutrients, potentially reducing plant available N (Diaz *et al.*, 1993). The response is hypothesized to be transitory, with subsequent decreased CO_2 flux from soil below the historical average over time due to long term effects of O_3 on standing root biomass. If there is a respite year when O_3 is low, the plant would be expected to compensate and perhaps reestablish conditions near a historical steady state.

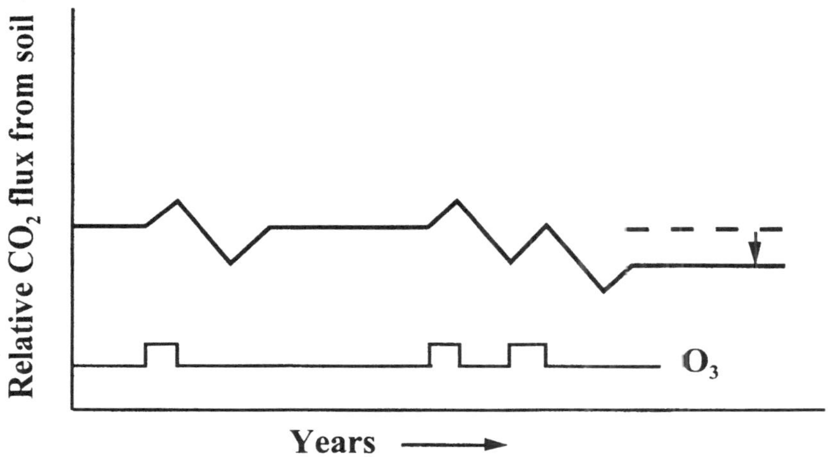

Figure 4. Hypothesized flux of CO_2 from soil in idealized ecosystem exposed to ozone during specific time periods. See text for details. Modified from Andersen, 2000.

If O_3 exposure is high year after year, soil CO_2 flux would be expected to decrease below the historical average due to the overall effect of O_3 on carbon allocation belowground to roots and mycorrhizal fungi. Over time, standing root biomass would decrease, and root respiration and carbon flux to soils would decrease. Similarly, accelerated needle senescence of older needle age classes would initially increase carbon flux to soils, stimulating heterotrophic respiration, however over time would lead to decreased litter inputs. Since O_3 varies from year to year, it is possible that soil ecosystems may remain in a state of nonequilibrium with respect to above and belowground carbon and N pools.

Although the effects of O_3 on individual plant species have been relatively well documented in many cases, the role that O_3 plays in altering carbon flux into and out of ecosystems is poorly understood. While not a substitute for field studies, controlled seedling studies provide the mechanistic basis for identifying the role O_3 may play in ecosystem carbon and nitrogen dynamics. It is important to understand both the short and long-term effects of O_3, and to understand potential interactions that may occur with other stresses in the field.

4. MODELING RESPONSES BELOWGROUND

Models provide a way of synthesizing and integrating below-ground complexity to better understand possible changes resulting from O_3 stress. Most process-level and ecosystem models include relatively detailed above ground subroutines and comparatively vague below-ground components. The slow development of models dealing with belowground ecosystems results in part from our general lack of knowledge about the complex interactions that occur belowground.

In any modelling effort, the hierarchical level of the question being addressed will determine the level of detail required in the model. Soil processes occur over time and spatial scales that vary by several orders of magnitude, therefore up-front identification of the scientific question(s) is pivotal to success. As one moves to lower levels, e.g., faster processes, the level of observable detail increases. As one moves to higher level processes, detail decreases. Currently there are few models that incorporate detailed low-level processes in soil, in part because of our relatively poor understanding of the controls and rates of these processes. Wullschleger *et al.* (1994) discussed the value of using models to follow below-ground response to climate change, and concluded that future modelling efforts should be encouraged because of the potential to better understand changes occurring below ground.

One approach to modelling belowground processes is to identify those key elements that affect higher order processes. Ideally, processes at lower levels of organization should be understood before higher level processes can be adequately modelled. O'Neill *et al.* (1991) discussed the importance of constraint in belowground systems when moving across levels. For example, if mycorrhizae are altered by the rate of carbon movement belowground, and mycorrhizae affect plant growth through water or nutrient uptake, then a change in mycorrhizae at a low-level may constrain higher level processes through changes in growth (O'Neill *et al.*, 1991). Consequently, factors that affect mycorrhizal associations can be expected to affect the linkage, which would be expected to affect higher order processes. Therefore, in a modelling context, understanding mycorrhizal response to stress would be an important factor to include when modelling stresses that affect the symbiotic association. Despite information suggesting O_3 alters the belowground system, there is currently insufficient information to identify all of the potential linkages affected by O_3.

The physiological process model TREGRO has been used to simulate the influence of O_3 on carbon allocation to roots in *Populus tremuloides* (Yun *et al.*, 2000), *Abies concolor* (Retzlaff *et al.*, 2000), *Quercus rubra* (Weinstein *et al.*, 1998), *Acer saccharum* (Retzlaff *et al.*, 1996), and *Pinus ponderosa* (Andersen *et al.*, 1992; Constable and Taylor, 1997). The model was developed specifically to address carbon-limiting stresses and includes detailed photosynthesis and allocation components which makes it particularly suited to simulate O_3 stress (Weinstein *et al.*, 1991). In simulations with *Acer saccharum*, O_3 exposure for 10 years depleted total nonstructural carbohydrate pools in coarse roots and reduced fine root production in a fashion similar to what was observed in the field (Retzlaff, *et al.*, 1996). A similar approach using white fir (*A. concolor*) resulted in no significant change in root biomass or TNC, probably due to the greater tolerance of white fir to O_3. In simulations with *Quercus rubra*, tree response was similar to observations in the field, however the model overestimated the magnitude of the response in roots. In this particular case, the model allowed the authors to identify where their input variables were incorrect, and they concluded that additional measurements of mature tree respiratory fluxes may be needed.

Constable and Taylor (1997) used the model to examine the response to O_3 of two varieties of ponderosa pine, a western and an eastern variety. Variety *Pinus ponderosa* var. *ponderosa*, the western variety, has higher stomatal conductance and tends to allocate less carbon belowground than the eastern variety, *P. ponderosa* var. *scopulorum*. When 3 years of O_3 exposure was simulated, fine root biomass of var. *ponderosa* was reduced by 117% compared to controls without O_3, while var. *scopulorum* showed fine root biomass reductions of only 10%. Total nonstructural carbohydrates also were reduced more by O_3 stress in the var. *ponderosa*. The simulations illustrate how models that incorporate physiological differences between plant varieties can be used to predict changes in carbohydrate allocation and growth belowground.

TREGRO was modified to simulate the presence of mycorrhizas on tree roots (Andersen,*et al.*, 1992). The goal was to determine if a model needed to include mycorrhizae as a discrete component of the root system, or whether changes in model input parameters could adequately simulate the different rates of respiration, storage and growth in mycorrhizas compared to roots. The results suggested that simply changing certain input parameters did not allow realistic simulation of ponderosa pine seedling growth over the three years. Because of the importance of mycorrhizas in plant uptake and metabolism, it was concluded that realistic models must include both roots and separate mycorrhizal components in order to accurately simulate growth.

The potential for using models to simulate the effects of O_3 on the soil ecosystem is tremendous. The few studies that have been done illustrate how relatively undetailed belowground subroutines in models can reasonably simulate plant response to O_3, including reduced root growth and TNC pools. As our understanding of the belowground ecosystem improves, models can incorporate more realistic subroutines to simulate interactions, and will be powerful tools for extrapolating across broader temporal and spatial scales.

5. SUMMARY

These examples illustrate how O_3 may lead to changes belowground, and how complex it may be to elucidate changes in response to O_3. Future emphasis should be on better understanding of the role stresses have, singly and in combination, on soil food web structure and function. Observing ecosystems in their natural state is necessary but restricts prediction of change beyond the conditions present at the time of measurement, i.e., in response to a stress. Controlled studies are useful in determining how fluxes and rates change under the conditions present in a particular experiment, but do not allow easy extrapolation of response to natural ecosystems. Mechanistic and field investigations need to proceed in parallel to address possible changes that may result from O_3 stress.

Until we have a better understanding of how stresses affect biotic relationships in the soil, it will be difficult to develop models that allow prediction of ozone response at higher levels of biological organization. Higher level investigations will require better mechanistic understanding of O_3 effects in order to identify what processes need to be included when predicting responses at the level of the ecosystem. There is compelling evidence that ozone alters soil food web structure and possibly function, and the challenge facing each of us is how to measure, analyze, and scale belowground responses in a context that is meaningful for answering important ecosystem questions. At the present time we do not fully understand how changes in the soil affect ecosystem processes, however results to date suggest that the impact of ozone may be greater belowground than aboveground.

ACKNOWLEDGEMENTS

I would like to thank Drs. John Laurence and Craig McFarlane for their constructive comments on an earlier version of this manuscript and my thanks to Virginia Robinson for editorial assistance in preparing this document. The research described in this article has been funded by the U.S. Environmental Protection Agency. It has been subjected to the Agency's peer and administrative review and approved for publication. Mention of trade names or commercial products does not constitute endorsement or recommendation for use.

REFERENCES

Adams M.B. and O'Neill E.G. Effects of ozone and acidic deposition on carbon allocation and mycorrhizal colonization of *Pinus taeda* L. Seedlings. *For. Sci.* **37(1)**, 5-16, 1991.

Andersen C.P. Ozone stress and changes below-ground: linking root and soil processes. *Phyton* **40**, 1-12, 2000.

Andersen C.P., Hogsett W.E., Wessling R. and Plocher M. Ozone decreases spring root growth and root carbohydrate content in ponderosa pine the year following exposure. *Can. J. For. Res.* **21**, 1288-1291, 1991.

Andersen C.P., Laurence J.A. and Hogsett W.E. To what extent to mycorrhizae need to be considered in process modelling efforts?. Poster presentation, The North American Forest Biology Workshop, Sault Ste. Marie, Canada, Aug. 17-20, 1992.

Andersen C.P. and Scagel C.F. Nutrient availability alters below-ground respiration of ozone exposed ponderosa pine. *Tree Physiol.* **17**, 377-387, 1997.

Andersen C.P., Wilson R., Plocher M. and Hogsett W.E. Carry-over effects of ozone on ponderosa pine root growth and carbohydrate concentrations. *Tree Physiol.* **17**, 805-81, 1997.

Andersen C.P. and Rygiewicz P.T. Stress interactions and mycorrhizal plant response: understanding carbon allocation priorities. *Environ. Pollut.* **73**, 217-244, 1991.

Andersen C.P. and Rygiewicz P.T. Allocation of carbon in mycorrhizal Pinus ponderosa seedlings exposed to ozone. *New Phytol.* **131**, 471-480, 1995.

Blum U. and Tingey D.T. A study of the potential ways in which ozone could reduce root growth and nodulation of soybean. *Atmos. Environ.* **11**, 737-739, 1977.

Brady N.C. The Nature and Properties of Soils. 8th Edition, MacMillan Publishing Co., Inc., New

York, 639 pp., 1974.

Cooley D.R and Manning W.J. The impact of ozone on assimilate partition in plants: A review. Environ. *Pollut.* **47**, 95-113, 1987.

Constable J.V.H. and Taylor Jr. G.E. Modeling the effects of elevated tropospheric O_3 on two varieties of *Pinus ponderosa*. *Can. J. For. Res.* **27**, 527-537, 1997.

Diaz S., Grime J.P., Harris J. and McPherson E. Evidence of a feedback mechanism limiting plant response to elevated carbon dioxide. *Nature* **364**, 616-617, 1993.

Edwards G.S. and Kelly J.M. Ectomycorrhizal colonization of loblolly pine seedlings during three growing seasons in response to ozone, acidic precipitation, and soil Mg status. *Environ. Pollut.* **76**, 71-77, 1992.

Edwards N.T. Root and soil respiration responses to ozone in *Pinus taeda* L. Seedlings. *New Phytol.* **118**, 315-321, 1991.

Fenn M. Increased site fertility and litter decomposition rate in high-pollution sites in the San Bernardino Mountains. *For. Sci.* **37(4)**, 1163-1181, 1991.

Fenn M.E. and Dunn P.H. Litter decomposition across an air-pollution gradient in the San Bernardino Mountains. *Soil Sci. Soc. Am. J.* **53**, 1560-1567, 1989.

Findlay S. and Jones C.G. Exposure of cottonwood plants to ozone alters subsequent leaf decomposition. *Oecologia* **82**, 248-250, 1990.

Findlay S., Carreiro M., Krischik V. and Jones C.G. Effects of damage to living plants on leaf litter quality. *Ecol. Appl.* **6(1)**, 269-275, 1996.

Gorissen A. and Van Veen J.A. Temporary disturbance of translocatoin of assimilates in Douglas-fir caused by low levels of ozone and sulfur dioxide. *Plant Physiol.* **88**, 559-563, 1988.

Gorissen A., Joosten N.A., Smeulders S.M. and Van Veen J.A. Effects of short-term ozone exposure and soil water availability on the carbon economy of juvenile Douglas-fir. *Tree Physiol.* **14**, 647-657, 1994.

Grulke N.E., Andersen C.P., Fenn M.E. and Miller P.R. Ozone exposure and nitrogen deposition lowers root biomass of ponderosa pine in the San Bernardino Mountains, California. *Environ. Pollut.* **103**, 63-73, 1998.

Hogsett W.E., Plocher M., Wildman V., Tingey D.T. and Bennett J.P. Growth response of two varieties of slash pine seedlings to chronic ozone exposures. *Can. J. Bot.* **63**, 2369-2376, 1985.

Kelting D.L., Burger J.A. and Edwards G.S. The effects of ozone on the root dynamics of seedlings and mature red oak (*Quercus rubra* L.). *For. Ecol. Manage.* **79**, 197-206, 1995.

Law B.E., Ryan M.G. and Anthoni P.M. Seasonal and annual respiration of a ponderosa pine ecosystem. *Global Change Biol.* **5**, 169-182, 1999.

Lippert M., Häberle K.H., Steiner K., Payer H.D. and Rehfuess K.E. Interactive effects of elevated CO_2 and O_3 on photosynthesis and biomass production of clonal 5-year-old Norway spruce (*Picea abies* (L.) Karst.) under different nitrogen nutrition and irrigation treatments. *Trees* **10**, 382-392, 1996.

Manning W.J., Feder W.A., Papia P.N. and Perkins I. Influence of foliar ozone injury on root development and root surface fungi of pinto bean plants. *Environ. Pollut.* **1**, 305-312, 1971.

Marshall J.D. and Waring R.H. Predicting fine root production and turnover by monitoring root starch and soil temperature. *Can. J. For. Res.* **15**, 791-800, 1985.

Matyssek R., Günthardt-Goerg M.S., Landolt W. and Keller T. Whole-plant growth and leaf formation in ozonated hybrid poplar (*Populus x euramericana*). *Environ. Pollut.* **81**, 207-21, 1993.

McCool P.M. and Menge J.A. Influence of ozone on carbon partitioning in tomato: potential role of carbon flow in regulation of the mycorrhizal symbiosis under conditions of stress. *New Phytol.* **94**, 241-247, 1983.

McCrady J.K. and Andersen C.P. The effect of ozone on below-ground carbon allocation in wheat. *Environ. Pollut.* **107**, 465-472, 2000.

McLaughlin S.B. and McConathy R.K. Effects of SO_2 and O_3 on allocation of ^{14}C-labelled photosynthate in *Phaseolus vulgaris*. *Plant Physiol.* **73**, 630-635, 1983.

Moore J.C, DeRuiter P.C., and Hunt H.W. Soil invertebrate/micro-invertebrate interactions: disproportionate effects of species on food web structure and function. *Veterinary Parasitology* **48**, 247-260, 1993.

Naeem S., Thompson L.J,. Lawler S.P., Lawton J.H. and Woodfin R.M. Declining biodiversity can alter the performance of ecosystems. *Nature* **368**, 734-737, 1994.

O'Neill E.G., O'Neill R.V. and Norby R.J. Hierarchy theory as a guide to mycorrhizal research on large-scale problems. *Environ. Pollut.* **73**, 271-284, 1991.

Pääkkönen E. and Holopainen T. Influence of nitrogen supply on the response of clones of birch (*Betula pendula* Roth.) to ozone. *New Phytol.* **129**, 595-603, 1995.

Patterson M.T. and Rundel P.W. Stand characteristics of ozone-stressed populations of *Pinus jeffreyi* (Pinaceae): Extent, development, and physiological consequences of visible injury. *Am. J. Bot.* **82**, 150-158, 1995.

Pimm S.L. and Lawton J.H. Number of trophic levels in ecological communities. *Nature* **268**, 329-331, 1977.

Pimm S.L . Food Webs. Chapman and Hall, London, 219 pp., 1982.

Reddy G.B., Reinert R.A. and Eason G. Enzymatic changes in the rhizosphere of loblolly pine exposed to ozone and acid rain. *Soil Biol. Biochem.* **23(12)**, 11115-1119, 1991.

Reich P.B., Schoettle A.W., Stroo H.W., Troiano J. and Amundson R.G. Effects of O_3, SO_2, and acidic rain on mycorrhizal infection in northern red oak seedlings. *Can. J. Bot.* **63**, 2049-2055, 1985.

Rennenberg H., Herschbach C. and Polle A. Consequences of air pollution on shoot-root interactions. *J. Plant Physiol.* **148**, 296-301, 1996.

Retzlaff W.A., Weinstein D.A., Laurence J.A. and Gollands B. Simulated root dynamics of a 160-year-old sugar maple (*Acer saccharum* Marsh.) tree with and without ozone exposure using the TREGRO model. *Tree Physiol.* **16**, 915-92, 1996.

Retzlaff W.A., Arthur M.A., Grulke N.E., Weinstein D.A. and Gollands B. Use of a single-tree simulation model to predict effects of ozone and drought on growth of a white fir tree. *Tree Physiol.* **20**, 195-202, 2000.

Rygiewicz P.T. and Andersen C.P. Mycorrhizae alter quality and quantity of carbon allocated below ground. *Nature* **369**, 58-60, 1994.

Scagel C.F. and Andersen C.P. Seasonal changes in root and soil respiration of ozone-exposed ponderosa pine (*Pinus ponderosa*) grown in different substrates. *New Phytol.* **136**, 627-64, 1997.

Shafer S.R. Influence of ozone and simulated acidic rain on microorganisms in the rhizosphere of *Sorghum*. *Environ. Pollut.* **51**, 131-152, 1988.

Simmons G.L. and Kelly J.M . Influence of O_3, rainfall acidity, and soil Mg status on growth and ectomycorrhizal colonization of loblolly pine roots. *Water, Air, Soil Pollut.* **44**, 159-171, 1989.

Spence R.D., Rykiel Jr. E.J. and Sharpe P.J.H. Ozone alters carbon allocation in loblolly pine:

Assessment with carbon-11 labeling. *Environ. Pollut.* **64**, 93-106, 1990.

Stow T.K., Allen H.L., and Kress L.W. Ozone impacts on seasonal foliage dynamics of young loblolly pine. *For. Sci.* **38**, 102-119,1992.

Stroo H.F,. Reich P.B. Schoettle A.W. and Amundson R.W. Effects of ozone and acid rain on white pine (*Pinus strobus*) seedlings grown in five soils. II. Mycorrhizal infection. *Can. J. Bot.* **66**, 1510-1516, 1988.

Temple P.J,. Riechers G.H., and Miller P.R. Foliar injury responses of ponderosa pine seedlings to ozone, wet and dry acidic deposition, and drought. *Environ. Exper. Bot.* **32**, 101-113, 1992.

Tingey D.T,. Wilhour R.G., and Standley C. The effect of chronic ozone exposures on the metabolite content of ponderosa pine seedlings. *For. Sci.* **22**, 234-241, 1976.

Turner N.C,. Rich S., and Waggoner P.E. Removal of ozone by soil. *J. Environ. Qual.* **2**:259-264, 1973.

U.S. Environmental Protection Agency. Air Quality Criteria for Ozone and related Photochemical Documents, Vol. II. EPA/600/P-93/004bf. Washington, D.C.,1996.

Van der Heijden M.G.A., Klironomos J.N., Ursic M., Moutoglis P., Streitwolf-Engel R., Boller T., Wiemken A. and Sanders I.R. Mycorrhizal fungal diversity determines plant biodiversity, ecosystem variability and productivity. *Nature* **396**, 69-72, 1998.

Weber J.A., Tingey D.T. and Andersen C.P. "Plant response to air pollution." In *Plant Response Mechanisms to the Environment*, R.E. Wilkinson ed. Marcel Dekker, Inc., New York, pp. 357-389, 1994.

Weinstein D.A., Beloin R.M. and Yanai R.D. Modeling changes in red spruce carbon balance and allocation in response to interacting ozone and nutrient stress. *Tree Physiol.* **9**:127-146, 1991.

Weinstein D.A., Samuelson L.J .and Arthur M.A. Comparison of the response of red oak (*Quercus rubra*) seedlings and mature trees to ozone exposure using simulation modeling. *Environ. Pollut.* **102**, 307-320, 1998.

Wullschleger S.D., Lynch J.P. and Bernston G.M. Modeling the belowground response of plants and soil biota to edaphic and climatic change- What can we expect to gain? *Plant Soil* **165**, 149-160, 1994.

Yun S.C., Park E.W. and Laurence J.A. Simulation of 1-year-old *Populus tremuloides* response to ozone stress at Ithaca, USA, and Suwon, Republic of Korea. *Environ. Pollut.* **111**, 1-8, 2000.

Zak D.R., Tilman D., Parmenter R.R,. Rice C.W., Fisher F.M,. Vose J., Milchunas D. and Martin C.W. Plant production and soil microorganisms in late-successional ecosystems: A continental-scale study. *Ecology* **75**:2333-2347, 1994.

OZONE-INDUCED CELL DEATH

Reactive oxygen species as signal molecules regulating cell death

JAAKKO KANGASJÄRVI, HANNELE TUOMINEN AND KIRK OVERMYER
Institute of Biotechnology, University of Helsinki, POB 56, FIN-00014 Helsinki, Finland

Key words: Reactive oxygen species, programmed cell death, ethylene, jasmonic acid, salicylic acid, ozone

Abstract: Ozone exposure induces oxidative stress in plant cells in a similar way in both trees and annual plants. As a result of exposure to ozone, several different responses, such as alterations in gene expression, and ultimately, cell death by the active and regulated process of oxidative cell death can be detected. Oxidative cell death takes place in plants also in response to other stress factors, such as pathogen attack. Signal transduction during oxidative cell death is poorly understood, but involvement of ethylene, jasmonic acid, and salicylic acid mediated signaling pathways have been implicated. Most of this knowledge is derived from annual model species, such as *Arabidopsis thaliana*, but is applicable to tree species as well. These results suggest that ozone damage, also in trees, is a result of deleterious firing by the reactive oxygen species of pathways normally associated with the plant hypersensitive response to pathogens, where reactive oxygen species have a central and integral role in regulating programmed cell death. Establishment of EST libraries and methods for functional genomics and proteomics in novel plant species will allow a more detailed elucidation of the mechanisms involved in ozone-induced cell death in trees as well.

1. INTRODUCTION

The growth, distribution, and productivity as well as product quality of forest trees are greatly influenced by the environmental stresses and other external stimuli to which they are continuously exposed. The potential for growth of any tree individual under optimal conditions is far above what in reality is achieved in forests. As sessile, long-lived organisms,

S. Huttunen et al. (eds.),
Trends in European Forest Tree Physiology Research, Cost Action E6: EUROSILVA, 81-92

trees have developed various and sophisticated ways to adapt in accordance with environmental demands by altering their metabolism, growth and development. Effective responses to both external and internal stimuli will ensure both optimal growth and survival in an environment where productivity and product quality are adversely affected by biotic stresses such as diseases and herbivores (wounding), and abiotic stresses, such as cold, heat and air pollutants. To an even higher degree than annual or short-lived plants, trees must have effective means of defending themselves against invading pathogens and adapting to changes in their environment. Both preformed and inducible defenses participate in these processes. The main features of such defense measures involve early recognition and perception of the developing stress, and subsequent activation of induced adaptive and defensive responses leading to both local and systemic resistance.

In this chapter, cell death related to ozone exposure is discussed in the context of cell death that is induced by reactive oxygen species within the perspective, framework, and information available from model systems. Other ozone-dependent responses in forest trees are discussed by Langebartels et.al. 2001, and the related signaling and antioxidant defense mechanisms by Karpinski et.al. 2001. Since only a limited amount of literature is available on cell death in trees, emphasis of this chapter is on the process of cell death itself and information available mostly from annual model plants is used. At the level of an individual cell, trees and annual plants are not expected to have major differences in signaling and communication processes, demonstrated for example in the ability of the regulators of *Arabidopsis* flower development to control flower development in *Populus* (Weigel and Nilsson 1995). Thus, information on mechanisms, obtained with model plants is applicable also in the same context in trees.

2. PROGRAMMED CELL DEATH IN PLANTS

Both plant and animal cells can die by two different processes: by necrotic and programmed cell death. Programmed cell death (PCD) is an active form of cell death that occurs during normal plant growth and development, and as a response to certain abiotic and biotic stresses (Greenberg 1997, Pennell and Lamb 1997). PCD differs from necrotic cell death; it is the controlled destruction of cells. Implicit in the name is that the cell itself, acting on cue of an intrinsic genetic program, is taking an active role in its own demise in an orderly and highly controlled manner. Apoptosis, a similar term often used synonymously with PCD, especially in context with animal cells, was originally coined to describe a very distinct series of morphological and biochemical changes first observed in dying mammalian cells. In plants usually the term PCD is preferred over apoptosis, since some of the processes defined with the term apoptosis can not take place in plant cells.

Necrotic cell death, on the other hand, is pathological, imposed by mechanical, chemical, or osmotic insults, which overwhelm the cell resulting in early loss of membrane integrity, and affects whole tissues. The characteristics of PCD are cell shrinkage, the formation of membrane protuberances or "membrane blebbing" as it is called, nuclear shrinkage, chromatin condensation, and the activation of endogenous enzymes resulting of the degradation of the cell from within. The best example of this internal degradation, and one of the hallmarks of PCD, is the activation of nucleases that cut nuclear DNA resulting in a characteristic ladder of oligonucleosome-sized fragments (Pennell and Lamb 1997).

In plants, PCD occurs in several developmental and inducible processes, such as carpel senescence after pollination (Orzáez and Granell 1997), root aerenchyma formation (Drew et al. 2000), cereal endosperm development (Young et al. 1997) and, especially during the hypersensitive response (HR) induced by invasion of an incompatible pathogen (Greenberg 1997, Pennell and Lamb 1997). Furthermore, leaf senescence, which was the meaning of the original Greek word 'apoptosis' (falling of leaves), displays several hallmarks of programmed cell death (Noodén et al. 1997, Pennell and Lamb 1997), though it progresses at a slower pace than cell death in the previous examples. In both HR and senescence, the formation of reactive oxygen species (ROS), such as H_2O_2, superoxide, and hydroxyl radicals, is one of the central factors regulating the process (Pennell and Lamb 1997). Further evidence of the regulatory role for ROS in these processes is, for example, that the genes that are up-regulated by ozone in young birch leaves (Tuomainen et al. 1996, Kiiskinen et al. 1997, Korhonen et al. 1998, Pääkkönen et al. 1998), are also induced by the developmental program active in senescing leaves without any external stress (Valjakka et al. 1999).

3. CELL DEATH AS A RESULT OF OXIDATIVE STRESS

3.1 Reactive oxygen species as regulators of cell death

Reactive oxygen species that are formed in plant cells by several stresses may be one of the factors that contribute to plant stress sensitivity/tolerance. Ozone, a gaseous air pollutant that is converted into reactive oxygen species inside the plant tissues, is one of the most phytotoxic and most widespread tropospheric air pollutants. It is estimated that ozone causes more damage to crops than all other air pollutants combined (Kley et al. 1999). It has been proposed that the ROS produced from ozone degradation in cell wall could induce uncontrollable chain reactions that damage proteins, DNA, and membrane lipids, cause reduced photosynthesis, electrolyte leakage and accelerated senescence of plant tissues (Heath and Taylor 1997). However, these intermediates act also as signal molecules that have a profound and central role in cellular signaling related to induction and regulation of programmed cell death. According to the current understanding (reviewed in Langebartels et al. 2000b), the necrotic lesions observed in plants treated with ozone are most likely caused by the activation of programmed cell death component of the HR (Kangasjärvi et al. 1994, Sharma and Davis 1997, Sandermann et al. 1998), not by uncontrolled necrotic processes. This chapter reviews how these processes are controlled and induced, a topic that is currently under active investigations in studies that are concentrating on both plant/pathogen and plant/ozone interactions.

3.2 Ozone induces production of reactive oxygen species

Trees have turned out to be especially sensitive to ozone (Kärenlampi and Skärby 1996) and the adverse effects can be carried out to the next growth seasons (Langebartels et al. 1997). Trees also exhibit extreme variation in their sensitivity to ozone. Extremely sensitive and very tolerant individuals/clones of broad-leaved tree species, such as birch and poplar have been used in studies of gene expression and cell death as a result of ozone exposure, to determine the mechanisms that are related to ozone-induced cell death in trees (Tuomainen et al. 1996, Koch et al. 1998, 2000). Ozone induces the death of a few individual cells or cell clusters also in leaves of resistant clones, accompanied by pathogen defense gene induction (Tuomainen et al. 1996). This pattern of cell death has a similar spatial distribution as the ozone-induced H_2O_2 accumulation that has been detected by subcellular localization (Pellinen et al. 1999). Similar correlation between H_2O_2 production and cell death was visible also in tobacco (Schraudner et al. 1998) and superoxide accumulation and cell death in ozone-sensitive *Arabidopsis* mutants (Overmyer et al. 2000). This can be regarded analogous to the recently described micro HR where oxidative burst and death of a few individual cells in systemic leaves is essential requirement for the development of systemic acquired resistance (SAR) and pathogen defense gene induction in the systemic leaves (Alvarez et al. 1998).

The ROS produced from degradation of ozone in cell wall include molecules such as H_2O_2, superoxide and hydroxyl radicals. These ROS, that are produced directly from ozone degradation are not, however, at least to major extent, directly responsible for the cell death observed in ozone-exposed plants, more probably they act as the trigger that activates ROS production by the cell itself. It has been shown that the plasma membrane bound NADPH oxidase, as well as other mechanisms for H_2O_2 accumulation are activated as a result of ozone exposure (Schraudner et al. 1998, Pellinen et al 1999, Rao and Davis 1999, Overmyer et al. 2000). These are responsible for the production of the oxidative burst in ozone-exposed plants. In birch, O_3 induces spatially and temporally separated oxidative bursts where H_2O_2 is produced by at least two different mechanisms (Pellinen et al. 1999). The first burst of H_2O_2 occurs in the cell walls and then later during the process of oxidative cell death H_2O_2 accumulation continues in the mitochondria and other internal compartments of the cell several hours after exposure to ozone is over. Furthermore, inhibition of the flavin containing oxidases, such as the plasma membrane NADPH oxidase, partially inhibited ozone-induced ROS production (Pellinen et al. 1999, Rao and Davis 1999), and also cell death (Overmyer et al. 2000).

3.3 Regulation of cell death

Plant cells are constantly aware of changes taking place in their immediate surroundings; various receptors are used for continuous sensing of the close perimeter of the cell. These receptors then relay information to the cellular signaling cascades, which regulate further responses. For example, during the hypersensitive response to pathogens, receptors encoded by the resistance genes are interacting with several downstream processes that are responsible for regulating the extent of HR cell death (Hammond-Kosack and Jones 1996). One of the earliest of these responses is the excessive accumulation of reactive oxygen species, often referred to as the 'oxidative burst', in the apoplast (Lamb and Dixon 1997). This apoplastic ROS accumulation is an important regulator of the subsequent cell death. Strong evidence has accumulated that ROS play a central role in the signaling resulting in the induction of plant defense responses (Lamb and Dixon 1997, Alvarez et al. 1998, Van Camp et al. 1998). The signaling pathways involved in the primary cellular responses to both biotic and abiotic stresses include rapid protein phosphorylation, induction of MAP kinase cascades, rapid ion fluxes, including K^+ and Cl^- efflux, and Ca^{2+} and H^+ influx leading to extracellular alkalinization. Rapid Ca^{2+} fluxes across the plasma membrane of ozone-exposed plants have also been observed (Heath and Taylor 1997) and these fluxes are required in ozone-induced gene induction (Clayton et al. 1999).

It was proposed recently that an oxidative cell death (OCD) cycle, where ROS and salicylic acid (SA) have a connected regulatory role, would be responsible for regulating the extent of PCD in HR (Van Camp et al. 1998). Three different processes, initiation, propagation and containment of lesion, all presumably regulated by different signaling pathways, can be separated in the 'oxidative cell death cycle' (Overmyer et al. 2000). Propagation of the lesion is most likely to involve generation of signals at the initiation site that regulate death of the neighboring cells. Similarly, containment of lesion is bound to be under similar strict regulation that involves cell-to-cell communication that determines the outer limit of the expanding lesion. The nature and identity of the signals and processes involved in lesion propagation and containment are, however, still more or less unknown but involvement of plant hormones, especially ethylene, jasmonic acid (JA), and salicylic acid is evident (Tuomainen et al. 1997, Dong 1998).

3.4 Hormonal control of cell death

Several abiotic or biotic stresses are experienced by plants much in the same way. Therefore, it seems obvious that the signal transduction pathways involved are, at least partially, identical. The primary cellular events in plant defense, ion fluxes and the oxidative burst, are followed by involvement of a small number of signal transduction pathways mediated by salicylic acid, ethylene and jasmonic acid (Dong 1998, Reymond and Farmer 1998). These signaling pathways modulate cellular responses, and result in specific transcriptional activation of several plant defense genes and *de novo* synthesis of defense-related proteins involved in resistance, and determine the extent of highly regulated cell death. The exact role of these pathways in intracellular and intercellular communication and, especially the cross talk between the pathways is, however, still mostly unknown even in the much studied model systems. Even less is known in trees.

Recent results on programmed cell death suggest a role for ethylene in coordination of plant-stress responses. Using accelerated cell death mutants of *Arabidopsis*, it has been shown that ethylene can modulate the spread of programmed cell death in response to pathogen attack (Greenberg and Ausubel 1993, Rate et al. 1999). Furthermore, ethylene has been shown to regulate programmed cell death during carpel senescence in pea (Orzáez and Granell 1997), during endosperm development in maize (Young et al. 1997), and during aerenchyma formation in maize roots (Drew et al. 2000). Recent results suggest also that ozone-induced cell death proceeds through PCD (Overmyer et al. 1998, Koch et al. 2000) and ethylene regulates or modulates this response (Kettunen et al. 1999, Overmyer et al. 2000). Taken together, ethylene seems to be involved in orchestrating plants' responses with ROS production in situations such as pathogen attack and ozone exposure (Overmyer et al. 2000). The presence of various isoforms of ethylene-biosynthetic enzymes, ethylene receptor and possible also the downstream components of ethylene signal transduction (Johnson and Ecker 1998, Stepanova and Ecker 2000) could offer the variability that is required for the proper spatial and temporal regulation of the stress responses.

3.5 Hormonal control of ROS production

Although much has been learned about the signaling molecules involved in the HR (Dong 1998), only a little is known about how different signaling pathways act, independently or in concert. Combinations of signal molecules might direct the activation of certain defense responses. For example, the presence of two different pathways in plant pathogen responses has been demonstrated in several experiments. One is mediated by SA and leads to the expression of acidic PR-proteins, and the other is mediated by jasmonate and ethylene and leads to the induction of, for example, defensins and basic PR proteins (Reymond and Farmer 1998). Similarly at least two independent signaling cascades are involved in defense gene induction and cell death by ozone (Rao and Davis 1999, Langebartels et al. 2000b, Overmyer et al. 2000), analogous to defense gene induction by pathogens.

Recent results indicate that the communication between plant cells plays a fundamental role in plant stress responses and, for example, determines the extent of cell death spread in the hypersensitive response to pathogens (HR). During the HR, lesion propagation continues, as proposed, by the oxidative cell death cycle (Van Camp et al. 1998) until the amplification of ROS production and subsequent cell death is contained. This could be due to increased antioxidative capacity induced in the cells surrounding the spreading lesion, effectively removing the ROS signal that might be responsible for recurring initiation events, or, by surrounding cells becoming insensitive to the message that regulates lesion propagation. It has been proposed that JA could act as a signaling molecule in plant ozone responses (Kangasjärvi et al. 1994). In tobacco, both wounding and treatment with jasmonates prior to exposure to ozone provided protection from injury (Örvar et al. 1997). Similarly, Koch et al. (1998, 2000) showed that an ozone-sensitive hybrid poplar clone has greatly attenuated ozone- and wound-induced responses, which is a result of JA-insensitivity of the clone. A more direct role and mechanism of JA-involvement in determining plant O_3-tolerance was shown with *Arabidopsis* mutants that are ozone sensitive, or JA insensitive (Overmyer et al. 2000, Rao and Davis 2000). In the ozone-sensitive *rcd1*, external application of JA prevented ROS-dependent lesion development (Overmyer et al. 2000), and in the JA-insensitive *jar1*, both ROS accumulation and lesion formation in O_3-exposed plants were predominant (Overmyer et al. 2000, Rao and Davis 2000). These results indicate that JA, either alone, or in interaction with other messengers, may be involved in regulating the propagation of ROS signal and therefore the degree of ozone resistance in both trees and annual plants.

JA, SA, and ethylene seem to control and potentiate each other's activities, indicating that cross talk between these pathways may be very common in defense gene regulation. For example, JA and ethylene co-operate to regulate the expression of many genes, and at least some JA-inducible genes are not inducible in plants unable to produce or sense ethylene (Xu et al. 1994, O'Donnell et al. 1996). Similarly, the sensitivity of *Arabidopsis* plants to SA-inducible pathogenesis related protein PR1 accumulation is enhanced by ethylene (Lawton et al. 1994). On the other hand, SA can often inhibit JA-inducible gene expression (Doares et al. 1995). Synergistic or inhibitory cross-talk between these regulators may help the plant to prioritize the activation of a particular signal pathway over another and fine-tune its defense by upregulating different sets of overlapping genes against different aggressors, such as various pathogens and herbivores, or toxic air pollutants (Reymond and Farmer 1998). These processes are supposed to take place by similar regulatory cascades and sequences in both annual plants and trees. However, experimental studies of these interactions are complicated when tree species are considered. Therefore, it is necessary to utilize the most recent developments in plant biology with model systems, and then to transfer this information to trees.

4. MODEL SPECIES AND MUTANTS AS TOOLS TO STUDY PROCESSES IN TREES

Plant science is currently undergoing a revolutionary phase with the completion of the *Arabidopsis* genome sequencing, generation of expressed sequence tag (EST) databases, and the possibility of gene expression analysis with array technologies. Even though most of the active EST sequencing processes are carried out in model- or crop species, also trees have been a subject to such an approach (Allona et al. 1998, Sterky et al. 1998). These EST sequencing programs yield thousands of novel genes without any similarity to previously known genes. Use of these EST clones in microarray-analysis with, for example, O_3-exposed trees will reveal genes that are up- or down regulated by ozone. However, with array hybridizations, a function, or a biological process to which the gene product is associated, can be assigned only by correlation; it will not reveal whether the changes in gene expression are the primary determinants of, in this case ozone sensitivity or resistance, or merely a secondary effect that acts downstream of the primary ones. To determine the actual function of the genes identified with array technologies, other approaches are needed.

4.1 From trees to model species

If gene expression analysis in trees with array hybridizations reveals a gene whose expression has a very strong correlation with, for example ozone-induced cell death, but has no similarity to any identified gene, its role and function in the process can be studied with two approaches: generation of transgenic trees, or utilizing *Arabidopsis*. The gene could be over-expressed or anti sensed in the tree species where it was discovered, if transformation technologies for the species exist. However, this process usually takes a long time. Most processes can be approached in *Arabidopsis* by identifying a transposon or T-DNA insertion to the orthologous *Arabidopsis* gene from the insertion-mutagenesis collections (Somerville 2000). This way a null-mutation in the gene in question can be obtained. If an ozone-related phenotype can be seen, this will provide an unequivocal evidence for a direct involvement of the gene in oxidative cell death.

As the next step, the insertion mutant can be crossed with other relevant mutant strains and these can be exposed to global gene expression analyses for elucidation of the processes affected. Finally, the backbones of the regulatory pathways identified in *Arabidopsis* this way can be used as starting points when addressing the same processes again in the tree species. Orthologs of the genes identified and characterized in *Arabidopsis* through this approach can then be identified in tree EST databases. In order to address their functional significance, they can be subjected to a global gene expression analysis in parallel to *Arabidopsis*. If a correlation is found, this will give way to more detailed molecular-genetic analyses, involving transgenic plants in both species.

4.2 From model species to trees

The approach of using plants like *Arabidopsis* as a model for example O_3-induced cell death in trees, should also involve isolation and characterization of mutants with visible phenotypes (Langebartels et al. 2000a), such as the O_3-sensitive *rcd1* (Overmyer et al. 2000). The power of this approach is the possibility for the use of genetics, which currently, and also in the near future, is impossible to the same extent in trees. The mutations can be mapped genetically and the gene that carries the mutation can be identified with fine mapping. In optimal cases, mutated gene can be identified already 6 to 8 weeks after the individual with the mutant phenotype has been discovered (Somerville 2000). The regulatory pathways involved can be further characterized by isolating suppressor mutations. Seeds of the 'primary' mutants can be mutagenized again and the M_2 generation is screened for loss of the phenotype. This way, both linear and parallel processes involved in oxidative stress-responses and tolerance can be identified. Parallel with the mutant and suppressor mutant cloning and identification, EST libraries are generated from the tree species of interest. Orthologs of genes identified in both *Arabidopsis* mutants and suppressor mutants, as well as with analysis of global gene expression with micro arrays in *Arabidopsis* can then be identified from the tree. Function and role of the tree genes identified with the information arising from the *Arabidopsis* mutant work can then be further studied in transgenic trees.

5. FUTURE PERSPECTIVES

The future research on ozone/oxidative stress-related processes and cell death in trees should include the use of molecular tools, EST sequencing, genomics, proteomics, and utilization of model systems. These will, without doubt, lead to a better understanding of the molecular mechanisms of stress tolerance in trees. In particular, the elucidation of signal processes employed for stress adaptation and resistance, fundamental features in plant life, will yield important discoveries. We can expect that novel ways of generating plants with enhanced stress tolerance and disease resistance can be designed. The focus of these efforts will be in engineering signal pathways that trigger stress responses and adaptation. These efforts will be greatly facilitated by EST sequencing projects allowing the identification of genes encoding components of signaling cascades. The practical benefit of this research will be in identification by molecular tools, and, ultimately, generation by genetic engineering of more tolerant plants to be used in practical forestry.

The studies on signal transduction will further lead to detailed understanding of several key signal transduction cascades involved in stress adaptation in plants. This should result in the identification of significant components of these cascades, their functional characterization, and ordering of the cascades. Elucidation of the molecular details of these cascades and the interplay of the different signal pathways will require more meticulous studies. The insights gained and the techniques developed and applied will allow expansion of this work to other aspects of plant signaling

The biochemical and genetic dissection of the signal transduction pathways will provide the basis for subsequent genetic engineering of stress responses by modulating the activity of entire signal transduction pathways. In addition, the methods used in modern cell biology will greatly facilitate the basic as well as applied research in tree biology.

REFERENCES

Allona I, Quinn M, Shoop E, Swope K, St.Cyr S, Carlis J, Riedl J, Retzel E, Campbell MM, Sederoff R, Whetten RW (1998) Analysis of xylem formation in pine by cDNA sequencing, *Proc. Natl. Acad. Sci. USA*. **95**, 9693-9698.

Alvarez ME, Pennell RI, Meijer P-J, Ishikawa A, Dixon RA, Lamb C (1998) Reactive oxygen intermediates mediate a systemic signal network in the establishment of plant immunity, *Cell*. **92**, 773-784.

Clayton H, Knight MR, Knight H, McAinsh MR, Hetherington AM (1999) Dissection of the ozone-induced calcium signature, *Plant J.* **17**, 575-579.

Doares SH, Syrovets T, Weiler EW, Ryan CA (1995) Oligogalacturonides and chitosan activate plant defensive genes through the octadecanoid pathway, *Proc. Natl. Acad. Sci. USA*. **92**, 4095-4098.

Dong X (1998) SA, JA, ethylene, and disease resistance in plants, *Curr. Opin. Plant Biol.* **1**, 316-323

Drew MC, He C-J, Morgan PW (2000) Programmed cell death and aerenchyma formation in roots, *Trends Plant Sci.* **5**, 123-127.

Greenberg JT (1997) Programmed cell death in plant-pathogen interactions, *Annu. Rev. Plant Physiol. Plant Mol. Biol.* **48**, 525-545.

Greenberg JT, Ausubel FM (1993) *Arabidopsis* mutants compromised for the control of cellular damage during pathogenesis and aging, *Plant J.* **4**, 327-341.

Hammond-Kosack KE, Jones JDG (1996) Resistance gene-dependent plant defense responses, *Plant*

Cell. **8**, 1773-1791.

Heath RL, Taylor GE Jr (1997) Physiological processes and plant responses to ozone exposure. *In* Sandermann H, Wellburn AR, Heath RL, eds, Forest decline and ozone, Vol 127. Springer-Verlag, Berlin,Heidelberg, pp 317-368.

Johnson,PR; Ecker,JR (1998) The ethylene gas signal transduction pathway: A molecular perspective. *Annu. Rev. Genet.* **32**, 227-254.

Kangasjärvi J, Talvinen J, Utriainen M, Karjalainen R (1994) Plant defense systems induced by ozone, *Plant Cell Environ.* **17**, 783-794.

Kärenlampi L, Skärby L (1996) Critical levels for ozone in Europe: Testing and finalizing the concepts. UN-ECE workshop report, University of Kuopio, Kuopio, Finland.

Kettunen R, Overmyer K, Kangasjärvi J (1999) The role of ethylene in the formation of cell damage during ozone stress. Does ozone-induced cell death require concomitant AOS and ethylene production. *In* Kanellis A, Chang C, Klee H, Bleecker AB, Pech JC, Grierson D, eds, Biology and biotechnology of the plant hormone ethylene II. Kluwer Academic Publishers, Dortreth, pp 299-306.

Kiiskinen M, Korhonen M, Kangasjärvi J (1997) Isolation and characterization of cDNA for a plant mitochondrial phosphate translocator (*Mpt1*). Ozone stress induces *Mpt1* mRNA accumulation in birch (*Betula pendula* Roth), *Plant Mol. Biol.* **35**, 271-279.

Kley D, Kleinman M, Sandermann H Jr, Krupa S (1999) Photochemical oxidants: State of the science, *Environ. Pollut.* **100**, 19-24.

Koch JR, Scherzer AJ, Eshita SM, Davis KR (1998) Ozone sensitivity in hybrid poplar is correlated with a lack of defense-gene activation, *Plant Physiol.* **118**, 1243-1252.

Koch,JR; Creelman,RA; Eshita,SM; Seskar,M; Mullet,JE; Davis,KR (2000) Ozone sensitivity in hybrid poplar correlates with insensitivity to both salicylic acid and jasmonic acid. The role of programmed cell death in lesion formation. *Plant Physiol.* **123**, 487-496.

Korhonen MS, Pellinen R, Kiiskinen M, Kangasjärvi J (1998) Cellular-molecular diagnostics of ozone injury in birch (*Betula pendula* Roth). *In* De Kok LJ, Stulen I, eds, Responses of Plant Metabolism to Air Pollution. Backhuys Publishers, Leiden, The Netherlands, pp pp. 351-354. In Press.

Lamb C, Dixon RA (1997) The oxidative burst in plant disease resistance, *Annu. Rev. Plant Physiol. Plant Mol. Biol.* **48**, 251-275.

Langebartels C, Ernst D, Heller W, Lütz C, Payer H-D, Sandermann H Jr (1997) Ozone responses of trees: Results from controlled chamber exposures at the GSF phytotron. *In* Sandermann H, ed, Ecological Studies, 127: Forest decline and ozone. Springer-Verlag, Heidelberg, pp 163-200.

Langebartels C, Ernst D, Kangasjärvi J, Sandermann H Jr (2000a) Ozone effects on plant defense. *In* Packer L, Sies H, eds, Singlet oxygen, UV-A and ozone: Methods in enzymology, Vol 319. Academic Press, San Dieco, CA, pp 520-535.

Langebartels C, Schraudner M, Heller W, Ernst D, Sandermann H Jr (2000b) Oxidative stress and defense reactions in plants exposed to air pollutants and UV-B radiation. *In* Inzé D, Van Montagu M, eds, Oxidative stress in plants. Harwood, Amsterdam, In Press.

Lawton KA, Potter SL, Uknes S, Ryals J (1994) Acquired resistance signal transduction in *Arabidopsis* is ethylene independent, *Plant Cell.* **6**, 581-588.

Noodén LD, Guiamét JJ, John I (1997) Senescence mechanisms, *Physiol. Plant.* **101**, 746-753.

O'Donnell PJ, Calvert C, Atzorn R, Wasternack C, Leyser HMO, Bowles DJ (1996) Ethylene as a signal mediating the wound response of tomato plants, *Science.* **274**, 1914-1917.

Örvar BL, McPherson J, Ellis BE (1997) Pre-activating wounding response in tobacco prior to high-level ozone exposure prevents necrotic injury, *Plant J.* **11**, 203-212

Orzáez D, Granell A (1997) DNA fragmentation is regulated by ethylene during carpel senescence in *Pisum sativum*, *Plant J.* **11**, 137-144.

Overmyer K, Kangasjärvi J, Kuittinen T, Saarma M (1998) Gene expression and cell death in ozone-exposed plants: Is programmed cell death involved in ozone damage in ozone sensitive *Arabidopsis* mutants? *In* De Kok L, Stulen I, eds, Responses of Plant Metabolism to Air Pollution and Global Change. Backhuys Publishers, Leiden, pp 403-406.

Overmyer,K; Tuominen,H; Kettunen,R; Betz,C; Langebartels,C; Sandermann, H,Jr; Kangasjärvi,J (2000) The ozone-sensitive *Arabidopsis rcd1* mutant reveals opposite roles for ethylene and jasmonate signalling pathways in regulating superoxide-dependent cell death. *Plant Cell* **12**, 1849-1862.

Pääkkönen E, Seppänen S, Holopainen T, Kokko H, Kärenlampi S, Kärenlampi L, Kangasjärvi J (1998) Induction of genes for the stress proteins PR-10 and PAL in relation to growth, visible injuries and stomatal conductance in birch (*Betula pendula*) clones exposed to ozone and/or drought, *New Phytol.* **138**, 295-305.

Pellinen R, Palva T, Kangasjärvi J (1999) Subcellular localization of ozone- induced hydrogen peroxide production in birch (*Betula pendula*) leaf cells, *Plant J.* **20**, 349-356.

Pennell RI, Lamb C (1997) Programmed cell death in plants, *Plant Cell.* **9**, 1157-1168

Rao M V, Davis K R (1999) Ozone-induced cell death occurs via two distinct mechanisms in *Arabidopsis*: the role of salicylic acid, *Plant J.* **17**, 603-614

Rao,MV; Lee,HI; Creelman,RA; Mullet,JA; Davis,KR (2000) Jasmonic acid signalling modulates ozone-induced hypersensitive cell death. *Plant Cell* **12**, 1633-1646.

Rate DN, Cuenca JV, Bowman GR, Guttman DS, Greenberg JT (1999) The gain-of-function Arabidopsis *acd6* mutant reveals novel regulation and function of the salicylic acid signaling pathway in controlling cell death, defenses, and cell growth, *Plant Cell.* **11**, 1695-1708.

Reymond P, Farmer EE (1998) Jasmonate and salicylate as global signals for defense gene expression, *Curr. Opin. Plant Biol.* **1**, 404-411.

Sandermann H Jr, Ernst D, Heller W, Langebartels C (1998) Ozone: an abiotic elicitor of plant defence reactions, *Trends Plant Sci.* **3**, 47-50.

Schraudner M, Moeder W, Wiese C, van Camp W, Inzé D, Langebartels C, Sandermann H Jr (1998) Ozone-induced oxidative burst in the ozone biomonitor plant, tobacco Bel W3, *Plant J.* **16**, 235-245.

Sharma YK, Davis KR (1997) The effects of ozone on antioxidant responses in plants, *Free Rad. Biol. Med.* **23**, 480-488.

Somerville C (2000) The twentieth century trajectory of plant biology, *Cell.* **100**, 13-25.

Stepanova,AN; Ecker,JR (2000) Ethylene signaling: from mutants to molecules. *Curr. Opin. Plant Biol.* **3**, 353-360.

Sterky F, Regan S, Karlsson J, Hertzberg M, Rohde A, Holmberg A, Amini B, Bhalerao R, Larsson M, Villarroel R, Van Montagu M, Sandberg G, Olsson O, Teeri TT, Boerjan W, Gustafsson P, Uhlén M, Sundberg B, Lundeberg J (1998) Gene discovery in the wood-forming tissues of poplar: Analysis of 5,692 expressed sequence tags, *Proc. Natl. Acad. Sci. USA.* **95**, 13330-13335.

Tuomainen J, Pellinen R, Roy S, Kiiskinen M, Eloranta T, Karjalainen R, Kangasjärvi J (1996) Ozone affects birch (*Betula pendula* Roth) phenylpropanoid, polyamine and active oxygen detoxifying pathways at biochemical and gene expression level, *J. Plant Physiol.* **148**, 179-188.

Tuomainen J, Betz C, Kangasjärvi J, Ernst D, Yin ZH, Langebartels C, Sandermann H Jr (1997) Ozone induction of ethylene emission in tomato plants: Regulation by differential transcript accumulation for the biosynthetic enzymes, *Plant J.* **12**, 1151-1162.

Valjakka M, Tuhkanen E, Kangasjärvi J, Vapaavuori E (1999) Expression of photosynthesis and senescence-related genes during leaf development and senescence in silver birch (*Betula pendula* Roth) seedlings, *Physiol. Plant.* **106**, 302-310.

Van Camp W, Van Montagu M, Inzé D (1998) H_2O_2 and NO: redox signals in disease resistance, *Trends Plant Sci.* **3**, 330-334

Weigel D, Nilsson O (1995) A developmental switch sufficient for flower initiation in diverse plants, *Nature* **337**, 495-500.

Xu Y, Chang P-FL, Liu D, Narasimhan ML, Ragothama KG, Hasegawa PM, Bressan RA (1994) Plant defense genes are synergistically induced by ethylene and methyl jasmonate, *Plant Cell.* **6**, 1077-1085.

Young TE, Gallie DR, DeMason DA (1997) Ethylene-mediated programmed cell death during maize endosperm development of wild-type and *shrunken2* genotypes, *Plant Physiol.* **115**, 737-751.

SIGNALLING AND ANTIOXIDANT DEFENCE MECHANISMS IN HIGHER PLANTS

STANISLAW KARPINSKI[1], BARBARA KARPINSKA[1], MICHAEL MELTZER[2], JAN-ERIK HÄLLGREN[3] AND GUNNAR WINGSLE[3]

[1]*Department of Botany, Stockholm University, Lilla Frescativ. 5, SE-106 91 Stockholm, Sweden.*
[2]*Department of Molecular Cell Biology, Institute of Plant Genetics and Crop Plant Research, 06466 Gatersleben, Germany.*
[3]*Umeå Plant Science Centre, Department of Forest Genetics and Plant Physiology, Faculty of Forestry, Swedish University of Agricultural Sciences, SE-901 83 Umeå, Sweden.*

Key words: antioxidants, excess excitation energy, low temperature, high light, oxidative stress, reactive oxygen species, redox status

Abstract: Major and rapid changes in environmental conditions can damage plant health and limit crop production. However, stress-sensing and subsequent regulation of the defence mechanisms in plants are poorly understood. The emphasis of this chapter will be to discuss: the evidence for the involvement of reactive oxygen species (ROS); the roles of the enzymatic and non-enzymatic ROS-scavenging and antioxidant systems; and avoidance mechanism of ROS production, during low-temperature and high light-induced oxidative stress. Furthermore the role of the redox status of the chloroplast will be discussed, particularly, quinone B and plastoquinone in regulation of expression of the nuclear- and the chloroplast-encoded genes that are necessarily for acclimatory processes.

1. INTRODUCTION

Two of the major factors that limit agricultural and forest productivity, and influence the plants distribution in natural ecosystems, are temperature and light. Temperatures and light are subject to wide fluctuations, both seasonal and diurnal, which have profound effects on plants in many habitats. The major, visible plant responses to changes in

S. Huttunen et al. (eds.),
Trends in European Forest Tree Physiology Research, Cost Action E6: EUROSILVA, 93-114

temperature are well known, but the primary, molecular effects of low temperature are much less well understood. Low temperatures induce complex changes at the whole plant level, on growth rate for instance, which are modulated by wide-ranging adjustments of both structural components and metabolic activities (Levitt 1980; Steponkus 1981; Sakai and Larcher 1987; Kacperska 1989; Guy 1990; Hällgren and Öquist 1990; Hällgren et al., 1991). The limits to our current knowledge of the physiological, biochemical and molecular mechanisms of low-temperature effects on phenomena such as growth inhibition, developmental impairment and other phenotypic changes have been described in several reviews (Palva 1994; Thomashow 1994; Wise 1995; Bohnert et al., 1995; Hughes and Dunn 1996; Nishida and Murata 1996; Huner et al., 1998).

Growth is one of the most readily monitored and informative indicators of the metabolic status of any plant, and changes in growth rate and habit can provide valuable early symptoms of stress. There are many environmental factors that can cause stress, and influence growth kinetics in plants, but three of the most important are light intensity, photoperiod and temperature. These factors also affect many other plant life processes, including developmental changes from germination, through bud break, flowering, seed formation and pigmentation, through to senescence. However, the sensitivity of the various plant organs differs markedly between different growth stages and different plant ages. Sensitivity also fluctuates considerably throughout the year. Furthermore, temperature affects plants both directly and indirectly: an important example of its direct effects being alterations in the rate of production of reactive oxygen species (ROS).

The key physiological process photosynthesis is a well known source of ROS. Several excellent reviews cover oxygen-photosynthesis-environmental interactions, focusing on features such as the role of oxygen in photoinhibition (Krause 1994), oxygen metabolism (Foyer 1997) and chilling stress (Baker 1994; Wise 1995). In addition, Asada (1999) has described the production of ROS and the role of water-water cycle in chloroplasts, and Horton et al., (1996) have discussed biophysical mechanisms believed to regulate light harvesting and energy dissipation in photosystem II (PSII), in which the Δ-pH across the thylakoid membranes is postulated to control energy dissipation. Low temperature (LT) and excess light (EL) cause excess excitation energy (EEE), leading to photooxidative stress. In extreme cases this can cause the light and oxygen-induced bleaching of photosynthetic pigments. Among conifers, the phenomenon of chlorophyll bleaching is naturally much more common, and potentially severe, in habitats exposed to the sun than in shaded locations (Karpinski et al., 1994). However, in Scots pine trees the chlorophyll concentration declines in the winter, while carotenoid levels remain constant or even show a moderate rise. Thus, at the end of the winter, when the quantum flux density has begun to rise, but the temperature is still too low to sustain coniferous growth, the photosynthetic pigments are at their lowest level (Karpinski et al., 1994). This phase coincides with a period of very low PSII efficiency (Karpinski et al., 1994; Ottander et al., 1995). It has been proposed that the reorganisations of the photosynthetic systems that occur in autumn and winter allow Scots pine to maintain a relatively high proportion of chlorophyll in a quenched, photo-protected state (Öquist et al., 1992). This would be adaptively advantageous, since it would allow rapid recovery of photosynthesis in the spring (Karpinski et al., 1993; 1994; Ottander et al., 1995).

The most sensitive components of the photosynthetic apparatus exposed to EEE-induced photooxidative stress are generally thought to be the PSII reaction centres, including the D1 protein. So-called "acceptor- and donor-side photoinhibition" which is associated with proteolytic degradation of the D1 protein and induction of photooxidative stress can ultimately lead to photodamage of leaf tissue (Aro et al., 1993; Barber 1995, Van Wijk et al., 1997, Karpinski et al. 1999). Photoinactivation of photosystem I (PSI) has been reported to be more severe than that of PSII during chilling stress (Sonoike, 1996). Chilling and EL lead to over-production of ROS through EEE. This is essentially because light reactions continue unabated in the chloroplast, but the rates of the key energy-consuming biochemical reactions are reduced in LT.

Despite their potential for destruction, the generation of ROS is indispensable for aerobic organisms. In sustainable environmental conditions, all aerobic cells maintain a dynamic equilibrium between production and use of ROS. Photosynthesis and respiration are the most efficient energy-harnessing mechanisms in such conditions. However, when EEE is generated, in fluctuating or extreme conditions for example (Karpinski et al., 1999), ROS production may exceed the capacity of leaf cells to scavenge them, and photooxidative stress is likely to occur (Karpinski et al., 1997, 1999). EEE may be generated in plants when there is excess light, or low temperature, or both (Krause 1988, Karpinski et al., 1993, 1994, Long et al., 1994, Russell et al., 1995, Foyer 1997, Noctor and Foyer 1998, Asada 1999). These stresses can significantly damage plants, and constrain crop production, but the stress-sensing mechanisms in plants, the subsequent acclimatory processes, and their regulation are generally not well understood.

However, studies we have recently published (Karpinski et al., 1997, 1999) present a new, more integrated view of immediate and acclimatory responses in plants to EEE-inducing environmental fluctuations. A wide range of responses are provoked by EEE in leaves, one of the most important being the induction of antioxidant defences. This step is controlled (at least in part) by the secondary quinone electron acceptor (Q_B) and/or plastoquinone (PQ) pools' redox states (Karpinski et al., 1997, 1999). Furthermore, it is now known that cells subjected to EEE-associated stresses also produce a systemic signal, a component of which is hydrogen peroxide (H_2O_2). We have shown that acclimatory responses to EEE and associated photooxidative stresses are induced by this signal in unstressed parts of the plant (Karpinski et al., 1999). We propose that systemic signals can promote redox changes in the QB and/or PQ pools of unstressed chloroplasts, prompting chloroplasts and cells elsewhere in the plant to initiate protective mechanisms. This process has been named systemic acquired acclimation (SAA, Karpinski et al., 1999).

The redox status of the Q_B and/or PQ pools has also been shown to regulate the transcriptional activity of chloroplast-encoded genes and, thus, modulate the stoichiometric balance between PSII and PSI in different light conditions (Pfannschmidt et al., 1999). Furthermore, this redox state appears to regulate the transcription of nuclear genes encoding light harvesting complex proteins (Escoubas et al., 1995, Maxwell et al., 1995). These results strongly indicate that the redox status of the Q_B and/or PQ pools is an important redox sensing mechanism which plays a key role in the regulation of the acclimatory responses of plants to EEE and associated photooxidative stress.

1.1 Acclimation, Oxidative Stress and ROS

It is essential for plants to be able to adjust to diurnal and seasonal fluctuations in light and temperature. Stable, long-term, genotypically-based responses to seasonal changes in the environment are regarded as "adaptations", while shorter term physiological responses are generally described as "acclimation". Both adaptation and acclimation induce a variety of physiological, biochemical and molecular changes that lead to significant phenotypic alterations. It should also be noted that mosses, lichens, algae and cyanobacteria can (and must) acclimate to EEE as well as higher plants. Many of our major crop plants, including staples like maize, rice, wheat and rye have low resistance to photooxidative stress caused by chilling and/or high light. Thus, their growth and development can be severely hampered by such conditions.

Injuries caused by low-temperature stress are similar to those caused by the strong oxidative stress inducer, methyl viologen (Bowler et al., 1992), and the recovery of thawed rice cells, following freezing, is limited by oxidative stress (Benson et al., 1992). These observations have led to suggestions that ROS play an active part in low-temperature related injuries (McKersie 1991; Bridger et al., 1994). It has been demonstrated that ROS tend to activate certain peptidases, and facilitate the action of proteases, thus inducing increased rates of proteolysis (Casano et al., 1994). It has also been suggested that ROS are involved in oxidative stress defence mechanisms (Del Corso et al., 1994) *via* oxidation of thiol groups in enzymes that regulate major metabolic pathways, and influence NADPH levels. Inactivation of thioredoxin, for instance, affects ribulose-1,5-bisphosphate carboxylase oxygenase (Byrd et al., 1995), amongst other metabolic enzymes, and thus has wide-ranging effects on metabolism (Wise 1995).

Most of the evidence suggesting that ROS production increases during low temperature stress in plants is indirect, centring on observations of changes associated with shifts in temperature in the levels of various ROS-scavengers and antioxidants. However, a few studies have also provided direct evidence of increases in ROS formation during low temperature stress. Electron-spin-resonance (ESR) analyses, for instance, have indicated that microsomes in winter wheat subjected to lethal freezing have significantly higher superoxide ($O_2^{\cdot-}$) levels than controls that are either unstressed or sub-lethally stressed (Kendall and McKersie 1989). In other ESR studies, signals attributed to unidentified organic free radicals in Scots pine needles increased following a rapid decrease in temperature to -40 °C (Tao et al., 1992). Furthermore, thylakoid membranes from cold-sensitive plants have been demonstrated to produce $O_2^{\cdot-}$ radicals more rapidly than thylakoids from cold-acclimated plants (Hodgson and Raison 1991).

The most thoroughly researched sites at which univalent reduction of O_2 to $O_2^{\cdot-}$ occurs in plants are the primary electron acceptors of the PSI complex, which are thylakoid-membrane-bound, and the peripheral reduced ferredoxin (Asada 1999). At neutral pH, the $O_2^{\cdot-}$ is produced at a rate of approximately 20 μmol mg chlorophyll (Chl)$^{-1}$ h^{-1} (Asada 1999). Whether $O_2^{\cdot-}$ can be generated at the PSII side is not known as yet (Chen et al., 1995). However, before it reaches the stromal space, the majority of the $O_2^{\cdot-}$ radicals produced in the thylakoid membrane are converted via O_2-mediated cyclic electron flow to O_2, or to H_2O_2 by non-catalytic dismutation (Asada 1999).

Despite being referred to as a ROS, H_2O_2 does not meet the chemical definition of a radical, since its two outer orbitals are completely filled. Furthermore, it is the most stable of the ROS, and it may act either as a reductant or as an oxidant (Salin 1987). Abundant evidence has shown the electron-transfer chains of chloroplasts, mitochondria and peroxisomes to be significant sources of H_2O_2 (Cadenas 1989, Del Rio et al., 1992, Asada 1999), and in low light (LL) acclimated *A. thaliana* leaves EEE can induce increased levels of the compound (Karpinski et al., 1997, 1999). The thiol groups of various systems are susceptible to H_2O_2, and fairly low concentrations (10 μM) have been shown to inhibit SH-enzymes in the Calvin cycle in isolated chloroplasts (Kaiser 1979). It has moreover been suggested that 10 to 12 mM of the compound can induce plant cell death (Levine et al., 1994, Lamb and Dixon 1997).

It has been thoroughly established that H_2O_2 and $O_2^{\cdot-}$ can react together in various biochemical systems to form the hydroxyl radical (OH·). However, *in vitro*, pure H_2O_2 and $O_2^{\cdot-}$ do not tend to react at significant rates, unless traces of iron or copper salts are added (Halliwell and Gutteridge 1993). O_2^- can reduce both Fe(III) and Cu(II), so acts as a source of H_2O_2 in the presence of these catalysts (Halliwell and Gutteridge 1993). However, these metals can also be reduced by a range of other agents, including semiquinone radicals, thiols and ascorbate (AsA). A number of alternative metal-catalysed reactions involving H_2O_2 also generate OH· radicals. Therefore, the damage caused by OH· in specific systems depends on the cellular location of metals and potential reductants such as thiols and AsA, together with the site of production (and abundance) of $O_2^{\cdot-}$ and H_2O_2. As yet, no specific agents that directly scavenge OH· radicals have been found in any organism (Asada 1999).

ROS also occur as various singlet species, for instance singlet ^{1}Chl*, generated via light excitation, and singlet oxygen ($^1O_2$, both of which appear to be quenched by carotenoids). Chloroplast membranes are highly susceptible to $^1O_2$-induced lipid peroxidation, since the unsaturated fatty acid-linolenate accounts for around 90 % of the fatty acids in the thylakoid glycolipids, phospholipids and sulpholipids (Knox and Dodge 1985).

Clearly, EEE generally causes ROS levels to rise in plants, and induces photooxidative stress, but the mechanisms involved are not yet fully understood. Similarly, it is clear that many other free radicals, such as AsA-, monodehydroascorbate (mdAsA)-, reduced glutathione (GSH)-, thiyl-radicals and other organic radicals and peroxides, play roles in the damage caused by EEE, and/or acclimation to it, but the precise nature of their action is not yet known.

Empirical data have been gathered suggesting that lipid peroxidation of the cellular membranes may be a significant component of radical-mediated cell injury, causing significant structural alterations, for example (Bhaumik et al., 1995). Free-radical-mediated peroxidation of lipids involves at least three steps. The first step, initiation, occurs when a free radical, such as OH· or extracts a proton from a polyunsaturated fatty acid, forming a fatty acid radical. The next step is propagation, in which the fatty acid radical reacts with O_2 to create a fatty acid peroxy radical, which can then react with other lipids or proteins, thus perpetuating the proton transfer process, and further oxidation of lipids and protein substrates (Bhaumik et al., 1995). Lipid peroxidation has been observed in several studies of chilling and freezing. It has been followed, for instance, in chilled *Arabidopsis* suspension cultures (by measuring thiobarbituric-acid-reactive substances), and noted in spruce during frosty conditions in the spring (Polle et al., 1996). Products of lipid peroxidation have also been detected in *Arabidopsis* plants kept in low temperatures for eight days (O'Kane et al., 1996) and in developing maize seedlings (Prasad et al., 1994; Prasad 1996). Similarly, the accumulation of fluorescent lipid pigments in both the phospholipid and glycolipid fractions of thylakoids has been observed in young cucumber seedlings chilled at 4°C in the dark (Haiyada and Parkin 1993).

The polyunsaturation of membrane lipids associated with low temperatures has both positive and negative consequences for plants. On the negative side, it is likely to increase the probability of oxidative stress damage. However, it also introduces alterations in viscosity of membranes (Kusters et al., 1991), which enhance the recovery of photosynthesis, especially regeneration of the D1-protein, following low-temperature-photooxidative stress (Nishida and Murata 1996).

1.2 Antioxidants

When conditions are optimal, a minor proportion of the O_2 used by the plant cell is converted to ROS (Asada 1999). A range of protective mechanisms against oxidative stress have evolved in plants, which can be divided into two classes: those involved in eliminating ROS, and those that limit their production. The relative importance of these two processes is not yet known, and may differ between plant species and cultivars. However, it is clear that plants regularly make "safe" use of O_2 (for example as a substrate in respiratory and oxygenase reactions) without generating any ROS. It is equally clear that plant cells can minimise oxidative injury by limiting contact between potentially reactive substances, e.g. reducing OH· formation by binding metals to various chelators or proteins, and O_2 to appropriate proteins (such as leghemoglobin). The anti-ROS systems in plants comprise a range of suppressing, scavenging and repair systems, which singly and interactively protect molecules and structures within plant cells. These systems are active in a wide range of physiological states, and in both aerobic and anaerobic conditions. However, if oxidative stresses exceed the maximum capacity of the defences, ROS can induce a lethal cascade of damage to plant cells.

Numerous ROS-scavengers and antioxidants are found in plants: enzymatic and non-enzymatic, water- and lipid- soluble, and located in various, discrete cellular compartments (Noctor and Foyer 1998; Asada 1999). The non-enzymatic antioxidants comprise a diverse range of compounds, including pigments, GSH, AsA, vitamin E and many others, but these will not be further discussed in this section. Membrane-bound antioxidants mainly scavenge $^1O_2$, whereas water-soluble antioxidants tend to scavenge $O_2^{\cdot-}$ and H_2O_2. The interactions found between different ROS-scavengers and antioxidants have been reviewed by Noctor and Foyer (1998) and Asada (1999).

There is also a wide array of enzymatic ROS scavengers including, in the chloroplast, superoxide dismutase (SOD), ascorbate peroxidase (APX), monodehydroascorbate reductase (MDAR), dehydroascorbate reductase (DHAR), glutathione peroxidase (GPX) and glutathione reductase (GOR) (Foyer 1997; Mullineaux and Creissen 1997, Mullineaux et al., 1998; Noctor and Foyer 1998; Asada 1999). The cytosol may also be well-endowed in this respect, with APX, SOD, catalase (CAT), GPX, GOR, MDAR and DHAR all present. ROS-scavenging enzymes are active in several important physiological processes. They contribute to the generation of proton gradients across the thylakoid membranes in the chloroplast and, thus, allow significant linear electron fluxes. They help maintain glutathione and ascorbate in the chloroplast in reduced states, thereby reducing excess production of electrons. They limit $O_2^{\cdot-}$ and H_2O_2 levels, helping to prevent formation of OH· radicals by metal-catalysed Haber-Weiss reactions, and thus protect macromolecules in the chloroplasts. It has been proposed that 10 to 30 % of all generated electrons in the chloroplast lumen flows through $O_2^{\cdot-}$ and H_2O_2 and are captured again within the water molecule by the ROS-enzymatic scavenging system in the chloroplast stroma (Asada, 1999).

However, other results suggest that Mehler reaction play a minor role in dissipating of EEE, and photorespiration is far more important component in this process (Badger et al., 2000).

SOD has evolved to be one of the enzymes with the fastest function (maximal velocity (V_{max}) = 2 x 10^9 M^{-1} s^{-1}) with an optimum close to the diffusion rate of $O_2^{\cdot-}$. SOD converts $O_2^{\cdot-}$ to H_2O_2 and constitutes the first link in the enzymatic scavenging system of ROS (McCord and Fridovich, 1969; Asada, 1999). The cumulative results of several decades of research into this enzyme show that SOD plays key roles in diverse physiological, biochemical and molecular processes in both aerobic and anaerobic organisms. In humans and various other animals, it has also been implicated in programmed neurone cell death (Kane et al., 1993; Raff et al., 1993).

Three different kinds of SOD have been found in prokaryotes and eukaryotes, which contain Mn, Fe or Cu and Zn as prosthetic metals (Asada 1994), and can be identified by differences in their sensitivity to KCN and H_2O_2. Furthermore, various studies have shown that different SOD isoforms tend to have different expression patterns in plants, and to be located in different compartments, either intra- or extra- cellular (Tsang et al., 1991; Wingsle et al., 1991; Bowler et al., 1992; Karpinski et al., 1992a; 1992b; Karpinski et al., 1993; Streller and Wingsle 1994; Wingsle and Karpinski 1996; Schinkel et al., 1998; B. Karpinska, unpublished data, Figure 1).

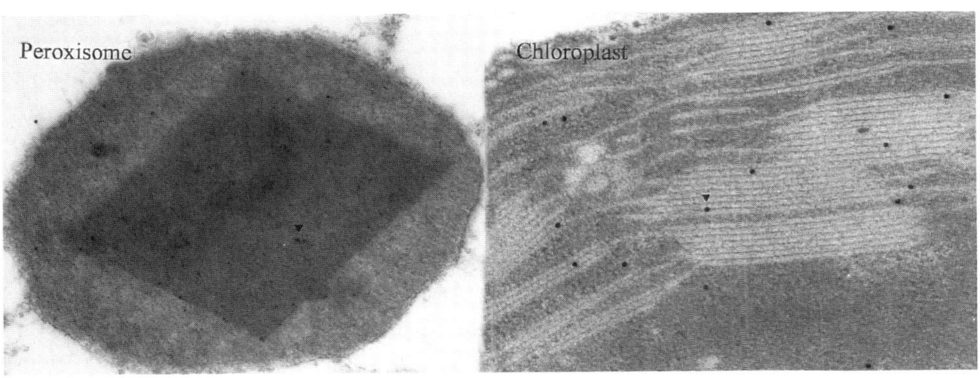

Figure 1. Peroxisomal and chloroplast localisation detected by immunogold staining of CuZnSOD in pine and tobacco, respectively.

In a study of pine needles recovering from natural winter conditions, a phase in which they may be exposed to both high light and low temperature stresses, Karpinski et al., (1993; 1994) recorded significant increases in *SOD* mRNA levels. Furthermore, in this study, higher levels of mRNA encoding both chloroplastic and cytosolic isoforms of CuZn-SOD were seen in needles protruding above snow than in needles covered by snow. However, corresponding increases in protein activities were not observed, and CuZn-SOD activities were similar in covered and protruding needles. Thus, CuZn-SOD turnover rates may be higher in needles that are not protected by snow. A similar lack of correlation between CuZn-SOD mRNA levels and protein activity during responses to oxidative stress, apparently due to compensatory changes in CuZn-SOD turnover rates, was also observed in an earlier study (Karpinski et al., 1992b). Further evidence has suggested that oxidative stress leads to increased rates of proteolytic degradation, initiated by the OH· radical (Casano et al., 1994). During high levels of oxidative stress, both CuZn- and Fe-SOD have the theoretical potential to produce OH· radicals, which may inactivate the enzyme.

The various isoforms of SOD show different patterns of expression during recovery from winter stress. In a study of SOD in Scots pine, for instance, mRNA transcripts of the chloroplastic CuZn-*SOD* where found to be four times higher than cytosolic mRNA levels until mid-May (Karpinski et al., 1993). The chloroplastic CuZn-SOD activity was also higher than the corresponding cytosolic activity. However, as the photosynthetic apparatus was repaired after the winter stress, transcript levels of both chloroplastic and cytosolic CuZn-*SODs* declined, falling to similar low levels (Karpinski et al., 1993; 1994). These observations indicate that chloroplasts in conifers tend to generate significant levels of ROS during low-temperature/high light related oxidative stress.

Earlier studies have also shown that SOD is an important component of responses to oxidative stresses induced by low temperatures in temperature-sensitive plants. For instance, significantly reduced CuZn-SOD activities have been observed in tomato leaves illuminated after storage in cold, dark conditions (Michalski and Kaniuga 1981), while increases in SOD activity have been recorded in wheat and maize following low temperature stress (Schöner and Krause 1990; Jahnke et al., 1991).

The key H_2O_2-scavenging enzyme in the light is APX, which catalyses the reaction: 2 AsA + H_2O_2 —> 2 monodehydroascorbate (mdAsA) + $2H_2O$. Catalases, which have a recently demonstrated protective role in photooxidative stress (Willekens et al., 1998), can also scavenge H_2O_2. These are peroxisomal enzymes that do not need a reducing substrate.

Having scavenged H_2O_2, AsA is regenerated from mdAsA or dehydroascorbate (dAsA) in the chloroplasts: mdAsA being converted to AsA either via the oxidation of reduced ferredoxin or NAD(P)H by the action of MDAR, while DHAR is believed to regenerate AsA using GSH as an electron donor. In both plants and animals, GPX has received considerable attention since it appears to be important in scavenging H_2O_2 and peroxidised lipids. Furthermore, a plastidial GPX has been identified recently (Mullineaux et al., 1998), suggesting that GSH can be enzymatically oxidised to oxidised glutathione (GSSG) in the chloroplast, without the involvement of DHAR. Elucidation of the role and function of the chloroplastic GPX during cold hardening and oxidative stress induced by low temperature in plants is awaited with interest.

Correlation's between the failure of antioxidant systems and damage to plant tissues in low temperatures have been observed in several investigations (Sagisaka 1985; Kuroda et al., 1992). Evergreens and deciduous trees are particularly vulnerable to late spring frosts during the initial flushing of buds, as shown in one such study, in which spruce seedlings were exposed to an artificial frost of -5°C for a single night in the spring (Polle et al., 1996). In this case, APX activity declined and lipid peroxidation increased before any discernible visual damage to the needles occurred. Furthermore, the entire antioxidant enzymatic system essentially collapsed in needles that were shown to be seriously injured. In contrast, a general stimulation of the enzymatic ROS-scavenging system was observed in surviving needles, transient increases being observed for all of the antioxidant enzymes. However, despite the transient increase in antioxidants seen in these needles, they still had lower levels than controls, indicating that the spring frost had left a "memory effect". The results clearly indicate that diverse elements of the enzymatic ROS-scavenging system need to be maintained at levels appropriate to the environmental conditions to ensure the survival and health of plant cells.

ROS-scavenging enzymes are important in both chilling-tolerant and chilling-sensitive plants during oxidative stress induced by low-temperatures. Nevertheless, sensitive and tolerant plants appear to differ in ability to increase their ROS-scavenging capacity during chilling, and to compensate for losses of capacity, as and when appropriate (Karpinski et al., 1993; Baker 1994; Wise 1995). Thus, it has been proposed that the high levels of ROS-scavenging enzymes often found in chilling-sensitive plants may indicate oxidative stress, rather than stress tolerance (Strassart et al., 1995).

1.3 H_2O_2 and Acclimation to EEE

A very wide array of compounds appear to be involved in signalling in biotic and abiotic stress responses, including: salicylic acid, H_2O_2 (Chen et al., 1993; Levine et al., 1994; Prasad et al., 1994; Bi et al., 1995; Neuenschwander et al., 1995; Alvarez et al., 1998; Karpinski et al., 1999; Karpinska et al., 2000), Nitric oxide (Delledonne et al., 1998), ethylene (Gan and Amasino 1997), $O_2^{\cdot-}$ (Doke 1983; Tsang et al., 1991, Jabs et al., 1996); GSH and GSSG (Wingate et al., 1988; Hérouart et al., 1993; Wingsle and Karpinski 1996; Karpinski et al., 1997, Creissen et al., 1999; Karpinska et al., 2000), Ca^{2+} (Price et al., 1994; Monroy and Dhindsa 1995; Knight et al., 1991; 1996), photoreceptors with Ca^{2+} (Neuhaus et al., 1993; Millar et al., 1995); abscisic acid (Leung and Giraudat 1998) and (via its redox status) the Q_B and/or PQ pool (Escoubas et al., 1995; Maxwell et al., 1995; Karpinski et al., 1997; 1999; Huner et al., 1998; Karpinski et al., 1999; Pfannschmidt et al., 1999). However, the signalling cascades initiated by these agents are not well understood. ROS have regulatory roles in diverse processes, such as resistance to various abiotic stresses (Bowler et al., 1992); acclimation to chilling (Prasad et al., 1994); the hypersensitive response and systemic acquired resistance (Lamb and Dixon 1997); photosynthetic electron transport and SAA (Karpinski et al., 1999).

There are also increasing indications that specific transcription factors regulate gene expression in oxidative stress responses (Lu et al., 1996; Cao et al., 1997; Dietrich et al.,

1998; Tamagnone et al., 1998). However, in the following section we focus on the roles of H_2O_2 and GSH in the acclimatory processes of plants, and in the influence of the redox status of the Q_B-PQ pools.

It has been repeatedly demonstrated that H_2O_2 acts as a signalling molecule in various secondary messenger systems in humans and in animals (Meyer et al., 1993; Ginnpease and Whisler 1996). The ability to control H_2O_2, $O_2^{\cdot-}$ and GSH levels is also clearly important to plants faced with either biotic or abiotic stresses. The results of experiments in which H_2O_2 or menadione, an $O_2^{\cdot}$ -generating compound, was exogenously applied to maize seedlings, suggest that mild oxidative stress at 27°C may induce chilling acclimation (Prasad et al., 1994). In these experiments, both compounds caused increases in *CAT3* and peroxidase (*POX)* transcript levels, and in corresponding protein activities, whereas SOD activity remained constant. In contrast, H_2O_2 did not increase the activity of these enzymes at 4°C, whether applied exogenously or endogenously accumulated. Therefore, it appears that elevated levels of H_2O_2, which would be potentially lethal at low temperatures, induced cold-acclimation at higher temperatures. It was proposed that the mechanism whereby H_2O_2 or menadione induce *CAT3* and *POX* gene expression in maize is chilling-sensitive, and that the chilling injury observed was due in part to excess ROS in non-acclimated seedlings (via protein and lipid oxidation). In contrast, the higher levels of ROS-scavenging enzymes induced in the chilling-acclimated seedlings are thought to have prevented ROS accumulation, and thus to have prevented subsequent damage to the lipids and proteins at 4°C (Prasad et al., 1994; Prasad 1996).

In tobacco, increased *Fe-SOD* mRNA levels have been shown to be induced by $O_2^{\cdot-}$ generated in the PSI vicinity by paraquat treatment (Tsang et al., 1991). It has also been shown that CuZn-*SOD4* and CuZn-*SOD4A* transcript levels rise in response to H_2O_2 treatment in maize (Kernodle and Scandalios 1996). Cytosolic and chloroplastic CuZn-*SOD* transcript levels show similar responses in Scots pine (S. Karpinski unpublished results).

Other studies have shown that the expression of two well-characterised plant genes, *APX1* and *APX2*, which encode cytosolic isoforms of APX, is swiftly induced during EEE generated by excess light (EL, Karpinski et al., 1997, 1999) or low temperature (S. Karpinski, data not shown). Critical primary events in the regulatory mechanism controlling *APX1* and *APX2* gene expression are redox changes in the QB and/or PQ pools (Karpinski et al., 1997, 1999)

We have also shown that treating *A. thaliana* leaves with 10 or 100 mM H_2O_2 prior to EEE protects chloroplasts, and the cell, from EEE-induced photooxidative damage (Karpinska et al., 2000). Concentrations of H_2O_2 found in both the cells and chloroplasts of leaves treated with 100 mM H_2O_2 (Figure 2A) are several-fold higher than levels previously believed to inhibit SH-enzymes involved in the Calvin cycle and to cause plant cell death (Kaiser 1979, Levine et al., 1994, Lamb and Dixon 1997). These treatments also induced large rises in *APX1* mRNA levels (Figure 2B).

Furthermore, the H_2O_2-treated leaves were more resistant to subsequent exposure to EL, photoinhibition of photosynthetic electron transport developed relatively slowly in them (Figure 2C and D), and their tissues showed no visual signs of photodamage. Photosynthetic parameters declined markedly faster in Low-Light (LL)-acclimated control leaves exposed to EL (for two hours or less), and the treatment caused permanent photodamage to their tissues: 65 % of these leaves never recovered (Figure 2C and D). The levels of H_2O_2 in the photodamaged, control leaves were much lower than those in the 100 mM H_2O_2- treated leaves (Figure 2A).

So, high H_2O_2 levels *per se* do not appear to be responsible for permanent EEE-induced photodamage, which must therefore be due to factors other than increased H_2O_2 production, for example lipid peroxidation. In other experiments, permanent photodamage was found to be accelerated in *Arabidopsis* leaves treated with 0.5 to 1 M H_2O_2 (Karpinska et al. 2000). In our experimental system, the lowest concentration of H_2O_2 observed to induce any significant changes in photosynthetic parameters in LL was 5 mM.

According to these results, the antioxidant capacities of active chloroplasts and photosynthetic tissues of mature, intact *A. thaliana* leaves cultivated under short day conditions (Karpinski et al., 1999) are several orders of magnitude higher than those of corresponding material from *in vitro* cell cultures. Thus, it appears that *in vitro* plant cell culture experimental systems, which tend to have very low levels of photosynthetic activity (Levine et al., 1994, Lamb and Dixon 1997), have limited relevance to the stress physiology of whole leaves.

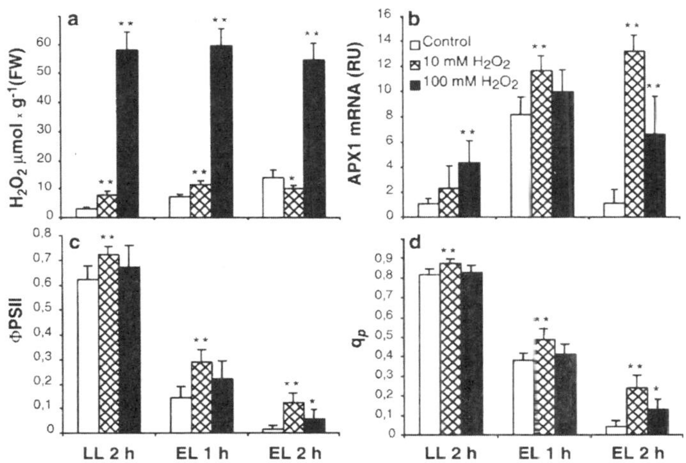

Figure 2. The role of hydrogen peroxide (H_2O_2) in acclimation to excess excitation energy (EEE) in *Arabidopsis thaliana* leaf tissue. A; H_2O_2 levels in detached leaves treated as described in Karpinska et al., (2000) with water (control), 10 mM H_2O_2 and 100 mM H_2O_2 for 2 hours (h) in low light (LL; 200 μmol of photons m^{-2} s^{-1}) and then exposed to excess light (EL, 2600 μmol of photons m^{-2} s^{-1}) for 1 or 2 h (n = 10±SD); B; *APX1* mRNA levels – relative units (RU) in leaves treated as above. The values were obtained after scanning autoradiograms from slot blot hybridisation experiments as described in Karpinski et al., (1997), (n = 3±SD). C and D; Photosynthetic electron transport efficiency in photosystem II (ΦPSII) and photochemical quenching (q_p) in detached leaves treated as above (n = 6±SD). Levels of significance were calculated from the ANOVA data (*$P < 0.05$, **$P < 0.01$).

1.4 Glutathione and Acclimation to EEE

Reduced glutathione is the most abundant thiol in higher plants, and it is recognised as one of the most important antioxidants in biological systems (Noctor and Foyer 1998, Asada 1999). Low levels of GSSG (Figure 3A) are generally found in plants: ca. 3% of the GSH levels, for instance, in *A. thaliana* leaves grown under optimal conditions (Figure 3A, Karpinski et al., 1997). The ratio of reduced to oxidised glutathione, i.e. the redox status [GSH/(GSSG +GSH)], is known to be affected by many environmental factors in diverse organisms (Karpinski et al., 1997, Noctor and Foyer 1998) and the accumulation of GSSG may indicate oxidative stress.

In a study of transgenic tobacco protoplasts (Hérouart et al., 1993), various thiolic compounds, such as GSH, cysteine and dithiothereitol, increased transcription of a reporter gene controlled by the cytosolic CuZn-*SOD* promoter. It has also been shown recently that transcript levels of genes encoding cytosolic and chloroplastic forms of CuZn-SOD in Scots pine (Wingsle and Karpinski 1996) and cytosolic *APX* in *Arabidopsis* (Karpinski et al., 1997) are regulated, at least in part, by changes in the glutathione levels and/or redox status of the glutathione pool.

The effects of glutathione on the expression of CuZn-*SOD* and *GOR* genes showed that GSH can decrease rates of expression of the cytosolic CuZn-*SOD* in plants (Wingsle and Karpinski 1996), and are consistent with the thiol-induced down-regulation of human CuZn-*SOD* and Mn-*SOD* genes observed by Suzuki et al., (1993), but they conflict with the findings of Hérouart et al., (1993). The differences may be due to differences in the experimental systems investigated. It has been proposed that either the absolute levels of GSH and GSSG, or the redox state of the glutathione pool, may be significant factors *in vivo* regulating genes that encode ROS-scavenging enzymes in plants. Our results suggest that, in Scots pine needles at least, altered levels of GSH and GSSG have different effects on the expression of *SOD* and *GOR* genes (Wingsle and Karpinski 1996). In our experiments, raised levels of GSSG induced increases in the activity of GOR, but not in *GOR* transcript levels. This suggests that the activity of the enzyme itself is modulated by redox events *in vivo*. However, GSH induced reductions in transcript levels of both cytosolic and chloroplastic CuZn-SOD.

We have shown that both photosynthetic electron transport and *APX1* gene expression are reduced in *A. thaliana* leaves treated with 10 mM GSH or GSSG, (Karpinska et al., 2000; Figure 3A-D). Extensive photodamage was apparent in leaves treated with 10mM GSH or GSSG and incubated in LL for 18 h. The damage was still greater in leaves incubated for 2 h in GSH or GSSG followed by EL treatment for 1.5 h. Combined treatments with 100 mM H_2O_2, followed by 10 mM GSH, greatly reduced these symptoms, and low levels of visual damage were seen in leaves incubated in the dark for 18 h with 10mM GSH.

The striking similarity of the effects induced by the GSH and GSSG treatments can be attributed to the increases in GSH levels (10-fold in the GSSG treatment and 50-fold in GSH treatment) caused by both treatments (Figure 3A): a similarity observed before in Scots pine needles (Wingsle and Karpinski 1996). Increases in GSH levels following both treatments have been reported to down-regulate the expression of genes encoding cytosolic and chloroplastic isoforms of copper-zinc superoxide dismutase in Scots pine needles (Wingsle and Karpinski 1996). The inhibition of *APX1* and *APX2* (Karpinski et al., 1997, Figure 3B) gene expression and photosynthetic electron transport in *A. thaliana* (Figure 3C and D) can be explained by the highly reduced environment in the leaf cells, caused by excess GSH. This, in turn, inhibits photosynthetic electron transport and redox signalling systems, and accelerates photooxidative damage, including bleaching of the chlorophyll (Åslund and Beckwith 1997, Karpinski et al., 1997, Åslund et al., 1999, Creissen et al., 1999). Excess AsA has been found to induce similar results to those caused by excess GSH (S. Karpinski, unpublished results. These results suggest that the correct antioxidants – oxidants balance is crucial for acclimation to EEE. Consequently, the above results strongly suggest that transiently higher H_2O_2 levels in the chloroplast or cytosol prior stress may prove to be much more important for the amelioration of photooxidative stress tolerance in plants than simply permanently higher levels of antioxidants.

1.5 Redox Regulation of Acclimation to EEE

We believe that the following mechanism may explain the phenomena described above. In the light, the extent of reduction of the Q_B-PQ pool depends largely on the trans-thylakoid ΔpH in the chloroplasts, and the electron transport efficiency in PSII (ΦPSII, Horton et al., 1996), which reflects the function and turnover of the D1 protein. The redox state of the primary electron acceptor, quinone A (Q_A), associated with the PSII complex, is indicated by photochemical quenching (q_p). In certain conditions (such as following treatment with 2.5 dibromo-3-methyl-6-isopropyl-p-benzoquinone, or DBMIB), this parameter may also reflect the redox states of the secondary electron acceptor, associated with the PSII (Q_B) and PQ pools. GSH/GSSG treatment tends to increase the degree of Q_A reduction and to reduce electron transport efficiency in PSII in LL (i.e. q_p and ΦPSII values decrease, Figure 3D). This reduction in FPSII is even more pronounced in EL conditions, since the excitation pressure on PSII is higher (Figure 3A-D). Following H_2O_2 treatment, in contrast, the oxidation of Q_A is enhanced and electron transport efficiency in PSII in LL rises (i.e. q_p and ΦPSII values increase, Figure 2D). Furthermore, in EL conditions Q_A is reduced to a lower extent than in control (H_2O-treated) leaves (Figure 2D).

These observations may suggest that the trans-thylakoid ΔpH gradient is decreased by excess glutathione, and increased by excess H_2O_2, so H_2O_2 tends to energise and GSH to de-energise thylakoid membranes. Thus, H_2O_2 and GSH appear to have antagonistic roles in regulation of the redox status of the Q_B-PQ pools and consequent acclimatory responses to EEE (Karpinski et al., 1997, 1999, Karpinska et al., 2000; Figures 2, 3 and 4). The proposed mechanism may also explain the chlorophyll bleaching that occurs in glutathione-treated leaves, a hypothesis supported by the results of treating leaves with excess GSH in the dark. Unlike leaves treated with GSH in LL, such leaves do not develop visual signs of chlorophyll bleaching. Moreover, leaves treated first with 100 mM H_2O_2 and then with10 mM GSH are more tolerant of EL than leaves treated solely with glutathione, or control leaves treated with water. A caveat that could be raised, is that no evidence was produced in these experiments proving that either GSH or GSSG reached the chloroplasts or leaf cells. Nevertheless, the observations demonstrate that excess GSH, either directly or indirectly, has potent inhibitory effects on photosynthetic electron transport and acclimation to EEE. Similar inhibitory effects also appeared in leaves treated with 1 mM GSH (S. Karpinski, unpublished data). The GSH concentration in active chloroplasts is estimated to be 5 to 25 mM (Noctor and Foyer 1998, Asada 1999). We have also shown that applied H_2O_2 readily penetrated to the chloroplast (Karpinska et al., 2000), and that the compound persisted in the leaf for the duration of the experiments, as demonstrated in other experimental systems used in previous investigations (Prasad et al., 1994, Willekens et al., 1998).

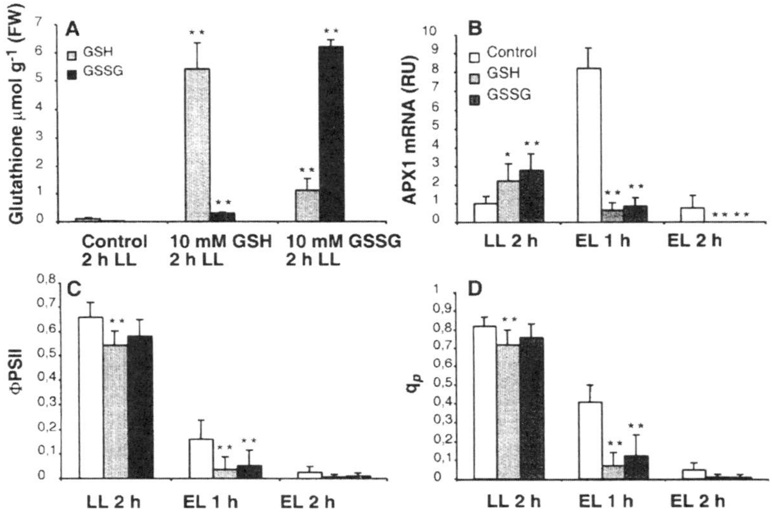

Figure 3. The roles of reduced (GSH) and oxidised (GSSG) glutathione in acclimation to excess excitation energy (EEE) in *Arabidopsis thaliana* leaf tissue. A; GSH and GSSG levels in detached leaves treated as previously described (Karpinski et al., 1997) with water (control), 10 mM GSH or 10 mM GSSG for 2 hours (h) in low light (LL; 200 µmol of photons m^{-2} s^{-1}, n = 6±SD). B; APX1 mRNA levels - relative units (RU) - in leaves treated as above and then exposed to excess light (EL, 2600 µmol of photons m^{-2} s^{-1}) for 1 or 2 h. The values were obtained after scanning autoradiograms from slot blot hybridisation experiments as previously described (Karpinski et al., 1997, n = 3±SD). C and D; Photosynthetic electron transport efficiency in photosystem II (ΦPSII) and photochemical quenching (q_p) in detached leaves treated as above (n = 6±SD). Levels of significance were calculated by ANOVA (*$P < 0.05$, **$P < 0.01$).

If detached *A. thaliana* leaves are treated with the photosynthetic electron transport inhibitor 3-(3,4-dichlorophenyl)-1,1-dimethylurea (DCMU) before being subjected to EL, the induction of *APX1* and *APX2* gene expression is diminished (Karpinski et al., 1997, 1999). However, DBMIB induces the expression of these genes in LL (Karpinski et al., 1997, 1999). The inductive effect of H_2O_2 observed in LL is also inhibited by DCMU treatment (Karpinski *et al.,* 1999) and in the dark (S. Karpinski, unpublished results).

DCMU and DBMIB, chemicals that are not naturally present in either plants or other organisms, block the reduction and oxidation, respectively, of the Q^B-PQ pool. However, DCMU and DBMIB clearly mimic, at a superficial level, the effects of glutathione and H_2O_2, respectively, on the redox status of the Q_A-Q_B-PQ pools and subsequent acclimatory processes (Figures 2 and 3, Karpinski et al., 1997, 1999, S. Karpinski, unpublished data). However, this does not necessarily mean that GSH has a similar mode of action to DCMU, or that H_2O_2 shares mechanistic similarities to DBMIB. A worthy challenge awaiting plant scientists now is to determine details of the signalling cascades and acclimatory processes influenced by the redox status of the Q_B-PQ pool.

1.6 Concluding Remarks

Different strategies of acclimation to photooxidative stress have evolved in different plant species. Improvements in our understanding of these natural strategies may help augment stress tolerance in target plants. As yet, however, the signal transduction events remain unknown that link changes in environmental factors such as day length, light intensity and temperature to changes in ROS, Ca^{2+}, antioxidant or phytohormone status, and ensuing adjustments to gene expression. Many researchers have attempted in recent years to enhance oxidative stress tolerance in plants by increasing their levels of antioxidants. Unfortunately, although transgenic plants with elevated GSH levels have been successfully generated in

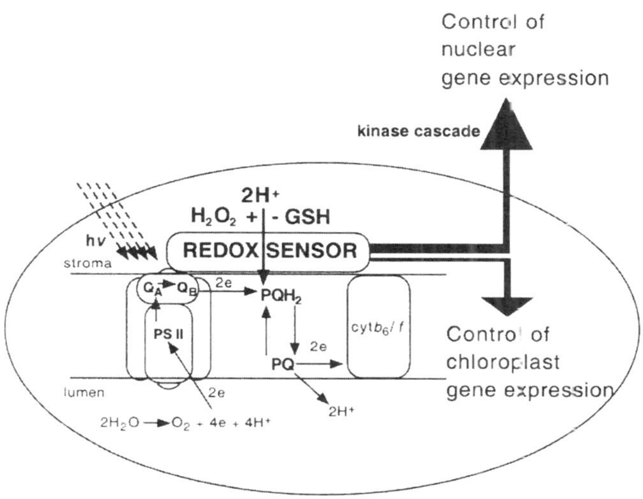

Figure 4. Redox control of nuclear and chloroplast gene expression mediated by a proposed sensor associated with plastoqunone (PQ). Hydrogen peroxide (H_2O_2) promote and glutathione (GSH) inhibits protonation of PQ. Thus H_2O_2 is a positive and GSH is a negative controller of the redox status of PQ and associated sensor, which initiate signalling cascade(s). This hypothetical model is proposed from the results presented here and those published before (Escoubas et al. 1995, Maxwell et al. 1995, Karpinski et al. 1997, 1999, Pfannschmidt et al. 1999).

these trials, some have proved to be more susceptible to EEE and photooxidative stress than the wild-type parental strains (Creissen et al., 1999). Our recent observations indicate that transiently raising H_2O_2 levels in the chloroplast or cytosol prior to stress may be more effective in enhancing the photooxidative stress tolerance of plants than simply raising baseline levels of antioxidants.

REFERENCES

Alvarez, M.E., Pennell, R. I., Meijer, P.J., Ishikawa, A., Dixon, R.A., Lamb, C. Reactive oxygen intermediates mediate a systemic signal network in the establishment of plant immunity. *Cell* 1998; **92**: 773-84.

Aro, E-M., Virgin, I., Andersson B. Photoinhibition of photosystem II. Inactivation, protein damage and turnover. Biochim. Biophys. *Acta* 1993; **1143**:113-34.

Asada, K. Production and action of active oxygen species in photosynthetic tissues. In: Foyer C.H., Mullineaux, P.M. eds. Causes of Photooxidative Stress and Amelioration of Defense Systems in Plants. CRC Press Inc., Boca Raton, FL, 1994; pp. 77-103.

Asada, K. The water-water cycle in chloroplasts: scavenging of active oxygens and dissipation of excess photons. Ann. Rev. Plant Physiol. *Plant Mol. Biol.* 1999; **50**: 601-39.

Åslund, F., Beckwith, Bridge over troubled water: Sensing stress by disulfite bond formation. *Cell* 1999; **96**: 751-3.

Åslund, F., Zheng, M., Beckwith, J., Storz, G. Regulation of OxyR transcription factor by hydrogen peroxide and the cellular thiol-disulfite status. *Proc. Natl. Acad. Sci.* USA 1999; **96**: 6161-5.

Badger, MR., von Caemmerer, S., Ruuska, S., Nakano, H. Electron flow to oxygen in higher plants and algae: rates and control of direct photoreduction (Mehler reaction) and rubisco oxygenase. Phil. *Trans. R Soc. Lond. B.* 2000; **355**: 1433-46.

Baker, N. Chilling stress and photosynthesis. In: Foyer CH. and Mullineaux PM eds. Causes of Photooxidative Stress and Amelioration of Defence Systems in Plants. CRC Press, Boca Raton FL, 1994; pp 127-54.

Benson, E.E., Lynch, P.T., Jones J. Detection of lipid peroxidation products in cryo-protected and frozen rice cells: consequences for post-thaw survival. *Plant Science* 1992; **85**: 107-14.

Bi, Y., M., Kenton, P., Mur, L., Darby, R., Draper, J. Hydrogen peroxide does not function downstream of salicylic acid in the induction of PR protein expression. *Plant J.* 1995; **8**: 235-45.

Bhaumik, G., Srivastava, K.K., Selvamurthy, W. The role of free-radicals in cold injuries. *Int. J. Biomet.* 1995; **38**: 171-81.

Bohnert, H.J., Nelson, D.E., Jensen, R.G. Adaptations to environmental stresses. *Plant Cell* 1995; **7**: 1099-111.

Bowler, C., Van Montagu, M., Inzé, D. Superoxide dismutase and stress tolerance. *Ann. Rev. Plant Physiol. Plant Mol. Biol.* 1992; **43**: 83-116.

Byrd, G.T., Ort, D.R. and Ogren, W.L. The effects of chilling in the light on ribulose-1,5-bisphosphate carboxylase/oxygenase activation in tomato (*Lycopersicon esculentum* Mill.). *Plant Physiol.* 1995; **107**: 585-91.

Cadenas, E. Biochemistry of oxygen toxicity. *Ann. Rev. Biochem.* 1989; **58**: 79-110.

Cao, H., Glazebrook, J., Clarke, J.D., Volko, S., Dong X.N. The *Arabidopsis NPR1* gene that controls systemic acquired resistance encodes a novel protein containing ankyrin repeats. *Cell* 1997; **88**: 57-63.

Casano, L.M., Lascano, H.E., Trippi, V.S. Hydroxyl radicals and a thylakoid-bound endopeptidase are involved in light- and oxygen-induced proteolysis in oat chloroplasts. *Plant Cell Physiol* 1994; **35**: 145-52.

Chen, G. X., Blubaugh, D. J., Hormann, P. H., Golbeck, J. H., Cheniae, G. M. Superoxide contributes to the rapid inactivation of specific secondary donors of the photosystem II reaction centre during photodamage of magnese-depleted photosystem II membranes. *Biochemistry* 1995; **34**: 2317-32.

Chen, Z., Silva, H., Klessig, D.F. Active oxygen species in the induction of plant systematic acquired resistance by salicylic acid. *Science* 1993; **262**: 1883-5.

Creissen, G., Firmin, J., Fryer, M., Kular, B., Leyland, N., Reynolds, H., Pastori, G., Wellburn, F., Baker, N., Wellburn, A., Mullineaux, P. Elevated glutathione biosynthesis capacity in the chloroplasts of transgenic tobacco plants paradoxically causes increased oxidative stress. *Plant Cell* 1999; **11**: 1277-91.

Delledonne, M., Xia, Y., Dixon, R.A., Lamb, C. Nitric oxide functions as a signal in plant disease resistance. *Nature* 1998; **394**: 585-7.

Del Corso, A., Cappiello, M., Mura, U. Thiol-dependent oxidation of enzymes: The last chance against oxidative stress. *Int. J. Biochem.* 1994; **26**: 745-50.

Del Rio, L.A., Sandalio, L.M., Palma, J.M., Bueno, P., Corpas, F.J. Metabolism of oxygen radicals in peroxisomes and cellular implications. *Free Radical in Biology and Medicine* 1992;**13**: 557-80.

Dietrich, R.A., Richberg, M.H., Schmidt, R., Dean, C., Dangl, J.L. A novel zinc finger protein is encoded by the *Arabidopsis* LSD1 gene and functions as a negative regulator of plant cell death. *Cell* 1998; **88**: 685-94.

Doke, N. Involvement of superoxide anion generation in hypersensitive response of potato tuber tissues to infection with an incompatible race of *Phytophthora infestans*. *Physiol. Plant Pathol.* 1983; **23**: 345-7.

Escoubas, J.M., Lomas, M., Laroche, J. Falkowski, P.G. Light-intensity regulation of cab gene-transcription is signaled by the redox state of the plastoquinone pool. *Proc. Natl. Acad. Sci.* USA 1995; **92**: 10237-41.

Foyer, C.H., Halliwell, B. The presence of glutathione and glutathione reductase in chloroplasts: A proposed role in ascorbic acid metabolism. *Planta* 1976; **133**: 21-5.

Foyer, C.H. Oxygen metabolism and electron transport in photosynthesis. In: Scandalios J.G. ed. Oxidative Stress and the Molecular Biology of Antioxidant Defences. Cold Spring Harbor Laboratory Press, 1997; pp. 587-622.

Gan, S.S. Amasino, R.M. Making sense of senescence - Molecular genetic regulation and manipulation of leaf senescence. *Plant Physiol.* 1997; **113**: 313-19.

Ginnpease, M.E. Whisler, R.L. Optimal NF-kB mediated transcriptional responses in Jurkat T-cells exposed to oxidative stress are dependent on intracellular glutathione and co-stimulatory signals. Biochem. *Biophys. Res. Com.* 1993; **226**: 695-702.

Gupta, A.S., Heinen, J.L., Holaday, S.A, Burke, J.J., Allen, R.D. Increased resistance to oxidative stress in transgenic plants that overexpress chloroplastic Cu/Zn superoxide dismutase. *Proc. Natl. Acad. Sci.* USA 1993; **90**: 1629-33.

Guy, C.L. Cold acclimation and freezing tolerance: role of protein metabolism. *Ann. Rev. Plant*

Physiol. Plant Mol. Biol. 1990; **41**:187-223.

Haiyada, P., Parkin, K.L. Chilling-induced oxidative stress in cucumber (*Cucumis sativus* L.cv.calypso) seedlings. *J. Plant Physiol.* 1993; **141**: 733-8.

Halliwell, B., Gutteridge, J.M.C. Biologically relevant metal ion-dependent hydroxyl radical generation. *FEBS Letters* 1993; **307**: 108-12.

Hérouart, D., Van Montagu, M., Inzé, D. Redox-activated expression of the cytosolic copper/zinc superoxide dismutase gene in *Nicotiana*. *Proc. Natl. Acad. Sci.* USA 1993; **90**: 3108-12.

Hodgson, R.A., Raison, J.K. Superoxide production by thylakoids during chilling and its implication in the susceptibility of plants to chilling-induced photoinhibition. *Planta* 1991; **183**: 222-28.

Horton, P., Ruban, A.V. Walters, R.G. Regulation of light harvesting in green plants. *Ann. Rev. Plant Physiol. Plant Mol. Biol.* 1996; **47**: 656-84.

Hughes, M.A., Dunn, A.M. The molecular biology of plant acclimation to low temperature. *J. Exp. Bot.* 1996; **47**: 291-305.

Huner, N.P.A., Öquist, G. Sarhan, F. Energy balance and acclimation to light and cold. *Trends Plant Sci.* 1998; **3**: 224-30.

Hällgren, J-E, Öquist, G. Adaptations to low temperatures. In: Alscher, R.G., Cummings, G.R., eds. Stress Responses in Plants: Adaptation and Acclimation Mechanisms. Wiley-Liss Inc., New York, 1990; pp. 265-93.

Hällgren, J-E, Strand, M, Lundmark, T. Temperature stress In: Raghavendra, A.S. ed. Physiology of Trees. John Wiley and Sons Inc. New York, 1991; pp 301-34.

Jabs, T., Dietrich, R.A., Dangl, J. Initiation of runaway cell death in an Arabidopsis mutant by extracellular superoxide. *Science* 1996; **273**: 1853-6.

Jahnke, L.S., Hull, M.R., Long, S.P. Chilling stress and oxygen metabolising enzymes in *Zea mays* and *Zea diploperennis*. *Plant Cell Environ* 1991; **14**: 97-104.

Kaiser, W.M. Reversible inhibition of the Calvin cycle and activation of oxidative pentose phosphate cycle in isolated intact chloroplasts by hydrogen peroxide. *Planta* 1979; **145**: 377-82.

Kacperska, A. Metabolic consequences of low temperature stress in chilling-insensitive plants. In: Li, P.H. ed. Low Temperature Stress Physiology in Crops. Boca Raton, FL, CRC Press Inc., 1989; pp. 27-40.

Kane, D.J. Srafian, T.A., Hahn, R-A.H., Gralla, E. B., Valentine, J.S., Örd, T., Bredesen, D.E. Bcl-2 inhibition of neural death: Decreased generation of reactive oxygen species. *Science* 1993: **262**: 1274-7.

Karpinska, B., Wingsle, G., Karpinski S. Antagonistic Effects of Hydrogen Peroxide and Glutathione on Acclimation to Excess Excitation Energy in *Arabidopsis*. *IUBMB Life* 2000 **50**: 21—26.

Karpinski, S., Wingsle, G., Olsson, O., Hällgren, J-E. Characterization of cDNAs encoding CuZn-superoxide dismutases in Scots pine. *Plant Mol. Biol.* 1992a; **18**: 545-55.

Karpinski, S., Wingsle, G., Karpinska, B., Hällgren, J-E. Differential expression of CuZn-superoxide dismutases in *Pinus sylvestris* (L.) needles exposed to SO_2 and NO_2. *Physiol. Plant.* 1992b; **85**: 689-96.

Karpinski, S., Escobar, C., Karpinska, B., Creissen, G. Mullineaux, P. Photosynthetic electron transport regulates the expression of cytosolic ascorbate peroxidase genes in *Arabidopsis* during excess light stress. *Plant Cell* 1997; **9**: 627-42.

Karpinski, S., Reynolds, H., Karpinska, B., Wingsle, G., Creissen, G., Mullineaux, P. Systemic signalling and acclimation in response to excess excitation energy in *Arabidopsis*. *Science* 1999;

284: 654-7.

Karpinski, S., Wingsle, G., Karpinska, B., Hällgren, J-E. Molecular responses to photooxidative stress in *Pinus sylvestris* (L.). II. Differential expression of CuZn-superoxide dismutases and glutathione reductase. *Plant Physiol.* 1993; **103**:1385-91.

Karpinski, S., Karpinska, B., Wingsle, G. Hällgren, J-E. Molecular responses to photooxidative stress in *Pinus sylvestris*. I. Differential expression of nuclear and plastid genes in relation to recovery from winter stress. *Physiol. Plant.* 1994; **90**: 358-66.

Kendall, E.J., McKersie, B.D. Free radical and freezing injury to cell membranes in winter wheat (superoxide, phospholipide-esterification). *Physiol. Plant.* 1989; **76**: 86-94.

Kernodle, S.P., Scandalios, J.G. A comparison of the structure and function of the highly homologous maize antioxidant Cu/Zn superoxide dismutase genes, *Sod4* and *Sod4A*. *Genetics* 1996 **143**: 317-28.

Knight, M.R., Campbell, A.K., Smith, S.M. Trewavas, A.J. Transgenic plant aquorin reports the effects of touch and cold-shock and elicitors on cytoplasmatic calcium. *Nature* 1991; **352**: 524-6.

Knight, H., Trewavas, A.J., Knight, M. Cold calcium signalling in *Arabidopsis* involves two cellular pools and change in calcium signature after acclimation. *Plant Cell* 1996; **8**: 489-503.

Knox, P.J., Dodge, A.D. Singlet oxygen and plants. *Phytochem.* 1985; **24**: 889-96.

Kranz, H.D., Denekamp, M., Greco, R., Jin, H., Leyva, A., Meissner, R. C , Petroni, K., Urzainqui, A., Bevan, M., Martin, C., Smeekens, S., Tonelli, C., PazAres, J., Weisshaar, B. Towards functional characterisation of the members of the R2R3-MYB gene family from *Arabidopsis thaliana*. *Plant J.* 1998; **16**: 263-76.

Krause, H. The role of oxygen in photoinhibition of photosynthesis. In Foyer, C.H., Mullineaux, P.M. eds. Causes of Photooxidative Stress and Amelioration of Defense Systems in Plants. Boca Raton, FL, CRC Press, 1994 pp. 43-76.

Kuroda, H., Sagisaka, S., Chiba, K. Collaps of peroxide-scavenging systems in apple flower-buds associated with freezing injury. *Plant Cell Physiol.* 1992; **33**: 743-50.

Kusters, R., Dowhan, W., de Kruijff, B. Negatively charged phospholipids restore prePhoE translocation across phosphatidylglycerol-depleted *Escherichia coli* inner membranes. *J. Biol. Chem.* 1991; **266**: 8659-62.

Lamb, C., Dixon, R.A. The oxidative burst in plant disease resistance. *Annu. Rev. Plant Physiol. Plant Mol. Biol.* 1997; **48**: 251-75.

Leung, J., Giraudat, J. Abscisic acid signal transduction. *Annu. Rev. Plant Physiol. Plant Mol. Biol.* 1998; **49**: 199-22.

Levitt, J. Responses of plants to environmental stress, 2nd ed. Academic Press, New York.1980.

Levine, A., Tenhaken, R., Dixon, R., Lamb, C. H_2O_2 from the oxidative burst orchestrates the plant hypersensitive disease resistance response. *Cell* 1994; **79**: 583-93.

Long, S.P., Humphries, S., Falkowski, P.G. Photoinhibition of photosynthesis in nature. *Annu. Rev. Plant Physiol. Plant Mol. Biol.* 1994; **45**: 633-62.

Lu, G., Paul, A.L., McCarty, D.R. Ferl, R.J. Transcription factor veracity: Is GBF3 responsible for ABA-regulated expression of *Arabidopsis* Adh? *Plant Cell* 1996; **8**: 847-57.

Maxwell, D.P., Laudenbach, D.E., Huner, N.P.A. Redox regulation of light-harvesting complex-II and CAB messenger-RNA abundance in *Dunaliella-salina*. *Plant Physiol.* 1995; **109**: 787-94.

McCord, J.M., Fridovich, I. Superoxide dismutase. An enzymic function for erythrocuprein (hemocuprein). *J. Biol. Chem.* 1969; **244**: 6049-55.

McKersie, B.D., Cen, Y., de Beus, M., Bowley, S.R., Bowler, C., Inzé, D., Dhalluin, K. Botterman,

J. Superoxide dismutase enhances tolerance of freezing stress in transgenic alfalfa (*Medicago sativa* L.). *Plant Physiol.* 1993; **103**: 1155-63.

Meyer, M., Schreck, R., Baeuerle, P.A. H_2O_2 and antioxidants have opposite effects on activation of NF-κB and AP-1 in intact cells: AP-1 as secondary antioxidant-responsive factor *EMBO J.* 1993; **12**: 2005-15.

Michalski, W.P., Kaniuga, Z. Photosynthetic apparatus of chilling sensitive plants. X. Relationship between superoxide dismutase activity and photoperoxidation of chloroplasts lipids. *Biochem. Biophys. Acta* 1981; **637**: 159-67.

Millar, A.J., Straume, M., Chory, J., Chua, N.H., Kay, S.A. The regulation of circadian period by phototransduction pathways in *Arabidopsis*. *Science* 1995; **267**: 1163-6.

Monroy, A.F., Dhindsa, R.S. Low-temperature signal transduction: Induction of cold acclimation-specific genes of alfalfa by calcium at 25 °C. *Plant Cell* 1995; **7**: 321-31.

Mullineaux, P.M. Creissen, G.P. Glutathione reductase: regulation and role in oxidative stress. In: Scandalios, J.G. ed. Oxidative Stress and the Molecular Biology of Antioxidant Defenses. 1997; pp. 667-714. Cold Spring Harbour Laboratory Press.

Mullineaux, P.M., Karpinski, S., Jimenéz, A., Cleary, S.P., Robinson, C., Creissen, G. Identification of cDNAs encoding plastid-targeted glutathione peroxidase. *Plant J.* 1998; **13**: 375-79.

Neuenschwander, U., Vernooij, B., Friedrich, L., Uknes, S., Kessmann, H., Ryals, J. Is hydrogen peroxide a second messenger of salicylic acid in systemic acquired resistance? *Plant J.* 1995; **8**: 227-33.

Neuhaus, G., Bowler, C., Kern, R,. Chua, N.H. Calcium/calmodulin-dependent and calcium/calmodulin-independent phytochrome signal-transduction pathways. *Cell* 1993; **73**: 937-52.

Nishida, I. and Murata, N. Chilling sensitivity in plants and cyanobacteria: The crucial contribution of membrane lipids. *Annu. Rev. Plant Physiol. Plant Mol. Biol.* 1996; **47**: 541-68.

Noctor, G., Foyer C.H. Ascorbate and glutathione: Keeping active oxygen under control. *Annu. Rev. Plant Physiol. Plant Mol. Biol.* 1998; **49**: 249-79.

O'Kane, D., Gill, V., Boyd, P., Burdon, B. Chilling, oxidative stress and antioxidant responses in *Arabidopsis thaliana* callus. *Planta* 1996; **198**: 371-7.

Ottander, C., Campbell, D., Öquist, G. Seasonal changes in photosystem II: organisation and pigment composition in *Pinus sylvestris*. *Planta* 1995; **197**: 176-83.

Öquist, G., Chow, W.S., Andersson, J.M. Photoinhibition of photosynthesis represents a mechanism for the long term regulation of photosystem II. *Planta* 1992; **186**: 450-60.

Palva, T.E. Gene expression under low temperature stress. In: Basra, A.S. ed. Stress-Induced Gene Expression in Plants. Hardwood Academic Publishers, 1994; pp. 103-30.

Pfannschmidt, T., Nilsson, A., Allen, J.F. Photosynthetic control of chloroplast gene expression. *Nature* 1999; **397**: 625-8.

Polle, A., Kroniger, W., Rennenberg, H. Seasonal fluctuations of ascorbate-related enzymes: acute and delayed effects of late frost in spring on antioxidative systems in needles of Norway spruce (*Picea abies* L.). *Plant Cell Phys.* 1996; **37**: 717-25.

Prasad, T.K., Anderson, M.D., Martin, B.A., Stewart, C.R. Evidence for chilling-induced oxidative stress in maize seedlings and regulatory role of hydrogen peroxide. *Plant Cell* 1994; **6**: 65-74.

Prasad, T.K. Mechanisms of chilling-induced oxidative stress injury and tolerance in developing maize seedlings: changes in antioxidant system, oxidation of proteins and lipids, and protease activities. *Plant J.* 1996; **10**: 1017-26.

Price, A.H., Taylor, A., Ripley, S.J., Griffiths, A., Trewavas, A.J. Knight, M.R. Oxidative signals in tobacco increase cytosolic calcium. *Plant Cell.* 1994; **6**: 1301-10.

Raff, M.C., Barres, B.A., Burne, J.F., Burne, J.F., Coles, H.S., Ishizaki, Y., Jacobson, M.D.. Programmed cell death and the control of cell survival: lessons from the nervous system. *Science* 1993: **262**: 695-700.

Russell, A.W., Critchley, C., Robinson, S.A., Franklin, L.A., Seaton, G.G.R., Chow, W-S., Anderson, J., Osmond, C.B. Photosystem II regulation and dynamics of the chloroplast D1 protein in *Arabidopsis* leaves during photosynthesis and photoinhibition. *Plant Physiol.* 1995; **107**: 943-52.

Salin, M.L. Toxic oxygen species and protective systems of the chloroplast. *Physiol. Plant.* 1987; **72**: 681-9.

Sagisaka, S. Injuries of cold acclimatized polar twigs resulting from enzyme inactivation and substrate depression during frozen storage at ambient temperatures for a long period. *Plant Cell Physiol.* 1985; **26**: 1135-45.

Sakai, A., Larcher, W. Frost Survival of Plants. Springer-Verlag. Berlin, Heidelberg, New York. 1987.

Schinkel, H., Streller, S., Wingsle, G. Multiple forms of extracellular superoxide dismutase in needles, stem tissues and seeds of Scots pine. *J. Exp. Bot.* 1998; **49**: 931-6.

Schöner, S., Krause, H. Protective systems against active oxygen species in spinach: Response to cold acclimation in excess light. *Planta* 1990; **180**: 383-9.

Sonoike, K. Photoinhibition of photosystem I; Its physiological significance in the chilling sensitivity of plants. *Plant Cell Physiol.* 1996; **37**: 239-47.

Steponkus, P.L. Responses to extreme temperatures; cellular and sub-cellular bases. In: Lange O.L., Nobel, P.S., Osmond, C.B., et al., eds. Encyclopaedia of Plant Physiology, New Series of Physiological Plant Ecology I, Responses to the Physical Environment Springer-Verlag Berlin, Heidelberg, New York. 1981; 12A: pp. 339-02.

Strassart, J-M., Slooten, L., Botterman, J., Van Breuseghem, F., Meray, N., Inzé, D. In: Mathis P. ed. Photosynthesis: from Light to Biosphere Cold-tolerance, oxidative stress tolerace and anti-oxidant enzyme levels in non-transgenic and Mn-Superoxide dismutase overexpressing maize, 1995; VI: 873-76.

Streller, S., Wingsle, G. *Pinus sylvestris* (L.) needles contain extracellular CuZn superoxide dismutase. *Planta* 1994; **192**: 195-201.

Suzuki, H, Matsumori, A., Matoba, Y., Kyu, B., Tanaka, A., Fujita, J., Sasayama, S. Enhanced expression of superoxide dismutase messenger RNA in viral myocarditis: an SH-dependent reduction of its expression and myocardial injury. *J. Clin. Invest.* 1993; **6**: 2727-33.

Tamagnone, L., Merida, A., Stacey, N., Plaskitt, K., Parr A., Chang, C. F., Lynn, D., Dow J. M., Roberts, K., Martin, C. Inhibition of phenolic acid metabolism results in precocious cell death and altered cell morphology in leaves of transgenic tobacco plants. *Plant Cell* 1998; **10**: 1801-16

Tao, D.L., Jin, Y.H., Du, Y.J. Response of organic-free-radicals production in overwintering conifer needles to low temperature and light. *Chinese J. App. Ecol.* 1992; **3**: 120-124.

Thomashow, M.F. *Arabidopsis thaliana* as a model for studying mechanisms of plant cold tolerance. In: Meyerowitz E.M., Somerville, C.R. eds. Arabidopsis. Cold Spring Harbor Laboratory Press. 1994; pp. 807-34.

Tsang, E.W.T., Bowler, C., Herouart, D., Van Camp, W., Villarroel, R,. Genetello, C., Van Montagu, M., Inzé, D. Differential regulation of superoxide dismutases in plants exposed to environmental stress. *Plant Cell* 1991; **3**: 783-92.

Van Wijk, K.J., Roobol-Boza, M., Kettunen, R., Andersson, B., Aro, E.M. Synthesis and assembly of the D1 protein into photosystem II: Processing of the C-terminus and identification of the initial assembly partners and complexes during photosystem II repair. *Biochem.* 1997; **36**: 6178-86.

Willekens, H., Chamnongpol, S., Davey, M., Schraudner, M., Langebartels, C., Van Montagu, M., Inzé, D., Van Camp, W. Catalase is a sink for H_2O_2 and is indispensable for stress defence in C-3 plants. *EMBO J.* 1998; **16**: 4806-16.

Wingate, V.P.M., Lawton, M.A., Lamb, C.J. Glutathione causes a massive and selective induction of plant defence genes. *Plant Physiol.* 1988; **87**: 206-10.

Wingsle, G., Gardeström, P., Hällgren, J-E., Karpinski, S. Isolation, purification, and subcellular localization of isozymes of superoxide dismutase from Scots pine (*Pinus sylvestris* L.) needles. *Plant Physiol.* 1991; **95**: 21-28.

Wingsle, G., Karpinski, S. Differential redox regulation by glutathione of glutathione reductase and CuZn superoxide dismutase genes expression in *Pinus sylvestris* (L.) needles. *Planta* 1996; **198**: 151-7.

Wise, R.R. Chilling-enhanced photooxidation: The production, action and study of reactive oxygen species produced during chilling in the light. *Photosynth. Res.* 1995; **45**: 79-97.

EFFECTS OF OZONE ON CONIFERS IN THE TIMBERLINE ECOTONE

GERHARD WIESER AND WILHELM M. HAVRANEK
Forstliche Bundesversuchsanstalt, Abt. Forstpflanzenphysiologie, Rennweg 1, A-6020 Innsbruck, Austria. E-mail: Gerhard.Wieser@uibk.ac.at

Key words: ozone, conifers, timberline ecotone, Europe, ozone flux, ozone uptake, threshold value, ontogeny, tree age

1. INTRODUCTION

Tropospheric ozone (O_3) is one of the most detrimental air pollutants known to affect forest trees. The negative impact of O_3 on biomass production and physiological functions have been investigated in a number of experiments (Matyssek et al. 1995, Sandermann et al. 1997, Skärby et al. 1998). Field observations in the Bavarian Forest, Germany (Baumgarten et al. 2000) and in Fennoscandia (Sellden et al. 1997) also implied that ambient O_3 might affect forest trees. For conifers at the alpine timberline however, unambiguous evidence of O_3 injury did not exist, although O_3 regimes are significantly higher when compared to those in Fennoscandia (Skärby et al. 1998) and to low elevation sites in Central Europe (Smidt 1998).

The following account concentrates on the timberline ecotone in central Europe. After providing a short overview of the O_3 regime, possible O_3 effects on conifers are examined. However, it is widely acknowledged that O_3 impacts are more closely related to cumulative O_3 uptake through the stomata (Wieser and Havranek 1993, Wieser 1997, Fuhrer and Achermann 1999, Musselman and Massman 1999) than to external exposure indices such as the AOT40 value in Europe (Fuhrer et al.1997) or the SUM06 in the USA (Lee et al. 1988). Therefore, the importance of the accumulated O_3 uptake will be highlighted with respect to a critical threshold uptake rate eliciting incipient O_3 effects. Finally, the focus will also be on the importance of ontogeny in relating potential O_3 impacts on trees.

S. Huttunen et al. (eds.),
Trends in European Forest Tree Physiology Research, Cost Action E6: EUROSILVA, 115-125

2. OZONE REGIMES AT HIGH ALTITUDE

During the last two decades annual mean O_3 concentrations at high altitude increased by about 0.5 to 0.8 ppb per year (Sladcovic et al. 1994, Smidt et al. 1999). Nowadays forest ecosystems in timberline ecotones in the central European Alps experience O_3 episodes above 120 ppb with to mean O_3 levels around 50 ppb during the growing season (Stockwell et al., 1997). In general, O_3 concentration exhibited a seasonal cycling with maxima during the summer and minima during the winter (Fig. 1).

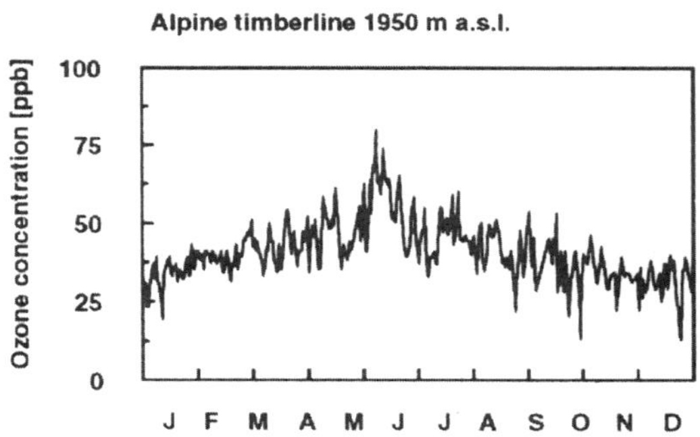

Figure 1. Seasonal course of daily mean ambient ozone concentration at the alpine timberline. (Austrian Timberline Research Station Mt. Patscherkofel, Innsbruck, Austria; 1996 data).

The high summer values can be attributed to an accumulation of O_3 precursors during hot, dry and sunny periods (Stockwell et al. 1997, Vecci and Valli 1999), the latter favouring the photochemical production of O_3 in the lower troposphere (Chameides and Lodge 1992). Consequently AOT40 values are widely exceeded and range from 20 to 70 ppm-h (Smidt 1998).

3. OZONE EFFECTS

3.1 Chamber studies

Evidence of negative impacts of O_3 on the physiology and biomass production has been provided by several chamber studies with 3 to 5-year old seedlings. Exposure to 100 ppb O_3 induced a significant decline in net photosynthesis of *Picea abies* (Kronfuß et al. 1998, Wieser et al. 1998) and *Larix decidua* (Wieser H. 1995) by 20 to 35 % when compared to O_3-free controls. The decline in net photosynthesis was accompanied by a reduction in the carboxylation efficiency (Kronfuß et al. 1998, Wieser H. 1995) strongly suggesting that carbon fixation was the site of O_3-induced injury, since the optimal quantum yield of PSII (Fv/Fm) was not affected neither in *Picea abies* (Wieser et al. 1998) nor in *Larix decidua* (authors unpublished observations).

O_3 induced reductions in the photosynthetic machinery of conifer seedlings are likely to influence carbon allocation including the translocation of assimilates in the plant (Adams et al. 1990, Spence et al. 1990), a decrease in starch concentration (Anttonen and Kärenlampi 1995, Meier et al 1990, Wieser et al. 1998) and an increase carbon partitioning into organic acids, lipids and pigments (Friend and Tomlinson 1992). Such a shift might suggest an allocation of carbon from compounds involved in storage to processes involved in repair and detoxification in order to counteract O_3 stress. As a result, O_3 effects that occur in one growing season probably carry over and might significantly affect growth in the following season as shown for seedlings of *Larix decidua* (Fig. 2), *Fagus sylvatica* (Pearson and Mansfield 1994), *Picea sitchensis* (Wellburn et al. 1997) and *Betula pendula* (Oksanen and Saleem 1999, Pääkkönen et al. 1999) post-cultivated in the absence of elevated O_3.

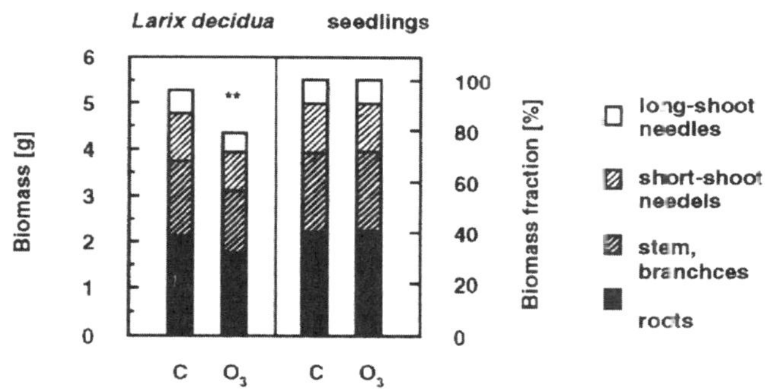

Figure 2. Carry-over effects of O_3 applied in 1993 on the absolute (left) and the relative (right) dry matter allocation in *Larix decidua* seedlings in 1994. Treatments : C: O_3-free control, O_3: 100 ppb O_3. Statistically different at: ** $p<0.01$; $n = 24$. (Data: Havranek and Wieser unpubl. 1994).

The leaf, shoot and root biomass ratios (dry mass in % of whole plant mass) of the *Larix decidua* seedlings however, showed no changes in relation to the previous year's O_3 treatment, as also found in *Pinus sylvestris* by Anttonen and Kärenlampi (1995). One explanation that there was no clear response in the carbon allocation patterns in *Larix decidua* seedlings (Fig. 2) might be, that at the leaf level there were no differences in net photosynthesis of the new flush which emerged in the year following O_3 exposure (G. Wieser and W.M. Havranek unpublished data).

3.2 Field studies

The physiological sensitivity of mature canopy trees to O_3 has been examined in a relatively small number of controlled field studies at the alpine timberline using cuvettes for exposing branches in tree crowns to defined O_3 regimes (Havranek and Wieser 1990, 1994). Such experiments indicated that even a 13 week exposure to ambient and twofold-ambient O_3 concentrations did not significantly affect net photosynthesis and the quantum yield of PSII neither in *Picea abies* (Havranek et al. 1989) nor in *Pinus cembra* needles (Fig. 3).

In *Pinus cembra* elevated O_3 also did not cause a depletion of chlorophyll and whole needle ascorbate (Wieser 1999). The only sign of O_3 stress was an increase in the amount of oxidised glutathione in one-year old needles of *Pinus cembra* (Wieser et al. 2001). Such a change in the redox state of glutathione is supposed to be an early indicator for early stages of hidden O_3 injury as recently shown for O_3 sensitive species of field grown *Pinus ponderosa* trees (Tausz et al. 1999).

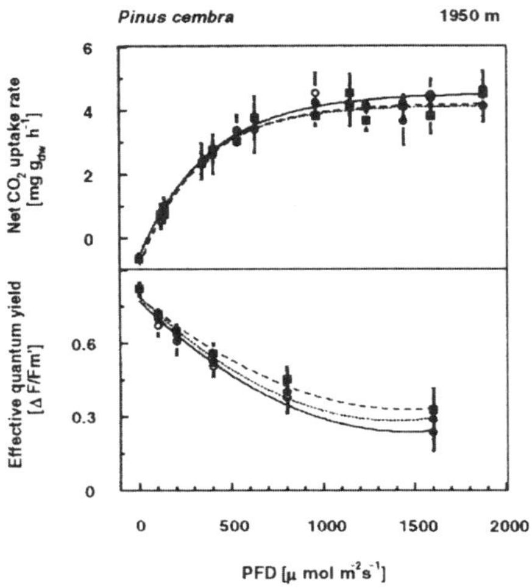

Figure 3. Net CO_2 uptake rate (top) and the effective quantum yield (bottom) in relation to photon flux density (PFD) in *Pinus cembra* needles after a 90 day exposure to charcoal filtered O_3 free air (○; solid line), ambient (mean 44 ppb O_3; ■; dashed line) and two-fold ambient air O_3 concentration (mean 87 ppb O_3; ●; dotted line). (Data: Havranek and Wieser unpubl. 1996).

4. OZONE UPTAKE

The use of twig chambers also provided valuable insights into the mechanisms underlying O_3 uptake of conifers under field conditions with regard to climate, crown position and tree age (Wieser 2001, Wieser et al. 2000). Total O_3 uptake by the foliage is the sum of adsorption onto external surfaces and absorption (flux) trough the stomata into the mesophyll (Fig. 4).

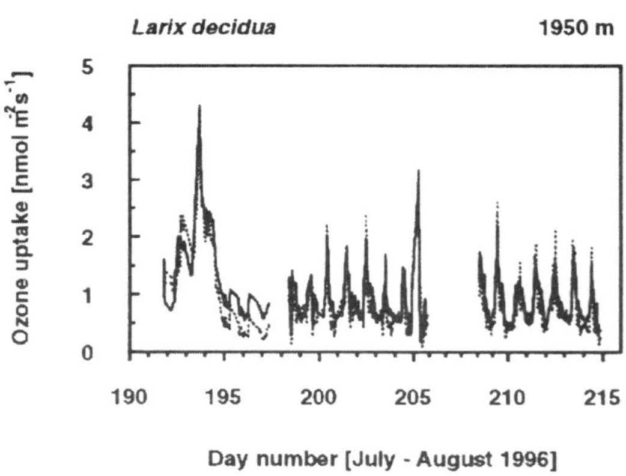

Figure 4. Time course of measured total ozone uptake (solid line) and calculated ozone flux (absorption) trough the stomata into the needle mesophyll (dotted line) of *Larix decidua*. (Data: Wieser unpubl. 1996).

At an average the amount of O_3 adsorbed on external surfaces was about 13 mmol/ m^{-1} s^{-1} of total twig surface area. Similar adsorption rates were also obtained for *Picea abies* by Havranek and Wieser (1990). The dominant pathway of O_3 uptake in conifers however, is the O_3 flux (absorption) through the stomata into the needle mesophyll (Wieser 2001, Wieser et al. 2000).

5. QUANTIFICATION OF TREE RESPONSE TO AMBIENT OZONE LEVELS

As shown above, elevated O_3 concentrations are sufficient to affect the physiology and growth of timberline associated conifers. A threshold relative to the accumulated O_3 uptake throughout one vegetation period under ambient O_3 concentrations however, has not been proposed for these species. Such a threshold can be determined from a scatter plot showing statistically significant initial effects on shoot photosynthesis of mature *Picea abies* trees exposed to various O_3 concentrations in controlled field studies (Fig. 5). The boundary line in Fig. 5 represents the threshold below which the combination of the exposure time

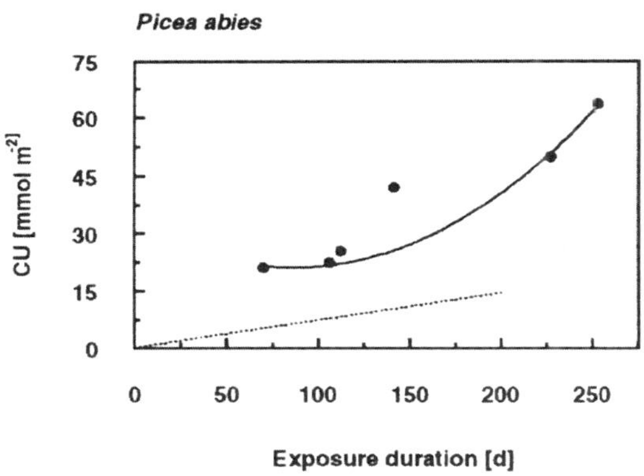

Figure 5. Boundary line analysis characterising the relationship between minimum ozone exposure [combination of exposure duration and cumulative O_3 uptake (CU)] and incipient effects on net photosynthesis of mature *Picea abies* trees. The boundary line defines the threshold below which the combination of exposure duration and cumulative O_3 uptake did not influence net photosynthesis when compared to O_3 free controls; n = 4 to 8; mean O_3 levels > 100 ppb. The dotted lines represents the cumulative O_3 uptake of evergreen mature conifers throughout one growing season at the alpine timberline at ambient O_3 concentration (Redrawn after data from Wieser et al. 2000).

and cumulative O_3 uptake did not cause a statistically significant reduction in shoot net photosynthesis when compared to O_3-free controls. Incipient effects only occurred at mean O_3 levels above 100 ppb when cumulative O_3 uptake exceeded 21 mmol m^{-2} of total needle surface area (Fig. 5). This threshold is about 45% above the accumulated O_3 uptake of

14.5 mmol m^{-2} of total needle surface area estimated for evergreen conifers throughout one vegetation period at the alpine timberline under ambient O_3 regimes (Wieser et al. 2000). By contrast, in *Picea abies* seedlings exposed to 100 ppb of O_3 in chambers statistically significant reductions in net photosynthesis already occurred when the accumulated O_3 uptake exceeded 4.5 mmol m^{-2} of total needle surface area (Kronfuß et.al. 1998).

When used by themselves cumulative O_3 uptake is a measure expressing the quantity rather than the intensity of exposure. Therefore, I also focused on O_3 flux as a rate-based measure in attempting to characterise age related differences in tree response to O_3. A redrawn scatter plot (Fig. 6) shows that the threshold flux inducing incipient effects on net photosynthesis approaches 2.45 nmol m^2 s^{-1} of total needle surface area for an exposure duration $\geq$ 70 in mature trees and $\geq$ 23 days in seedlings, respectively. This threshold is two times above the average O_3 flux during the growing season estimated for mature conifers at high altitude under ambient O_3 regimes (Wieser et al. 2000). Although based on *Picea abies* only, these results strongly suggest that tropospheric O_3 presently does not represent a dominating stress factor to conifers in the timberline ecotone. Nevertheless, chronic effects on the allocation of carbon, due to carbon demands in the foliage for repair and detoxification, with consequences for competitiveness and predisposition to other stress, cannot be ruled out completely.

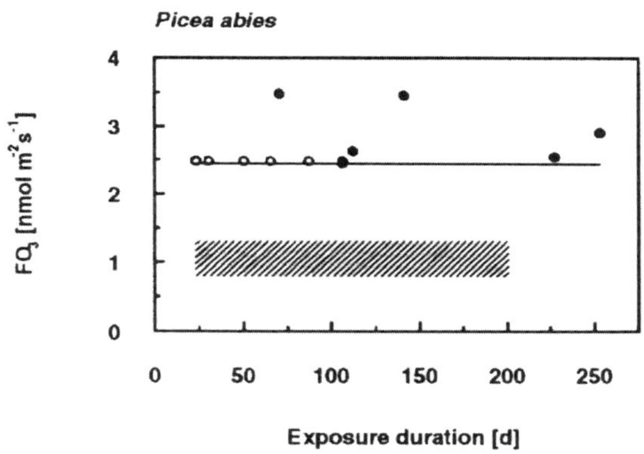

Figure 6. Boundary line analysis characterising the relationship between minimum ozone exposure [combination of exposure duration and average O_3 flux (FO_3) during the exposure period] and incipient effects on net photosynthesis of seedlings (○) and mature *Picea abies* trees (●). The boundary line defines the threshold below which the combination of exposure duration and mean average O_3 flux did not influence net photosynthesis when compared to O_3 free controls; n = 4 to 8; mean O_3 levels > 100 ppb in mature trees and 100 ppb in seedlings, respectively. The dotted band represents the mean average O_3 flux estimated for evergreen mature conifers at high altitude under ambient O_3 concentrations. (After data from Kronfuß et al. 1998, Wieser et al. 1998 and Wieser et al. 2000).

Regarding this "threshold O_3 flux" (Fig. 6) there seems to be a consistency between mature field grown trees and chamber exposed seedlings of *Picea abies*. Such a consistency between juvenile and mature trees is further corroborated by findings from Baumgarten et al. (2000). Although O_3 uptake was based on a model approach (Emberson et al. 1998, 2000); Baumgarten and co-workers (2000) observed that in *Fagus sylvatica* an accumulated O_3 uptake of 3.3. to 3.6 mmol m^{-2} initiated visual leaf injury in both mature canopy trees and chamber exposed seedlings. In both studies however, filed grown trees showed a significantly delayed response to O_3 uptake when compared to chamber exposed seedlings.

How ontogeny modifies tree response to O_3 is not known and the shortcoming in scaling O_3 effects from chamber to stand conditions and from seedlings to mature trees is well documented (Hogsett and Anderssen 1998). Age related differences in the response to O_3 have been attributed to a decline in leaf conductance (c.f. Kolb et al. 1997) and O_3 uptake with increasing tree age and tree size (Wieser 2001). However, the results obtained in *Picea* (Fig. 6) and *Fagus* (Baumgarten et al. 2000) clearly indicate that variations in the ontogenetic response are not only modified by O_3 flux and cumulative uptake, respectively. There are several other factors than O_3 flux that might be important in determining symptom expression including mechanisms of maintenance and repair as well as antioxidant defence mechanisms. Although detoxification is suggested to be less energy demanding than repair mechanisms (Musselman and Massman 1999) there is a lack of data on the control of carbon allocation between both metabolic pathways which attempt to compensate for O_3 stress.

In the timberline ecotone however, there were no significant differences in whole needle ascorbate concentrations of *Picea abies* trees of different ages beyond the early juvenile stage (Fig. 7). Furthermore, the redox state of the ascorbate pool (reduced ascorbate expressed as % of ascorbate and dehydroascorbate) also remained constant and was 70 ± 10% in all the tree age classes investigated (Wieser and Tegischer unpublished).

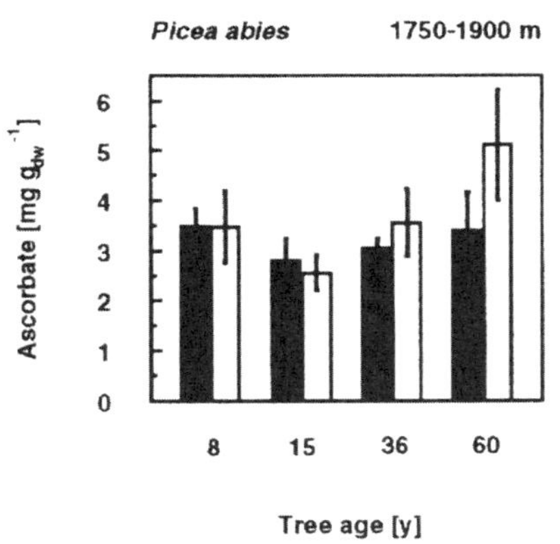

Figure 7. Contents of whole needle ascorbate in current (solid bars) and 1-year-old needles (open bars) of *Picea abies* trees of different age. n = 5 ± SD. Samples of 8, 15 and 36 year-old trees were collected at 1750 m a.s.l. and of 60 year-old trees at 1900 m a.s.l., respectively. (Data: Wieser and Tegischer unpubl. 1999).

These properties however, are under genetic control and are modified by prevailing site conditions (Polle et al. 1999); as there is evidence that *Picea abies* trees from different sites with similar exposure to oxidative pollutants clearly exhibited different levels of foliar ascorbate contents (Tausz and Grill 1995, Tausz et al. 1996).

Another reason for ontogenetic differences in O_3 sensitivity might be related to contrasts in carbon allocation because the trees defence against O_3 must be fuelled from the internal resource pool (Matyssek and Innes 1999). Laurence et al. (1994) observed that fast growing trees with a high metabolic activity (as for example reflected in a high photosynthetic performance) have a higher capacity to allocate carbon for stress compensation while slow growing trees with a low metabolic activity rely more carbon on storage. When comparing trees of different age beyond the early juvenile stage net photosynthesis was found to decrease with increasing tree age in *Pinus contorta, Pinus ponderosa* (Yoder et al. 1994) and *Picea abies* (Wieser and Tegischer unpublished). Such observations support the suggestion that during stages of high metabolic activity trees, within and between species, are more susceptible to O_3 stress. A low metabolic activity by contrast, might enhance tolerance to O_3. In the long term however, it might be possible that the energy costs for defence mechanisms might be exceeded by chronic O_3 exposure (Langebartels et al. 1997). There is a need to examine age related differences in the trees response to O_3 in combination with other biotic and abiotic site factors under field conditions

6. CONCLUSIONS

The findings presented for the timberline ecotone show that tropospheric O_3 presently does not represent a dominating stress factor to evergreen conifers in the timberline ecotone because cumulative uptake rates below 21 mmol m^{-2} of total needle surface area (i.e. about 45% above the cumulative O_3 uptake estimated throughout one vegetation period) did not cause any impairment in the gas exchange of mature evergreen conifers. Nevertheless, elevated O_3 levels and flux rates above 100 ppb and above 2.45 nmol m^2s^{-1} of total needle surface area, respectively cannot be excluded to be a risk for seedlings and mature trees. However, given site (and even chamber) conditions strongly can predispose the responses to O_3. Although there is evidence from chamber experiments that *Picea abies* seedlings have the possibility to acclimate to slowly increasing O_3 (Wieser et al. 1998) uncertainty remains about the extent how future O_3 concentrations in combination with other stress factors (temperature, irradiance, water and nutrient availability) might influence tree metabolism in the field, and hence also influence the development of high altitude forest ecosystems. Linked to controlled chamber and cuvette studies in the field, "free-air" canopy fumigations (Häberle et al. 1999) are necessary in order to determine the risks O_3 might present to forest trees in a changing environment.

REFERENCES

Adams M.B., Ewards N.T., Taylor G.E. and Skaggs B.L. 1990. Whole-plant carbon-14 photosynthate allocation in *Pinus taeda*, seasonal patterns at ambient and elevated ozone levels. *Canadian Journal of Forest Research* **20**: 152-158.

Anttonen S. and Kärenlampi L. 1995. Fatty acids, starch and biomass of scots pine needles and roots in open-air ozone exposure. *Trees* **10**: 74-82.

Baumgarten M., Werner H., Häberle K.-H., Emberson L.D., Fabian P. and Matyssek R. 2000. Seasonal ozone response of mature beech trees (*Fagus sylvatica*) at high elevation of the Bavarian Forest (Germany) in comparison with young beech in field and phytotron. *Environmental Pollution* **109**: 431-442.

Chameides, W.L., Lodge, J.P. 1992. Tropospheric ozone. Formation and fate. In: Lefohn, A.S. (ed.). Surface level ozone exposures and their effects on vegetation. Lewis, Chelsea, Mi, pp 5-30.

Emberson L.D., Ashmore M.R. and Cambridge H.M. 1998. Development of methodologies for mapping Level II critical levels for Ozone. Imperial College of London, DETR Report no. EPG 1/3/82.

Emberson L.D., Wieser G. and Ashmore M.R. 2000. Comparison of measured and modelled stomatal conductance and ozone flux for Norway spruce. *Environmental Pollution* **109**: 393-402.

Friend A.L. and Tomlinson P.T. 1992. Mild ozone exposure alters 14C dynamics in foliage of *Pinus taeda*. *Tree Physiology* **11**: 215-227.

Fuhrer J. and Achermann B. 1999. Critical Levels for ozone –Level II. Environmental Documentation No. 115. Swiss Agency for Environment, Forest and Landscape, Bern, Switzerland, 333p.

Fuhrer J., Skärby L. and Ashmore M.R. 1997. Critical levels for ozone effects on vegetation in Europe. *Environmental Pollution* **97**: 91-106.

Häberle, K.H., Werner H., Fabian P., Pretzsch H., Reiter I. and Matyssek R. 1999. "Free-air" ozone fumigation of mature forest trees: a concept for validating AOT40 under stand conditions. In: Fuhrer J. and Achermann B. (eds.): Critical levels for ozone – Level II. *Environmental Documentation* **No. 115**. Swiss Agency for Environment, Forest and Landscape, Bern, Switzerland. pp. 133-137.

Havranek W.M. and Wieser G. 1990. Research design to measure ozone uptake and its effects on gas exchange in spruce in the field. In: Payer H.D., Pfirrmann T. and Mathy P (eds.): Environmental Research with plants in closed chambers. *Air pollution research report* **26**. Commission of the European Communities, Brussels, pp. 148-152.

Havranek W.M. and Wieser G. 1994. Design and testing of twig chambers for ozone fumigation and gas exchange measurements in mature trees, *Proceedings of the Royal Society of Edinburgh* **102b**:541-546.

Havranek W.M., Wieser G. and Bodner M. 1989. Ozone fumigation of Norway spruce at timberline. *Annales des Sciencies Forestieres* **46 suppl.**, 581s-585s.

Hogsett W.E. and Andersen C.P. 1998: Ecological effects of tropospheric ozone: a U.S. Perspective – past, present and future. In: Schneider T. (ed.): Air pollution in the 21st century Priority issues and policy. *Studies in Environmental Science* **72**, Elsevier, Amsterdam, Lausanne, New York, Oxford, Shannon, Tokyo, pp.419-437.

Kolb T.E., Fredricksen T.S., Steiner, K.C. and Skelly, J.M. 1997. Issues in scaling tree size and age responses to ozone: a review. *Environmental Pollution* **98**: 195-208.

Kronfuß G., Polle A., Tausz M., Havranek W.M. and Wieser G. 1998. Effects of ozone and mild drought stress on gas exchange, antioxidants and chloroplast pigments in current-year needles of young Norway spruce [*Picea abies* (L.) Karst.]. *Trees* **12**: 482-489.

Langebartels C., Ernst D., Heller W., Lütz, C, Payer H.D. and Sandermann H.jr. 1997. Ozone response of trees: results from controlled chamber exposures at the GSF phytotrones. In: Sandermann H.jr., Wellburn A.R. and Heath R.L. (eds.): Forest decline and ozone: a comparison of controlled chamber and field experiments. *Ecological Studies* **Vol. 127**, Springer, Berlin, Heidelberg, New York, pp. 163-200.

Laurence J.A., Amundson R.G., Friedl A.L., Friend A.L., Pell E.J. and Temple P.J. 1994. Allocation of carbon in plants under stress: an analysis of the ROPIS experiments. *Journal of Environmental Quality* **23**: 412-417.

Lee E.H., Tingey D.T. and Hogsett W.E. 1988. Evaluation of ozone exposure indices in exposure-response modelling. *Environmental Pollution* **53**: 43-62.

Matyssek R. and Innes J.L. 1999. Ozone- a risk for forest trees and forests in Europe? *Water, Air, and Soil Pollution* **116**: 199-226.

Matyssek R., Reich P.B., Oren R. and Winner W.E. 1995. Response mechanisms of conifers to air pollutants. In: Smith W.E. and Hinckley T.M. (eds.): Ecophysiology of coniferous forests. Academic Press, San Diego. pp. 255-308.

Meier S., Grand L.F., Schoenberger M.M, Reinert R.A. and Bruck R.I. 1990. Growth, ectomycorrhizae and non structural carbohydrates of loblolly pine seedlings exposed to ozone and soil water deficit. *Environmental Pollution* **64**: 11-27.

Musselman R.C. and Massman W.J. 1999. Ozone flux to vegetation and its relationship to plant response and ambient air quality standards. *Atmospheric Environment* **33**: 65-73.

Oksanen E. and Saleem A. 1999. Ozone exposure results in various carry-over effects and prolonged reduction in biomass in birch (*Betula pendula* Roth.). *Plant Cell and Environment* **22**: 1401-1411.

Pääkkönen E., Holopainen T. and Kärenlampi L. 1999. Ozone impact remains in birch (*Betula pendula* Roth.) one and two seasons after exposure. In: Fuhrer J. and Achermann B. (eds.): Critical Levels for Ozone –Level II. *Environmental Documentation* **No 115**, Swiss Agency for Environment, Forest and Landscape, Bern. pp 139-143.

Pearson M. and Mansfield T.A. 1994. Effects of exposure to ozone and water stress on the following season's growth of beech (*Fagus sylvatica* L.). *New Phytologist* **126**: 511-515.

Polle A., Baumbusch L.O., Oschinski C., Eiblmeier M., Kuhlenkamp V., Vollrath B., Scholz F. and Rennenberg H. 1999. Growth and protection against oxidative stress in young clones and mature spruce trees (Picea abies L.) at high altitude. *Oecologia* **121**: 149-156.

Sandermann H.jr., Wellburn A.R. and Heath R.L. 1997. Forest decline and ozone: a comparison of controlled chamber and field experiments. *Ecological Studies* **Vol. 127**, Springer, Berlin, Heidelberg, New York, 398p.

Sellden G., Sutinen S. and Skärby L. 1997. Controlled ozone exposures and field observations in Fennoscandia. In: Sandermann H.jr., Wellburn A.R. and Heath R.L. (eds.): Forest decline and ozone: a comparison of controlled chamber and field experiments. *Ecological Studies* **Vol. 127**, Springer, Berlin, Heidelberg, New York, pp. 249-276.

Skärby L., Ro-Poulse H., Wellburn A.A.M. and Sheppard L.J. 1998, Impacts of ozone on forests: a European perspective. *New Phytologist* **139**: 109-122.

Sladovic R., Scheel H.E. and Seiler W. 1994. Ozone climatology at the mountain sites Wank and Zugspitze. In: Borrel P.M., Borrel P., Cvitas T. and Seiler W. (eds.): Proc. EUROTRAC Symp. 94. SPB The Hague, pp 253-258.

Smidt S. 1998. Risk assessment of air pollutants for forested areas in Austria, Bavaria and Switzerland. *Environmental Science and Pollution Research* **Spec Issue 1**: 25-31.

Smidt S., Block J., Jandl R. and Gehrmann J. 1999. Trends of pollutant concentrations and inputs at forest stations in Austria and Germany. *Centralblatt für das gesamte Forstwesen* **116**: 193-209.

Spence R.D., Rykiel E.J. and Sharpe P.J.H. 1990. Ozone alters carbon allocation in loblolly pine assessment with carbon-11 labelling. *Environmental Pollution* **64**: 93-106.

Stockwell, W.R., Kramm, G., Scheel, H-E., Mohnen, V.A., Seiler, W. 1997. Ozone formation, destruction and exposure in Europe and in the United States. In: Sandermann, H., Wellburn, A.R., Heath, R,L. (eds.). Forest decline and ozone. A comparison of controlled chamber and field experiments. *Ecological Studies* **Vol. 137**. Springer, Berlin, Heidelberg, New York, Tokyo, pp 1-38.

Tausz M. and Grill D. 1995. Physiological reactions of spruce trees to environmental stress – field study results from various locations in Austria. Proc. Int. Workshop: Exceedances of critical loads and levels, spatial and temporal interpretation for elements in landscapes sensitive to atmospheric pollutants (background papers). Nov 22-24, 1995, Vienna.

Tausz M., Batic F. and Grill D. 1996. Bioindication at forest sites – concepts, practice, and outlook. *Phyton* **36**: 7-14.

Tausz M., Bytnerowicz A., Weidner W., Arbaugh M.J., Padgett P. and Grill D. 1999. Changes in free radical scavengers describe the susceptibility of Pinus ponderosa to ozone in southern Californian forests. *Water, Air and Soil Pollution* **116**: 249-254.

Vecci, R., Valli, G. 1999. Ozone assessment in the southern part of the Alps. *Atmospheric Environment* **33**: 97-109.

Wellburn A.R., Barnes J.D., Lucas P.W., McLeod A.R. and Mansfield T.A. 1997. Controlled O_3 exposure and field observations of O_3 effects in the UK. In: Sandermann H.jr., Wellburn A.R. and Heath R.L. (eds.): Forest decline and ozone: a comparison of controlled chamber and field experiments. Ecological Studies Vol. 127, Springer, Berlin, Heidelberg, New York, pp. 201-247.

Wieser G. 1997. Ozone impact on photosynthetic capacity of mature and young Norway spruce (Picea abies (L.) Karst): external versus internal exposure. *Phyton* **37**: 297-302.

Wieser G. 1999. Evaluation of the impact of ozone on conifers in the Alps: a case study in spruce, pine and larch in the Austrian Alps. *Phyton* **39**: 241-252.

Wieser G. 2001. Exchange of trace gases at the tree-atmosphere interface – Ozone. In: Papen H., Gasche R. and Rennenberg H. (eds.): Trace gas exchange in forest ecosystems. Kluwer, in press.

Wieser G. and Havranek W.M. 1993. Ozone uptake in the sun and shade crown of spruce: quantifying the physiological effects of ozone exposure *Trees* **7**: 227-232.

Wieser G., Havranek W.M., Loidoldt-Nagele M., Kronfuß G. and Polle A. 1998. Response of photosynthesis, carbohydrates and antioxidants in needles of Norway spruce to slow and rapid changes in ozone. *Botanica Acta* **111**: 35-41.

Wieser G., Häsler R., Götz B., Koch W. and Havranek W.M. 2000. Role of climate, crown position, tree age and altitude on calculated ozone flux into needles of Picea abies (L.) Karst and Pinus cembra L.: a synthesis. *Environmental Pollution* **109**: 415-422.

Wieser G., Tausz M., Wonisch A. and Havranek W.M. 2001 Free radical scavengers and photosynthetic pigments in needles of a cembran pine (*Pinus cembra* L.) tree growing at the alpine timberline as affected by ozone exposure. *Biologia Plantarum* **44**: in press.

Wieser H. 1995. Auswirkungen chronischer Ozonbelastung auf die Gaswechselphysiologie junger Hoch- und Tieflagenherkünfte der Lärche (*Larix decidua* Mill.) in Klimakammern. Diploma Thesis University Innsbruck, 104 p.

Yoder B.J., Ryan M.G., Waring R.H., Schoettle A.W. and Kaufmann M.R. 1994. Evidence of reduced photosynthetic rates in old trees. *Forest Science* **40**: 513-527.

AGEING AS AN INFLUENCE ON TREE RESPONSE TO OZONE: THEORY AND OBSERVATIONS

THOMAS E. KOLB
School of Forestry, Northern Arizona University, Flagstaff, Arizona 86011-5018, USA.
E-mail:tom.kolb@nau.edu

Key words: aging, air pollution, maturation, nutrient, ontogeny, ozone, photosynthesis, physiology, water relations, stomatal conductance, scaling

1. INTRODUCTION

Tropospheric ozone (O_3) is considered to be a likely contributing factor to tree and forest decline at several forested areas in North American and Europe (Sandermann et al. 1997, Matyssek and Innes 1999). This hypothesis is largely based on O_3 exposure-response relationships from studies of seedlings. Ozone exposure-response data for the growth of mature forest trees are not yet widely available, and will be difficult to obtain for many species in the future because of logistical constraints.

Several approaches have been used to estimate impacts of O_3 on the growth of mature forest trees. Tree-ring studies have been used to examine relationships between stem radial growth and climate, competition, and unexplained factors that may include air pollutants (Zahner et al. 1989, Bartholomay et al. 1997). Correlation between temporal variation in stem radial growth and ambient O_3 concentration has provided evidence that current O_3 levels can reduce mature tree growth with the magnitude of impact dependent on seasonal climatic conditions (McLaughlin and Downing 1996). Leaf physiological response data from controlled exposures of mature tree branches have been used to estimate effects of O_3 on tree carbon balance (Teskey et al. 1991, Grulke et al. 1996, Wieser 2000). Tree physiological simulation models are another approach whose use is growing. Such

S. Huttunen et al. (eds.),
Trends in European Forest Tree Physiology Research, Cost Action E6: EUROSILVA, 127-155

models often use O_3 exposure-response data from leaves of seedlings to predict impacts on mature forest trees (Reich et al. 1990, Taylor et al. 1994, Constable and Retzlaff 1997, Constable and Taylor 1997, Ollinger et al. 1997, Retzlaff et al. 1997).

Simulation models used to predict effects of O_3 on mature forest trees often make the simplifying assumption that leaf-level responses of seedlings to O_3 can be scaled directly to leaf-level responses of mature forest trees. The accuracy of this assumption is challenged by growing evidence that seedlings and mature trees can differ in physiological characteristics and leaf-level response to O_3 (e.g., Samuelson and Edwards 1993, Grulke and Miller 1994, Samuelson 1994, Fredericksen et al. 1995, Samuelson and Kelly 1996, Fredericksen et al. 1996b, Kolb et al. 1997a, Samuelson and Kelly 1997, Momen et al. 1997, Grulke 1999, Bond 2000). Differences in leaf physiological response to O_3 between seedlings and mature trees could lead to large errors in predicting responses to O_3 for mature forests unless such differences are considered. Clearly, more information on how tree physiological characteristics change as trees age is needed to guide efforts to scale O_3 responses to mature forests.

In this paper, I first review changes during tree ageing in physiological, growth, and environmental characteristics that have been predicted from ecophysiological theories or observed in previous studies. The focus of this section is on changes that are relevant to attempts to scale O_3 impacts from seedlings to mature forest trees. Next, I summarize the results of previous studies that have compared O_3 response between seedlings and mature forest trees to synthesize current findings and to emphasize sources of variation that can influence the results of O_3 scaling studies. Last, I conclude with recommendations and precautions about future studies that are designed to measure O_3 exposure-response relationships of small trees to predict responses of mature trees, and highlight several research questions that need to be addressed to further such scaling efforts.

2. CHANGES DURING TREE MATURATION: PATTERNS AND MECHANISMS

2.1 Water relations

The cohesion theory of xylem water movement predicts differences in leaf water potential (Ψ_L) among different size trees of –0.01 MPa m^{-1} under conditions of no flow (Scholander et al. 1965, Zimmermann 1983). Evidence from older literature for this hydrostatic gradient in trees was mixed, and has led some to question the accuracy of the cohesion theory (Zimmermann et al. 1994, Canny 1995). However, several recent studies have strongly supported the cohesion theory for trees (Holbrook et al. 1995, Pockman et al. 1995, Tyree 1997). Moreover, recent careful measurements in old-growth *Pseudotsuga menziesii* trees showed that vertical gradients in predawn Ψ_L conformed closely to the hydrostatic gradient (Bauerle et al. 1999). For *Prunus serotina* in Pennsylvania, open-grown seedlings had lower predawn Ψ_L than the upper canopy of mature trees and the magnitude of the difference was also close to the gravitational hydrostatic gradient (Fig. 1, Fredericksen et al. 1996c).

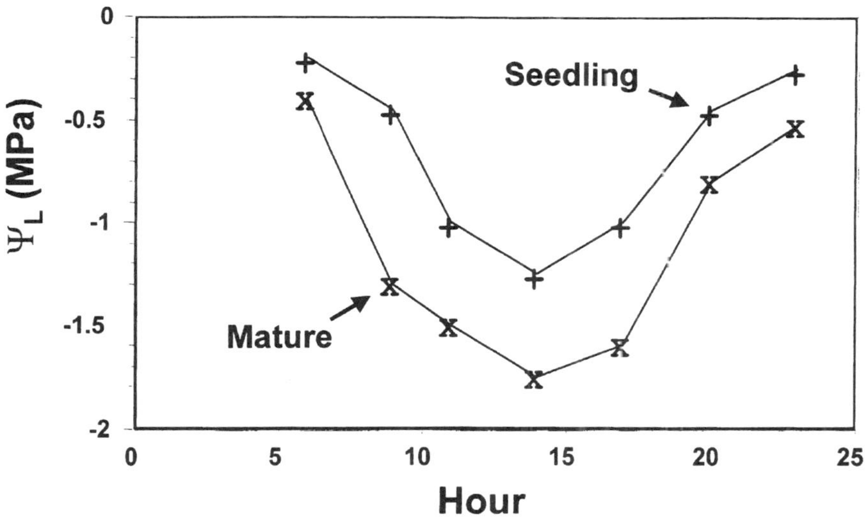

Figure 1. Diurnal variation in leaf xylem water potential (Ψ_L) in June for open-grown seedlings and the upper canopy of mature *Prunus serotina* trees in Pennsylvania. N = 8 leaves for each of 4 mature trees, and 2 leaves from each of 8 seedlings. (After data in Fredericksen et al. 1996c).

Differences in predawn Ψ_L between small and large *Pinus ponderosa* in Arizona were close to the hydrostatic gradient early in the growing season, but often exceeded the hydrostatic gradient later in the season (Fig. 2, Kolb and Stone 2000). This result suggests that low soil-to-leaf hydraulic conductance of large trees may limit overnight equilibration of leaves with the water potential of the soil in some environments.

Stomatal conductance (G_W) often differs between young and old trees. In studies of several species (*Juniperus occidentalis*, *Pinus aristata*, *Pinus ponderosa*, *Pinus taeda*, *Prunus serotina*, *Pseudotsuga menziesii*, *Sequoiadendron giganteum*) that compared unshaded, small trees with large trees that were at least 4 m taller, large trees had lower G_W than small trees (Kolb et al. 1997a, Bauerle et al. 1999, Hubbard et al. 1999, Kolb and Stone 2000, Kolb and Matyssek in press). For example, for *Pinus ponderosa* in Arizona, saplings had greater G_W than old-growth trees on most mornings and afternoons during the growing season (Fig. 3, Kolb and Stone 2000). For *Prunus serotina* in Pennsylvania, open-grown seedlings had greater G_W than the upper canopy of mature trees at all times of the day and night, and showed less stomatal closure during mid day (Fig. 4, Fredericksen et al. 1996c). *Quercus rubra* is an exception to the general pattern for unshaded juveniles: large trees had greater G_W than container-grown seedlings (Samuelson and Edwards 1993, Hanson et al. 1994, Samuelson 1994a). The mechanisms that underlie this result for *Quercus rubra* are not clear, but may include stimulation of leaf gas exchange in large trees by strong carbohydrate sinks such as seed development, reduction of leaf gas exchange in container-grown seedling because of restrictions on root growth, and lower soil-to-leaf hydraulic conductance of seedlings than large trees of this species. Another exception is

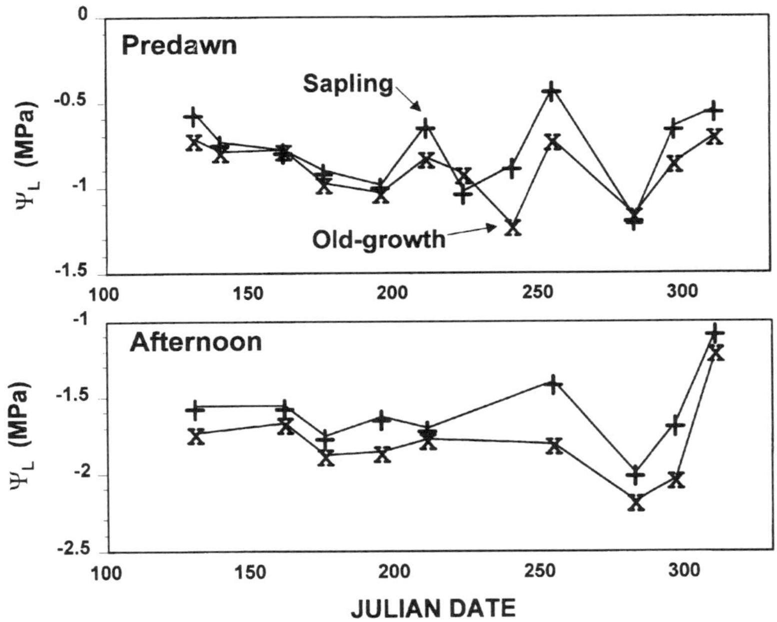

Figure 2. Mean predawn and afternoon leaf xylem water potential (Ψ_L) over the growing season for open-grown saplings and the middle canopy of old-growth *Pinus ponderosa* trees in Arizona. N = 1 leaf from each of 5 trees for each size class. (After data in Kolb and Stone 2000).

Pinus ponderosa in California where G_W of 3-8 year-old seedlings declined in response to seasonal drought earlier in the season than G_W of older trees because of shallower roots, resulting in lower G_W for seedlings than larger trees in late summer (Fig. 5, Grulke 1999).

Lower G_W and Ψ_L of large trees compared with small trees under unshaded conditions is consistent with the 'hydraulic limitation' hypothesis that water movement to the top of tall trees is restricted because of the gravitational hydrostatic gradient and cumulative frictional resistance (Ryan and Yoder 1997, Ryan et al. 1997, Bond and Ryan 2000). Although tapered xylem conduits may buffer hydraulic resistance from path-length effects (West et al. 1999, Becker et al. 2000a), such buffering appears to be overridden by hydraulic constraints in some tall trees. The tissues most responsible for frictional resistance to water flow in xylem are not understood well, but current evidence suggests that branch junctions and bud scars are important (Zimmermann 1983, Tyree and Ewers 1991, Rust and Huttl 1999). Moreover, other buffering factors such as greater water storage in sapwood (Becker et al. 2000b) and increased xylem permeability to water flow with tree ageing (Mencuccini and Magnani 2000) may compensate to some extent for long path lengths for water movement in tall trees.

The hydraulic limitation hypothesis also predicts that xylem water at the top of tall trees is close to the Ψ that causes xylem cavitation, consequently stomates close more

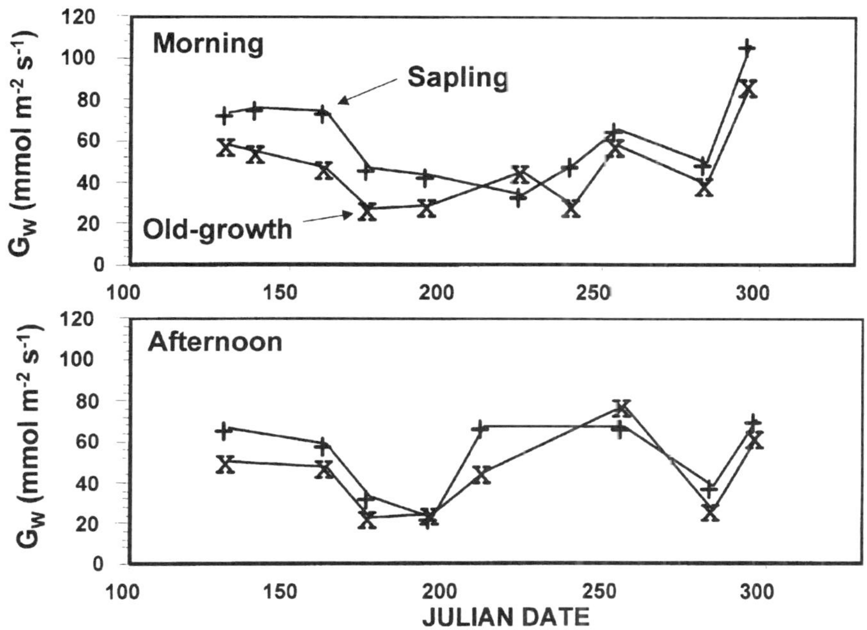

Figure 3. Mean morning and afternoon stomatal conductance (G_W) over the growing season for open-grown saplings and the middle canopy of old-growth *Pinus ponderosa* trees in Arizona. N = 1 leaf from each of 5 trees for each size class. (After data in Kolb and Stone 2000).

often in old trees than in young trees to prevent such cavitation (Ryan and Yoder 1997). It is not clear yet whether a common mechanism explains limitations to G_W in old trees of all species and in all environments. In some studies, stomatal closure of old trees was accompanied by lower Ψ_L compared with young trees, suggesting closure in response to a decrease in leaf water status (Fig. 6, Bauerle et al. 1999, Kolb and Stone 2000). Stomatal closure of old trees has been associated with lower soil-to-leaf hydraulic conductance compared with young trees both in the presence (Kolb and Stone 2000) and absence of differences in Ψ_L between tree sizes (Yoder et al. 1994, Hubbard et al. 1999). This latter finding suggests regulation of stomatal closure by a component(s) of the soil-plant hydraulic system. Another question relevant to scaling is whether the threshold Ψ_L for stomatal closure differs during tree ageing. A study of *Pinus ponderosa* in Arizona suggested the onset of stomatal closure of both juvenile and mature trees at a daytime Ψ_L of approximately –1.8 MPa for unshaded leaves (Fig. 6, Kolb and Stone 2000).

Studies of tree ageing that have compared scions from branches of mature and juvenile trees after grafting onto seedling rootstocks have shown that stem growth rate and physiological traits such as G_W can also vary with tissue age independent of differences in tree height (Greenwood 1995). For example, G_W was lower for scions from old *Picea rubens* trees compared with scions from seedlings when both were grafted onto similar-

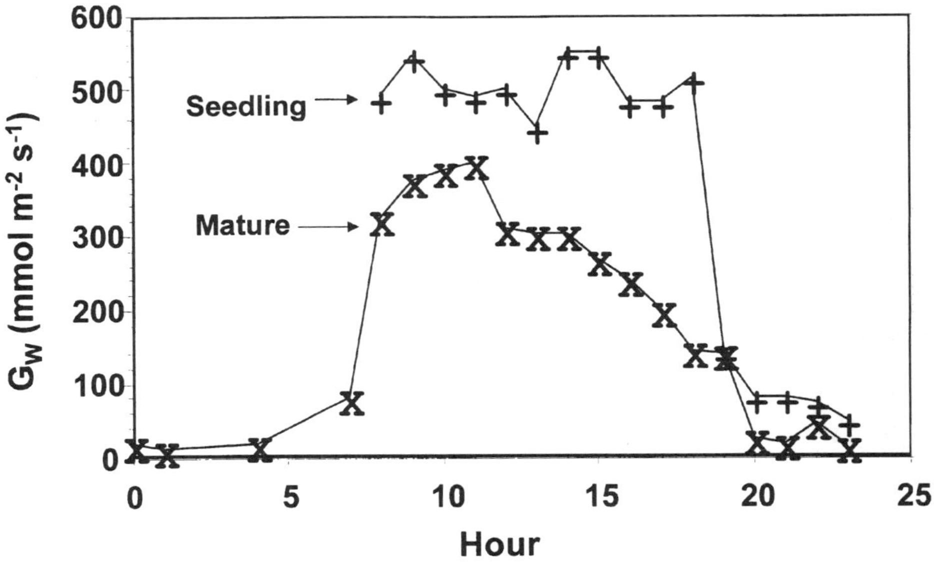

Figure 4. Diel variation in stomatal conductance (G_W) in June for open-grown seedlings and the upper canopy of mature *Prunus serotina* trees in Pennsylvania. N = 8 leaves for each of 4 mature trees, and 2 leaves from each of 8 seedlings. (After data in Fredericksen et al. 1996c).

sized seedling rootstocks and grown in a common environment (Rebbeck et al. 1993). Such results demonstrate that differences in G_W between young and old trees can arise from tree maturation processes. Limited evidence suggests that some changes in morphological and physiological traits during tree maturation result from changes in phytohormones and gene expression (Hutchison et al. 1990, Greenwood 1995).

Resource competition can limit G_W of small, suppressed trees more than larger, dominant trees. In particular, shaded seedlings and saplings often have lower G_W than unshaded dominant trees despite limitations to water movement in large trees (Kostner et al. 1992, Fredericksen et al. 1995, Fredericksen et al. 1996a, Martin et al. 1997, Samuelson and Kelly 1997). There are several mechanisms that explain low G_W of shaded seedlings and saplings. First, G_W of shaded leaves is often limited by low light intensity when measurements are made in ambient light. Second, shade leaves often have lower G_W than sun leaves when both are compared at photosynthetically saturating light intensity, and this difference has been associated with differences in leaf nitrogen concentration per leaf area between sun and shade leaves (Samuelson and Kelly 1997). Lower leaf nitrogen concentration per leaf area for shaded seedlings may result from competition for nitrogen with larger trees, or it may be a consequence of the greater specific leaf area (area per mass) of leaves that develop in shade (Abrams and Kubiske 1990, Ellsworth and Reich 1993).

Dramatic differences in soil water available to juvenile and mature trees in some

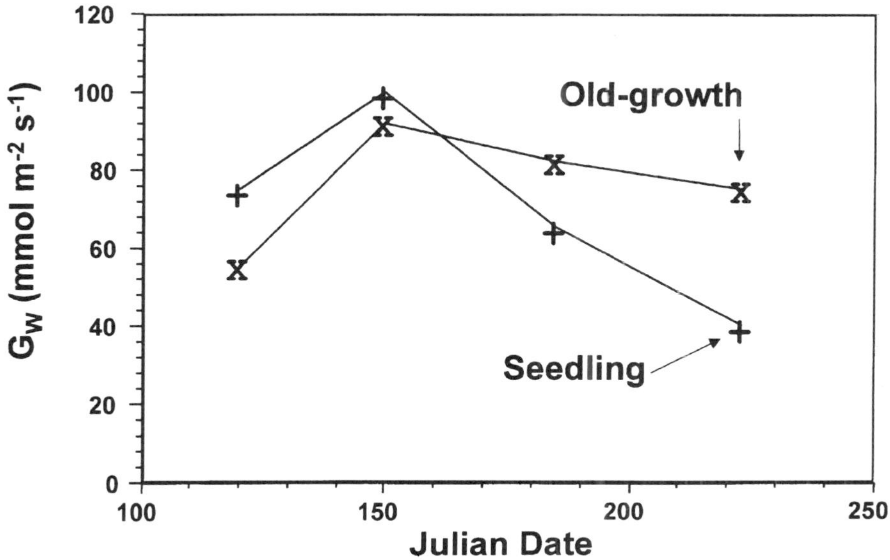

Figure 5. Mean mid day stomatal conductance (G_W) during the growing season for open-grown seedlings and old-growth *Pinus ponderosa* trees at a clean-air site California. (After data in Grulke 1999).

environments also can cause large differences in G_W. In particular, dry soil can impose greater constraints on G_W for shallow-rooted small trees than for deep-rooted large trees that use deep soil or ground water that is not available to small trees (Fig. 5, Dawson and Ehleringer 1991, Donovan and Ehleringer 1991, Dawson and Ehleringer 1993, Dawson 1996, Grulke 1999). However, there is evidence that water availability to small trees that originate from sprouting by large trees does not differ from water availability to large trees. For example, in a riparian forest with dry surface soil but shallow ground water, seedling-sized sprouts from mature trees of *Acer negundo* had similar G_W and Ψ_L to mature trees when measured in a similar light environment because both sprouts and mature trees used ground water (Kolb et al. 1997b).

There is growing evidence that chronic exposure to ambient O_3 levels increases the sensitivity of mature tree growth to drought (Zahner et al. 1989, McLaughlin and Downing 1996) because of reduced stomatal control of water loss (Maier-Maercker 1989, Maier-Maercker and Koch 1991, Maier-Maercker 1997, Mansfield 1998). This evidence suggests that stomatal closure measured in seedlings following short-term exposure to high O_3 concentrations may not adequately represent the expected response of large trees to ambient exposures in the field (McLaughlin and Percy 1999).

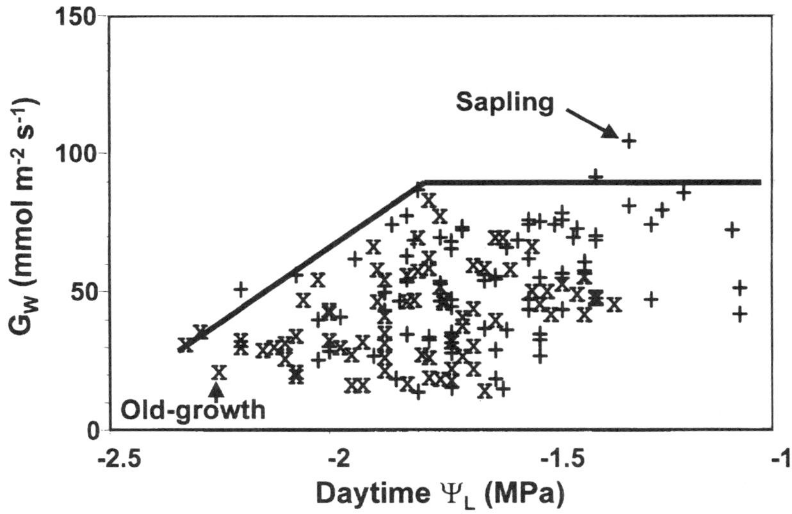

Figure 6. Stomatal conductance (G_W) versus daytime leaf xylem water potential (Ψ_L) pooled over morning and afternoon measurements during the growing season for open-grown saplings and the middle canopy of old-growth *Pinus ponderosa* trees in Arizona. The line shows the boundary line for the onset of stomatal closure for both size classes. (After data in Kolb and Stone 2000).

2.2 Carbon relations

Most forests worldwide exhibit a decrease in net primary productivity (NPP) in old compared with young or middle-aged stands (Gower et al. 1996, Ryan et al. 1997). Although the factors that limit NPP in old stands may differ among forest types (Ryan and Waring 1992, Hunt et al. 1999), low P_N of old trees is a favored hypothesis (Ryan et al. 1997). Indeed, there is evidence from several studies that sunlit leaves of old trees have lower P_N than sunlit leaves of young trees under comparable environmental conditions because of reduced G_W caused by limitations to long-distance water transport through xylem and against the hydrostatic gradient in old trees (Yoder et al. 1994, Fredericksen et al. 1996c, Ryan and Yoder 1997, Hubbard et al. 1999, Niinemets et al. 1999, Bond 2000, Kolb and Stone 2000). The P_N of old trees can match the P_N of young trees under favorable environmental conditions for photosynthesis such as times of the day when vapor pressure deficit is low and Ψ_L is high (Yoder et al. 1994, Fredericksen et al. 1996c). However, large trees appear to close stomates earlier in the day because of lower Ψ_L (Fredericksen et al. 1996c, Kolb and Stone 2000) and greater stomatal sensitivity to vapor pressure deficit (Hubbard et al. 1999).

2.3 Nitrogen relations

Leaf nitrogen concentration is a good predictor of light-saturated P_N (Field and Mooney 1986, Reich et al. 1997) and G_W (Samuelson and Kelly 1997), thus information on variations in leaf nitrogen concentration with tree ageing might improve understanding of changes in leaf gas exchange with ageing. Previous studies have shown that the pattern of change in leaf nitrogen concentration with tree ageing varies among species and environments. Decreases (Kull and Koppel 1987, Schoettle 1994, Carey et al. 1998), increases (Carey et al. 1998, Kolb and Stone 2000), and no change (Schoettle 1994) in leaf nitrogen concentration with increasing tree age have been reported.

Changes in leaf nitrogen concentration with tree ageing have been linked to age-related changes in other morphological or physiological traits in some studies. For example, the pattern of change in leaf nitrogen concentration is quite different for two co-occurring pines in the Rocky Mountains, North America. Old trees of *Pinus aristata* had lower leaf nitrogen concentration compared with young trees, but also had longer leaf longevity and more leaf biomass per shoot (Schoettle 1994). In contrast, old and young trees of *Pinus contorta* did not differ in leaf nitrogen concentration, and plasticity in leaf longevity and leaf biomass per shoot was lower for this species than for *Pinus aristata* (Schoettle 1994). This finding suggests that decreases in leaf nitrogen concentration with tree ageing are compensated by greater leaf longevity and maintenance of greater leaf biomass in some species (e.g., *Pinus aristata*) to ameliorate reductions in carbon uptake that would occur by changes in leaf nitrogen concentration alone. Clearly, prediction of changes in photosynthetic capacity with tree ageing based on leaf nitrogen concentration must also consider such compensating factors.

Environment can also influence how leaf nitrogen concentration varies with tree size. Carey et al. (1998) compared seasonally integrated P_N and leaf nitrogen concentration among *Pinus ponderosa* that differed in age in montane and desert environments in Nevada. The environments differed in temperature and water availability, but not soil type. Whole-tree relative growth rate declined with increasing age in both environments. Interesting, leaf nitrogen concentration and seasonally integrated P_N increased with increasing age in the desert environment. In contrast, in the montane environment leaf nitrogen concentration decreased with increasing age, but P_N did not vary. Carey et al. (1998) concluded that old *Pinus ponderosa* trees growing at the desert site allocated more biomass to sapwood and less to leaves during ageing to increase water supply to leaves, and that this pattern of allocation was facilitated by increases in leaf nitrogen concentration and P_N with tree ageing.

2.4 Ozone exposure

Variations in O_3 concentration within forests, or between forests and areas where test seedlings are grown, could cause in differences in O_3 exposure between mature trees and seedlings. There is considerable evidence that dense vegetation may act as a O_3 sink, and thus reduce concentrations of O_3 moving from atmosphere to the forest floor (Kozlowski and Constantidou 1986, Fuentes et al. 1992, Taylor and Hanson 1992, Coe et al. 1995, Skelly et al. 1996, Samuelson and Kelly 1997, Baumgarten et al. 2000). In addition, nitric oxide (NO) produced by soil microorganisms, which converts O_3 to oxygen, may reduce O_3 concentration near the soil surface (Baumback and Baumann 1989). On the other hand, volatile hydrocarbons produced by trees might stimulate the formation of O_3 in the forest because such hydrocarbons are precursors to O_3 formation (Enders et al. 1989).

The interplay among these factors that influence local O_3 concentration could result in important spatial variations in O_3 concentration within forests that would need to be considered in scaling studies. For example, Samuelson and Kelly (1997) reported that average daytime O_3 concentration was about 40% lower at 2 m above the forest floor compared with above the forest canopy in a dense, cove-hardwood forest in Tennessee. Similar results were found by Baumgarten et al. (2000) in a *Fagus sylvatica* forest in Germany, where average O_3 concentration was about 33% lower near the forest floor compared with in the upper canopy. However, there are also reports of little spatial variation in O_3 concentration within local forest areas. For example, Fredericksen et al. (1995, 1996b) reported small and inconsistent differences in ambient daytime O_3 concentration within *Prunus serotina* forests, and between the forest canopy and adjacent open fields where test seedlings were grown in central Pennsylvania.

2.5 Leaf production phenology

A difference in the phenology of leaf production between young and old trees has implications concerning O_3 exposure to leaves. Free or indeterminate growth, where new leaves are produced continuously during the growing season, is a trait of certain tree species including some members of the genera *Acer*, *Betula*, *Liriodendron*, *Populus*, *Prunus*, *Robinia*, and *Salix* (Kozlowski and Pallardy 1997). Moreover, free growth occurs more often for seedlings of these species than for older trees, whose leaf production is often limited to the expansion of those primordia contained in the overwintering bud, resulting in fixed or determinate growth. Likewise, seedlings of species with recurrently flushing shoots typically produce more flushes annually than mature trees (Kozlowski and Pallardy 1997).

These differences in leaf production phenology can cause large differences in the longevity of individual leaves, and thus cumulative exposure to O_3, between juvenile and mature tree leaves. Leaves of mature trees produced by fixed growth are exposed to O_3 for the entire growing season. In contrast, free-growing seedlings have leaves that vary in age. In such seedlings, leaves formed early in the growing season are exposed to a cumulative O_3 dose similar to leaves of mature trees produced by fixed growth, whereas seedling leaves formed later in the season are exposed to a lower cumulative O_3 dose because of their shorter longevity.

Environment must also be considered when comparing leaf production phenology between young and old trees, as expression of free growth in small trees can be strongly regulated by resource availability. Cuttings of *Betula pendula* grown with high nitrogen availability had a high rates of leaf production and turnover typical of free growth, whereas cuttings grown with low nitrogen availability had low rates of leaf production and turnover, characteristics more similar to fixed growth (Maurer and Matyssek 1997). Thus, young plants of some species may show characteristics of either free or fixed growth depending on resource availability.

Differences in leaf production phenology have been associated with differences in plasticity of carbon allocation patterns in response to stress in comparisons among species. For example, Laurence et al. (1994) concluded that plant species with high inherent growth rates and free growth are more likely to change carbon allocation patterns in response to stress than are slower growing species with fixed growth. Species with fixed growth often store more carbohydrates than species with free growth, which helps them maintain more constant carbon allocation patterns. In contrast, low carbohydrate storage by species with free growth requires large adjustments in carbon allocation among absorbing tissues (i.e., leaves, fine roots) in response to stress in order to maintain adequate resource uptake.

3. RESULTS OF OZONE SCALING STUDIES

3.1 Leaf carbon balance

Comparisons between juvenile and mature trees of several species of the effects of O_3 exposure on leaf carbon balance have highlighted the importance of differences in G_W in understanding scaling patterns (Kolb et al. 1997a, Kolb and Matyssek in press). Such comparisons have been made for *Picea rubens* (Rebbeck et al. 1992, Rebbeck et al. 1993), *Pinus ponderosa* (Momen et al. 1996, Momen et al. 1997), *Prunus serotina* (Fredericksen et al. 1995, Fredericksen et al. 1996a, Fredericksen et al. 1996b), *Pseudotsuga menziesii* (Smeulders et al. 1995), *Quercus rubra* (Samuelson and Edwards 1993, Hanson et al. 1994, Samuelson 1994a), and *Sequoiadendron giganteum* (Grulke and Miller 1994) using a variety of research approaches and response variables. In all studies where characteristics of leaf carbon balance (P_N, dark respiration rate, visible foliage injury) were compared for leaves exposed to full sun for much of the day, the tree size with the greatest G_W exhibited the largest negative response to O_3 exposure (Kolb et al. 1997a, Kolb and Matyssek in press). Interestingly, this pattern of leaf-level response to O_3 is the same as that reported for comparisons among species (Reich 1987). In most cases when unshaded juveniles were compared to mature trees, juveniles had larger G_W, and exhibited the largest negative leaf-level response to O_3 exposure. This pattern occurred for *Picea rubens*, *Pinus ponderosa*, *Prunus serotina*, and *Sequoiadendron giganteum* (Kolb et al. 1997a, Kolb and Matyssek in press). In contrast, the opposite pattern of response occurred for *Quercus rubra*: mature trees had greater G_W and a larger negative leaf-level response to O_3 exposure than unshaded, container-grown seedlings. These results show a consistent scaling pattern of O_3 impact on the carbon balance of unshaded leaves with respect to differences in G_W among different-aged trees.

Recent studies from Europe have shown a similar response to O_3 for sun leaves of juvenile and mature trees when leaf injury is expressed as a function of cumulative O_3

uptake. O_3 uptake, which is the product of O_3 concentration at the leaf surface and stomatal conductance to O_3, implicitly includes differences in G_W that may occur among tree sizes. For *Picea abies*, incipient effects of O_3 on P_N of chamber-grown seedlings and field-grown mature trees started at about the same cumulative O_3 uptake, and could be described by the same boundary-line function (Wieser 2001). For *Fagus sylvatica*, the onset of visible foliar injury of sun leaves started at about the same cumulative O_3 uptake for chamber-grown seedlings and field-grown mature trees, and did not differ between years for mature trees (Baumgarten et al. 2000). These results support the use of exposure indices based on O_3 uptake into the leaf mesophyll, rather than indices based on O_3 concentration alone, in efforts to scale leaf-level O_3 effects from juvenile to mature trees.

3.2 Effects of light environment

Scaling patterns of leaf-level response to O_3 exposure appear to be different for unshaded leaves than for heavily shaded leaves of some species (Kolb and Matyssek in press). Despite lower rates of O_3 uptake for shade leaves because of lower G_W and, in some cases, lower exposure to O_3 concentration, heavily shaded leaves often exhibit greater foliar injury symptoms to O_3 than sun leaves. This pattern has been reported for several species, including *Acer saccharum* (Tjoelker et al. 1993, Tjoelker et al. 1995), *Fraxinus americana* (Hildebrand 1994, Hildebrand et al. 1996), *Liriodendron tulipifera* (Hildebrand 1994, Hildebrand et al. 1996), and *Prunus serotina* (Hildebrand 1994, Hildebrand et al. 1996, Fredericksen et al. 1995, Fredericksen et al. 1996b, Lee et al. 1999). For *Acer saccharum*, this pattern has been shown in controlled O_3 exposures for both seedlings (Tjoelker et al. 1993) and branches of mature trees (Tjoelker et al. 1995). The low palisade/spongy mesophyll ratio of shade leaves may increase exposure of palisade cells to O_3 compared with sun leaves (Bennett et al. 1992). For *Acer saccharum* and *Prunus serotina*, exposure of shaded leaves to O_3 resulted in uncoupling of P_N and G_W (Volin et al. 1993, Tjoelker et al. 1995, Fredericksen et al. 1996a, Fredericksen et al. 1996b), which reduced instantaneous water use efficiency. After prolonged exposure to O_3, shaded leaves of these species often had low P_N and high G_W; this condition resulted in a high rate of O_3 uptake per P_N. These findings support the hypothesis that low allocation of photosynthate to repair and defensive processes in shade leaves because of low P_N predisposes them to O_3 injury despite low rates of O_3 uptake.

Whereas most studies have shown greater sensitivity of shaded leaves to O_3 exposure than unshaded leaves, there is evidence that the direction of the shading influence on leaf-level response to O_3 may differ among species. O_3 exposure reduced growth and P_N of *Acer saccharum* seedlings more for plants grown at a low light intensity compared with plants grown at a high light intensity, whereas the opposite pattern of response occurred for hybrid *Populus* (Tjoelker et al. 1993, Volin et al. 1993). Moreover, foliar discoloration attributed to O_3 injury was lower for shade leaves than sun leaves of *Fagus sylvatica* (Baumgarten et al. 2000). These differential responses to O_3 and shading are not clearly related to differences in shade tolerance among species. For example, greater O_3 sensitivity of shaded versus unshaded leaves has been shown by both shade intolerant (*Prunus serotina*) and shade tolerant (*Acer saccharum*) species.

The existence of such light environment and species x light environment interaction effects on leaf-level scaling responses to O_3 exposure will complicate predictions of O_3 impacts to mature trees. A larger number of shade leaves in large, forest-grown trees compared with open-grown seedlings may increase the potential for greater exposure of palisade cells to O_3, and result in more O_3 injury per leaf. On the other hand, a larger number of shade leaves in understory seedlings compared with large canopy trees could produce the opposite result. To complicate the issue further, the magnitude and direction of such interactions may vary among species.

An additional aspect of the light environment that might be important in scaling leaf-level response to O_3 from juvenile to mature trees is the possibility of differences in nighttime O_3 uptake among different size trees. Although low light intensity is often assumed to severely limit G_W at night, G_W can be surprising high at night because of low vapor-pressure deficit and perhaps other factors (Tobiessen 1982, Wieser and Havranek 1993). Matyssek et al. (1995) showed high rates of G_W and O_3 uptake at night in young, well-watered *Betula pendula* cuttings exposed to O_3 in outdoor chambers. Further, nighttime plus daytime exposure to O_3 reduced biomass growth and altered carbon allocation patterns of cuttings about two-fold more than daytime exposure alone. This finding has implications regarding calculations of cumulative O_3 uptake for juvenile and mature trees if nighttime G_W varies with tree age and high concentrations of O_3 occur at night. In a study of different-sized *Prunus serotina* trees growing in a Pennsylvania forest without supplemental watering (Fredericksen et al. 1996c), nighttime G_W of mature trees was about 6% of maximum daytime values, whereas nighttime G_W of seedlings was about 18% of maximum daytime values. Seedlings of *Prunus serotina* had higher G_W than mature trees in both the daytime and nighttime (Fredericksen et al. 1996c). Given that O_3 concentrations in forests can be high at night under some conditions (Lefohn and Jones 1986, Krupa and Manning 1988, Wieser and Havranek 1993), these findings suggest that seedlings may take up more O_3 at night than mature trees.

3.3 Effects of leaf production phenology

Differences in phenology of leaf production between juvenile and mature forest trees have been shown to influence cumulative leaf exposure to O_3. In a comparison of unshaded *Prunus serotina* seedlings growing in an open field and mature forest trees of this species growing in an adjacent forest (Fredericksen et al. 1995), seedlings had free growth and produced leaves continuously between May and August (Fig. 7). The largest monthly increase in leaf production of seedlings occurred between July and August. In contrast, mature trees had fixed growth and finished leaf production in June (Fig. 7). Despite greater instantaneous rates of O_3 uptake for seedlings because of greater G_W (Fig. 8A), many seedling leaves were exposed to O_3 for only part of the growing season, resulting in lower cumulative uptake per leaf for seedlings than mature trees (Fig. 8B).

Changes in leaf production phenology with tree ageing also may influence carbon allocation responses to O_3 exposure. As mentioned earlier, Laurence et al. (1994) hypothesized that fast-growing plants with free growth are more likely to alter carbon

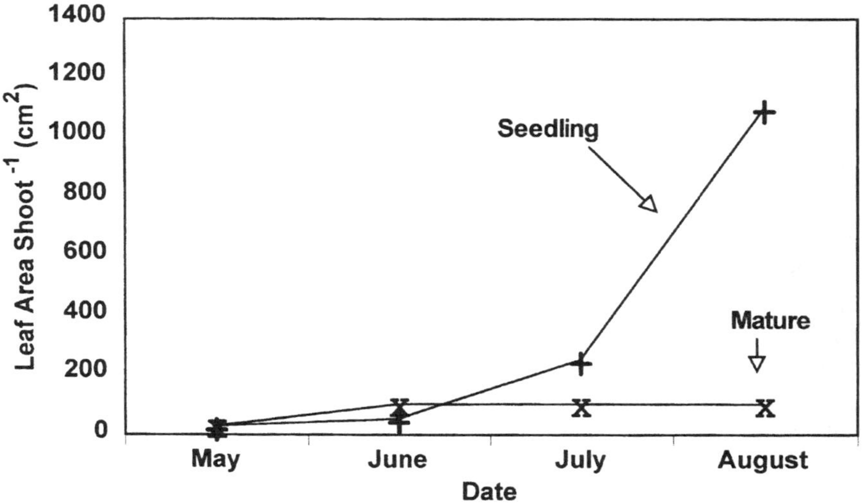

Figure 7. Average leaf area per terminal shoot for seedlings and mature forest trees of *Prunus serotina* averaged over upper and lower crown positions during 1993 in central Pennsylvania. The seedlings were growing in an open field, and the mature trees were growing in an adjacent, natural forest. N = 20 shoots per tree for each of 4 mature trees and 12 seedlings. (After data in Fredericksen et al. 1995.)

allocation among absorbing organs in response to stress than slow-growing plants. This hypothesis is consistent with results from comparisons of response to O_3 exposure among species in the ROPIS experiments (Laurence et al. 1994), between fertilized and unfertilized *Betula pendula* cuttings (Maurer and Matyssek 1997), and between fast- and slow-growing genotypes of *Prunus serotina* (Kouterick et al. 2000). For example, slow-growing, unfertilized *Betula pendula* cuttings had a low rate of leaf production and maintained O_3-injured leaves, whereas fast-growing, fertilized cuttings had fast senescence of O_3-injured leaves because of a high rate of leaf turnover (Maurer and Matyssek 1997, Kolb and Matyssek in press). Further, the respiratory costs for maintenance of injured leaves in the whole-plant carbon balance were higher for slow-growing cuttings than fast-growing cuttings (Maurer and Matyssek 1997, Kolb and Matyssek in press). Extrapolation of the Laurence et al. (1994) hypothesis to the responses of juvenile and mature trees to O_3 exposure suggests that slow-growing mature trees with fixed growth may use a greater proportion of their whole-tree carbon balance to maintain O_3-injured leaves than free-growing seedlings. Moreover, high respiratory demands of maintaining O_3-injured leaves in mature trees with fixed growth would be expected to reduce carbon allocation to other sinks, especially to root growth, which could lead to low rates of nutrient and water uptake, and to storage, which could increase vulnerability to stress. On the other hand, rapidly growing juveniles with indeterminate growth may have lower respiration costs for maintaining O_3-injured leaves because of fast leaf turnover and greater plasticity in carbon allocation.

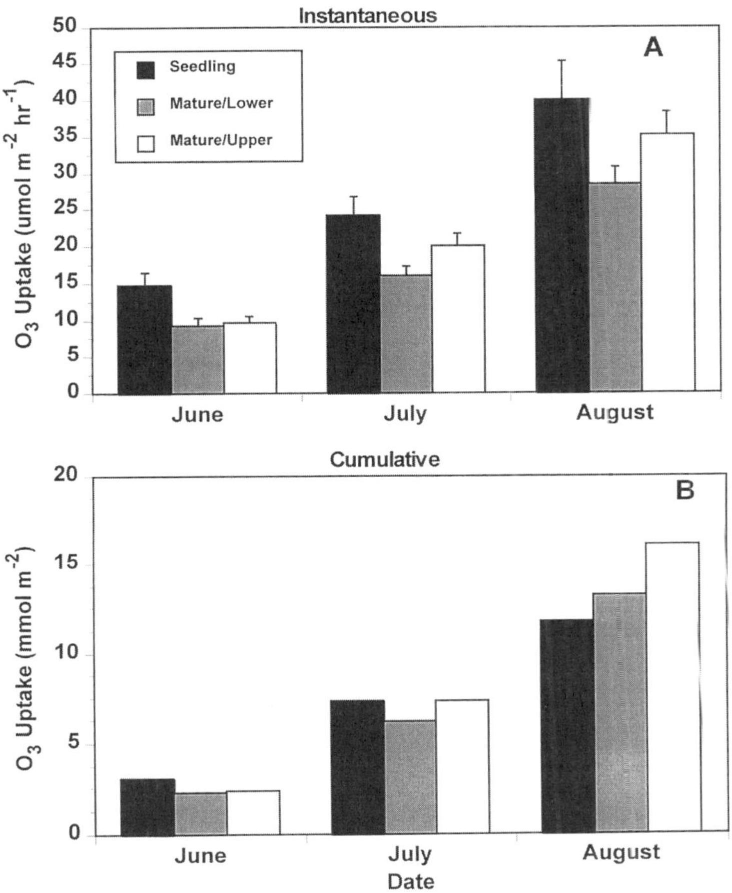

Figure 8. Instantaneous O_3 uptake rate (A) and cumulative O_3 uptake (B) for unshaded seedlings and lower and upper canopy positions of mature forest trees of *Prunus serotina* during June-August 1993 in central Pennsylvania. The seedlings were growing in an open field, and the mature trees were growing in an adjacent, natural forest. Instantaneous O_3 uptake rate was calculated from monthly averages of diurnal measurements of O_3 concentration in the canopy and stomatal conductance for each tree size. Cumulative uptake was calculated by summing instantaneous rates of uptake during the study. Because of indeterminate growth of seedlings, cumulative uptake was weighted for this tree age by the proportion of leaf area (monthly mean leaf area/total end-of-season leaf area) exposed to O_3 each month. N = 30 to 60 leaves per date for each of 5 seedlings and 4 mature trees. The bars in (A) estimate one standard error of the mean based on pooled variances for O_3 concentration and stomatal conductance. (After data in Fredericksen et al. 1995.)

3.4 Avoidance versus tolerance

A difference in G_W between juvenile and mature forest trees results in a difference in O_3 uptake given similar leaf production phenology and O_3 exposure. Because O_3 uptake is often calculated from G_W and ambient O_3 concentration (Reich 1987, Wieser and Havranek 1993, Wieser and Havranek 1995, Fredericksen et al. 1995, Samuelson and Kelly 1997), most previous scaling studies have shown that the tree age with lower G_W also exhibited greater avoidance of O_3 uptake (Kolb et al. 1997a, Kolb and Matyssek in press).

Plant tolerance of O_3, which includes repair of injured tissues, compensation in carbon allocation to minimize injury, and defense against injury (Taylor and Hanson 1992, Laurence et al. 1994, Pell et al. 1994), is more difficult to measure. Limited evidence suggests that mature trees often avoid O_3 uptake into the leaf mesophyll more than juveniles because of lower G_W, but mature trees also may have lower tolerance to O_3 injury (Fredericksen et al. 1996b). In a study of *Prunus serotina*, forest-grown mature trees showed less visible O_3 foliar injury than seedlings grown in an adjacent open field as a function of O_3 exposure to ambient concentrations in Pennsylvania (Fig. 9A). This result suggest greater depression of P_N in seedlings as a function of O_3 exposure because visible foliar injury is strongly and negatively related to P_N in *Prunus serotina* (Samuelson 1994b, Fredericksen et al. 1996b). In contrast, seedlings had less O_3 foliar injury than mature trees at low amounts of cumulative uptake (<7 mmol m^{-2}) (Fig. 9B), suggesting greater tolerance and capacity for compensation, defense, or repair for seedlings than mature trees. However, high cumulative uptake of seedlings may have eventually overwhelmed tolerance mechanisms, resulting in more foliar injury for seedlings than mature trees by the end of the growing season (Fig. 9B).

3.5 Modeled predictions of growth

Stomatal characteristics are a key parameter in many process models used to predict O_3 impacts because O_3 uptake into the leaf depends on G_W and the O_3 concentration gradient between atmosphere and leaf mesophyll (Weinstein et al. 1991, Weinstein and Yanai 1994, Rubin et al. 1996, Retzlaff et al. 1997). Further, G_W has been correlated with photosynthetic response to O_3 exposure in comparisons among species (Reich 1987) and among different age trees of the same species (Kolb et al. 1997a); tree species or tree ages with high G_W typically show the greatest negative photosynthetic response to O_3 for unshaded leaves. A prediction of the effect of O_3 on young and old trees of *Quercus rubra* using the process model TREGROW (Weinstein et al. 1998) was consistent with this pattern. Mature trees of *Quercus rubra* had higher G_W than seedlings in the data set used to parameterize the model, which resulted in greater O_3 uptake and a larger depression of P_N and biomass growth by O_3 exposure in mature trees compared with seedlings (Weinstein et al. 1998). Extension of this result to different-age trees of other species is tenuous given the many factors that can influence the direction and magnitude of changes in G_W with tree ageing.

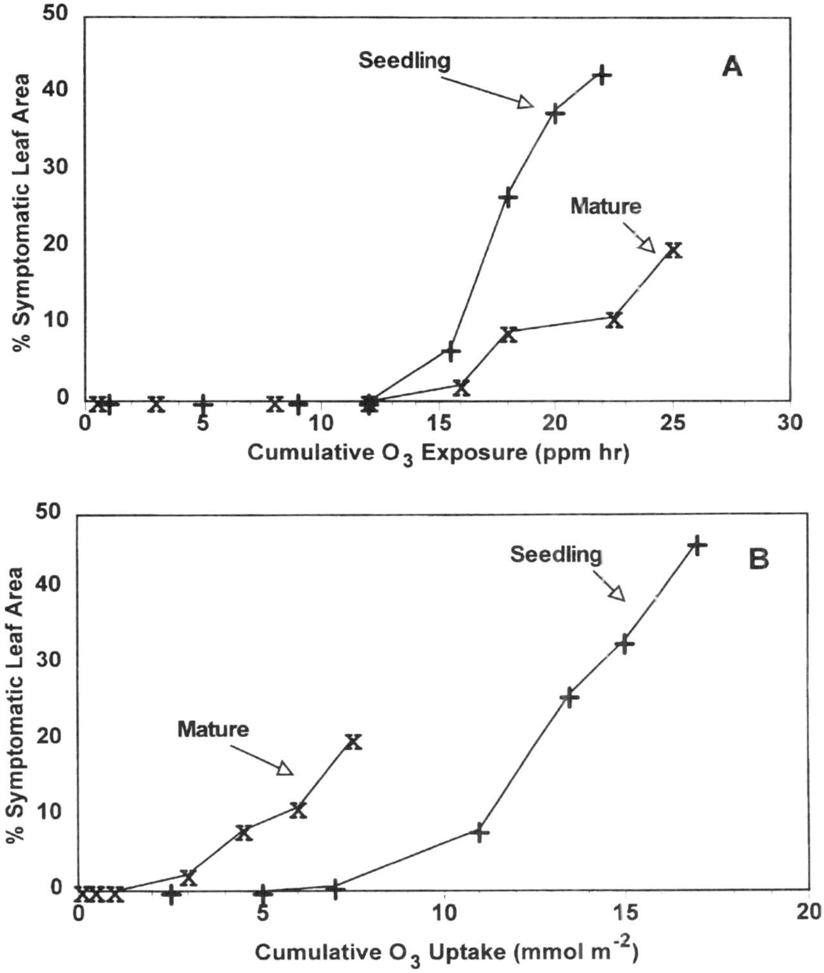

Figure 9. Average percent of leaf area with visible symptoms of O_3 injury (dark adaxial stipple) on shoots of unshaded seedlings and the upper canopy of mature forest trees of *Prunus serotina* during 1994 in central Pennsylvania versus cumulative O_3 exposure (A) and cumulative O_3 uptake (B). The seedlings were growing in an open field, and the mature trees were growing in an adjacent, natural forest. Cumulative O_3 exposure was calculated as the sum of all average hourly O_3 concentrations > 60 ppb (Sum60). Cumulative O_3 uptake was calculated as the product of stomatal conductance to O_3 and O_3 concentrations using hourly concentrations > 60 ppb. N = all leaves on 3 shoots for each of 5 seedlings, and all leaves on 5 shoots for each of 4 mature trees. (After data in Fredericksen et al. 1996b.)

4. FRAMEWORK FOR UNDERSTANDING SCALING STUDIES

Table 1 shows a framework to help in understanding the results of O_3 scaling studies between juvenile and mature forest trees. This framework is based on hypothesized changes in physiology and morphology as trees age, differences in resources available to juvenile and mature forest trees, and morphological and physiological responses to these differences in resource availability.

Several factors likely influence all scaling studies where juvenile trees are compared to mature forest trees (Table 1). First, gravity causes differences in Ψ_L between short and tall trees, and within the canopy of tall trees, that may influence G_W, O_3 uptake, and P_N, but only when differences in leaf height are large because of the magnitude of the gravitational gradient in Ψ (0.01 MPa m^{-1}). An additional influence on Ψ_L that likely affects differences among tree sizes in G_W, O_3 uptake, and P_N under all study conditions is xylem resistance to water flow (Table 1: A3, B1, C2, D2, E2). This influence should increase in importance as the difference in path length for water flow from soil to leaf increases between young and old trees. Next, leaf morphology likely influences tree size comparisons under all conditions where the proportion of leaves with shade morphology versus sun morphology differs between juvenile and mature trees. Such differences in leaf morphology should have the largest influence on comparisons between shaded, understory juveniles and mature, dominant forest trees (Table 1: A1). The influence of shade leaf morphology on scaling studies performed under other study conditions in Table 1 (B, C, D, E) will vary depending on differences between tree ages in the proportion of shade- and sun-leaves. The proportion of seedling leaves with shade morphology will vary from close to none for open-grown seedlings to 100% for seedlings in shaded, understory environments. Likewise, the proportion of mature tree leaves with shade morphology will vary depending on shading within the canopy and from other trees.

The influence of other factors on scaling studies likely depends on study conditions. One such factor is a difference in O_3 exposure between juvenile and mature trees caused by spatial variation in O_3 concentration within forests or other test sites. Limited data suggests that such variations are most pronounced in dense forests where understory seedlings may be exposed to lower O_3 dose than mature forest trees (e.g., Samuelson and Kelly 1997). Thus, spatial variation in O_3 concentration is hypothesized to be most important in studies that compare shaded, understory seedlings with mature, dominant trees in forests with high foliage density (Table 1: A4).

Light environment is another factor that likely has a greater influence on comparisons between shaded, understory juveniles and mature forest trees (Table 1: A2) compared with other study conditions in Table 1 (B, C, D, E). Juveniles growing in low light often have lower G_W, O_3 uptake, and P_N than unshaded leaves of mature forest trees. Further, uncoupling of P_N and G_W in heavily shaded leaves damaged by O_3 may create conditions where G_W occurs at moderate rates at the same time that P_N is very low, which reduces water-use efficiency and promotes a high rate of O_3 uptake per P_N that may overwhelm leaf repair and defensive processes.

Table 1. Hypothesized factors that are important influences on ozone (O_3) scaling studies under different study conditions, and the mechanisms of influence.

Study Conditions	Important Influencing Factors	Mechanisms of Influence
A. Shaded, understory seedlings vs. mature trees	1. Shade- vs. sun-leaf morphology 2. Light environment 3. Ψ_L (xylem resistance, gravity) 4. O_3 concentration	1. O_3 exposure to palisade cells 2. O_3 uptake via G_W, P_N available for tolerance processes, coupling of P_N to G_W 3. O_3 uptake via G_W, P_N available for tolerance processes via G_W 4. O_3 exposure and uptake
B. Unshaded seedlings vs. mature trees	1. Ψ_L (xylem resistance, gravity) 2. Shade- vs. sun-leaf morphology	1. O_3 uptake via G_W, P_N available for tolerance processes via G_W 2. O_3 exposure to palisade cells
C. Container-grown, unshaded seedlings vs. mature trees	1. Sink regulation of carbon relations 2. Ψ_L (xylem resistance, gravity) 3. Shade- vs. sun-leaf morphology	1. O_3 uptake via P_N and G_W 2. O_3 uptake via G_W, P_N available for tolerance processes via G_W 3. O_3 exposure to palisade cells
D. Unshaded seedlings using shallow soil water vs. mature trees using deep soil or ground water	1. Water availability in rooting zone 2. Ψ_L (xylem resistance, gravity) 3. Shade- vs. sun-leaf morphology	1. O_3 uptake via G_W, P_N available for tolerance processes via G_W 2. O_3 uptake via G_W, P_N available for tolerance processes via G_W 3. O_3 exposure to palisade cells
E. Unshaded seedlings with free shoot growth vs. mature trees with fixed shoot growth	1. Leaf production phenology 2. Ψ_L (xylem resistance, gravity) 3. Shade- vs. sun-leaf morphology	1. Cumulative O_3 uptake/leaf, compensation by new leaf production, plasticity in carbon allocation, respiration costs of maintaining O_3-injured leaves 2. O_3 uptake via G_W, P_N available for tolerance processes via G_W 3. O_3 exposure to palisade cells

Sink regulation of P_N may be an important influence on scaling comparisons where growing conditions or tree maturation cause large differences in carbon sink strength between juveniles and mature trees. There is considerable evidence that strong sinks for photosynthate can stimulate P_N (Maier and Teskey 1992, Luxmoore et al. 1995), and consequently increase G_W and O_3 uptake. One study condition that may lead to such differences is when container-grown juveniles having low sink activity because of container limitations to root growth are compared with larger trees that are not growing in containers (Table 1: C1). Further, mature trees may have periods of high sink activity caused by heavy flowering and fruiting that do not occur in juveniles. These differences in source-sink relationships could cause greater G_W and O_3 uptake by mature forest trees compared with juveniles.

Soil water availability likely has a strong influence on scaling comparisons made in environments that have seasonally dry surface soil, and wetter deep soil or shallow ground water (Table 1: D1). This environment is common in some arid regions, for example riparian areas in western North America (Kolb et al. 1997b). In such environments, G_W, O_3 uptake, and P_N of shallow-rooted juveniles should be limited more by low availability of soil water in the rooting zone than for more deeply rooted mature trees.

Phenology of leaf production can strongly influence scaling comparisons between seedlings with free growth and mature trees with fixed growth (Table 1: E1). Seedlings with free growth may be better able to compensate for O_3 injury to older leaves by early senescence of injured leaves and by producing new leaves that have high P_N and lower cumulative O_3 exposure. In contrast, compensatory ability is limited for mature trees with fixed growth; their old leaves are not replaced in the same season, thus they should have greater cumulative O_3 uptake per leaf than leaves of free-growing seedlings. In addition, greater capacity of juveniles with free growth to quickly alter carbon allocation among absorbing tissues or towards repair and compensation processes in response to O_3 exposure compared with mature trees with fixed growth may be an important influence on biomass and carbon balance responses to O_3. Finally, mature trees with fixed growth may incur high respiration costs of maintaining O_3-injured leaves that juvenile trees with free growth avoid through rapid leaf turnover.

5. CONCLUSIONS

5.1 Considerations in the design of future scaling studies

My review of theory and observations related to scaling effects of O_3 from small, easily manipulated trees to mature trees in forests suggests that future attempts at such scaling can be improved by consideration of the following factors.

Leaf morphology, light environment in the canopy, and leaf production phenology can differ between young and old trees and strongly influence G_W, O_3 uptake, and leaf-level responses to O_3 exposure. Thus, the design of scaling studies should be guided by information on these factors in mature tree canopies, a suggestion also made by others (Hinckley et al. 1999). For example, if mature forest trees of the target species contain leaves with both sun- and shade-morphology, exposure-response data should be obtained for both leaf types. If leaves of mature forest trees are always produced by fixed growth, the most useful exposure-response data will be from trees with fixed growth. In this case, O_3 response data from seedlings with free growth may not be useful for scaling projects.

Comparison of leaf-level response to O_3 between juvenile and mature trees can vary depending on the method used to calculate O_3 exposure. Comparisons based on O_3 concentration alone do not include regulation of O_3 flux into the leaf mesophyll by stomatal aperature, which often differs during tree ageing because of changes in resource availability and intrinsic changes in tree water relations with tree ageing. In contrast, comparisons based on O_3 uptake into the leaf implicitly include tree size-related changes in G_W. The only current reports of similar leaf-level response to O_3 between juvenile and mature trees are based on O_3 uptake into the leaf rather than O_3 concentration external to the leaf (Baumgarten et al. 2000, Wieser 2001).

Branches are usually autonomous in their production and use of photosynthate for much of the growing season (Sprugel et al. 1991), and thus are potentially useful for scaling of physiological responses to the canopy of large trees. Experience with *Prunus serotina* in Pennsylvania indicated that open-grown saplings had similar leaf production phenology to mature forest trees (Fredericksen et al. 1996b). Response functions of P_N to O_3 from sapling branches exposed to controlled concentrations of O_3 in large open-top chambers or with branch chambers should be useful for scaling to larger trees.

Shading treatments could be used on small trees exposed to controlled concentrations of O_3 to produce the range of leaf morphology present among sun- and shade-leaves in the canopy of mature trees. Exposure-response functions measured for both leaf types on small trees could be used to scale physiological responses to the canopy of mature trees based on the proportion of sun- and shade-leaves in mature tree canopies. Furthermore, juvenile plants grown at low nutrient levels often have low rates of leaf production and stem growth (Maurer and Matyssek 1997), characteristics also exhibited by mature trees with fixed growth. Thus, nutrition may be another factor that can be manipulated to create mature-tree characteristics on juvenile trees used for O_3 exposure-response relationships.

It may be possible for some species to create small trees having branch and leaf characteristics of mature trees by grafting mature shoots onto seedling rootstocks for use in O_3 exposure experiments. This approach is used routinely in tree genetic improvement programs (Wright 1976), and it has been used in O_3 scaling studies of *Picea rubens* (Rebbeck et al. 1992, Rebbeck et al. 1993), *Pinus ponderosa* (Momen et al. 1996, Momen et al. 1997), *Prunus serotina* (J.M. Skelly, K.C. Steiner, T.E. Kolb, unpublished data). Leaf-level physiological responses to O_3 exposure from such grafted trees may be more useful for scaling to mature forest trees than responses from seedlings if seedlings and mature trees differ substantially in leaf morphology and leaf production phenology.

The different research approaches that have been used to estimate O_3 impacts on mature forest trees have desirable attributes and all have furthered understanding of O_3 impacts. Yet, there are limitations to all approaches. For example, correlative tree-ring and growth analyses can provide only weak evidence of cause and effect relationships and are often complicated by the potential influence of other stresses. Morphological and biochemical symptoms of O_3 damage to tree leaves or crowns are not always easily related to other important physiological responses or growth, and assignment of cause and effect can be difficult. Carbon relations of mature-tree branches that are exposed to controlled concentrations of O_3 are not always autonomous, and represent only a portion of the canopy. Finally, the accuracy of simulation models used to predict O_3 impacts on mature trees is not known.

Exposure of whole trees or forest areas to O_3 with free-air systems is a favored approach that overcomes many of the limitations of earlier approaches (Matyssek and Innes 1999). While large-scale, free-air O_3 experiments are currently limited by logistical constraints and funding, their future use is recommended if society desires more accurate data on O_3 impacts to mature forests. The simultaneous use of research approaches that address responses to O_3 at different levels of organization (cellular, leaf, branch, whole tree, simulation modeling) with free-air O_3 exposure experiments in the future should provide considerable insight into the feasibility of accurately predicting forest-level responses from responses measured at lower hierarchical levels.

5.2 Important research questions

Despite considerable progress in understanding how resource availability, water relations, and carbon relations change as trees age in the last decade, major knowledge gaps still limit the design and interpretation of scaling studies for O_3 and other air pollutants. My review has highlighted the following research questions that need to be addressed to further such understanding.

- Under what environmental conditions do G_W and O_3 uptake differ between juvenile and mature trees, and what are the mechanisms of such differences?
- Do small and large trees differ in nighttime G_W, and are these differences large enough to cause important differences in O_3 uptake?
- What factors influence the nature and pattern of changes in leaf nitrogen concentration with leaf age and tree maturation?
- Can gradients in O_3 concentration in forests be predicted reliably from variations in foliage density?
- How does the proportion of leaves with shade morphology and leaves with sun morphology change with tree ageing, and does environment affect this change?
- How does leaf production phenology change over environments and during tree ageing?
- How well does the Laurence et al. (1994) hypothesis about linkages among growth rate, leaf production phenology, and capacity to allocate carbon to compensate for O_3 injury apply to young and old trees of the same species?
- How important are tolerance mechanisms (defense, repair, compensation) in ameliorating O_3 injury?
- How important and common are species x light environment interactions in leaf-level O_3 responses, and what mechanisms underlie such interactions?

ACKNOWLEDGEMENTS

Special thanks to my collaborators on the project "Scaling seedling to mature black cherry responses to ambient ozone in forests of north central Pennsylvania," especially T.S. Fredericksen, K.C. Steiner, and J.M. Skelly, for providing ideas and data for this paper. Cooperative Agreement CR820417-01-0, Environmental Research Laboratory, U.S. Environmental Protection Agency supported a portion of the work reported in this paper. The manuscript benefited from review comments by R. Matyssek.

REFERENCES

Abrams, M.D. and Kubiske, M.E. 1990. Leaf structural characteristics of 31 hardwood and conifer species in central Wisconsin: Influence of light regime and shade-tolerance rank. *Forest Ecology and Management* **31**: 245-253.

Bartholomay, G.A., Eckert, R.T., and Smith, K.T. 1997. Reductions in tree-ring widths of white pine following ozone exposure at Acadia National Park, Maine, U.S.A. *Canadian Journal of Forest Research* **27**: 361-368.

Bauerle, W.L., Hinckley, T.M., Cermak, J., and Kucera, J. 1999. The canopy water relations of old-growth Douglas-fir trees. *Trees* **13**: 211-217.

Baumbach, G. and Baumann, K. 1989. Ozone in forest stands - examinations to its occurrence and degradation. In: Georgii, H.W. (ed.): Mechanisms and Effects of Pollutant Transfer into Forests. Kluwer Academic Publishers, pp. 37-44.

Baumgarten, M., Werner, H., Haberle, K.-H., Emberson, L.D., Fabian, P., and Matyssek, R. 2000. Seasonal ozone response of mature beech trees (*Fagus sylvatica*) at high altitude in the Bavarian forest (Germany) in comparison with young beech grown in the field and in phytotrons. *Environmental Pollution*: **109**: 431-442.

Becker, P., Gribben, R.J., and Lim, C.M. 2000a. Tapered conduits can buffer hydraulic conductance from path length effects. *Tree Physiology* **20**: 965-968.

Becker, P., Meinzer, F.C., and Wullschleger, S.D. 2000b. Hydraulic limitation of tree height: A critique. *Functional Ecology* **14**: 4-11.

Bennett, J.P., Rassat P., Berrang P., and Karnosky, D.F. 1992. Relationships between leaf anatomy and ozone sensitivity of *Fraxinus pennsylvanica* Marsh. and *Prunus serotina* Ehrh. *Environmental and Experimental Botany* **32**: 33-41.

Bond, B.J. 2000. Age-related changes in photosynthesis of woody plants. *Trends in Plant Science* **5**: 349-353.

Bond, B.J., and Ryan, M.G. 2000. Comment on 'Hydraulic limitation of tree height: A critique' by Becker, Meinzer, and Wullschleger. *Functional Ecology* **140**: 135-140.

Canny, M.J. 1995. A new theory for the ascent of sap – cohesion supported by tissue pressure. *Annuals of Botany* **75**: 343-357.

Carey, E.V., Callaway, R.M., and DeLucia, E.H. 1998. Increased photosynthesis offsets costs of allocation to sapwood in an arid environment. *Ecology* **79**: 2281-2291.

Coe, H., Gallagher, M.W., Choularton, T.W., and Dore, C. 1995. Canopy scale measurements of stomatal and cuticular O_3 uptake by Sitka spruce. *Atmospheric Environment* **29**: 1413-1423.

Constable, J.V.H., and Retzlaff, W.A. 1997. Simulating the response of mature yellow popular and loblolly pine trees in peak ozone periods during the growing season using the TREEGRO model. *Tree Physiology* **17**: 627-635.

Constable, J.V.H., and Taylor, G.E., Jr. 1997. Modeling the effects of elevated tropospheric O_3 on two varieties of *Pinus ponderosa*. *Canadian Journal of Forest Research* **27**: 527-537.

Dawson, T.E. 1996. Determining water use by trees and forests from isotopic, energy balance and transpiration analyses: The role of tree size and hydraulic lift. *Tree Physiology* **16**: 263-272.

Dawson, T.E. and Ehleringer, J.R. 1991. Streamside trees that do not use stream water. *Nature* **350**: 335-337.

Dawson, T.E. and Ehleringer, J.R. 1993. Gender-specific physiology, carbon isotope discrimination, and habitat distribution in boxelder, *Acer negundo*. *Ecology* **74**: 798-815.

Donovan, L.A. and Ehleringer, J.R. 1991. Ecophysiological differences among juvenile and reproductive plants of several woody species. *Oecologia* **86**: 594-597.

Ellsworth, D.S., and Reich, P.B. 1993. Canopy structure and vertical patterns of photosynthsis and related leaf traits in a deciduous forest. *Oecologia* **96**: 169-178.

Enders, G., Teichmann, U., and Kramm, G. 1989. Profiles of ozone and surface layer parameters over a mature spruce stand. In: Georgii, H.W. (ed.): Mechanisms and Effects of Pollutant Transfer into Forests. Kluwer Academic Publishers, pp. 21-35.

Field, C., and Mooney, H.A. 1986. The photosynthesis-nitrogen relationship in wild plants In: Givnish, T.J (ed.): On the Economy of Plant Form and Function. Cambridge University Press, London, UK, pp. 25-55.

Fredericksen, T.S., Joyce, B.J., Skelly, J.M., Steiner, K.C., Kolb, T.E., Kouterick, K.B., Savage, J.E., and Snyder, K.R. 1995. Physiology, morphology, and ozone uptake of leaves of black cherry seedlings, saplings, and canopy trees. *Environmental Pollution* **89**: 273-283.

Fredericksen, T.S., Kolb, T.E., Skelly, J.M., Steiner, K.C., Joyce, B.J., and Savage, J.E. 1996a. Light environment alters ozone uptake per net photosynthetic rate in black cherry trees. *Tree Physiology* **16**: 485-490.

Fredericksen, T.S., Skelly, J.M., Steiner, K.C., Kolb, T.E., Kouterick, K.B. 1996b. Size-mediated foliar injury response in black cherry trees. *Environmental Pollution* **91**: 53-63.

Fredericksen, T.S., Steiner, K.C., Skelly, J.M., Joyce, B.J., Kolb, T.E., Kouterick, K.B., and Ferdinand, J.E. 1996c. Diel and seasonal patterns of leaf gas exchange and xylem water potentials in different-sized *Prunus serotina* Ehrh. trees. *Forest Science* **42**: 359-365.

Fuentes, J.D., Gillespie, T.J., den Hartog, G., and Neumann, H.H. 1992. Ozone deposition onto a deciduous forest during dry and wet conditions. Agricultural and Forest Meterology **62**: 1-18.

Gower, S.T., McMurtrie, R.E., and Murty, D. 1996. Above-ground net primary production decline with stand age: Potential causes. *Trends in Ecology and Evolution* **11**: 378-382.

Greenwood, M.S. 1995. Juvenility and maturation in conifers: Current concepts. *Tree Physiology* **15**: 433-438.

Grulke, N.E. 1999. Physiological responses of ponderosa pine to gradients of environmental stressors. In: Miller, P.R., and McBride, J.R. (eds.): Oxidant Air Pollution Impacts in the Montane Forests of Southern California. Springer, Berlin, pp. 127-163.

Grulke, N.E. and Miller, P.R. 1994. Changes in gas exchange characteristics during the life span of giant sequoia - implications for response to current and future concentrations of atmospheric ozone. Tree Physiology **14**: 659-668.

Grulke, N.E., Miller, P.R., and Scioli, D. 1996. Response of giant sequoia foliage to elevated concentrations of atmospheric ozone. *Tree Physiology* **16**: 575-581.

Hanson, P.J., Samuelson, L.J., Wullschleger, S.D., Tabberer, T.A., and Edwards, G.S. 1994. Seasonal patterns of light-saturated photosynthesis and leaf conductance for mature and seedling *Quercus rubra* L. foliage - differential sensitivity to ozone exposure. *Tree Physiology* **14**: 1351-1366.

Hinckley, T.M., Keyes, M.R., Martin, T.A., and Sprugel, D.G. 1999. Lessons from montane forests of the Pacific Northwest. *Phyton* **39(4)**: 73-84.

Hildebrand, E.S. 1994. Incidence of ozone-induced injury on sensitive hardwood species in Shenandoah National Park, VA. M.S. Thesis, The Pennsylvania State University, University

Park, 198 pp.

Hildebrand, E.S., Skelly, J.M., Fredericksen, T.S. 1996. Foliar response of ozone-sensitive hardwood tree species from 1991 to 1993 in the Shenandoah National Park, Virginia. *Canadian Journal of Forest Research* **26**: 658-669.

Holbrook, N.M., Burns, M.J., and Field, C.B. 1995. Negative xylem pressures in plants: A test of the balancing pressure technique. *Science* **270**: 1193-1194.

Hubbard, R.M., Bond, B.J., and Ryan, M.G. 1999. Evidence that hydraulic conductance limits photosynthesis in old Pinus ponderosa trees. *Tree Physiology* **19**: 165-172.

Hunt, Jr., E.R., Lavigne, M.B., Franklin, S.E. 1999. Factors controlling the decline of net primary production with stand age for balsam fir in Newfoundland assessed using an ecosystem simulation model. *Ecological Modelling* **122**: 151-164.

Hutchison, K.W., Sherman, C.D., Weber, J., Smith, S.S., Singer, P.B., and Greenwood, M.S. 1990. Maturation in larch II. Effects of age on photosynthesis and gene expression in developing foliage. *Plant Physiology* **94**: 1308-1315.

Kolb, T.E., Fredericksen, T.S., Steiner, K.C., and Skelly, J.M. 1997a. Issues in scaling tree size and age responses to ozone: A review. *Environmental Pollution* **98**: 195-208.

Kolb, T.E., Hart, S.C., and Amundson, R. 1997b. Boxelder water sources and physiology at perennial and ephemeral stream sites in Arizona. *Tree Physiology* **17**: 151-160.

Kolb, T.E., and Matyssek, R. in press. Limitations and perspectives about scaling ozone impacts in trees. *Environmental Pollution*.

Kolb, T.E., and Stone, J.E. 2000. Differences in leaf gas exchange and water relations among species and tree sizes in an Arizona pine-oak forest. *Tree Physiology* **20**: 1-12

Kouterick, K.B., Skelly, J.M., Fredericksen, T.S., Steiner, K.C., Kolb, T.E., and Ferdinand, J.A. 2000. Foliar injury, leaf gas exchange and biomass responses of black cherry (*Prunus serotina* Ehrh.) half-sibling families to ozone exposure. *Environmental Pollution* **107**: 117-126.

Kostner, B.M.M., Schulze, E-.D., Kelliher, F.M., Hollinger, D.Y., Byers, J.N., Hunt, J.E., McSeveny, T.M., Meserth, R., and Weir, P.L. 1992. Transpiration and canopy conductance in a pristine broad-leaved forest of Nothofagus: An analysis of xylem sap flow and eddy correlation measurements. *Oecologia* **91**: 350-359.

Kozlowski, T.T., and Constantinidou, H.A. 1986. Environmental pollution and tree growth. Part II. Factors affecting responses to pollution and alleviation of pollution effects. *Forestry Abstracts* **47**: 105-132.

Kozlowski, T.T., and Pallardy, S.G. 1997. Physiology of Woody Plants. Academic Press, NY. 411 pp.

Krupa, S.V., and Manning, W.J. 1988. Atmospheric ozone: formation and effects on vegetation. *Environmental Pollution* **50**: 101-137.

Kull, O. and Koppel, A. 1987. Net photosynthetic response to light intensity of shoots from different crown positions and age in *Picea abies* (L.) Karst. *Scandinavian Journal of Forest Research* **2**: 157-166.

Laurence, J.A., Amundson, R.G., Friend, A.L., Pell, E.J., and Temple, P.J. 1994. Allocation of carbon in plants under stress: An analysis of the ROPIS experiments. *Journal of Environmental Quality* **23**: 412-417.

Lee, J.C., Skelly, J.M., Steiner, K.C., Zhang, J.W., Savage, J.E. 1999. Foliar response of black cherry (*Prunus serotina*) clones to ambient ozone exposure in central Pennsylvania. *Environmental Pollution* **105**: 325-331.

Lefohn, A.S., and Jones, C.K. 1986. The characterization of ozone and sulfur dioxide air quality for assessing possible vegetation effects. *Journal of Air Pollution Control Association* **36**: 1123-1129.

Luxmoore, R.J., Oren, R., Sheriff, D.W., and Thomas, R.B. 1995. Source-sink-storage relationships of conifers. In: Smith, W.K. and Hinckely, T.M. (eds.): Resource Physiology of Conifers: Acquisition, Allocation, and Utilization. Academic Press, NY, pp. 179-216.

Maier, C.A. and Teskey, R.O. 1992. Internal and external control of net photosynthesis and stomatal conductance of mature eastern white pine (*Pinus strobus*). *Canadian Journal of Forest Research* **22**: 1387-1394.

Maier-Maercker, U. 1989. Delignification of subsidiary and guard cell walls of *Picea abies* (L.) Katst. By fumigation with ozone. *Trees* **3**: 57-64.

Maier-Maercker, U. 1997. Experiments on the water balance of individual attached twigs of *Picea abies* (L.) Karst. In pure and ozone-enriched air. *Trees* **11**: 229-239.

Maier-Maercker, U., and Koch, W. 1991. Experiments on the control capacity of stomata of Picea abies (L.) Karst. after fumigation with ozone and in environmentally damaged material. *Plant, Cell, Environment* **14**: 175-184.

Mansfield, T.A. 1998. Stomata and plant water relations: Does air pollution create problems? *Environmental Pollution* **101**: 1-11.

Martin, T.A., Brown, K.J., Cermack, J., Ceulemans, R., Kucera, J., Meinzer, F.C., Rombold, J.S., Sprugel, D.G., and Hinckley, T.M. 1997. Crown conductance and tree and stand transpiration in a second-growth Abies amabilis forest. *Canadian Journal of Forest Research* **27**: 797-808.

Matyssek, R., Gunthardt-Goerg, M.S., Maurer, S., and Keller, T. 1995. Nighttime exposure to ozone reduces whole-plant production in *Betula pendula*. *Tree Physiology* **15**: 159-165.

Matyssek, R., and Innes, J.L. 1999. Ozone – A risk factor for trees and forests in Europe? *Water, Air, and Soil Pollution* **116**: 199-226.

Maurer, S., and Matyssek, R. 1997. Nutrition and the ozone sensitivity of birch (*Betula pendula*), II. Carbon balance, water-use efficiency and nutritional status of the whole plant. *Trees* **12**: 11-20.

McLaughlin, S.B., and Downing, D.J. 1996. Interactive effects of ambient ozone and climate measure on growth of mature loblolly pine trees. *Canadian Journal of Forest Research* **26**: 670-681.

McLaughlin, S.B., and Percy, K. 1999. Forest health in North America: Some perspectives on actual and potential roles of climate and air pollution. *Water, Air, and Soil Pollution* **116**: 151-197.

Mencuccini, M., and Magnani, F. 2000. Comment on 'Hydraulic limitation of tree height: A critique' by Becker, Meinzer, and Wullschleger. *Functional Ecology* **14**: 135-140.

Momen, B., Anderson, P.B., Helms, J.A., and J.L.J. Houpis. 1997. Acid rain and ozone effects on gas exchange of *Pinus ponderosa*: A comparison between trees and seedlings. *International Journal of Plant Science* **158**: 617-621.

Momen, B., Helms, J.A. and Criddle, R.S. 1996. Foliar metabolic heat rate of seedlings and mature trees of *Pinus ponderosa* exposed to simulated acid rain and elevated ozone. *Plant, Cell, Environment* **19**: 747-753.

Niinemets, U., Kull, O., and Tenhunen, J.D. 1999. Variability in leaf morphology and chemical composition as a function of canopy light environment in coexisting deciduous trees. *International Journal of Plant Science* **160**: 837-848.

Ollinger, S.V., Aber, J.D., and Reich, P.B. 1997. Simulating ozone effects on forest productivity: interactions among leaf-, canopy, and stand-level processes. *Ecological Applications* 7: 1237-1251.

Pell, E.J., Temple, P.J., Friend, A.L., Mooney, H.A. and Winner, W.E. 1994. Compensation as a plant response to ozone and associated stresses: An analysis of ROPIS experiments. *Journal of Environmental Quality* **23**: 429-436.

Pockman, W.T., Sperry, J.S., and O'Leary, J.W. 1995. Sustained and significant negative water pressure in xylem. *Nature* **378**: 715-716.

Rebbeck, J., Jensen, K.F., and Greenwood, M.S. 1992. Ozone effects on the growth of grafted mature and juvenile red spruce. *Canadian Journal of Forest Research* **22**: 756-760.

Rebbeck, J., Jensen, K.F., and Greenwood, M.S. 1993. Ozone effects on grafted mature and juvenile red spruce: photosynthesis, stomatal conductance, and chlorophyll concentration. *Canadian Journal of Forest Research* **23**: 450-456.

Reich, P.B. 1987. Quantifying plant response to ozone: A unifying theory. *Tree Physiology* **3**: 63-91.

Reich, P.B., Ellsworth, D.S., Kloeppel, B.D., Fownes, J.H., and Gower, S.T. 1990. Vertical variation in canopy structure and CO2 exchange of oak-maple forests: Influences of ozone, nitrogen, and other factors on simulated canopy carbon gain. *Tree Physiology* **7**: 329-345.

Reich, R.B., Walters, M.B. and Ellsworth, D.S. 1997. From tropics to tundra: global convergence in plant functioning. *Proceedings of the National Academy of Sciences USA* **94**: 13730-13734.

Retzlaff, W.A., Weinstein, D.A., Laurence, J.A., and Gollands, B. 1997. Simulating the growth of a 160-year-old sugar maple (*Acer saccharum*) tree with and without ozone exposure using the TREGROW model. *Canadian Journal of Forest Research* **27**: 783-789.

Rubin, G., McCullouch, C.E. and Laurence, J.A. 1996. A model for estimating daily ozone dose for plants from atmospheric ozone concentration and vapor pressure deficit. *Journal of Agricultural, Biological, and Environmental Statistics* **1**: 1-16.

Rust, S. and Huttl, R.F. 1999. The effect of shoot architecture on hydraulic conductance in beech (*Fagus sylvatica* L.). *Trees* **14**: 39-42.

Ryan, M.G., Binkley, D., and Fownes, J.H. 1997. Age-related decline in forest productivity: pattern and process. *Advances in Ecological Research* **27**: 213-262.

Ryan, M.G. and Waring, R.H. 1992. Maintenance respiration and stand development in a subalpine lodgepole pine forest. *Ecology* **73**: 2100-2108.

Ryan, M.G. and Yoder, B.J. 1997. Hydraulic limits to tree height and tree growth. *Bioscience* **47**: 235-242.

Samuelson, L.J. 1994a. The role of microclimate in determining the sensitivity of *Quercus rubra* L. to ozone. *New Phytologist* **128**: 235-241.

Samuelson, L.J. 1994b. Ozone-exposure responses of black cherry and red maple seedlings. *Environmental and Experimental Botany* **34**: 355-362.

Samuelson, L.J. and Edwards, G.S. 1993. A comparison of sensitivity to ozone in seedlings and trees of *Quercus rubra* L. *New Phytologist* **125**: 373-379.

Samuelson, L.J. and Kelly, J.M. 1996. Carbon partitioning and allocation in northern red oak seedlings and mature trees in response to ozone. *Tree Physiology* **16**: 853-858.

Samuelson, L.J. and Kelly, J.M. 1997. Ozone uptake in *Prunus serotina, Acer rubrum* and *Quercus rubra* forest trees of different sizes. *New Phytologist* **136**: 255-264.

Sandermann, Jr., H., Wellburn, A.R., Heath, R.L. 1997. Forest decline and ozone: Synopsis. In: Sandermann, Jr., H., Wellburn, A.R., and Heath, R.L. (eds.), Forest Decline and Ozone: A Comparison of Controlled Chamber and Field Experiments. Springer-Verlag, Berlin, pp. 369-377.

Schoettle, A.W. 1994. Influence of tree size on shoot structure and physiology of *Pinus contorta* and *Pinus aristata*. *Tree Physiology* **14**: 1055-1068.

Scholander, P.F., Hammel, H.T., Bradstreet, E.D., and Hemmingsen, E.A. 1965. Sap pressure in vascular plants. *Science* **148**: 339-346.

Skelly, J.M., Fredericksen, T.S., Savage, J.M., and Snyder, K.R. 1996. Vertical gradients of ozone and carbon dioxide in a Pennsylvania deciduous forest. *Environmental Pollution* **94**: 235-240.

Smeulders, S.M., Gorissen, A., Joosten, N.N., and Vanveen, J.A. 1995. Effects of short-term ozone exposure on the carbon economy of mature and juvenile Douglas firs [*Pseudotsuga menziesii* (Mirb) Franco]. *New Phytologist* **129**: 45-53.

Sprugel, D.G., Hinckley, T.M., Schaap, W. 1991. The theory and practice of branch autonomy. *Annual Review of Ecology and Systematics* **22**: 309-334.

Taylor, G.E., Jr. and Hanson, P.J. 1992. Forest trees and tropospheric ozone: Role of canopy deposition and leaf uptake in developing exposure-response relationships. *Agriculture, Ecosystems and Environment* **42**: 255-273.

Taylor, G.E., Jr., Johnson, D.W., and Andersen, C.P. 1994. Air pollution and forest ecosystems: A regional to global perspective. *Ecological Applications* **4**: 662-689.

Teskey, R.O., Dougherty, P.M., and Wiselogel, A.E. 1991. Design and performance of branch chambers suitable for long-term ozone fumigation of foliage in large trees. *Journal of Environmental Quality* **20**: 591-595.

Tjoelker, M.G., Volin, J.C., Oleksyn, J., Reich, P.B. 1993. Light environmental alters response to ozone stress in seedlings of *Acer saccharum* Marsh. and hybrid *Populus* L. I. In situ net photosynthesis, dark respiration and growth. *New Phytologist* **124**: 627-636.

Tjoelker, M.G., Volin, J.C., Oleksyn, J., Reich, P.B. 1995. Interaction of ozone pollution and light effects on photosynthesis in a forest canopy experiment. Plant, Cell, Environment **18**: 895-905.

Tobiesson, P. 1982. Dark opening of stomata in successional trees. *Oecologia* **52**: 356-359.

Tyree, M.T. 1997. The cohesion-tension theory of sap ascent: *Journal of Experimental Botany* **48**: 1753-1765.

Tyree, M.T., and Ewers, F.W. 1991. The hydraulic architecture of trees and other woody plants. *New Phytologist* **119**: 345-360.

Volin, J.C., Tjoelker, M.G., Oleksyn, J., Reich, P.B. 1993. Light environmental alters responses to ozone stress in seedlings of *Acer saccharum* Marsh. and hybrid *Populus* L. II. Diagnostic gas exchange and leaf chemistry. *New Phytologist* **124**: 637-646.

Weinstein, D.A, Beloin, R.M. and Yanai, R.D. 1991. Modeling changes in red spruce carbon balance and allocation in response to interacting ozone and nutrient stress. *Tree Physiology* **9**: 127-146.

Weinstein, D.A., Samuelson, L.J. and Arthur, M.A. 1998. Comparison of the response of red oak (*Quercus rubra*) seedlings and mature trees to ozone exposure using simulation modeling. *Environmental Pollution* **102**: 307-320.

Weinstein, D.A. and Yanai, R.D. 1994. Integrating the effects of simultaneous multiple stresses on plants using the simulation model TREGROW. *Journal of Environmental Quality* **23**: 418-428.

Wieser, G. 2001. Effects of ozone on conifers in the timberline ecotone. In: Huttunen, S. et.al. (eds.): Trends in European Forest Tree Physiology Research, Cost Action E6: EUROSILVA, Kluwer Academic Publishers, Netherlands.

Wieser, G. and Havranek, W.M. 1993. Ozone uptake in the sun and shade crown of spruce: Quantifying the physiological effects of ozone exposure. *Trees* **7**: 227-232.

Wieser, G. and Havranek, W.M. 1995. Environmental control of ozone uptake in *Larix decidua* Mill.: A comparison between different altitudes. *Tree Physiology* **15**: 253-258.

West, G.B., Brown, J.H., and Enquist, B.J. 1999. A general model for the structure and allometry of plant vascular systems. *Nature* **400**: 664-667.

Wright, J.W. 1976. Introduction to Forest Genetics. Academic Press, NY. 463 pp.

Yoder, B.J., Ryan, M.G., Waring, R.H., Schoettle, A.W., and Kaufmann, M.R. 1994. Evidence of reduced photosynthetic rates in old trees. *Forest Science* **40**:513-527.

Zahner, R., Saucier, J.R., and Meyers, R.K. 1989. Tree-ring model interprets frowth decline in natural stands of loblolly pine in southeastern United States. *Canadian Journal of Forest Research* **19**: 612-621.

Zimmermann, M.H. 1983. Xylem structure and the ascent of sap. Springer, NY. 143 pp.

Zimmermann, U., Meinzer, F.C., Benkert, R., Zhu, J.J., Scheider, H., Goldstein, G., Kuchenbroad, E., and Haase, A. 1994. Xylem water transport: Is the available evidence consistent with the cohesion theory? *Plant, Cell, Environment* **17**: 1169-1181.

OZONE AND UV-B RESPONSES OF TREES AND THE QUESTION OF FOREST SUSTAINABILITY

CHRISTIAN LANGEBARTELS, SABINE ANEGG, KLAUS HAHN[1], HÉLÈNE CHIRON[2], ALAIN DROUET[2], DIETER ERNST, WERNER HELLER, JÖRG-PETER SCHNITZLER[3], ROLAND SCHUBERT[4], CHRISTIAN ZINSER AND HEINRICH SANDERMANN

GSF National Research Center for Environment and Health, Institute of Biochemical Plant Pathology and [1]Institute of Biomathematics and Biometry, D-85764 Oberschleißheim, Germany;
[2]University of Orléans, Laboratory of Plant Physiology, BP6749, F-45067 Orléans, France;
[3]Fraunhofer Institute for Atmospheric Environment Research, D-82467 Garmisch - Partenkirchen, Germany;
[4]Technical University of Munich, Section of Forest Genetics, D - 85354 Freising, Germany

Key words: Beech, birch, pine, spruce, biomarker, critical level for ozone, ethylene, memory effect for ozone, pathogenesis-related protein, stilbene, UV-screening pigment.

Abstract: Forest trees respond to increased ozone levels and UV-B radiation by reduced photosynthesis, allocation and growth, but also by increased expression of plant defence systems. Ozone effects usually are cumulative and affect in particular deciduous trees such as beech and birch. Current concepts of forest succession, climax tree species and timber production are thereby made questionable.

1. INTRODUCTION

Forest trees are perennial plants that have developed many mechanisms of stress resistance during evolution. This concerns natural stress factors such as climatic extremes, pathogens and marginal supplies of mineral nutrients. The present chapter will concentrate on two anthropogenic stress factors that have over the past decades already increased (ozone; Kley et al., 1999) or are predicted to increase (UV-B radiation; Krupa et al., 1998). Thorough

S. Huttunen et al. (eds.),
Trends in European Forest Tree Physiology Research, Cost Action E6: EUROSILVA, 157-166

ozone measurements of the last century have been re-calibrated. The average concentration of around 10-20 nl/l ozone showed a slight annual maximum in spring time. Average ozone concentrations nowadays are two- to five-fold higher and are typically maximal in the main growth season of forest trees and agricultural crop plants (Kley et al., 1999). National ozone values and in particular AOT40 (accumulated ozone dose over a threshold of 40 nl/l) values are exceeded every year in Europe, the US and many other countries. An increase of UV-B radiation is expected due to the loss of stratospheric ozone caused by anthropogenic chlorofluorocarbons (Krupa et al., 1998). This loss is most dramatic in the Antarctic and the occasional Arctic ozone holes, but also extends over wide continental areas including Central Europe. Ground-level UV-B radiation has not yet increased in Central Europe or the US. However, measures taken currently to reduce air pollution are likely to lead to higher tropospheric transmittance for UV-B radiation. Ozone and UV-B radiation are both known to potentially decrease photosynthesis and growth, and to induce signal molecules and plant defence systems. The present article will not try to review the vast existing literature but will focus on selected recent results.

2. BIOMARKERS FOR OZONE AND UV-B RADIATION IN TREES

It has been well established for herbaceous plants that ozone can act as an abiotic elicitor of the basic types of defence reactions, i.e. pathogenesis-related (PR) proteins, antioxidant responses, cellular barriers, biosynthesis of signal molecules and phytoalexin production (Pell et al., 1997; Sandermann, 1996; Sandermann et al., 1998; Sharma and Davis, 1997). A similar situation appears to exist in tree species (Langebartels et al., 1997) where information on birch, European beech, hybrid poplar (deciduous trees) and Norway spruce as well as Scots pine (conifers) has been obtained so far. Research in trees has focussed on genes involved in phenylpropanoid metabolism and cell wall strengthening. In contrast to herbaceous plants, no ozone-affected antioxidant genes have been analyzed until now. An up-to-date general review of ozone-induced plant responses is available (Langebartels et al., 2001).

It was postulated that, in general, ozone acts as an abiotic elicitor of plant defence reactions that are known from the hypersensitive reaction (HR) and systemic acquired resistance (SAR) in plant-pathogen interactions (Sandermann, 1996; Sandermann et al., 1998). Ozone-induced lesions often resemble those of HR (J. Kangasjärvi, 2001). They typically develop in resistant plants at the site of pathogen or elicitor entry, and may prevent the pathogen from spreading to other parts of the plant. SAR, on the other hand, is a broad spectrum resistance that typically occurs in distal leaves after the attack of necrotizing pathogens. The similarity of ozone and pathogen responses in plants have led to the concept that ozone (or ozone-derived reactive oxygen species, ROS) and pathogens trigger gene expression via common signaling pathways. Aside ROS, ethylene, jasmonate and salicylic acid have been implicated in mediating ozone responses in plants (Langebartels et al., 2001; Rao et al., 2000; Sandermann et al., 1998).

Ozone triggers an oxidative burst, an enzymatic production of reactive oxygen species,

Ozone triggers an oxidative burst, an enzymatic production of reactive oxygen species, in sensitive plants such as the bioindicator tobacco Bel W3 (Schraudner et al., 1998) and the ozone-sensitive *Arabidopsis* cultivar Cvi-0 (Rao et al., 2000). The accumulation of hydrogen peroxide has recently been demonstrated at the subcellular level in birch leaves (Pellinen et al., 1999). H_2O_2 accumulation was first observed in the apoplast and at the plasmalemma of palisade and spongy mesophyll cells, while intracellular accumulation, especially in mitochondria, was a late event and obviously occurred together with cell death (Pellinen et al., 1999).

With regard to biomarkers, stilbene synthase gene expression in pine and grapevine had previously been found to be one of the most sensitive early ozone responses. These findings, and the analysis of regulatory elements in the stilbene synthase promoter of grapevine have been summarized (Ernst et al., 1999). Transcripts for porin- and metallothionein-like proteins were induced by ozone in various clones of spruce (Buschmann et al., 1998). Transcripts of PR-10, PAL and the LEA-group protein BP8 were elevated in ozone-treated birch (Pääkkönen et al., 1998). A comparison of ozone-sensitive and -tolerant poplar lines has revealed that sensitivity correlated with a lack of defence gene activation (Koch et al., 1998). In contrast, several stress-related transcripts increased in both ozone-sensitive or tolerant birch clones, although many ecophysiological parameters differed (Pääkkönen et al., 1998). A multifunctional *O*-methyl transferase has recently been cloned from Scots pine (Chiron et al., 2000a). The ozone induction of the transcript activity could be attributed to distinct regulatory sequence elements of the gene promotor (Chiron et al., 2000b).

The first gene of conifer wood formation (cinnamyl alcohol dehydrogenase) had been isolated, and it was shown in our laboratory to respond to ozone (Galliano et al., 1993). Subsequent research on this enzyme and on the structure of stress lignin has been summarized in a previous EUROSILVA volume (Boudet et al., 1995). Three genomic clones of cinnamyl alcohol dehydrogenase have recently been analyzed (Schubert et al., 1997, 1998). The evolution of these genes appeared to be significantly slower in conifers than in angiosperms. Stilbene synthase and cinnamyl alcohol dehydrogenase transcripts have been shown to be differentially induced by ozone in pine seedlings (Zinser et al., 1998). Multiple ozone treatment led to a prolonged (12 days) increase of cinnamyl alcohol dehydrogenase transcripts. In addition, there was an interaction between UV-B radiation and ozone in the induced gene expression (Zinser et al., 2000). Exposure experiments with ozone and UV-B indicated that increased UV-B radiation had an ameliorating effect with regard to ozone sensitivity (Schnitzler et al., 1999). Complex interactions between ozone and UV-B were also detected at the level of needle antioxidants (Baumbusch et al., 1998). Birch showed a number of detrimental responses, but also synthesized protective flavonoids under high UV-B radiation (Wulff et al., 1999).

3. UV-B PROTECTIVE PIGMENTS IN TREES

A detailed phytochemical analysis of UV-B-screening pigments was performed in the coniferous species, Scots pine and Norway spruce (Fischbach et al., 1999; Jungblut et al., closely related diacylated flavonol monoglucosides. These molecules contained either two coumaroyl or one coumaroyl and one feruloyl residue attached to the glucose moiety, shifting the absorption maximum from 350 nm (UV-A range, typical for flavonols) to 315 nm in the UV-B range (Jungblut et al., 1995). Norway spruce also accumulated diacylated flavonol glucosides, in particular kaempferol 3-*O*-(3",6"-*O*-di-*p*-coumaroyl)-glucoside, during needle development (Fischbach et al., 1999). The soluble acylated flavonol glucosides were induced by UV-B radiation and were exclusively found in the epidermal layer of pine needles as demonstrated with isolated epidermal strips (Schnitzler et al., 1996). When the UV-screening efficiency of the soluble (acylated flavonol glucosides) and the wall-bound metabolites (p-coumaric and ferulic acids, kaempferol glucoside) was calculated at 300 nm, UV-B radiation was found shielded by the epidermal layer by more than 99.99% (Fischbach et al. 1999; Schnitzler et al., 1996). In the meantime, these studies on UV protection in conifers have been extended to subarctic regions (Turunen et al., 1999). In addition, analysis of beech leaves has revealed that structurally slightly modified compounds, namely 2'',4''-di-p-coumaroyl-kaempferol 3-arabinopyranoside and 3-rhamnopyranoside, are UV-B-induced (J. Rothenburger, W. Heller, unpublished results).

The low calculated epidermal transmittance values indicate that adult conifer needles, and possibly beech leaves, should be quite resistant to UV-B radiation so that attention should be focused on sensitive growth stages in seedling and pollen development. It needs to be clarified whether UV-B can penetrate into leaves even in the presence of screening pigments, or whether indirect mechanisms are responsible for UV-B responses in leaves protected by screening pigments.

4. THE OZONE 'MEMORY' OF TREES

Ozone-induced needle symptoms and needle loss of spruce and pine occurred only in the year following treatment (Langebartels et al., 1998). In parallel, certain ozone-responsive secondary compounds remained elevated over many months, so that the concept of a 'memory' of conifers for ozone was generated (Langebartels et al., 1998). The deciduous tree species, beech, also showed 'memory' or 'carry-over' effects into the year after ozone treatment (delayed bud break, increased lamma growth; Leonardi and Langebartels, 1991; Lippert et al., 1996; Pearson and Mansfield, 1994). 'Carry-over' ozone effects have also been found for pine (Andersen et al., 1997) and birch (Oksanen and Saleen, 1999). These various observations may indicate that trees sense a long-lasting cumulative ozone dose. It has been shown some time ago that the ozone sensitivity of deciduous and conifer trees becomes very similar when ozone lifetime dose per leaf or needle is considered (Reich, 1987). Memory effects are also in the centre of a recently proposed epidemiological model to study ozone/biotic disease interactions (Sandermann, 2000). As depicted in *Figure 1*, a single episode of elevated ozone can suffice to change the host status of a plant for pathogen attack. In contrast, it had been stated repeatedly in the literature that ozone and infection periods are non-coincident, implying negligible ozone/biotic disease interaction. The methods needed to elucidate ozone/biotic disease relationships have recently been summarized (Langebartels et al., 2001).

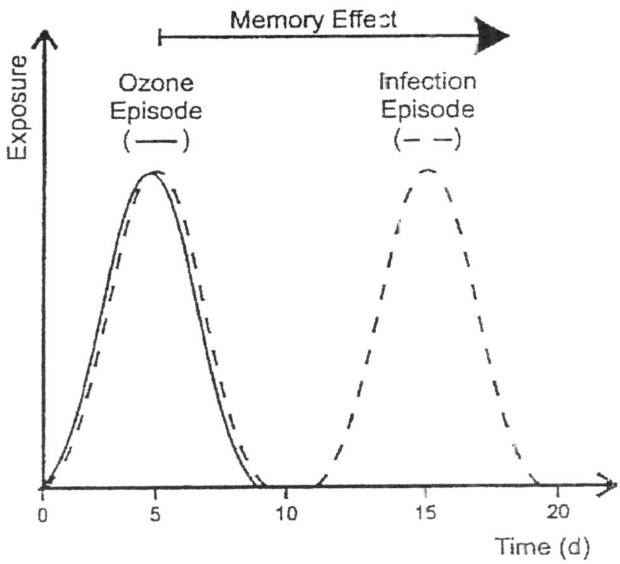

Figure 1. The 'memory' effect as the basis for the epidemiological analysis of ozone/ biotic disease interactions. Episodes of high ambient ozone are usually non-coincident with weather conditions favoring biotic disease. This implies negligible interaction. However, 'memory' or 'carry-over' effects can transmit ozone effects into the infection period (from Sandermann, 2000).

5. OZONE SENSITIVITY OF BEECH AND BIRCH MAY AFFECT SUSTAINABILITY

Beech saplings were intermediately sensitive in exposure experiments with ozone. There was a decrease of photosynthesis, and necrotic leaf lesions appeared already at twice ambient ozone (Leonardi and Langebartels, 1991). The initial data were confirmed in more detail, including thorough photosynthetic measurements leading to a critical AOT40 value of about 25 ppm x h (Lippert et al., 1996). A multidisciplinary subsequent study confirmed the relative ozone sensitivity of beech, and furthermore revealed a protective effect of increased carbon dioxide concentration (Grams et al., 1999).

In more recent studies we tested beech saplings for the induction of ethylene biosynthesis. Ethylene emission had earlier been established as being closely related to the ozone sensitivity of herbaceous and tree species (Sandermann, 1996, Wellburn and Wellburn, 1996). Increased ethylene and ACC had also been good biomarkers for ozone in spruce seedlings (Langebartels et al., 1998). It has now been found, that ethylene emission from beech leaves was in close correlation to the symptomatic area (*Figure 2*), suggesting that ethylene plays a role as regulator of cell death also in tree species. Ethylene development, as well as leaf contents of the ethylene precursor, ACC and its conjugated form, malonyl-ACC, appear to be promising early biomarkers of ozone damage in beech. In contrast to herbaceous plants, the peak maxima for ethylene emission are delayed in trees and occur during the night, during ozone minima. It has been proposed that ozone exerts its toxicity through a chemical reaction with ethylene, yielding toxic products that initiate a self-propagating lipid peroxidation cycle (Elstner et al. 1985). In trees, however, the peak episodes of ozone and ethylene emission can be separated, so that other, plant-based mechanisms have to be taken into accout to explain ozone toxicity. Recent studies in herbaceous plants indicate that ethylene is an active component enhancing the damage process. When ethylene biosynthesis, perception or signalling was blocked, by inhibitors in ozone-treated tomato (Bae et al., 1996; Tuomainen et al., 1997) or in the *Arabidopsis* signaling mutant *ein2* (Overmyer et al., 2000), ROS accumulation and subsequent cell death were both repressed. These results suggest that ethylene promotes ROS accumulation and plays a role in the initiation and propagation phases of lesion development. Similar results have not yet been obtained in trees. It was shown, however, that ozone-sensitive poplar and loblolly pine proveniences emitted significantly more ethylene than ozone-tolerant lines (Wellburn and Wellburn, 1996). Ozone-induced acceleration of foliar senescence was found to be highest in black cherry and poplar, and much less in sugar maple and red oak (Pell et al., 1997).

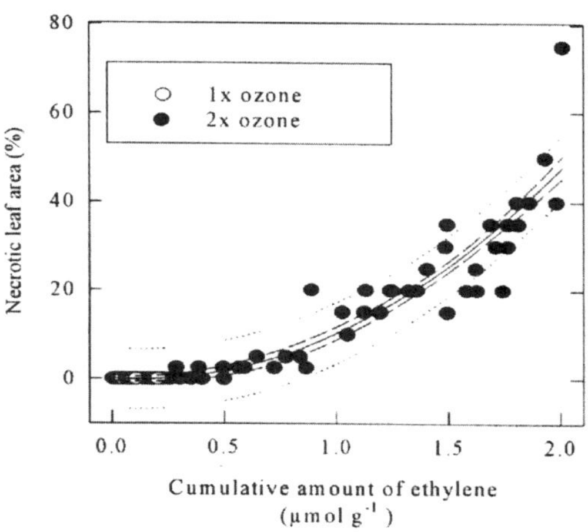

Figure 2. Correlation between cumulative ethylene dose and necrotic area of beech leaves. Ozone-induced ethylene emission is plotted against symptoms in percent of total leaf area. The various data points at ambient (1x) ozone are not resolved and are all located near the zero-effect origin. Ethylene emission peaked on average 15 to 20 days prior to lesion development (from Anegg, 1999).

In Germany and other European countries, a large scale re-forestration with beech was initiated after spruce as the predominant tree species was affected by decline phenomena as well as fungal diseases. The above results, however, make it necessary to reconsider current concepts of forest sustainability that neglect ozone phytotoxicity. It can be concluded that certain birch, beech and poplar proveniences accumulate an ozone dose that may decrease photosynthesis and growth and perturb defence systems. Conifers seem to be affected in the chronic ozone regime through 'memory' effects (Langebartels et al., 1998). Chronic effects of ozone on tree ecophysiological parameters have recently been discussed

(Matyssek and Innes, 1999). These authors concluded that knowledge on the long-term effects of chronic ozone is a prerequisite for silvicultural measures. This conclusion is emphasized by the results summarized above and by the existence of ozone-sensitive and ozone-tolerant genotypes of many trees species. As documented in Southern California (Miller and McBride, 1998), elevated ozone concentrations may be detrimental for certain tree species of forest successions or even for climax tree species, so that timber yields are affected. There is a need to clarify whether elevated ozone (and, potentially, UV-B radiation) could reduce forest sustainability.

ACKNOWLEDGEMENTS

This work has been supported by the Bayerisches Staatsministerium für Landesentwicklung und Umweltfragen (BayForKlim), and in part by Deutsche Forschungsgemeinschaft (SFB 607). Valuable cooperations with Dr. J. Kangasjärvi, University of Helsinki, Finland, and Prof. R. Matyssek, Technical University, München, are gratefully acknowledged.

REFERENCES

Andersen, C.P., Wilson, R., Plocher, M. and Hogsett, W.E. (1997) Carry-over effects of ozone on root growth and carbohydrate concentrations of ponderosa pine seedlings. *Tree Physiol* **17**, 805-811.

Anegg, S. (1999) Untersuchungen zur Wirkung von CO_2 und Ozon auf die Polyamin- und Ethylenbiosynthese sowie auf Wachstum und Symptomentwicklung bei Buchen und Fichten. PhD Thesis, Univ. of Munich.

Bae, G.Y., Nakajima, N., Ishizuka, K. and Kondo, N. (1996) The role in ozone phytotoxicity of the evolution of ethylene upon induction of 1-aminocyclopropane-1-carboxylic acid synthase by ozone fumigation in tomato plants. *Plant Cell Physiol*, **37**: 129-134.

Baumbusch, L., Eiblmeier, M., Schnitzler, J.P., Heller, W., Sandermann, H. and Polle, A. (1998) Joint exposure to enhanced ozone and moderate UV-B: Responses of antioxidants in spruce (*Picea abies*) and pine (*Pinus sylvestris*) needles under a simulated alpine climate. *Physiol Plant*, **104**: 248-254.

Boudet, A.M., Hawkins, S., Cabané, M., Ernst, D., Galliano, H., Heller, W., Lange, M., Sandermann, H. and Lapierre, C. (1995) Developmental and stress lignification. In Sandermann, H. and Bonnet-Masimbert, M. (eds.), *Eurosilva. Contribution to Forest Tree Physiology*. INRA Editions, Paris 76, pp. 13-34.

Buschmann, K., Etscheid, M., Riesner, D. and Scholz, F. (1998) Accumulation of a porin-like mRNA and a metallothionein-like mRNA in various clones of Norway spruce upon long-term treatment with ozone. *Eur J For Pathol*, **28**: 307-322.

Chiron, H., Drouet, A., Claudot, A.C., Eckerskorn, C., Trost, M., Heller, W., Ernst, D. and Sandermann, H. (2000a) Molecular cloning and functional expression of a stress-induced multifunctional O-methyltransferase with pinosylvin methyltransferase activity from Scots pine (*Pinus sylvestris* L.). *Plant Mol Biol*, **44**: 733-745.

Chiron, H., Drouet, A., Lieutier, F., Payer, H.D., Ernst, D. and Sandermann, H. (2000b) Gene induction of stilbene biosynthesis in Scots pine in response to ozone treatment, wounding, and fungal infection. *Plant Physiol*, **124**: 865-872.

Elstner, E.F., Osswald, W. and Youngman, R.J. (1985) Basic mechanisms of pigment bleaching and loss of structural resistance in spruce (*Picea abies*) needles: Advances in phytomedical diagnostics. *Experientia*, **41**: 591-597.

Ernst, D., Grimmig, B., Heidenreich, B., Schubert, R. and Sandermann, H. (1999) Ozone-induced genes: Mechanisms and biotechnological applications. In Smallwood, M.F., Calvert, C.M. and Bowles, D.J. (eds.), *Plant responses to environmental stress*. BIOS, Oxford, pp. 33-41.

Fischbach, R.J., Kossmann, B., Panten, H., Steinbrecher, R., Heller, W., Seidlitz, H.K., Sandermann, H., Hertkorn, N. and Schnitzler, J.P. (1999) Seasonal accumulation of ultraviolet-B screening pigments in needles of Norway spruce (*Picea abies* (L.) Karst.). *Plant Cell Environ*, **22**: 27-37.

Galliano, H., Cabané, M., Eckerskorn, C., Lottspeich, F., Sandermann, H. and Ernst, D. (1993) Molecular cloning, sequence analysis and elicitor-/ozone-induced accumulation of cinnamyl alcohol dehydrogenase from Norway spruce (*Picea abies* L.). *Plant Mol Biol*, **23**: 145-156.

Grams, T.E.E., Anegg, S., Häberle, K.H., Langebartels, C. and Matyssek, R. (1999) Interactions of chronic exposure to elevated CO_2 and O_3 levels in the photosynthetic light and dark reactions of European beech (*Fagus sylvatica*). *New Phytol*, **144**: 95-107.

Jungblut, T. P., Schnitzler, J. P., Heller, W., Hertkorn, N., Metzger, J. W., Szymczak, W. and Sandermann, H (1995) Structure of UV-B induced sunscreen pigments of Scots pine (*Pinus sylvestris* L.). *Angew Chem Int Ed Engl* **34**: 312-314.

Kley, D., Kleinman, M., Sandermann, H. and Krupa, S. (1999) Photochemical oxidants: State of the science. *Environ Pollut*, **100**: 19-42.

Koch, J.R., Scherzer, A.J., Eshita, S.M. and Davis, K.R. (1998) Ozone sensitivity in hybrid poplar is correlated with a lack of defense-gene activation. *Plant Physiol*, **118**: 1243-1252.

Krupa, S.V., Kickert, R.N. and Jäger, H.J. (1998) *Elevated Ultraviolet (UV)-B Radiation and Agriculture*. Springer, Berlin

Langebartels, C., Ernst, D., Heller, W., Lütz, C., Payer, H.D. and Sandermann, H. (1997) Ozone responses of trees. Results from controlled chamber exposures at the GSF phytotron. In Sandermann, H., Wellburn, A.R. and Heath, R.L. (eds.), *Forest decline and ozone: A comparison of controlled chamber and field experiments*. Ecol. Studies, Vol. 127, Springer, Berlin, pp. 163-200.

Langebartels, C., Heller, W., Führer, G., Lippert, M., Simons, S. and Sandermann, H. (1998) Memory effects in the action of ozone on conifers. *Ecotox Environ Safe*, **41**: 62-72.

Langebartels, C., Ernst, D., Kangasjärvi, J. and Sandermann, H. (2000) Ozone effects on plant defense. *Meth Enzymol*, **319**: 520-535.

Langebartels, C., Schraudner, M., Ernst, D., Heller, W. and Sandermann, H. (2001) Oxidative stress and defense reactions in plants exposed to air pollutants and UV-B radiation. In Inzé, D. and Van Montagu, M. (eds.), *Oxidative stress in plants*. Harwood, Amsterdam, pp. 105-135.

Leonardi, S. and Langebartels, C. (1991) Fall exposure of beech saplings (*Fagus sylvatica* L.) to ozone and simulated acidic mist: Effects on gas exchange and leachability. *Water Air Soil Pollut*, **54**: 143-153.

Lippert, M., Steiner, K., Payer, H.-D., Simons , S., Langebartels , C. and Sandermann, H. (1996) Assessing the impact of ozone on photosynthesis of European beech (*Fagus sylvatica* L.) in environmental chambers. *Trees*, **10**: 268-275.

Matyssek, R. and Innes, J.L. (1999) Ozone – A risk factor for trees and forests in Europe? *Water Air*

Soil Pollut **116**: 199-226.

Miller, P.R. and McBride, J.R. (1998) *Oxidant Air Pollution Impacts in the Montane Forests of Southern California. A Case Study of the San Bernardino Mountains.* Ecol. Studies, Springer, New York.

Oksanen, E. and Saleem, A. (1999) Ozone exposure results in various carry-over effects and prolonged reduction in biomass in birch (*Betula pendula* Roth) *Plant Cell Environ*, **22**: 1401-1411

Overmyer, K., Tuominen, H., Kettunen, C., Betz, C., Langebartels, C., Sandermann, H. and Kangasjärvi, J. (2000) The ozone-sensitive *Arabidopsis rcd1* mutant reveals opposite roles for ethylene and jasmonate signaling pathways in regulating superoxide-dependent cell death. *Plant Cell*, **12**: 1849-1862.

Pääkkönen, E., Seppänen, S., Holopainen, T., Kokko, H., Kärenlampi, S., Kärenlampi, L. and Kangasjärvi, J. (1998) Induction of genes for the stress proteins PR-10 and PAL in relation to growth, visible injuries and stomatal conductance in birch (*Betula pendula*) clones exposed to ozone and/or drought. *New Phytol*, **138**: 295-305.

Pearson, M. and Mansfield, T.A. (1994) Effects of exposure to ozone and water stress on the following season's growth of beech (*Fagus sylvatica* L.). *New Phytol*, **126**: 511-515.

Pell, E.J., Schlagnhaufer, C.D. and Arteca, R.N. (1997) Ozone-induced oxidative stress: Mechanisms of action and reaction. *Physiol Plant*, **100**: 264-273.

Pellinen, R., Palva, T. and Kangasjärvi, J. (1999) Subcellular localization of ozone-induced hydrogen peroxide production in birch (*Betula pendula*) leaf cells. *Plant J*, **20**: 349-356.

Rao, M.V., Koch, J.R. and Davis, K.R. (2000) Ozone: A tool for probing programmed cell death in plants. *Plant Mol Biol*, **44**: 345-358.

Reich, P.B. (1987) Quantifying plant response to ozone: A unifying theory. *Tree Physiol*, **3**: 63-91.

Sandermann, H. (1996) Ozone and plant health. *Annu Rev Phytopathol*, **34**: 347-366.

Sandermann, H. (2000) Ozone/biotic disease interactions: Molecular biomarkers as a new experimental tool. *Environ Pollut*, **108**: 327-332.

Sandermann, H., Ernst, D., Heller, W. and Langebartels, C. (1998) Ozone: An abiotic elicitor of plant defense reactions. *Trends Plant Sci*, **3**: 47-50.

Schnitzler, J.-P., Jungblut, T.P., Heller, W., Köfferlein , M., Hutzler, P., Heinzmann, U., Schmelzer, E., Ernst, D., Langebartels, C. and Sandermann , H. (1996) Tissue localization of UV-B screening pigments and of chalcone synthase mRNA in needles of Scots pine (*Pinus sylvestris* L.) seedlings. *New Phytol*, **132**: 247-258.

Schnitzler, J.P., Langebartels, C., Heller, W., Liu, J., Lippert, M., Döhring, T., Bahnweg, G. and Sandermann, H. (1999) Ameliorating effect of UV-B radiation on the response of Norway spruce and Scots pine to ambient ozone concentrations. *Global Change Biol*, **5**: 83-94.

Schraudner, M., Möder, W., Wiese, C., Van Camp, W., Inzé, D., Langebartels, C. and Sandermann, H. (1998) Ozone-induced oxidative burst in the ozone biomonitor plant, tobacco Bel W3. *Plant J*, **16**: 235-245.

Schubert, R., Müller-Starck, G., Sandermann, H., Ernst, D. and Häger, K.P. (1997) The molecular structure and evolutionary relationships of a 16.9 kDa heat shock protein from Norway spruce. *For Genetics* **4**: 131-138.

Schubert, R., Sperisen, S., Müller-Starck, G., La Scala, S., Ernst, D., Sandermann, H. and Häger, K.P. (1998) The cinnamyl alcohol dehydrogenase gene structure in Picea abies (L.) Karst.: Genomic sequences, Southern hybridization, genetic analysis and phylogenetic relationships. *Trees* **12**:

453-463.

Sharma, Y.K. and Davis, K.R. (1997) The effects of ozone on antioxidant responses in plants. *Free Radical Biol Med*, **23**: 480-488.

Tuomainen, J., Betz, C., Kangasjärvi, J., Ernst, D., Yin, Z.H., Langebartels, C. and Sandermann, H. (1997) Ozone induction of ethylene emission in tomato plants: Regulation by differential accumulation of transcripts for the biosynthetic enzymes. *Plant J*, **12**: 1151-1162.

Turunen, M., Heller, W., Stich, S., Sandermann, H., Sutinen, M.L. and Norokorpi, Y. (1999) The effects of UV exclusion on the soluble phenolics of young Scots pine seedlings in the subarctic. *Environ Pollut*, **106**: 219-228.

Wellburn, F.A.M. and Wellburn, A.R. (1996) Variable patterns of antioxidant protection but similar ethene emission differences in several ozone-sensitive and ozone-tolerant plant selections. *Plant Cell Environ*, **19**: 754-760.

Wulff, A., Anttonen, S., Pellinen, R., Savonen, E.M., Sutinen, M.L., Heller, W., Sandermann, H. and Kangasjärvi, J. (1999) Birch (Betula pendula Roth.) reponses to high UV-B radiation. *Boreal Environ Res*, **4**: 77-88.

Zinser, C., Ernst, D. and Sandermann, H. (1998) Induction of stilbene synthase and cinnamyl alcohol dehydrogenase mRNAs in Scots pine (*Pinus sylvestris* L.) seedlings. *Planta*, **204**: 169-176.

Zinser, C., Jungblut, T., Heller, W., Seidlitz, H.K., Schnitzler, J.P., Ernst, D. and Sandermann, H. (2000) The effect of ozone in Scots pine (*Pinus sylvestris* L.): Gene expression, biochemical changes and interactions with UV-B radiation. *Plant Cell Environ*, **23**: 975-982.

INVESTIGATIONS ON THE ACTION OF FUNGAL TOXINS ON PLANTS

W. Oßwald[1], I. Heiser[2], J. Fromm[3], A. Albrecht[1] and M. Brummer[1]
[1]*Forest Botany, Section Phytopathology, Technische Universität München, 85354 Freising, Am Hochanger 13*
[2]*Institute of Phytopathology, Technische Universität München, 85350 Weihenstephan, Am Hochanger 2*
[3]*Institute of Wood Research, Technische Universität München, 80797unich, Winzererstr. 45*

Key words: *Fusarium solani*, Dihydrofusarubin, Reactive oxygen species, *Phytophthora quercina*, Elicitin, Quercinin, Cryptogein

Abbreviations: DHF = dihydrofusarubin; MV = methylviologen; DCMU = 3-(3', 4'-dichlorophenyl)-1,1-dimethylurea; fd = ferredoxin; AM = ammonium chloride; DCC = dicyclohexylcarbodiimid; SOD = superoxide dismutase

Abstract: In this contribution we characterise *Phytophthora quercina* peptides and demonstrate their effects on tobacco leaves in comparison with the *Fusarium solani* toxin dihydrofusarubin.

We demonstrated that dihydrofusarubin interacted with the electron transport chain of chloroplasts, thus forming reactive oxygen species. The inhibition by DCMU proved that oxygen was reduced on photosystem I by the toxin in a light dependent reaction. The toxin reduced NADPH-formation, whereas ATP-formation was significantly accelerated in the presence of dihydrofusarubin. The reactive oxygen species finally induced chlorophyll degradation and pigment cooxidation shown by ethane formation of toxin treated tobacco leaf discs.

The *Phytophthora quercina* toxins were characterised as three small, heat stable peptides with molecular masses of 6.8 kDa and basic as well as acidic isoelectric points. These peptides caused chlorosis and necrosis of tobacco leaves in the light. Transmission electron microscopy studies showed that membrane structures especially of chloroplasts were damaged. Western-blot studies with the cryptogein antibody proved that the *P. quercina* peptides are similar to cryptogein, the toxin produced by *Phytophthora cryptogea*.

S. Huttunen et al. (eds.),
Trends in European Forest Tree Physiology Research, Cost Action E6: EUROSILVA, 167-180

1. INTRODUCTION

Many plant pathogenic fungi and bacteria were shown to produce toxins which contribute to symptom development of infected plants (Kern, 1956; Yoder, 1980). Many of these toxins belong to the group of non-selective toxins, which are produced by a lot of fungi and bacteria and cause damage not only to the host plant but also to other plant species. In many cases disease symptoms caused by non-selective toxins are chlorotic lesions and wilt (Oßwald, 1995). Often reactive oxygen species (ROS), such as superoxide ($O_2^{\bullet-}$), hydrogenperoxide (H_2O_2) or hydroxyl radicals ($OH^{\cdot}$) are involved in symptom development causing pigment degradation and the destruction of membranes in toxin treated plants at the subcellular level (Heiser *et al.*, 1998).

A characteristic feature of *Phytophthora* species is the release of non-selective toxins, called elicitins, into the culture medium (Yu, 1995). These small peptides are produced by many *Phytophthora* species and consist of 98 amino acids. Elicitins are synthesised within the mycelium as pre-proteins carrying a signal peptide which is removed co-translationally before they are abundantly secreted (Terce-Laforgue *et al.*, 1992). When injected into the stem of tobacco plants, these proteins migrate towards the leaf lamina, where they induce leaf necrosis in combination with a rapid chloroplast breakdown and a collapse of parenchyma cell tissue (Zanetti *et al.*, 1992). A characteristic feature of the *Phytophthora* elicitins besides their leaf necrotic activity is that they also induce resistance towards different pathogens. Cryptogein, the elicitin produced by *Phytophthora cryptogea*, was shown to induce the accumulation of pathogenesis-related proteins, the synthesis of the phytoalexin capsidiol and the release of the phytohormone ethylene in tobacco leaves. Currently, the biological role of elicitins is still puzzling. Elicitins share sequence similarity with cerato-ulmin toxin, the protein thought to be involved in Dutch elm disease caused by *Ophiostoma novo-ulmi* (Yu, 1995). Recently, Mikes et al. (1997) demonstrated that cryptogein has one binding site with a strong affinity to dehydroergosterol. Moreover, this protein catalyzes the transfer of sterols between phospholipidic artificial membranes. This unique property could contribute to the understanding of the role of cryptogein in host-pathogen interactions knowing that *Phytophthora* fungi need sterols for vegetative growth and for the formation of sexual structures.

Non-selective toxins are also synthesised by *Fusarium solani* (Mart.) Sacc. a common soil-born fungus that causes root rot symptoms in peas, beans, lentils and cucurbits (Baywater, 1959; Synder *et al.*, 1975). On citrus in Florida, *Fusarium solani* is a primary coloniser of healthy fibrous roots causing fibrous and scaffold root rots (Graham *et al.*, 1983). This fungus produces a large number of phytotoxic naphthazarin pigments that may be synthesised in invaded roots and are toxic to citrus (Nemec *et al.*, 1988). When grown on shake cultures, *F. solani* synthesises dihydrofusarubin in a range of 36 to 71% of the total quantity of naphthoquinones produced (Baker and Tatum, 1983). Achor et al. (1993) demonstrated that leaves of rough lemon seedlings grown in a solution containing 100µg/L dihydrofusarubin showed veinal chlorosis, similar to symptoms in the field of blight trees with root rot disease. *Fusarium solani* naphthazarin toxins were detected by competitive enzyme-linked immunosorbent assay (ELISA) analysis in xylem fluid of citrus roots rotted by *F. solani* (Nemec *et al.*, 1991). Significantly more toxin was also present in branches of blight trees than in healthy trees. The authors therefore drew the conclusion that *in situ* toxin concentration may be required to cause blight in susceptible rootstocks.

In this paper we characterise and compare the actions of non-selective toxins of *Phytophthora quercina* and *Fusarium solani* on chloroplasts and on tobacco leaves.

2. MATERIAL AND METHODS

2.1 Chloroplast and ferredoxin preparation and chlorophyll detection

Chloroplast lamellae were isolated from spinach leaves according to Walker (1971) and Kobza and Edwards (1984).

Ferredoxin was prepared from 5 kg spinach leaves as described by Buchanan and Arnon (1971) and Hall et al. (1972). The concentration was calculated using the extinction coefficient of efd = 9700 M-1 x cm-1 at 420 nm.

Chlorophyll was extracted from tobacco leaves and spinach chloroplasts with 80% acetone and was determined according to the method of Arnon (1949).

2.2 Chloroplast experiments in the presence of dihydrofusarubin and methylviologen

A typical reaction mixture for measuring superoxide formation and NADPH- as well as ATP-production of spinach chloroplasts contained the following compounds in 2.5 ml thylakoid reaction medium: 30 mM tricine buffer pH 7.7, 200 mM sucrose, 20 mM sodium chloride, 3 mM magnesium chloride, 5 μM ferredoxin, 100 μM $NADP^+$, 500 μM ADP, 50 mM sodium di-hydrogen phosphate and chloroplasts thylakoids corresponding to 100 μg chlorophyll. For the different experiments dihydrofusarubin, methylviologen, DCMU, DBMIB or superoxide dismutase were added.

The following experiments with spinach chloroplasts were done with 10 ml glas tubes which were illuminated with a Philips Agrarphoto-BM-lamp equipment (335μmole E x m-2 x sec-1). A water barrier was placed between the reaction tube and the light source in order to keep a constant temperature of 20°C within the test tube.

Superoxide formation was determined as described earlier (Elstner and Heupel, 1976). In this assay hydroxylamine (50 μM) is oxidized to nitrite by superoxide. After the incubation time nitrite was quantified by adding sulfanilamide (10 mM in the final volume) and α-naphthylethylenediamine (1.5 mM in the final volume) to the reaction mixture. The formed dye was detected photometrically at 540 nm after 30 minutes incubation at room temperature in the dark.

In case of measuring NADPH the reaction tube was first centrifuged at 8000 g for 5 minutes and the extinction of the supernatant was monitored at 340 nm. After adding 10 μl

1M HCl the extinction was measured again and substracted from the first value. The concentration of NADPH formed was calculated with the extinction coefficient of ϵNADPH = 6211 M-1 x cm-1.

ATP-formation was measured with the Sigma Bioluminescence Somatic Cell Assay Kit. After centrifugation as mentioned above and after a dilution step of 1:5000 the assay kit was used as described by the manufacturer. For quantification a ATP calibration curve was used.

2.3 Ethane detection of toxin treated tobacco leaf discs

Ethane formation of tobacco leaf discs was determined gas chromatographically as described by Elstner and Konze (1976). Incubations were done in gas-tight vessels sealed with serum rubber stoppers. After 24 hours of incubation 1 ml of gas was withdrawn from the headspace with gas-tight syringes and was injected into the gas chromatograph. Each vessel contained one tobacco leaf disc (diameter of 2 cm) which floated on 2 ml toxin-buffer solution (0.1 M citrate buffer pH 6.0). The vessels were illuminated from the top with lamps mentioned above. Each vessel was placed in a water bath in order to guarantee a constant temperature of 30°C.

2.4 Isoelectric focussing of *Phytophthora* proteins

Isoelectric focusing (IEF) was performed with the Mini IEF Cell (Modell 111) from Biorad according to the manufacturer instructions using acrylamide 5%, glycerol 5,22% and a 2% ampholite solution (pI- range 4.45 - 9.6). IEF was run 15 min at 100V, 15 min at 200V and 1 h at 450V.

2.5 Transmission electron microscopy

About 1-2 mm^2 segments of leaf tissue were immersed in 6% glutaraldehyde in 50 mM cacodylate buffer pH 7.0, briefly evacuated and fixed for 90 min at 4°C. Following fixation, the tissue was washed in four changes of cold buffer and postfixed in 2% OsO_4 in 50 mM buffer overnight at 4°C. For electron microscopy, the sections were dehydrated in a graded ethanol series and embedded in Spurr's epoxy resin (Spurr, 1969). Sections were cut with a diamond knife on a Reichert ultramicrotome and stained with 3% uranyl acetate and lead citrate. Photographs were taken using a Zeiss 10C electron microscope.

2.6 Pronase and heat treatment of *Phytophthora* proteins

The pronase treated samples were incubated with 5µg pronase/µg protein in a 100mM Tris-buffer pH 8 for two days at 37°C in 2 ml eppendorf caps. The enzymatic reaction was stopped by boiling the samples for 5 min. For heat treatment the samples were boiled for 15 min. Before use all samples were dialysed against water for 8 hours.

3. RESULTS

3.1 Nitrite formation of spinach chloroplasts treated with dihydrofusarubin and methylviologen in the presence or absence of different electron transport blockers

Table 1 shows that nitrite formation from hydroxylamine (a test for superoxide) was significantly increased compared to the control, when spinach thylacoid membranes were incubated with DHF in the light. Similar results were obtained with the herbizid methylviologen. Light dependent nitrite formation of control as well as of dihydrofusarubin and methylviologen treated spinach chloroplasts was totally blocked after adding the electron transport blocker DCMU. Nitrite formation was also reduced to almost zero with superoxide dismutase (SOD), proving that superoxide was formed by chloroplasts.

3.2 Formation of NADPH and ATP by illuminated spinach chloroplasts in the presence of dihydrofusarubin and methylviologen

In table 2 the effects of dihydrofusarubin and methylviologen on ATP- and NADPH-formation by illuminated spinach chloroplasts are compared. The data show that dihydrofusarubin as well as methylviologen accelerated the electron transport of chloroplast membranes, measured as an increase in ATP-formation. Simultaneously $NADP^+$-reduction was inhibited by both compounds. Both compounds did not act as an uncoupler (ammonia) or an ATPase inhibitor (dicyclohexylcarbodiimid) on spinach chloroplasts.

3.3 Pigment degradation and ethane release of tobacco leaf discs treated with dihydrofusarubin in the light

Figure 1 shows the light dependent chlorophyll degradation and the simultaneously detected ethane production of dihydrofusarubin (DHF) treated tobacco leaf discs. A significant increase in ethane formation and pigment degradation was measured for 50 µmol/L DHF. Gas formation and chlorophyll degradation increased with increasing DHF concentration. No pigment loss and no significant ethane production was measured for DHF treated tobacco discs kept in the dark. Pigment degradation of illuminated tobacco leaves incubated with 10 µmol/L dihydrofusarubin started at the bottom of the leaf and close to the veins. Control leaves, treated with the same toxin solution, however kept in the dark, did not show any symptoms (data not shown).

Table 1. Hydroxylamine oxidation of illuminated spinach chloroplasts in the presence of methylviologen (MV), dihydrofusarubin (DHF), superoxide dismutase (SOD) and the electron transport blocker DCMU.

	nitrite (µmol / 5 min x mg chl)	
	mean	SD
C (control)	128	11
C + 10 µmol/L DHF	239	28
C + 10 µmol/L MV	348	51
C + DCMU	14	4
C + DCMU + 10 µmol/L DHF	21	1
C + DCMU + 10 µmol/L MV	12	8
C + SOD	5	4
C + SOD + 10 µmol/L DHF	21	1
C + SOD + 10 µmol/L MV	25	15

Spinach thylakoid membranes corresponding to 100 µg chlorophyll were suspended in 2.5 mL thylakoid reaction medium and illuminated in the presence of 10 µmol/L dihydrofusarubin, 10 µmol/L methylviologen, 50 µmol/L DCMU and DBMIB as well as 100U superoxide dismutase. In addition 50 µmol/L hydroxylamine was added to each experiment. Finally distilled water was added to give a final volume of 3 mL. The reaction tubes were centrifuged after an incubation time of 10 minutes and nitrite was measured as mentioned under material and methods. The control assay represents the thylakoid reaction medium including thylakoids. The data are means of six independent determinations.

Table 2. NADPH- and ATP-production of illuminated chloroplasts in the presence of methylviologen, dihydrofusarubin, ammonium chloride and dicyclohexylcarbodiimid.

	ATP (nmol/mg chl x h)		NADPH (µmol/mg chl x h)	
	mean	SD	mean	SD
C (control)	424.3	24.4	3.67	0.15
C + DCC	56.9	18.7	2.23	0.72
C + AM	45.0	17.3	4.32	0.14
C + 10 µmol/L DHF	817.4	45.8	2.21	0.06
C + 100 µmol/L DHF	982.1	141.2	0.47	0.32
C + 10 µmol/L MV	964.3	46.1	1.38	0.23
C + 100 µmol/L MV	752.3	124.7	0.08	0.05

Spinach thylakoid membranes corresponding to 100 µg chlorophyll were suspended in 2.5 mL thylakoid reaction medium and illuminated in the presence of 10 and 100 µmol/L dihydrofusarubin or methylviologen respectively. Ammonium chloride and dicyclohexylcarbodiimid, both 1 mmol/L, were used as indicated. Distilled water was added to give a final volume of 3 mL. The reaction tubes were centrifuged after an incubation time of 10 minutes and NADPH as well as ATP were detected as mentioned under material and methods. The control assay represents the thylakoid reaction medium including thylakoids. The data are means of six independent replications.

3.4 Transmission electron microscopy studies of tobacco leaves treated with protein solutions from *Phytophthora quercina*

In order to investigate the effect of *Phytophthora quercina* proteins on tobacco, leaves were put into different concentrated protein solutions and kept in the light. After an incubation time of three days, the protein solution of 0.5 µg x ml^{-1} caused severe wilt and necrosis on leaves. However, tobacco leaves treated with 1.0 µg x ml^{-1} of protein started to break into pieces showing severe wilt symptoms (data not shown).

Chloroplasts of the photosynthetic mesophyll of control leaves appear well developed with stromal as well as stacked granal membranes (G), intact outer envelopes (E), starch grains (S), peroxysomes (Pe) and some plastoglobuli (Pl, fig. 2a). Surprisingly tobacco leaf cells treated with *Phytophthora quercina* proteins for one day in the light were severely destroyed (data not shown). Especially, chloroplasts showed massive membrane degradation and less regularly stacked grana (fig. 2b). Figure 2c demonstrates the grana destruction in more detail.

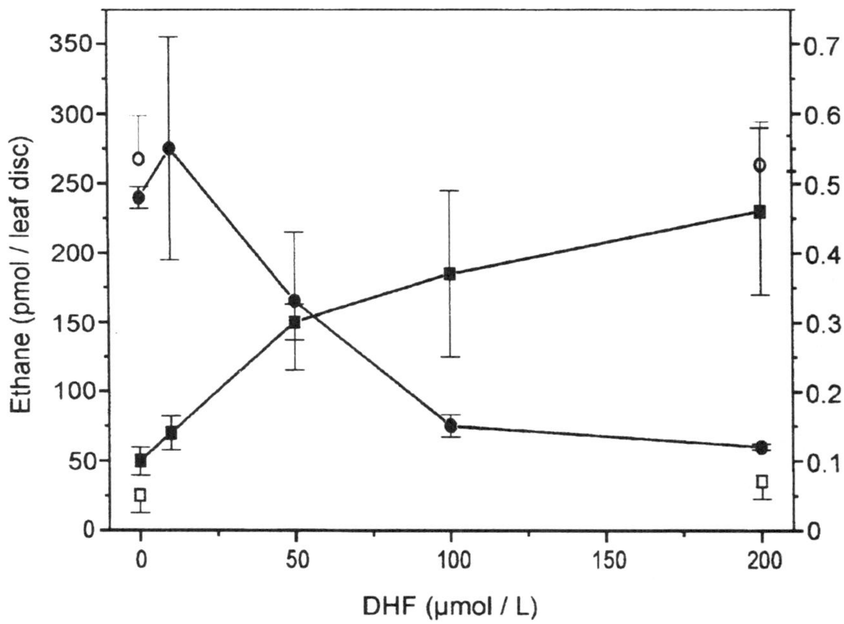

Figure 1. Ethane production and chlorophyll degradation of tobacco leaf discs treated with different dihydrofusarubin concentrations.

Tobacco leaf discs were floated on 0.1 mol/L citrate buffer pH 6.0 containing different dihydrofusarubin concentrations. After 24 hours of incubation in the light at 30°C, 1 mL of the gaseous headspace was analysed in a gaschromatograph and the chlorophyll content of the leaf discs was determined. The data are means of five independent determinations. For details see materials and methods.

● chlorophyll (in the light); ■ ethane (in the light);

O chlorophyll (in the dark); □ ethane (in the dark)

3.5 Pronase and heat treatment of the *Phytophthora quercina* proteins

Table 3 shows that the *Phytophthora quercina* proteins did not loose their activity towards tobacco leaves after heat treatment. Severe symptoms were still visible after boiling the protein solution for 15 minutes. However, no symptoms were seen on leaves after the protein solution was preincubated with pronase, indicating that the toxic effect is due to proteins released by *Phytophthora quercina* into the M1-medium.

Table 3. The influence of a pronase and heat treatment on the toxicity of the *Phytophthora quercina* proteins on tobacco leaves.

Sample number	Sample	Symptoms on tobacco leaves	Medium (ml) taken up by the leaf
1	*P. quercina* M1-medium *	++	4.2
2	*P. quercina* M1-medium * boiled for 15 min	++	5.0
3	*P. quercina* M1-medium * + pronase (active) and boiled for 15 min after incubation	-	5.0
4	*P. quercina* M1-medium * + pronase (inactive#) and boiled for 15 min after incubation	++	4.8

* The protein concentration was set up to 1.0µg/ml. The incubation was done for two days in the light using the conditions mentioned under material and methods.
++ = severe wilt symptoms on leaves and about 75% of the leaf area shows necrotic lesions.
– = no symptoms on leaves after treatment
= the pronase was inactivated during a 5 minute heat treatment before use

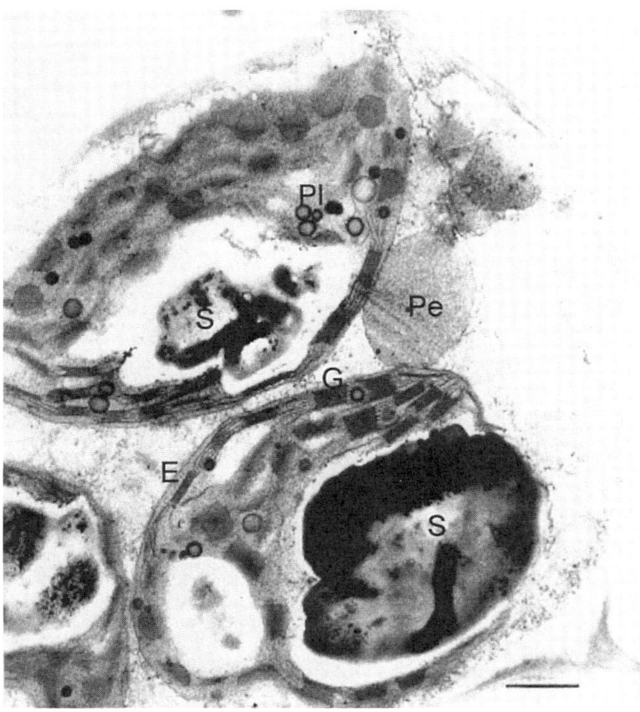

Figure 2. Transmission electron microscopy studies of tobacco leaves treated with *Phytophthora quercina* proteins in the light.

A: Cross section of a mesophyll cell from an untreated mature leaf with well developed chloroplasts. E, Outer envelope; G, grana membranes; Pl, plastoglobuli; Pe, peroxysome; S, starch grains. Bar = 1µm.

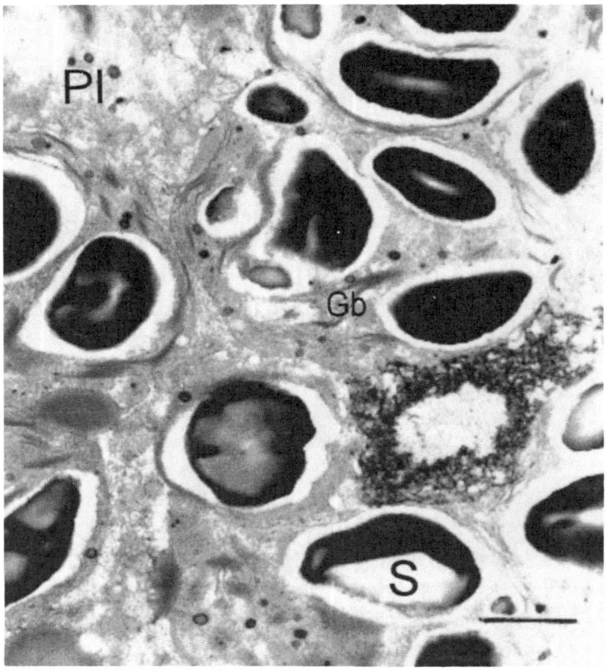

B: Cross section of a tobacco leaf treated with *P. quercina* proteins ($1\mu g \times ml^{-1}$) for two days in the light. Only starch grains (S) and broken grana structures (Gb) can be clearly identified. There are almost no intact membrane systems in the cytoplasm. Bar = 1 μm.

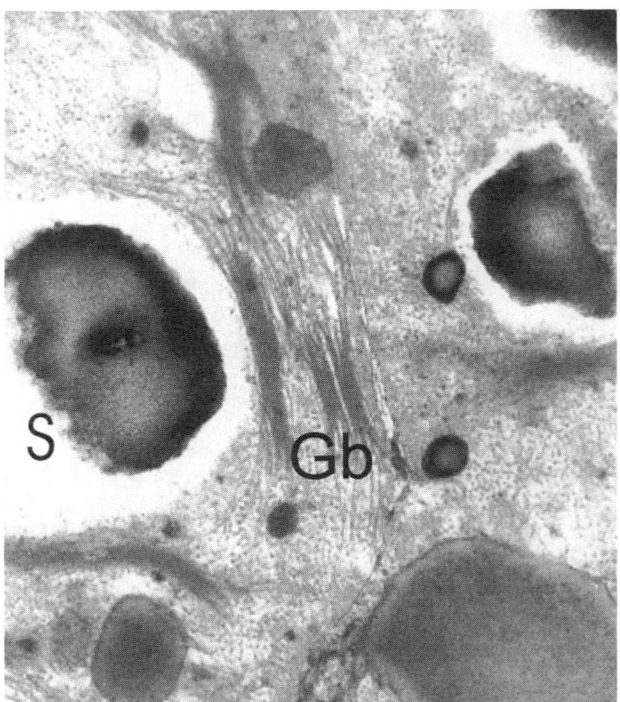

C: Another cross section of a tobacco leaf treated with *P. quercina* proteins ($1\mu g \times ml^{-1}$) showing the degradation of grana membranes (Gb) within chloroplasts in more detail. Bar = 1 μm.

3.6 Characterization and isoelectric focussing of the *Phytophthora* proteins released into the M1 -medium

The isoelectric points of the proteins of the two *Phytophthora quercina* isolates are summarized in table 4. The two *Phytophthora quercina* isolates produced three small peptides with isoelectric points of 3.6, 3.9 and 8.3. Only one single protein band was found after silver staining, when the concentrated M1-media of the *Phytophthora quercina* isolates were run on a 20% SDS gel. The identified molecular masses of the proteins (6.8 kDa) was almost identical with that of cryptogein, the phytotoxic peptide produced by *P. cryptogea*. Furthermore the antibody raised against cryptogein nicely crossreacted with the *Phytophthora quercina* peptide (data not shown).

Table 4. The isoelectric points of the *Phytophthora quercina* isolates.

Phytophthora quercina isolates	pH-value	
	acidic	basic
Phytophthora quercina (Higa1)	3.6; 3.9	8.3
Phytophthora quercina (Eger 3)	3.6; 3.9	8.3

The isoelectric points were determined with the Mini Protean II cell from Biorad as mentioned under material and methods.

4. DISCUSSION

4.1 The effect of the *Fusarium solani* toxin on spinach chloroplasts

Nemec et al. (1991) demonstrated that *Fusarium solani* naphthazarin toxins could be detected in the xylem fluid of scaffold roots and branches of infected citrus trees, although the fungus has not been detected microscopically in vessels. Furthermore Achor et al. (1993) showed that leaves of rough lemon seedlings grown in a solution of dihydrofusarubin revealed veinal chlorosis, similar to symptoms in the field of blight trees with root rot disease. On the basis of these findings we performed experiments to investigate the action of dihydrofusarubin on spinach chloroplasts.

We demonstrated that the *F. solani* toxin interacted with the electron transport chain of illuminated chloroplasts, causing the formation of superoxide. Experiments with the specific electron transport inhibitor DCMU proved that superoxide was generated on photosystem I. Furthermore, we found that NADPH-formation was reduced, whereas ATP-production was enhanced in the presence of dihydrofusarubin compared to untreated controls. Concentrations as low as 10 μmol/L of dihydrofusarubin were sufficient to induce these reactions. Surprisingly, the same results were obtained, when dihydrofusarubin was substituted by the well known herbizide methylviologen (MV). The generation of superoxide by the *F. solani* toxin can be explained by its redox potential of E_0 = -203 mV (Albrecht, 1996). The formation of superoxide by dihydrofusarubin and further reactive oxygen species, such as hydroperoxy radicals ($HO_2^\bullet$), formed at the Gouy-Chapman-Stern-layer of lipophilic membranes (Elstner, 1990), and electron donor-hydrogen peroxide complexes (crypto-OH) or free $OH^\bullet$-radicals might explain the findings of Achor et al. (1993). The destruction of membrane structures, primarily those of chloroplasts, the plasmalemma and the tonoplast, might be caused by lipid peroxidation initiated by reactive oxygen species, mentioned above. The detected chlorophyll degradation of tobacco leaf discs treated with dihydrofusarubin in the light, can be explained by the cooxidation of chlorophyll by peroxyradicals formed via lipid peroxidation (Elstner and Oßwald, 1980). The simultaneously measured ethane production proves that lipid peroxidation of membrane bound α−linolenic acid occurred.

The lack of NADPH within the stroma of chloroplasts will impair the ascorbate-glutathione cycle, where superoxide and hydrogen peroxide are detoxified at the expense of NADPH catalysed by ascorbate peroxidase, dehydroascorbic acid reductase and glutathione reductase (Foyer and Halliwell, 1976). The consequences would be an uncontrolled accumulation of different reactive oxygen species within chloroplasts of leaves, containing dihydrofusarubin, so that lipid peroxidation and pigment cooxidation might be accelerated (Albrecht et al., 1998).

4.2 Characterisation of the *Phytophthora quercina* peptides and their effect on tobacco leaves

In another set of experiments we characterised *Phytophthora quercina* peptides and investigated their effect on tobacco leaves. Recently we demonstrated that several root rot pathogens, including *Phytophthora quercina* (Jung *et al.*, 1998), are the primary cause of oak decline in the field in certain stands (Blaschke, 1994; Jung *et al.*, 1996). Declining trees exhibit wilted, chlorotic and necrotic leaves in the crown. We demonstrated that *P. quercina* released proteins into the M1-medium during a 2-week growth in the dark. This is in good agreement with Terce-Laforgue et al. (1992), who showed that the fungus *Phytophthora cryptogea* also released huge amounts of a small peptide, called cryptogein, into the medium. The protein solution of *P. quercina* caused severe wilt symptoms and pigment degradation on tobacco leaves in the light. A protein concentration as low as 0.5 µg protein x ml^{-1} was sufficient to induce symptoms. Pronase treatment of the *P. quercina* culture medium destroyed its activity, indicating that the damage is due to proteins released by that fungus. Ricci et al. (1989) also reported that the toxic components of *P. cryptogea* and *P. capsici* are small peptides which were inactivated after pronase treatment. After separation on a 20 % SDS-gel the *P. quercina* proteins showed up as one single band with a molecular mass of 6.5 kDa, identical with that of cryptogein. Surprisingly, the antibody raised against cryptogein showed a strong cross reaction with the *P. quercina* proteins in

Western-blot experiments (Heiser et al., 1999). Furthermore isoelectric focussing revealed that the *P. quercina* protein solution consists of one basic and two acidic peptides. The isoelectric points of the *P. quercina* proteins were very similar with that of *P. cryptogea*, reported by Le Berre et al. (1994). Transmission electron microscopy studies demonstrated that membrane structures, especially those of chloroplasts, were destroyed. Finally cells of treated leaves collapsed, leading to a disorganization of the parenchama tissue.

Taken all together, our results demonstrate that the characterised proteins, released by *P. quercina*, can be grouped into the family of *Phytophthora* leaf necrotic proteins called elicitins.

Investigations on the mode of action of the basic, highly purified *P. quercina* toxin Quercinin on chloroplasts and on root tissue are under way.

ACKNOWLEDGEMENTS

We thank Prof. Dr. E. F. Elstner (Institute of Phytopathology, Technical University of Munich) for his helpful discussions. The skillfull technical work on electron microscopy of Mrs. Maria Wenzkowski and of Mrs. Angelika Muhr for some biochemical analysis is greatly appreciated. We also thank Prof. Dr. Ricci and Dr. Pernollet (Institut National de la Recherche Agronomique, Cedex, France) for the cryptogein antibody and a cryptogein sample.

REFERENCES

Achor, D.S., Nemec, S., Baker, R.B.: 1993, *Mycopathologia* **123**, 117-126.

Albrecht, A., Heiser, I., Baker, R., Nemec, S., Elstner, E.F., Oßwald W.F.: 1998, *J. Plant Physiol.* **153**, 462-468.

Albrecht, A.: 1996, PhD thesis. Technical University, Munich

Arnon, D. I.: 1949, *Plant Physiol.* **24**, 1 – 15.

Baker, R. A. and J. H. Tatum, J.H.: 1983, *Proc. Fla. State Hort. Soc.* **96** 53-55.

Baywater, J.: 1959, *Trans. Br. Mycol. Soc.* **42**, 201-212.

Blaschke, H.: 1994. *Eur. J. For. Path.*, **24**, 386-398.

Buchanan, B. B., D. I. Arnon: 1971, *Methods Enzymol.,* **23a**, 413-440.

Elstner, E. F. and A. Heupel, A.: 1976, *Biochem.* **70**, 616-620.

Elstner, E. F. and J. R. Konze, J.K.: 1976, *Nature* **263,** 251-352.

Elstner, E. F. and W. F. Oßwald, W.F.: 1980, *Z. Naturforsch.* **35c**, 129-135.

Elstner, E. F.: 1990, Der Sauerstoff: Biochemie, Biologie, Medizin. BI-Wissenschafts-verlag Mannheim, Wien, Zürich

Foyer, C. H. and Halliwell, B.: 1976, *Planta* **133**, 21 – 25.

Graham, J. H., Timmer, L.W., Young, R.H.: 1983, *Plant Disease* **67**, 1273-1276.

Hall, D. O., Rao, K.K., Cammack, R.: 1972, *Biochem. Biophys. Res. Commun.* **47**, 798-802.

Heiser, I., Oßwald, W. F., Elstner, E. F.: 1989, *Plant Physiol. Biochem.* **36**, 703-713.

Heiser, I., Fromm, J., Giefing, M., Koehl, J., Jung, T., Oßwald, W.F.: 1999, *Plant Physiol. Biochem.*

37, 73-81.

Jung T., Cooke, D., Blaschke, H., Duncan, J., Oßwald W.F.: 1998, *Mycol. Research* **103**, 785-798.

Jung T., Blaschke, H., Neumann, P.: 1996, *Eur. J. For. Pathol.* **26**, 253-272.

Kern H.: 1956, *Umschau* **19**, 548-551.

Kobza, J. and Edwards, G. E.: 1984, Isolation of organells: Chloroplasts. In: Cell, cultures and somatic cell genetics of plants. Vol 1, Academic Press London / New York, pp 471-482.

Le Berre, J-Y., Panabieres, F., Ponchet, M., Denoroy, L., Bonnet, P., Marais, A., Ricci, P.: 1994. *Plant Physiol. Biochem.* **32**, 251-258.

Mikes, V., Milat, M-L., Ponchet, M., Ricci, P., Blein, J-P.: 1997, *FEBS Lett.* **416**, 190-192.

Nemec, S., Baker, B., Tantum, J.H.: 1988, *Soil Biol. Biochem.* **20**, 493-499.

Nemec, S., Jabaji-Hare, S., Charest, P.M.: 1991, *Phytopathol.* **81**, 1497-1503.

Oßwald, W. F.: 1995, Pages 315-369, In: *Schadwirkungen auf Pflanzen*, B. Hock, E.F. Elstner., Spektrum akademischer Verlag, Heidelberg, Berlin, Oxford,.

Ricci, P., Bonnet, P., Huet, J.C., Sallatin, M., Beauvais-Cante, F., Bruneteau, M., Billard, V., Michel, G., Pernollet, J.C.: 1989, *Eur. J. Biochem.* **183**, 555-563.

Spurr, A. R.: 1969, *J. Ultrastruct. Research* **26**, 31-43.

Synder, W. C., Georgopoulos, S.G., Webster, R.K., Smith, S.N.: 1975, *Hilgardia* **43**, 161-185.

Terce-Laforgue, T., Huet, T., Pernollet, J.C.: 1992, *Plant Physiol.* **98**, 936-941.

Walker, D. A.: 1971, *Methods Enzymol.* **23**, 211-220.

Yoder, O. C.: 1980, *Annu. Rev. Phytopathol.* **18**, 103-129.

Yu, L. M.:1995, *Proc. Natl. Acad. Sci. USA* **92**, 4088-4094.

Zanetti, A., Beauvais, F., Huet, C.J., Pernollet, J.C.: 1992, *Planta* **187**,163-170.

SECTION III

Trees and their Environment

EFFECTS OF ATMOSPHERIC CO_2 ON GROWTH AND BRANCHINESS OF DIFFERENT POPLAR (*Populus*) GENOTYPES IN THE POPFACE EXPERIMENT

B. GIELEN[1], C. CALFAPIETRA[2], I.A. JANSSENS[1], G. SCARASCIA-MUGNOZZA[2] AND R. CEULEMANS[1]

[1] *University of Antwerpen, UIA, Department of Biology, Universiteitsplein 1, B-2610 Wilrijk, Belgium, email: bgielen@uia.ua.ac.be,*

[2] *Universita degli Studi della Tuscia, Department of Forest Environment and Resources (DISAFRI), Via San Camillo de Lellis, I-01100 Viterbo, Italy*

Key words: FACE, *Populus*, elevated CO_2, growth, SRFC, branches, nitrogen

Abstract: A FACE facility has been installed in a high density poplar plantation to investigate the effects of increasing atmospheric CO_2 on three different *Populus* genotypes. First-year height growth was stimulated significantly and a faster growth rate was observed in the beginning of the growing season. At the end of the first season, stem volume index was significantly larger in the FACE treatment and an increased number of sylleptic branches was observed. Leaf nitrogen concentration remained unchanged or increased under elevated CO_2. Significant genotype differences in the response to elevated CO_2 were observed.

1. INTRODUCTION

One of the main components of global climatic change is the rising atmospheric CO_2 concentration, that is expected to reach 550 ppm after the middle of this century (Schimel *et al.* 1996). Since C_3 photosynthesis is carbon limited in the present atmosphere, an increased atmospheric CO_2 concentration will enhance photosynthesis and consequently growth of individual plants (Bowes 1996). Forests cover 3.5 billion ha (FAO 1995) of the

S. Huttunen et al. (eds.),
Trends in European Forest Tree Physiology Research, Cost Action E6: EUROSILVA, 183-196

earth and make up 85% of plant and 35% of soil carbon (Kirschbaum and Fischlin 1996). Therefore, a major emphasis on the study of the carbon sequestration capacity of forests is justified in the framework of global climatic change research (Mooney *et al.* 1999; Wullschleger *et al.* 1995).

A compilation of the studies of tree responses to an elevated atmospheric CO_2 concentration under different experimental conditions can be found in recent review papers (Eamus and Jarvis 1989; Ceulemans and Mousseau 1994; Saxe *et al.* 1998; Norby *et al.* 1999). Despite the numerous studies, it has by now become evident that a number of questions remain unanswered. This is due to the fact that the majority of experiments last only a fraction of a trees life span and are realised in open top chambers, involving a change of the microclimate around the trees (Norby *et al.* 1999; Janous *et al.* 1996). To overcome this limitation, the recently developed FACE (Free Air Carbon Enrichment) technology has been applied to complete ecosystems (Hendrey *et al.* 1993; Norby *et al.* 1999). Sofar, FACE studies on trees are restricted to a loblolly pine stand in North Carolina, USA (DeLucia *et al.* 1999), to a high-density aspen and birch plantation (AspenFACE) in Wisconsin, USA (Dickson *et al.* 2000), to a sweetgum canopy in Tennessee, USA and to a prototype FACE on rainforest species in Panama.

In the recently initiated POPFACE experiment (Scarascia-Mugnozza *et al.* 2000; http://www.unitus.it/dipartimenti/disafri/progelti/Popface), an intensively cultured high-density poplar plantation with three *Populus* genotypes is being treated to investigate CO_2 effects at an ecosystem level. An additional objective is to quantify the carbon sequestration capacity of short rotation forestry systems. Poplars are the most promising trees for short rotation intensive culture (SRIC), a kind of agriculture in which trees are intensively managed with the aim of maximising biomass production for pulp, timber or energy (Beaton *et al.* 1991; Galinski *et al.* 1991; Ceulemans *et al.* 1992). Since managed forest plantations are of major global significance, it is crucial to know how these ecosystems will perform in a changing climate (Janssens *et al.* 2000). All studies on the effects of atmospheric CO_2 on poplars (Bosac *et al.* 1995; Ceulemans *et al.* 1995; Curtis *et al.* 1994; Gardner *et al.* 1995; Radoglou and Jarvis 1990; Taylor and Frost 1992; Pregitzer *et al.* 1995; Kalina and Ceulemans 1997) have been carried out at individual plant or tree levels only, and have shown pronounced responses to the CO_2 treatment. No studies thusfar have been made at the larger ecosystem scale. In this contribution, tree growth in the establishment year of the POPFACE experiment will be reported. The aim is to illustrate effects of CO_2 enrichment and genotype on the initial growth of a high-density, intensively cultivated poplar plantation.

2. MATERIAL AND METHODS

2.1 Site and treatment description

In late spring 1999, a plantation with a FACE design was established in Central Italy, near Viterbo (Tuscania; 42°22' N, 11°48' E, alt 150 m). Following detailed soil analysis, six experimental plots (30 m x 30 m) were selected within a 9 ha former agricultural land. In three of these plots, a FACE facility was installed while the three remaining plots, representing the control treatment, were left under untreated conditions. The carbon enrichment is realised through a polyethylene octagonal ring (diameter = 22 m) mounted on telescopic poles. These are extendable up to 12 m to adjust for height growth during the course of the study. Pure CO_2 is released through laser drilled holes in the horizontal pipes of the ring, in contrast to other FACE designs where CO_2 enriched air is emitted through ports in vertical vent pipes (Hendrey *et al.* 1993; Lewin *et al.* 1994). The target CO_2 concentration inside the FACE treatment is 550 ppm. In order to avoid cross contamination between FACE and control plots, the minimum distance between plots is 120 m. A meteorological station in the centre of each ring collects data that are used to control the directional release of gas along the ring. The released quantity of gas was established, according to wind speed, using an algorithm developed for the facility and based on a 3-D gas dispersion model. A detailed description of the set-up and performance of the POPFACE facility is given by Scarascia-Mugnozza *et al.* (2000).

2.2 Plant material and plantation lay-out

After the land was ploughed and crumbled twice using a miller, the poplar plantation was set up using hardwood cuttings (length 25 cm) selected for size, bud status and vigour uniformity. The entire field was planted with *Populus x euramericana* clone I-214 at a planting density of 5000 trees per ha (spacing 2 m x 1 m). The six experimental plots were planted with three different poplar genotypes at a planting density of 10000 trees per ha (spacing of 1 m x 1 m) in order to have a sufficient number of experimental trees and a full canopy already after the first year. The three genotypes were *P. x euramericana* (clone I-214), a clone of *P. nigra* L. (Jean Pourtet) and a local selection of *P. alba* L. (clone 2AS11). Each ring, which contains 350 trees, is divided into two parts by a physical resin-glass barrier (1 m deep in the soil) for future nitrogen treatments. Each halve ring is further divided in three sectors for the different genotypes. No nitrogen treatments were applied during the first year of the experiment.

A drip irrigation system was installed both in the experimental plots and in the surrounding plantation to avoid drought stress. Herbs were removed manually or mechanically; a limited use of insecticides was indispensable to control stem boring insects.

2.3 Growth measurements

From August 1999 onward, stem height and basal stem diameter at 20 cm above the soil were measured, and sylleptic branches were counted, every two weeks. Tree height was measured with a graduated pole and measurements of stem diameter were made using a digital caliper (Mitutoyo, type CD-15DC, U.K.). Sylleptic branches on the main stem, defined as branches that develop from axillary buds not undergoing a rest period (Remphrey and Powell 1985), were counted.

All measurements were made on a sample of six adjacent trees selected within each sector of the experimental plots. These trees, surrounded by at least one row of the same genotype (to avoid possible border effects) represent the permanent growth plot (PGP) which is being left undisturbed during the course of the study. This means that there are six PGPs per ring, two per genotype. Since no nitrogen treatment was applied during the first year, all growth parameters were measured on a sample of twelve trees per genotype in each ring.

Near the end of the growing season stem height (H) and stem diameter (D) at 20 cm above the soil were assessed on 48 trees per genotype in each FACE ring and control plot. Stem volume index (VI) was estimated as D^2H.

For each of the different genotypes, correlation coefficients were calculated between the stem VI and the total number of sylleptic branches per tree of all the PGP trees.

At the two-weekly intervals, absolute growth rate (AGR) of tree height was calculated according to (Kvèt *et al.* 1971) and relative growth rate (RGR) of tree height after (Evans 1972):

$$AGR = (H_2 - H_1)/(t_2 - t_1)$$

$$RGR = (\ln H_2 - \ln H_1)/(t_2 - t_1)$$

where H_1 and H_2 are heights of the main stem (cm) at times t_1 and t_2 (days).

2.4 Leaf nitrogen

In August 1999, two recently matured leaves per tree in the PGP and 8 leaves of surrounding trees per genotype were harvested in each sector of the FACE rings and control plots. Leaves were oven dried for 48 h (50 °C), ground in a mill (Cyclotec 1093 Sample Mill, Sweden) and analysed for nitrogen (N) and carbon (C) by a Dynamic Flush Combustion Method with a NC 2100 Soil Analyser (Carlo Erba Strumentazione, Rodano, Italy). Nitrogen concentrations were expressed on a dry mass basis.

2.5 Statistical analysis

To determine the main effects of CO_2 treatment and genotype, both fixed factors, data were analysed by a nested Analysis of Variance (ANOVA). Ring, nested within CO_2 treatment, and the interaction between ring and genotype, were included as random factors in the design to account for between ring variation. Significance of this interaction was tested with the Likelihood ratio test. Analysis of stem height was performed separately for each measuring date. All statistical analyses were performed in SAS (SAS Institute, Cary, NC) using the mixed procedure. Satterthwaite's procedure was used to obtain the denominator degrees of freedom. Where the ANOVA F-test highlighted an interaction between CO_2 treatment and genotype, *a posteriori* comparison of means was done, using parameter estimates as given by SAS. The Bonferroni method was applied to correct for the multiple comparisons. Differences between parameter means were considered significant when $p < 0.05$.

2.6 Growth

No significant treatment effect on tree height was found at the start of measurements but from September onwards, differences became significant (Table 1) with p = 0.0297 at the end of the season. Trees of *P. alba*, *P. nigra* and *P. x euramericana* were respectively 8%, 11% and 10% higher in the FACE rings. Significant differences among genotypes were found, but diminished towards the end of season (Table 1, Figure 1). Due to its fast initial growth rate, *P. x euramericana* was significantly higher than *P. alba,* while no significant difference between *P. x euramericana* and *P. nigra* was observed in the beginning of the growing season (Table 1, Figure 1). *P. x euramericana*, however, stopped growing in September whereas *P. alba* and *P. nigra* continued growth. This resulted in non-significant differences between *P. x euramericana* and *P. alba*, and significant differences between *P. nigra* and *P. alba*, respectively, *P. x euramericana* at the end of the first growing season (Table 1).

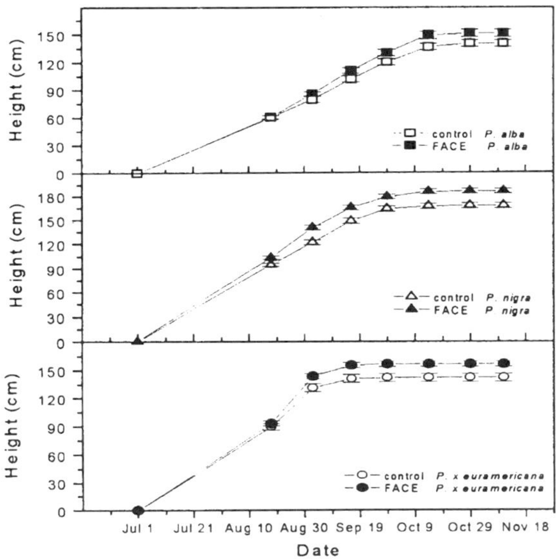

Figure 1. Evolution of tree height during the first growing season (1999) of three *Populus* genotypes in FACE rings and control plots. Symbols represent the mean ± SE (n = 36). ANOVA results are shown in Table 1.

The differences in growth rate became evident from AGR and RGR, calculated over the two-weekly intervals. During the initial growing period, both AGR and RGR were higher under the FACE treatment (Figure 2), but ANOVA showed only a significant treatment effect on AGR (p = 0.0241, respectively 0.0789). Afterwards, no CO_2 effect was observed neither on AGR nor on RGR of the different genotypes; only during the second half of September a higher AGR was observed in the FACE treatment (Figure 2, p = 0.0143). The effect of genotype on AGR and RGR was significant throughout the entire measuring period (p < 0.001). In August, *P. x euramericana* had the fastest growth rate (AGR = 3.5 cm day^{-1}) but due to a quick decline, and at the same time a sustained growth of *P. alba*, highest values of growth rate near the end of the season were observed for *P.*

alba (Figure 2). No interaction between the effects of treatment and genotype on AGR and RGR was observed (p of the interaction ranged between 0.19 and 0.99).

Table 1. Analysis of variance for the main effects of CO_2 treatment, genotype and their interaction (X) on tree height at different times during the first season, stem volume index (D^2H) and number of sylleptic branches per unit stem length of three *Populus* genotypes in FACE rings and control plots. Genotypes that are significantly ($p < 0.05$) different are shown in the last column. P-values of the ANOVA F-test are indicated as: ns: $p > 0.05$, *: $p < 0.05$, **: $p < 0.01$, ***: $p < 0.001$, ****: $p < 0.0001$.

Parameter	Date	Source of variation			Genotype Differences
		Treat-ment	Genotype	X	
Height	August	ns	****	ns	*P. alba-P. nigra; P. alba-P. x eur*
	September	*	****	ns	*P. alba-P. nigra; P. alba-P. x eur*
		ns	****	ns	*P. alba-P. nigra; P. alba-P. x eur*
		*	***	ns	*P. alba-P. nigra; P. nigra-P. x eur*
	October	*	**	ns	*P. alba-P. nigra; P. nigra-P. x eur*
		*	*	ns	-
	November	*	*	ns	-
D^2H	November	**	****	*	*P. alba-P. nigra; P. alba-P. x eur, P. nigra- P. x eur*
Sylleptic branches	November	ns	****	ns	*P. alba-P. nigra; P. alba-P. x eur, P. nigra- P. x eur*

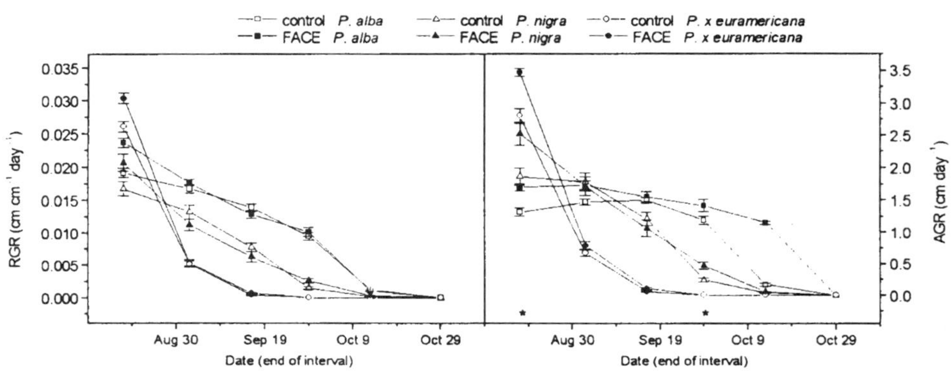

Figure 2. Evolution of absolute growth rate (AGR; right panel) of tree height and relative growth rate (RGR; left panel) of tree height for two-weekly intervals in the first growing season of three *Populus* genotypes in FACE rings and control plots. Symbols represent the mean ± SE (n = 36). When the effect of CO_2 treatment was significant, the p-value of the ANOVA F-test is indicated in the graph as: *: $p < 0.05$.

As a result of this initial stimulation of both height and stem diameter (data not shown), the stem volume index was significantly (p = 0.0058) larger in the FACE rings at the end of the first growing season (Table 1, Figure 3). Due to the small stimulation of *P. alba* in the FACE treatment, a significant interaction between CO_2 treatment and genotype was found (p = 0.0123). Therefore, *a posteriori* comparison showed significant differences between treatments only for *P. nigra* and *P. x euramericana* that doubled their stem volume index under FACE, reaching 1.17 $dm^3\ m^{-2}$ and 0.85 $dm^3\ m^{-2}$ respectively (Figure 3). The stem volume index of *P. alba* was significantly smaller (0.36 $dm^3\ m^{-2}$) compared to the two other genotypes and showed an enhancement under FACE of 64% of the end-of-season VI. Genotype had a significant effect on VI (Table 1) but comparison of parameter estimates showed that the mean VI of *P. x euramericana* was only significantly different from *P. nigra* and *P. alba* in the FACE treatment.

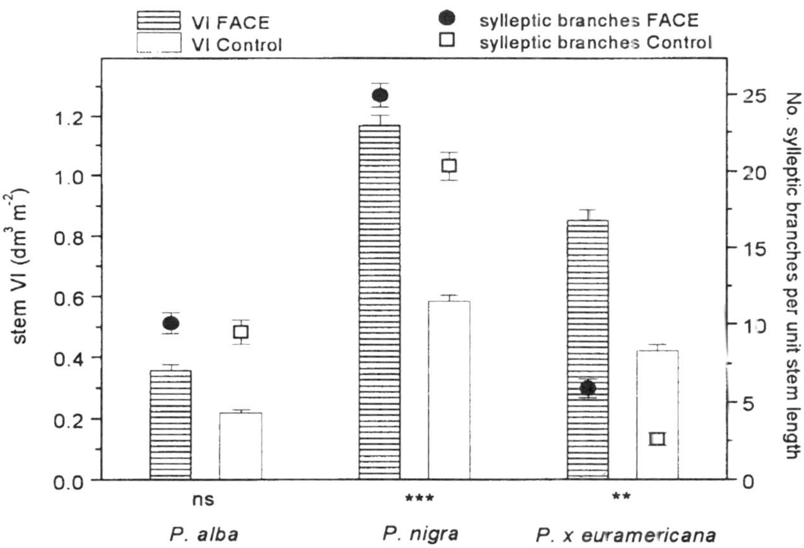

Figure 3. Volume index (VI = D^2H = stem diameter^2 x tree height) per unit ground area (1 m^2) and number of sylleptic branches per unit (1 m) stem length of three different *Populus* genotypes in FACE rings and control plots at the end of the first growing season. Columns represent the mean VI $\pm$ SE (n = 144) and symbols indicate the number of sylleptic branches $\pm$ SE (n = 36) for each genotype and treatment. ANOVA results for VI and sylleptic branches are given in Table 1. Significant differences in stem VI between treatments for each genotype are indicated below the columns as: ns: $p > 0.05$, * : $p < 0.05$, ** $p < 0.01$, *** $p < 0.001$.

P. nigra produced considerably more sylleptic branches in the FACE rings with a 23% increase in the number of branches per unit stem length (Figure 3). Due to the small numbers, a relative stimulation of 127% was observed for trees of *P. x euramericana*. However, no significant overall treatment effect was shown (p = 0.1184, Table 1). Genotype had a high significant effect (p = 0.0001) on sylleptic branching (Table 1); *P. nigra* is characterized by numerous sylleptic branches whereas the number of sylleptic branches of *P. x euramericana* is four times smaller in the FACE rings, and seven times smaller in the control plots than this of *P. nigra* (Figure 3).

Correlation coefficients between stem VI and the number of sylleptic branches per tree were not very strong. Values for *P. alba*, *P. nigra* and *P. x euramericana* were 0.65, 0.74 and 0.69 respectively.

2.7 Nitrogen

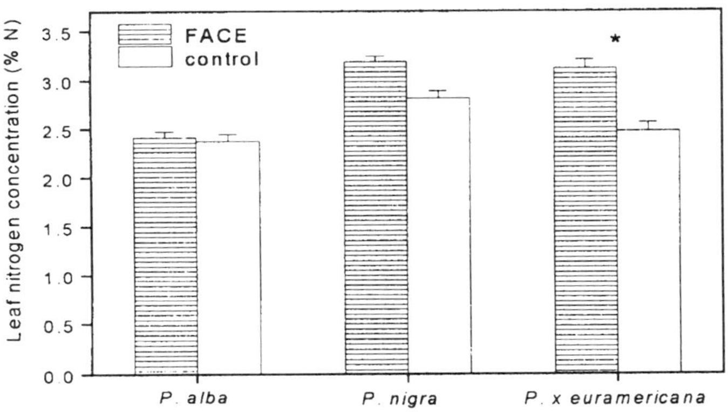

Figure 4. Leaf nitrogen concentration on a mass basis (% N) of three *Populus* genotypes in FACE rings and control plots. Columns represent the mean ± SE (n = 29-39) of leaves sampled in August, 1999. Significant differences between treatments for each genotype are indicated as * : $p < 0.05$.

Leaf nitrogen concentrations, obtained from the August samples, were in general relatively high, ranging between 1.17% and 3.43% (on a dry mass basis) for *P. alba*, from 1.79% to 3.88% for *P. nigra* and 1.27% to 4.22% for *P. x euramericana*. On a mass basis, leaves of *P. x euramericana* contained significantly (p=0.0125) more nitrogen in the FACE treatment than in the ambient control. The relative difference was 26% and 13% for *P. x euramericana* and *P. nigra* respectively. Analysis of *P. alba* leaves showed no significant response to the CO_2 enrichment. Within the FACE rings, nitrogen concentrations of *P. alba* were significantly lower compared to *P. x euramericana* and *P. nigra*. Genotypic differences between *P. nigra* and *P. alba*, respectively *P. x euramericana* were significant in the control treatment.

3. DISCUSSION

In the present study, where an entire forest plantation is exposed to a future atmospheric CO_2 concentration, tree height was significantly enhanced in the FACE treatment. At the end of the growing season the canopy was 11% and 10% higher under CO_2 enrichment for *P. nigra* and *P. x euramericana* respectively. Minor differences were observed for *P. alba*, due to ring variation and initial rooting difficulties. Although these results agree with open top chamber observations on single trees (Lee *et al.* 1998; Norby *et al.* 1999), absolute values might be more reliable since limitations of chamber experiments are excluded and CO_2 is enriched up to 550 ppm. Most of previous studies have used doubled ambient CO_2 concentrations, which could have overestimated effects in view of the most recent predictions.

Height differences could be attributed to a faster absolute growth under elevated CO_2 in the beginning of the measuring period. From September onwards only differences between genotypes were present; *P. x euramericana* stopped growing already in late summer while *P. alba* still had a mean growth rate of 1.4 cm day^{-1} under FACE at that time. Consequently, not only growth rate but also length of the growing season are main factors that determine plant productivity, and that influence the ability of different genotypes to take advantage of a carbon enrichment to a varying degree. Since growth rate was also enhanced on a relative basis, these observations suggest that trees grew faster in the FACE treatment and that the growth enhancement was not only a size effect with bigger trees with a larger leaf area in the FACE treatment.

Poplars are characterized by an indeterminate growth habit, and produce sylleptic branches during the growing season until September. Since it has often been suggested that CO_2 responses of plants are dependent on the availability to create new sinks (Webber *et al.* 1994; Oechel and Strain 1985; Ward and Strain 1999), poplars might profit even more from a carbon enrichment. Few studies thusfar have examined the number of branches under elevated CO_2 but they all reported an increased branch number (Ceulemans *et al.* 1996; Idso *et al.* 1991). In the present FACE experiment we observed a stimulation of branch production since the number of sylleptic branches per unit length on the main stem was higher, although not significant. Trees with a lot of sylleptic branches, were characterized by a high volume index; correlation was, however, relatively poor. Larson *et al.* (1976) also observed a poor correlation, and Scarascia-Mugnozza *et al.* (1997) found no significant differences in stem biomass between branchy clones versus less branchy clones. Moreover, it has been shown that main stem leaves provide most of the photosynthetic products for stem height and diameter growth (Isebrands and Nelson 1983).

Stem volume index of *P. x euramericana* and *P. nigra* was doubled significantly in the FACE rings. Although, this approximation of stem volume gives an overestimation of real volume production (Pontailler *et al.* 1997), relative comparison between genotypes and treatments can give valuable information about genotype specific performance in future SRIC. Because of the importance of managed forests from an economic viewpoint and their possible role as a carbon sink, understanding and predicting their response is crucial (Janssens *et al.* 2000). Tree responses to global climatic change have been shown to differ among species and among clones e.g. (Mousseau and Enoch 1989; Ceulemans *et al.* 1996). This is particularly relevant for forest crops as has already been emphasized by Centritto *et al.* (1999b). In their study on Sitka spruce, clonal provenance affected dry mass production only under elevated CO_2 concentrations. In the present study genotype had a significant effect on all measured tree characteristics and moreover, an interaction of genotype with CO_2 treatment was shown for stem volume index. *P. nigra* produced significantly more biomass (as estimated by the stem volume index) in the FACE treatment compared to *P. x euramericana*, but differences in mean volume index in the control treatment were not significant.

Because of the tight relation between leaf nitrogen concentration and photosynthesis, leaves have been analysed for nitrogen concentration in a number of studies. Overall, these have shown a reduction in nitrogen concentration on a mass basis of 19% (Cotrufo *et al.* 1998). In the present experiment however, a higher nitrogen concentration was observed in leaves of the FACE treatment. Since leaves were sampled during the initial period of the experiment, it is probable premature to conclude that these findings contradict previous studies. In an open top chamber experiment with *Pinus sylvetris*, for example (Jach and Ceulemans 1997), an increase of needle nitrogen was observed in the first year whereas the opposite was found in the following years.

Our first-year results show a significant enhancement of volume production of fast-growing poplar genotypes under free air CO_2 enrichment. It has been suggested that forests will not produce more biomass because complex ecosystem feedbacks, including intra- and interspecific competition, or limited nutrient supply would decrease the initial biomass stimulation (Hättenschwiler and Körner 1998). Our study is a typical intensively managed forest ecosystem without limitations of water and nutrient supply. However, since plant spacing is very small, competition might influence whole-canopy responses to CO_2 enrichment in the following years. Furthermore, it has been suggested that a growth stimulation under carbon enrichment is mainly an initial effect resulting in trees that grow bigger more quickly (Lee *et al.* 1998; Centritto *et al.* 1999a). Longer-term investigations will yield relevant information about the impact of an increase in atmospheric CO_2 on the productivity and carbon sequestration potential of forest plantations and contribute to our understanding of climate change responses beyond the individual tree level. This type of ecosystem studies provides valuable data to improve and validate modelling efforts, which are necessary to make reliable predictions of ecosystem responses in a changing climate.

ACKNOWLEDGEMENTS

Funding has been provided by the EC Fourth Framework Programme, Environment and Climate RTD Programme, research contract ENV4-CT97-0657 within the Terrestrial Ecosystems Research Initiative (TERI). The authors acknowledge Dr T. Crowe (University of Southampton) for his recommendations concerning the statistical analysis and E. De Bruyn for nitrogen analysis. We thank M. Sabatti for useful discussions and S. Van Dongen for practical advice about SAS. B. Gielen is a Research Assistant of the Fund for Scientific Research-Flanders (Belgium).This study is a core project within the GCTE (Gobal Change & Terrestrial Ecosystems) Elevated CO_2 Consortium of the IGBP (International Geosphere Biosphere Programme).

REFERENCES

Beaton, A., Thompson, D. and Webber, J. (1991) British poplar-a 21st century challenge. *For. Br. Timber,* **august,** 19-21.

Bosac, C., Gardner, S. D. L., Taylor, G. and Wilkins, D. (1995) Elevated CO_2 and hybrid poplar: a detailed investigation of root and shoot growth and physiology of *P. euramericana* 'Primo'. *Forest Ecology and Management,* **74,** 103-116.

Bowes, G. (1996) Photosynthetic responses to changing atmospheric carbon dioxide concentration. In: *Advances in Photosynthesis. Photosynthesis and the Environment,* vol. 5 (ed. N. R. Baker). Kluwer Academic Publishers, the Netherlands.

Centritto, M., Lee, H. S. J. and Jarvis, P. G. (1999a) Increased growth in elevated [CO_2]: an early, short-term response? *Global Change Biology,* **5,** 623-633.

Centritto, M., Lee, H. S. J. and Jarvis, P. G. (1999b) Long-term effects of elevated carbon dioxide concentration and provenance on four clones of Sitka spruce (*Picea sitchensis*). I. Plant growth, allocation and ontogeny. *Tree Physiology,* **19,** 799-806.

Ceulemans, R., Impens, I., Mau, F., van Hecke, P. and Chen, S. G. (1992) Dry mass production and solar radiation conversion efficiency of poplar clones. In: *Biomass for Energy, Industry and Environment: 6th E.C. Conference* (ed. G. Grassi, A. Collina and H. Zibetta), pp. 157-163. Elsevier, London.

Ceulemans, R., Jiang, X. N. and Shao, B. Y. (1995) Effects of elevated atmospheric CO_2 on growth, biomass production and nitrogen allocation of two *Populus* clones. *Journal of Biogeography,* **22,** 261-268.

Ceulemans, R. and Mousseau, M. (1994) Tansley Review No. 71. Effects of elevated atmospheric CO_2 on woody plants. *New Phytologist,* **127,** 425-446.

Ceulemans, R., Shao, B. Y., Jiang, X. N. and Kalina, J. (1996) First- and second-year aboveground growth and productivity of two *Populus* hybrids grown at ambient and elevated CO_2. *Tree Physiology,* **16,** 61-68.

Cotrufo, M. F., Ineson, P. and Scott, A. (1998) Elevated CO_2 reduces the nitrogen concentration of plant tissues. *Global Change Biology,* **4,** 43-54.

Curtis, P. S., Zak, D. R., Pregitzer, K. S. and Teeri, J. A. (1994) Above- and belowground response of *Populus grandidentata* to elevated atmospheric CO_2 and soil N availability. *Plant and Soil,* **165,** 45-51.

DeLucia, E. H., Hamilton, J. G., Naidu, S. L., Thomas, R. B., Andrews, J. A., Finzi, A., Lavine, M., Matamala, R., Mohan, J. E., Hendrey, G. R. and Schlesinger, W. H. (1999) Net primary production of a forest ecosystem with experimental CO_2 enrichment. *Science,* **284,** 1177-1179.

Dickson, R. E., Lewin, K. F., Isebrands, J. G., Coleman, M. D., Heilman, W. E., Riemenschneider, D. E., Sober, J., Host, G. E., Hendrey, G. R., Pregitzer, K. S. and Karnosky, D. F. (2000). Forest atmosphere carbon transfer and storage (FACTS II)-The Aspen free-air CO_2 and O_3 enrichment (FACE) project: An overview: USDA Forest Service North Central Forest Experiment Station General Technical Report NC-214.

Eamus, D. and Jarvis, P. G. (1989) The direct effects of increase in the global atmospheric CO_2 concentration on natural and commercial temperate trees and forests. *Advances in Ecological Research,* **19,** 2-55.

Evans, G. C. (1972) Relative growth rate. In: *The Quantitative Analysis of Plant Growth* (ed. D. J.

Anderson, P. Greigh-Smith and F. A. Pitelka), pp. 246-254. Blackwell Scientific Publications, Oxford, U.K.

FAO (1995) Forest products 1991-1995. FAO Forestry series No. 30, FAO statistic series No. 137, FAO, Rome, 57 pp.

Galinski, W., Goosens, R., Ceulemans, R. and Impens, I. (1991) The wood productivity of two poplar clones (*Populus trichocarpa x Populus deltoides*) as affected by stocking and age. *Biomass Bioenergy,* **1,** 233-239.

Gardner, S. D. L., Taylor, G. and Bosac, C. (1995) Leaf growth of hybrid poplar following exposure to elevated CO_2. *New Phytologist,* **131,** 81-90.

Hättenschwiler, S. and Körner, S. (1998) Biomass allocation and canopy development in spruce model ecosystems under elevated CO_2 and increased N deposition. *Oecologia,* **113,** 104-114.

Hendrey, G. R., Lewin, K. F. and Nagy, J. (1993) Free air carbon dioxide enrichment: development, progress, results. *Vegetatio,* **104/105,** 17-31.

Idso, S. B., Kimball, B. A. and Allen, S. G. (1991) CO_2 enrichment of sour orange trees: 2.5 years into a long-term experiment. *Plant, Cell and Environment,* **14,** 351-353.

Isebrands, J. G. and Nelson, N. D. (1983) Distribution of ^{14}C-labeled photosynthates within intensively cultured *Populus* clones during the establishment year. *Physiologia Plantarium,* **59,** 9-18.

Jach, M. E. and Ceulemans, R. (1997) Impact of elevated CO_2 on physiology and needle morphology of Scots pine (*Pinus sylvestris*) seedlings. In: *Impacts of Global Change on Tree Physiology and Forest Ecosystems* (ed. G. M. J. Mohren, K. Kramer and S. Sabaté), pp. 57-73. Kluwer Academic Publishers, Dordrect.

Janous, D., Dvorák, V., Oplustilová, M. and Kalina, J. (1996) Chamber effects and responses of trees in the experiment using open top chambers. *Journal of Plant Physiology,* **148,** 332-338.

Janssens, I. A., Mousseau, M. and Ceulemans, R. (2000) Crop ecosystem responses to global climate change: tree crops. In: *Climate Change and Global Crop Productivity* (ed. K. R. Reddy and H. F. Hodges). CAB International, Wallingford, UK.

Kalina, J. and Ceulemans, R. (1997) Clonal differences in the response of dark and light reactions of photosynthesis to elevated atmospheric CO_2 in poplar. *Photosynthetica,* **33,** 51-61.

Kirschbaum, M. U. F. and Fischlin, A. (1996) Climate change impacts on forests. In: *Climate Change 1995 Impacts, Adaptations and Mitigation of Climate Change: Scientific-Technical Analysis* (ed. R. T. Watson, M. C. Zinyowera and R. H. Ross), pp. 99-125. Cambridge University Press, Cambridge.

Kvèt, J., Ondok, J. P., Necas, J. and Jarvis, P. J. (1971) Methods of growth analysis. In: *Plant Photosynthetic Production* (ed. Z. Sesták, Z. Catsky and P. J. Jarvis), pp. 343-391. Dr. W. Junk N.V. Publishers, The Hague.

Larson, P. R., Dickson, R. E. and Isebrands, J. G. (1976) Some physiological applications for intensive culture. In: *Intensive plantation culture: five years research*, pp. 10-18. USDA For. Serv. Gen. Tech. Rep. NC-21,

Lee, H. S. J., Overdieck, D. and Jarvis, P. G. (1998) Biomass, growth and carbon allocation. In: *European Forests and Global Change: the Likely Impacts of Rising CO_2 and Temperature* (ed. P. G. Jarvis), pp. 126-191. Cambridge University Press, Cambridge.

Lewin, K. F., Hendrey, G. R., Nagy, J. and LaMorte, R. L. (1994) Design and application of a free-air carbon dioxide enrichment facility. *Agricultural and Forest Meteorology,* **70,** 15-29.

Mooney, H. A., Canadell, J., Chapin, F. S., Ehleringer, J., Körner, C., McMurtrie, R., Parton, W. J.,

Pitelka, L. and Schulze, E.-D. (1999) The terrestrial biosphere and global change: Ecosystem physiology responses to global change. In: *Implications of Global Change for Natural and Managed Ecosystems: A Synthesis of GCTE and Related Research* (ed. B. H. Walker, J. Canadell and J. S. I. Ingram). Cambridge University Press, Cambridge.

Mousseau, M. and Enoch, Z. H. (1989) Carbon dioxide enrichment reduces shoot growth in sweet chestnut seedlings (*Castanea sativa* Mill.). *Plant, Cell and Environment,* **12,** 927-934.

Norby, R. J., Wullschleger, S. D., Gunderson, C. A., Johnson, D. W. and Ceulemans, R. (1999) Tree responses to rising CO_2 in field experiments: implications for the future forest. *Plant, Cell and Environment,* **22,** 683-714.

Oechel, W. C. and Strain, B. R. (1985) Native species responses to increased atmospheric carbon dioxide concentration. In: *Direct Effects of Increasing Carbon Dioxide on Vegetation* (ed. B. R. Strain and J. D. Cure), pp. 117-154. Department of Energy, Office of Basic Energy Sciences, Carbon Dioxide Research Division, Springfield, VA, Washington D.C.

Pontailler, J. Y., Ceulemans, R., Guittet, J. and Mau, F. (1997) Linear and non-linear functions of volume index to estimate woody biomass in high density young poplar stands. *Annual Science of Forestry,* **54,** 335-345.

Pregitzer, K. S., Zak, D. R., Curtis, P. S., Kubiske, M. E., Teeri, J. A. and Vogel, C. S. (1995) Atmospheric CO_2, soil nitrogen and turnover of fine roots. *New Phytologist,* **129,** 579-585.

Radoglou, K. M. and Jarvis, P. G. (1990) Effects of CO_2 enrichment on four poplar clones. I. Growth and leaf anatomy. *Annals of Botany,* **65,** 617-626.

Remphrey, W. R. and Powell, G. R. (1985) Crown architecture of *Larix laricina* saplings: sylleptic branching on the main stem. *Canadian Journal of Botany,* **63,** 1296-1302.

Saxe, H., Ellsworth, D. S. and Heath, J. (1998) Tansley Review No. 98. Tree and forest functioning in an enriched CO_2 atmosphere. *New Phytologist,* **139,** 395-436.

Scarascia-Mugnozza, G. E., Ceulemans, R., Heilman, P. E., Isebrands, J. G., Stettler, R. F. and Hinckley, T. M. (1997) Production physiology and morphology of *Populus* species and their hybrids grown under short rotation. II. Biomass components and harvest index of hybrid and parental species clones. *Canadian Journal of Forest Research,* **27,** 285-294.

Scarascia-Mugnozza, G., DeAngelis, P., Sabatti, M., Calfapietra, C., Ceulemans, R., Peressotti, A. and Miglietta, F. (2000) A FACE experiment on a short-rotation, intensive poplar plantation: objective and experimental set-up of POPFACE. In: *Terrestrial Ecosystem Research in Europe: Successes, Challenges and Policy. Final Conference of the Terrestrial Ecosystem Research Initiative-Concerted Action (TERICA)* (ed. M.A. Sutton, J.M. Moreno, W van der Putten, and S. Struwe), pp. 136-140. Ecosystem Research Report, European Commission, Luxembourg.

Schimel, D., Alves, D., Enting, D., Heimann, M. and Joos, F. (1996) Radiative forcing of climate change. In: *Climate Change 1995: the Science of Climate Change* (ed. J. T. Houghton, L. G. M. Filho, B. A. Callander, N. Harris, A. Kattenberg and K. Maskell), pp. 65-132. Cambridge University Press, Cambridge.

Taylor, G. and Frost, D. L. (1992) Impact of gaseous air pollution on leaf growth of hybrid poplar. *Forest Ecology and Management,* **51,** 151-162.

Ward, J. K. and Strain, B. R. (1999) Elevated CO_2 studies: past, present and future. *Tree Physiology,* **19,** 211-220.

Webber, A. N., Nie, G.-Y. and Long, S. P. (1994) Acclimation of photosynthetic proteins to rising atmospheric CO_2. *Photosynhtesis Research,* **39,** 413-425.

Wullschleger, S. D., Post, W. M. and King, A. W. (1995) On the potential for a CO_2 fertilization effect in forests: estimates of the biotic growth factor based on 58 controlled-exposure studies. In: *Biotic Feedbacks in the Global Climatic System* (ed. G. M. Woodwell and F. T. Mackenzie), pp. 85-107. Oxford University Press, New York.

BIOTIC INTERACTIONS

WOLFRAM BEYSCHLAG
Universität Bielefeld, Lehrstuhl für experimentelle Ökologie und Ökosystembiologi W4-107, Universitätstr. 25, D 33615 Bielefeld, Fax: +49 521 106 6038
E-mail: w.beyschlag@biologie.uni-bielefeld.de

Key words: ecophysiology, biotic interactions, hydraulic lift, forest decline, leaf cutting ants

Abstract: Using three examples covering the whole range of biotic interactions (hydraulic lift in desert shrubs, forest decline in middle European spruce stands and leaf cutting ants in the neotropical rainforests), it is shown, that ecophysiological research can provide valuable information for the understanding of the mechanisms behind population dynamics and particularly competition. It is emphasized that such interactions have mostly been neglected during the past but will increasingly move into the focus of the plant ecophysiology of the 21. century.

1. INTRODUCTION

During the last decades the new research objective "plant ecophysiology" became well established within the context of the botanical disciplines. Based on the results of plant physiology, plant ecophysiologists try to understand the causal background of existence and niche formation of plant species under natural field conditions. The development of high quality field gas exchange measurement systems in the late 1960s and the increasing availability of field suitable chlorophyll fluorometers as well as mathematical simulation models in the 1990s stimulated a vast number of rather interesting studies on the effect of the various abiotic habitat (and particularly stress) conditions on the behavior of plant organs and/or whole plant individuals (for reviews see Schulze and Caldwell, 1994).

Of course, plant behavior under natural field conditions can not fully be explained solely by the effects of abiotic habitat conditions on isolated plants, because most of these conditions are considerably dependent on biotic influences (including mutualistic, competitive and parasitic interactions) from other members of the community.

S. Huttunen et al. (eds.),
Trends in European Forest Tree Physiology Research, Cost Action E6: EUROSILVA, 197-205

Consequently, based on the enormous available knowledge about the behavior of single plants, plant ecophysiologists now increasingly expand their activities towards the study of biotic interactions and which includes the development of links to population biology and systems ecology. Thus, plant ecophysiology becomes more and more a connecting element within botany and ecology.

The present chapter presents three examples from this new direction of plant ecophysiology. All three interactions have in common that they were either not known at al or that their role for the ecosystem was greatly underestimated. thus indicating that this new field of ecophysiological research is much larger than expected.

2. HYDRAULIC LIFT

The term "hydraulic lift" was coined by Richards and Caldwell (1987) and describes "the process of water movement from relatively moist to dry soil layers using plant root systems as a conduit" (Caldwell et al. 1998). In dry habitats, plants normally take up water in the deeper soil layers close to the ground water table. Under normal daylight conditions this water moves passively upward within the plant along an osmotic gradient between the roots in the moist soil and the atmosphere and is finally transpired by the aboveground plant organs. When transpiration ceases at night, the prevailing gradient would be from the roots in the moist soil layers to the relatively dry upper soil layers. Thus, water would be lifted from deeper to upper soil layers using the root system as a bridge. The water is then released via the fine roots present in these layers (Fig. 1).

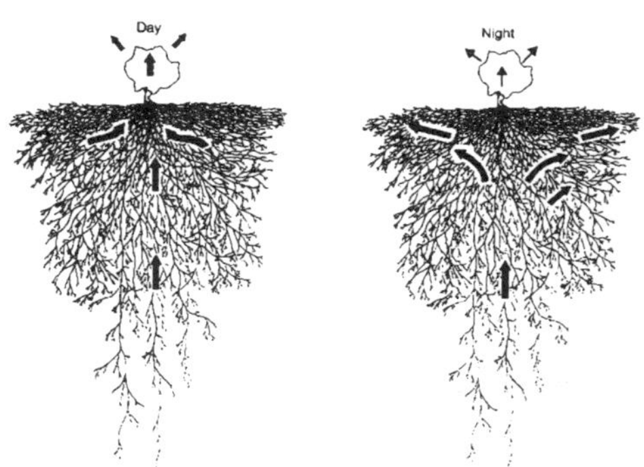

Figure 1. Pattern of water movement in the plant according to the hydraulic lift hypothesis during day and night periods (from Caldwell 1988).

There is plenty of evidence for this phenomenon including the results of continuous soil moisture measurements, direct transpiration measurements of isolated fine roots from the top soil layer (Caldwell et al. 1991) as well as from deuterium labelling experiments (for review see Caldwell et al. 1998). Studies on the quantity of overnight lifted water revealed values up to approx. 33% of daily total evapotranspiration (e.g. Richards and Caldwell (1987)). Hydraulic lift seems to be a rather widespread phenomenon which up to now has been reported for 19 woody taxa as well as for 8 species of herbs and grasses

(Caldwell et al. 1998). The water lifted at night can be reabsorbed during the following day by the same fine root system in the top soil layers through which it was released. However, it may also be possible that root systems of neighbouring shallow rooted plant species which lack direct contact to the ground water benefit from this lifted water. First evidence for this hypothesis was found by Caldwell and Richards (1989) in a deuterium labelling experiment (Fig. 2).

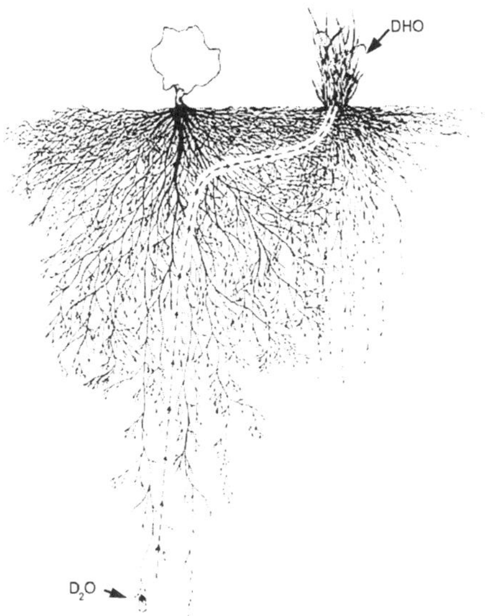

Figure 2. Deuterium labelling experiment designed to determine if water absorbed by deep roots of *Artemisia tridentata* would appear in the stem water of neighbouring Agropyron desertorum tussock grasses (slightly modified from Caldwell et al. 1991).

The magnitude of water utilization by neighbours was analyzed in a study of Dawson (1993) with *Acer saccharum* as lifting species. Depending on the species, neighbouring plants were shown to utilize proportions between 3 and 60% of the lifted water.

The question arises why would a plant give its water away to the dry soil or even share it with its neighbours? (1) Since a considerable part of the released water can be reacquired by the same plant during the next day, hydraulic lift provides an excellent way for temporary external water storage. Having this additional water available during the following day, increases the time where the stomata of the lifting plant can be kept open and, therefore, increases the daily carbon gain. Caldwell and Richards (1989) could show for *Artemisia* shrubs, that hydraulic lift increased the daily transpiration rate up to 50%. This is particularly important for shrubs. While trees are typically tall and voluminous enough to store a considerable amount of water within their own biomass (i.e. out of reach of competitors), the smaller shrubs need much more to rely on external water storage. In both life forms, however, the diurnal pattern of using the stored water is rather similar i.e. refilling the storage at night and using the stored water during the morning hours. (2) Hydraulic lift facilitates nutrient acquisition of the fine roots in the top soil layers, where nutrient concentrations are normally higher than in deeper zones. This is explained by the well known fact that mineralization rates, nutrient mobility, mycorrhizal activity and longevity of fine roots and root hairs are higher in humid than in dry soil. (3) An intact vegetation cover consisting of shallow rooted species around the water lifting plant continously protects the top soil from complete dryout and overheating and, therefore, enhances all of the above advantages in a way that the costs for the water lifting species of sharing nutrients and water seem to be overcompensated.

3. PARTIAL SHEDDING OF FOLIAGE INCREASES BIOMASS PRODUCTION IN DENSE FOREST STANDS

Needle chlorosis and shedding of older needles are the most prevalent symptoms of conifer forest decline, a phenomenon observed within the last 20 years throughout the entire northern hemisphere. Most of the decline symptoms could be successfully related to various air pollutants (Schulze et al. 1989a) However, under moderate stress conditions and on a short term basis, the loss of older needle age classes does not seem to be as deleterious as might be expected from the visual appearance of the tree crowns. There are reports that an adult spruce tree may temporarily lose up to 20 % of its crown without a remarkable negative effect on its wood production (Mitscherlich and von Gadow 1968, Gruber 1990) and foresters frequently report that wood production of the trees even slightly increases during the first years of decline.

Photosynthetic capacity of spruce needles decreases with age (Lange et al. 1989a) but they still maintain considerably high transpiration rates (Lange et al. 1989b). So the water use efficiency of a tree would certainly benefit from the loss of older needle age classes. In a situation where a tree grows in a dense forest canopy, where major parts of its crown are shaded by neighbouring trees, this needle loss may have an even more beneficial effect on overall tree photosynthesis. Under these conditions thinning of typical adult spruce crowns (i.e. with hanging second order branches) due to shedding of older needles will possibly lead to a better illumination of the remaining young and photosynthetically highly active needles. So it can be hypothesized that this effect may lead to a compensation or even an overcompensation of the loss of foliage in terms of carbon assimilation which would explain the observation of the foresters.

In order to test this hypothesis Beyschlag et al. (1994) employed a modified version of a three-dimensional whole-plant canopy photosynthesis model (Ryel et al. 1993, see also Beyschlag and Ryel 1999). After thorough validation of the model, structural and photosynthetic data from young Norway spruce trees (height 140 cm, diameter 100 cm) and climatic data from a typical middle European summer day were used for parametrization. Effects of increasing needle loss (simulated removal of needle age classes, starting always with the oldest needle age) were calculated for isolated single trees as well as for trees growing in simulated monotypic forest stands as a function of planting density.

The top curve in Fig. 3 shows the effect of simulated needle removal on whole tree assimilation (A_{Tree}) for an isolated tree. As expected, both A_{Tree} and E_{Tree} showed a pronounced decline with increasing foliage removal. However, if the target tree is regarded as part of a plantation, among neighboring trees, the result looks quite different. Since the model tree

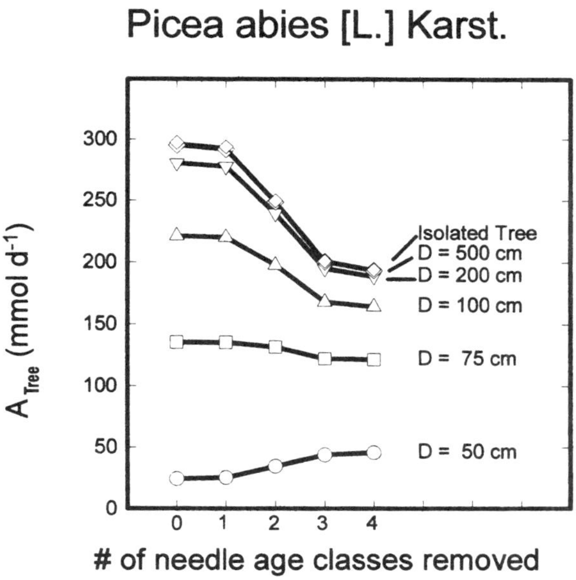

Figure 3. Calculated absolute changes in daily overall tree photosynthesis (A_{Tree}) of the experimental 7-year-old spruce tree plotted as a function of the number of needle age classes removed. The figure compares the situation of an isolated tree with the situation of this tree being the center of a simulated canopy of identical trees. Stand density is varied by changing the distance D between the equidistantly placed stems (from Beyschlag et al. 1994).

had a diameter of 100cm, a distance D of 500cm between the trees resulted in an identical relationship as found for the isolated tree. However, the negative effect of needle removal on A_{Tree} became less pronounced at lower D-values. At D = 75 cm the relative effect of needle removal was minimal and in the extreme overlap situation at D = 50 cm A_{Tree} of the target tree actually increased with increasing needle removal. Thus, the above hypothesis can be accepted. Of course, the total extent of stimulation for a particular tree is determined by its particular crown architecture.

However, using a 7-year-old potted spruce tree as the basis of the simulations, it is clear that the goal of this analysis was to outline a potential mechanism rather than making quantitative statements on its effect on adult forest stands. Care should, therefore, be taken when extrapolating these results to mature forest stands since crown structure as well as needle physiology is known to respond adaptively to the surrounding environment (Tucker et al. 1987, Gruber 1990, Wang and Jarvis 1993).

Finally the question arises: Why do healthy spruce trees keep so many needle age classes while they could achieve similar primary production with much less foliage? Answering this question concerns more than just photosynthesis. (1) There may be negative effects of crown thinning on the water-use-efficiency of trees because of a greater wind exposure of the remaining needles (Whitehead and Jarvis, 1981 but see also Werk et al. 1988). (2) Trees frequently utilize old needles as storage volumes for Nitrogen, carbohydrates and mineral nutrients (e.g. Schulze et al. 1989b), but may also use them as "garbage cans" i.e. load them up with toxic substances and subsequently shed them. (3) There may have been a strong evolutionary pressure to keep foliage density high, because this increases the competitive ability of an individual tree by shading out potential competitors in the understory (e.g. Lieffers and Stadt 1994). Comparing the frequently observable abundant understory vegetation of declining conifer forests with the rather poor situation in healthy stands illustrates the importance of this strategy.

4. A FUNGUS AS HERBIVORE

Restricted to the New World, leaf cutting ants (Genus *Atta* and *Acromyrmex*) are among the most visible and abundant harvesters of foliage in lowland tropical wet forests. Mature colonies consist of several million workers (Hölldobler and Wilson, 1990) with foragers cutting and gathering vegetation along multiple trails. Removal of foliage can be extensive and affect vegetation growth in the vicinity of the colony. The ants do not feed on the harvested biomass (although harvesters can utilize the cell sap extravasating during the cutting process as a major source of carbohydrates) but use it as a substrate for the cultivation of a fungus (*Leucocoprinus gongyliphorus*, Basidiomycetes) which represents their major food source. Thus, the fungus acts as herbivore using the ants just as a vector.

There are direct and indirect influences of leaf cutting ants on the vegetation. Indirect effects include soil nutrient enrichment from nest refuse dumps as well as soil turnover during construction of the nests which can reach a total volume of up to 5 m^3. The ants create understory gaps by clearing all vegetation on top of the colonies. Leaf cutting ants may also influence vegetation succession by destroying numerous flowers of forest tree species and also contribute to the seed dispersal of certain forest plants (Dalling and Wirth 1998).

Leaf cutting ants further have a pronounced effect on the light microclimate within the tree canopy. Since the tropical rainforest is a light limited ecosystem any organism that opens up gaps in the canopy should have a pronounced effect on the surrounding vegetation. In fact, Weigelt and Wirth (unpublished data) could show that the activity of the ants leads to a significant increase in the number of sunflecks in the upper and middle canopy. The importance of sunflecks for the primary production of tropical forest plants and epiphytes has been demonstrated repeatedly (e.g. Pearcy 1990). All this additional light penetrating into the forest canopy due to the cutting activities of the ants is already intercepted in the upper and middle parts of the canopy. In the understory Wirth (1996) could not find any significant ant related increase in light penetration.

In 1997, Wirth et al. presented the first long term study on the harvesting activities of leaf cutting ants which revealed pronounced seasonal patterns, particularly in the ratio between the green and non-green fraction of the harvest (Fig. 4).

The annual harvest of an adult *Atta* colony in Panama amounted to 3,800 m^2 foliage area or respectively 370 kg dry weight. At the level of the particular foraging territory of this colony, this represents a herbivory rate of approx. 12.5%. From a second study of 45 colonies distributed over an area of 100 ha it can be derived that the overall herbivory rate at the ecosystem level amounts to values less than 2.5% (Herz unpublished data).

The most striking direct effect of the ants on the vegetation is certainly their role as generalist herbivores. Up to 80% of the apparent leaf damage in some Panamenian rain forests may be caused by *Atta* species (Wint 1989, Cherrett, 1989). Massive defoliation of trees in natural forests can occasionally induce tree mortality. Possible impacts of large-

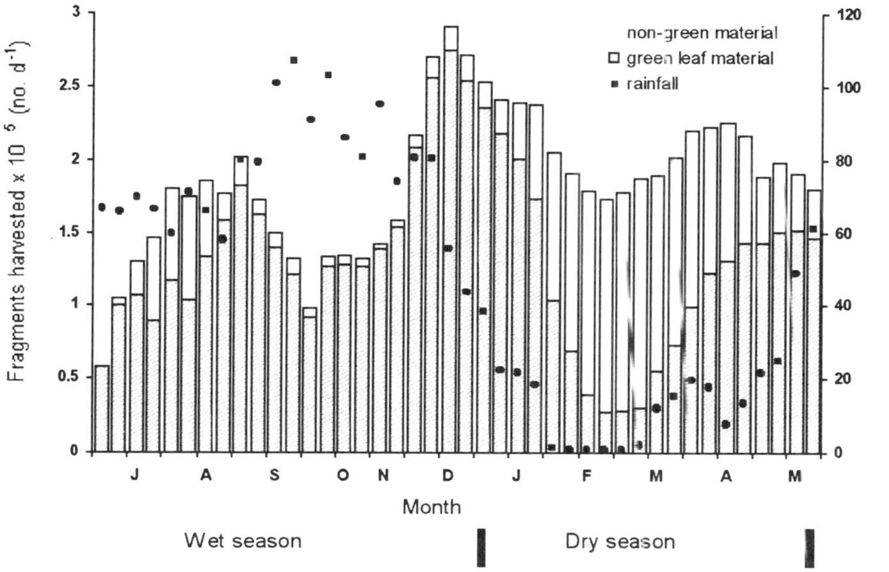

Figure 4. Annual pattern of daily biomass intake harvested by a colony of leaf-cutting ants and weekly precipitation totals in the study area of Barro Colorado Island, Panama, June 1993-June 1994 (from Wirth et al. 1997).

scale herbivory could be shifts in plant species composition or changes in competitive balanees between species. Through these influences, leaf cutting ants may play an important role for development and maintenance of biodiversity (Marquis 1984, 1992).

5. CONCLUSIONS

The above examples cover the whole range of biotic interactions from intra- and interspecific effects to interactions between plants, and other systematic kingdoms like fungi and animals. They further illustrate the fact that biotic interactions and particularly the unexpected and frequently neglected phenomena can sometimes be of essential importance for the existence and survival of a particular plant species. It also becomes clear that working in this exciting new field requires a broader range of scientific abilities than the classical plant ecophysiology. Thus, additional to the botanical skills, profound knowledge in Zoology, Microbiology, Population Biology, Ecosystem theory and Modelling will become increasingly important for the plant ecophysiologist of the 21. century.

ACKNOWLEDGEMENTS

Parts of the work presented here have been supported by the Alexander-von-Humboldt-Foundation, Bonn (Feodor-Lynen-Fellowship to the author) and the Deutsche Forschungsgemeinschaft (SFB 251, Universität Würzburg).

REFERENCES

Beyschlag W, Ryel RJ (1999) Canopy Modelling. In: Pugnaire F, Valladares, F (eds). Handbook of Plant Functional Ecology. New York, Marcel Dekker, pp. 771-804.

Beyschlag W, Ryel RJ, Dietsch C (1994) Shedding of older needle age classes does not necesssarily reduce photosynthetic primary production of Norway spruce. Analysis with a 3-dimensional canopy photosynthesis model. *Trees, Struct. Funct.* **9**: 51-59.

Caldwell MM (1988) Plant root systems and competition. In: Greuter W, Zimmer B (eds) Proceedings of the XIV International Botanical Congress, Königstein, Koelz, pp. 385-404.

Caldwell MM, Richards JH (1989) Hydraulic lift: water efflux from upper roots improves effectiveness of water uptake by deep roots. *Oecologia* **79**: 1-5.

Caldwell MM, Richards JH, Beyschlag W (1991) Hydraulic lift: Ecological implications of water efflux from roots. In: Atkinson D. (ed) Plant Root Systems: Their Effects on Ecosystem Composition. Special Publication Number 10 of the British Ecological Society; Oxford, Blackwell, pp. 423-436.

Caldwell MM, Dawson TE, Richards JH (1998) Hydraulic lift: consequences of water efflux from the roots of plants. *Oecologia* **113**:151-161.

Cherret JM (1989) Leaf-cutting ants. in: Lieth H. Werger MJA (eds) Ecosystems of the World Vol. 14 B, Amsterdam, Elsevier, pp. 473-486.

Dalling JW, Wirth R (1998) Dispersal of *Miconia argentea* seeds by the leaf-cutting ant *Atta colombica. J. Trop. Ecol.* **14**: 705-710.

Dawson TE (1993) Hydraulic lift and water use by plants: implications for water balance, performance and plant-plant interactions. *Oecologia* **107**:13-20.

Gruber F (1990) Aufbau und Anpassungsfähigkeit der Krone von *Picea abies* (L.) Karst. *Flora* **181**:205-242.

Hölldobler B., Wilson EO (1990) The Ants. Cambridge MA (Harvard University Press).

Lange OL, Weikert RM, Wedler M, Gebel J, Heber U (1989a) Photosynthese und Nährstoffversorgung von Fichten aus einem Waldschadensgebiet auf basenarmem Untergrund. *Allg. Forstz.* **3**: 55-65.

Lange, OL, Heber U, Schulze ED, Ziegler H (1989b) Atmospheric pollutants and plant metabolism. In: Schulze ED, Lange OL, Oren R (eds) Forest decline and air pollution. A study of spruce (*Picea abies*) on acid soils. Ecological Studies Vol. 77, Berlin Heidelberg New York, Springer-Verlag, pp. 238-273.

Lieffers VJ, Stadt KJ (1994) Growth of understory *Picea glauca*, *Calamagrostis canadensis*, and *Epilobium angustifolium* in relation to overstory light transmission. *Can. J. For. Res.* **24**: 1193-1198.

Marquis RJ (1984) Leaf herbivores decrease fitness of a tropical plant. *Science* **226**: 537-539.

Marquis RJ (1992) Selective impact of herbivores. In: Fritz RS, Simms EL (eds) Ecology and

Evolution of Plant Resistance, Chicago, University of Chicago Press, pp. 301-325.

Mitscherlich G, Gadow K von (1968) Über den Zuwachsverlust bei der Ästung von Waldbäumen. *Allg. Forst Jagdz.* **139**: 135-184.

Pearcy RW (1990) Sunflecks and photosynthesis in plant canopies. *Annu. Rev. Plant Physiol. Plant Mol. Biol.* **41**: 421-453.

Richards JH, Caldwell MM (1987) Hydraulic lift: substantial nocturnal water transport between soil layers by *Artemisia tridentata* roots. *Oecologia* **73**: 486-489.

Ryel RJ, Beyschlag W, Caldwell MM (1993) Foliage orientation and carbon gain in two tussock grasses as assessed with a new whole-plant gas-exchange model. *Funct. Ecol.* **7**: 115-124.

Schulze ED, Caldwell MM (eds) (1994) Ecophysiology of photosynthesis. Ecological Studies Vol. 100, Berlin Heidelberg New York, Springer Verlag.

Schulze ED, Lange OL, Oren R (eds) (1989a) Forest decline and air pollution. A study of spruce (*Picea abies*) on acid soils. Ecological Studies Vol. 77, Berlin Heidelberg New York, Springer-Verlag.

Schulze ED, Oren R, Lange OL (1989b) Nutrient relations of trees in healthy and declining Norway spruce stands. In: Schulze ED, Lange OL, Oren R (eds) Forest decline and air pollution. A study of spruce (*Picea abies*) on acid soils. Ecological Studies Vol. 77, Berlin Heidelberg New York, Springer-Verlag, pp. 392-417.

Tucker GF, Hinckley TM, Leverenz JM, Jiang SM (1987) Adjustments of foliar morphology in the acclimation of understory Pacific silver fir following clearcutting. *For. Ecol. Manage.* **21**: 249-268.

Wang YP, Jarvis PG (1993) Influence of shoot structure on the photosynthesis of Sitka spruce (*Picea sitchensis*). *Funct. Ecol.* **7**: 433-453.

Werk KS, Oren R, Schulze ED, Zimmermann R, Meyer J (1988) Performance of two *Picea abies* (L.) Karst. stands at different stages of decline. III. Canopy transpiration of green trees. *Oecologia* **76**: 519-524.

Whitehead D, Jarvis PG (1981) Coniferous forests and plantations. In: Kozlowski TT (ed) Water deficits and plant growth, Vol. VI, New York, Academic Press.

Wint GRW (1983) Leaf damage in tropical rainforest canopies. in: Sutton SL, Whitmore, TC, Chadwick AC (eds) Tropical Rainforest, Ecology and Management. Oxford (Blackwell Scientific Publishers), pp. 229-239.

Wirth R (1996) Der Einfluß von Blattschneiderameisen der Gattung *Atta* auf einen tropischen halbimmergrünen Regenwald in Panama: Herbivorieraten, Effekte auf das Lichtklima im Bestand und Ökologie der Futtersuche. PhD Dissertation, University of Würzburg.

Wirth R, Beyschlag W, Ryel RJ, Hölldobler B (1997) Annual foraging of the leaf-cutting ant *Atta colombica* in a semideciduous rain forest in Panama. *J. Trop. Ecol.* **13**: 741-757.

THE INFLUENCE OF THE ENVIRONMENT DURING SEXUAL REPRODUCTION ON ADAPTATIONS OF CONIFERS ALONG LATITUDINAL AND ALTITUDINAL GRADIENTS

ØYSTEIN JOHNSEN AND TORE SKRØPPA
Norwegian Forest Research Institute, Høgskoleveien 12, 1432 Ås, Norway.
Tel. 47 64 949000, Fax. 47 64 942980,
E-mail: oystein.johnsen@nisk.no, E-mail: tore.skroppa@nisk.no

Key words: Climatic adaptation, adaptive traits, adaptive plasticity, phenotypic plasticity, environmental preconditioning, *Picea abies*.

Abstract: Climatic adaptation seems to be the most important component in the evolutionary process of boreal forest tree species. Recent evidence has been accumulating that this process may be less clear-cut than earlier assumed. In *Picea abies*, several independent tests have shown that the climate and weather conditions during sexual reproduction influence the adaptive properties of the progenies. The phenomenon is expressed in Norway spruce seed orchards established by moving parent trees, propagated as grafts, from north to south, from high to low elevation, or from outdoor to indoor greenhouse conditions. The progenies exhibit a changed phenology and a delayed development of frost hardiness in the autumn compared to their siblings reproduced in a colder environment. Observations on seedlings from seeds harvested in northern stands of provenances transferred from the south to the north, show that they have tuned their photoperiodic responses towards that the of the local provenances. Results with other conifer species indicate that they express similar effects. A possible explanation for the phenomenon is the existence of a regulatory mechanism, sensing the changes in temperature and/or photoperiod at some stages during the reproductive process in the female flower, and then modifying the expression of genes controlling adaptive traits in the progeny. Regardless of the type of mechanisms involved, these observations have changed our understanding of ecotype formation in Norway spruce, and given us some ideas why this species

S. Huttunen et al. (eds.),
Trends in European Forest Tree Physiology Research, Cost Action E6: EUROSILVA, 207-221

express strong clinal variation in combination with a large within population variation for the same adaptive traits. The maternal influence will often alter the phenotypic performance of the progenies in the same direction as natural selection. It may be one of the reasons why some conifer species under boreal conditions seem so highly adaptable.

1. ADAPTATION

Adaptedness is the degree to which an organism is able to live and reproduce in a given set of environments. Adaptation is the process of becoming adapted or more adapted (Dobzhansky, 1968a,b), the process by which a population responds to a given environment by different forms of beneficial genetic changes. An adaptive trait is an aspect of the developmental pattern of an organism that enables or enhances the probability of that organism to survive and reproduce (Dobzhansky, 1956, 1968a). Turesson (1922, 1925) introduced the ecotype concept, that plant populations will express adaptations to the environments from which they originated, even when cultivated in common garden experiments. Preadapted genotypes, which already from birth carry the genotypic constitution to support life in certain habitats, will have a higher probability to survive and breed than lesser-fit genotypes in these habitats. This process of natural selection is believed, among neo-Darwinian supporters, to entirely account for the observed differences among ecotypes in common garden experiments, especially when the species have been present for many generations in the habitats. The principles developed from observations of herbaceous annuals and biennials have been extended to forest trees. Langlet (1936) demonstrated that phenotypic variation among natural populations of *Pinus sylvestris* was arranged along clines that paralleled Scandinavian climates. Common garden plantations have since then provided data for studies of the variability in quantitative adaptive traits among and within wild populations of forest tree species. The distribution of variability among and within populations, its patterns across the landscape and relationships to environmental variables have been used to interpret the ecological basis for genetic differentiation and evolution of the species (Rehfeldt, 1988, 1989, 1995; Campbell et al., 1989; Hannerz, 1998). The studies have shown that climatic adaptation seems to be the most important component in the evolutionary process of temperate and boreal conifers. Most species show substantial variability in adaptive traits both within and among populations, even at the climatic margin of their range (Savolainen, 1997). The main evolutionary forces influencing the variability patterns are natural selection, genetic drift, gene flow and mutations, and phenotypic plasticity will affect the outcome of natural selection by increasing the within population variation and counteract the population differentiation (Eriksson, 1998a). The many types of variability patterns shown by different species and their degree of adaptedness to the sites suggest that species occupying the same environment may be influenced by different evolutionary forces; see examples discussed by Eriksson (1998b). Several selective forces may even operate on the same population, some simultaneously and some at different life stages or during different seasons of the year (Eriksson, 1998a). The variability patterns are therefore most likely under the influence of several evolutionary factors interacting in rather complex ways. This conclusion is supported by the rather high levels of genetic variation in survival and annual growth rhythm traits maintained in the boreal forest (Savolainen, 1997), in spite of a strong directional selection against the traits.

2. GROWTH VERSUS FROST HARDINESS

Temperature is probably the major selecting agent causing population differentiation along latitudinal and altitudinal gradients. A heritable trait is often interpreted as being adaptive if it fits closely these clines, which frequently are believed to be the product of natural selection. Temperature may operate indirectly by selections against genotypes with a delayed onset of growth in the spring and premature dormancy in the autumn. These genotypes will under-utilise a given site's resources, and may be categorised as having insufficient growth capacity (Heide, 1985), and could possibly be outcompeted by genotypes expressing longer-lasting growth within the growing seasons (Rehfeldt et al., 1999). On the other hand, low temperature extremes may affect mortality directly by killing genotypes with premature onset of vegetative growth in the spring and delayed growth cessation in the autumn, i.e. plants will have insufficient ability to survive at freezing temperatures (Heide, 1985). In general, however, one cannot assume that native populations are optimally adapted to minimise frost damage and yet utilise the full growing season in the face of varying temperatures from year-to-year and within years (see Rehfeldt et al., 1999 and Rehfeldt, 2000 for a very good discussion on this matter). The relationships between frost hardiness and growth depend on many factors: how many generations the species has been present, genetic correlations among growth, phenology, dormancy and timing of hardening and dehardening, genetic variation in these adaptive traits, the selection pressure from temperature extremes, the magnitude of gene flow among populations, the degree of phenotypic plasticity, and the degree of inter- and intra-specific competitions.

A large number of studies have shown that temperate and boreal conifer species express a rather close phenotypic relationship between bud phenology and frost hardiness among populations (e.g. Rehfeldt, 1979; Johnsen & Apeland, 1988; Hurme et al., 1997). The length of the seasonal frost-free period within a provenance region often correlates well with the duration of growth of the provenance in common garden experiments (e.g. Rehfeldt, 1988, 1989). Thus, at the provenance level, fairly high correlations exist between bud phenology and hardening/dehardening during autumn and spring. However, when similar studies are made with families within populations, the same strong relationships do not always occur. Weak or no relationships between bud-set and frost hardiness in the autumn have been found among full-sib- or open-pollinated families from seed orchards or natural stands of Norway spruce (Johnsen & Apeland, 1988; Skrøppa, 1991; Johnsen & Skrøppa, 2000) and within populations of Douglas fir (Aitken et al., 1996). Nevertheless, a physiological and genetic conflict exists between being in a state of active growth and being frost hardy (Levitt, 1980; Olsen et al., 1997). Growth cessation and frost hardiness development could still be regulated by some independently segregating genes, producing families or clones that combine late growth cessation with rapid development of frost hardiness after growth has terminated. In spring the situation seems less complicated. Quite high correlations between frost hardiness and time of flushing have been found also among families within populations (e.g. Dietrichson, 1993; Aitken & Adams, 1997).

3. FAST POPULATION DIFFERENTIATION IN CONIFERS

After the last ice age conifer species migrated at an average rate of 10–40 km per century, with maximum migration rate of 200 km per century for white spruce (see Schwartz, 1991 and references therein). Norway spruce may serve as an example. It is a relatively young species in the Scandinavian Peninsula. It migrated from the Russian refugium, advanced through the forest area at the present border between Russian and Finland and passed through Finland and northern Sweden into Norway in the period 3500-500 BC (Moe, 1970; Schmidt-Vogt, 1977). Some new findings have suggested an earlier establishment of the species in some areas (Kullmann, 1995). The border area between Sweden and Central Norway was crossed approx. 500 BC. In the southern part of Norway the earliest immigration took place in the area between the present border between Norway and Sweden in the years 400-200 BC. Eight hundred years later the species was established in the lowland south-eastern area. The migration into the valleys to its present altitudinal boundaries was not completed until the period AD 1000-1300. Thus, the species is young in Norway and has had a short time to adapt to local climatic conditions, taking the long generation interval (minimum 30 years) into account. Nevertheless, the Norwegian provenances show remarkably distinct and strong clinal variation in both bud-set (Kohmann, 1996) and frost hardiness in the autumn (Dæhlen et al., 1995; Fig. 1). This provenance differentiation has thus evolved in less than 100 generations of natural selection in Norway, and a large within-stand variation in e.g. frost hardiness during autumn (Skrøppa, 1991; Johnsen & Østreng, 1994, Fig. 2) has been maintained.

Seedlings from seed harvested in planted stands with transferred provenances perform differently from seedlings of their original provenance. In Denmark, seedlings from seeds harvested in first generation stands of *Picea sitchensis* have better survival and are less injured by climate than seedlings of the same provenances from direct seed import from North America (Nielsen, 1994). Similar experiences are found with *Abies normaniana* in Denmark (Nielsen, 1999). In Scotland, seedlings from first generation of stands of *Abies grandis* perform far better than direct import of the same provenance (Ennos et al., 1998). In Germany, Douglas fir seedlings from German stands have performed better than the other provenances tested (Kleinschmit et al., 1974). Thus, a fast formation of seemingly adapted populations (from one generation to the next) may to occur in several conifer species.

Norway spruce provenances have been extensively transferred from southern to northern latitudes. Seeds have been collected in stands in West Norway (lat. 61°N) and in Central Norway (lat. 64°N) being established with seedlings of origin Harz, Germany (lat. 52°N). In one of the stands included in the study, the mean diameter of 15 trees of Harz origin was 24% larger than that of the 15 trees sampled in the adjacent stand of local

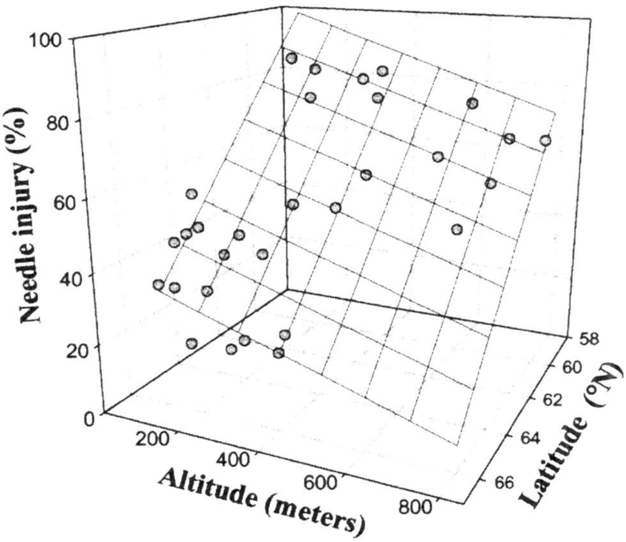

Figure 1. The relationship between needle injury (% of needles brown or discoloured) after freezing tests from -6 – -15°C of Norway spruce (*Picea abies*) provenances and their origin (latitude and altitude) within Norway. Provenances express increased hardiness during cold acclimation from south to north and from low to high elevation, and 87 % of the provenance variation could be "explained" by latitude and altitude of provenance origins (data from Dæhlen et al., 1995).

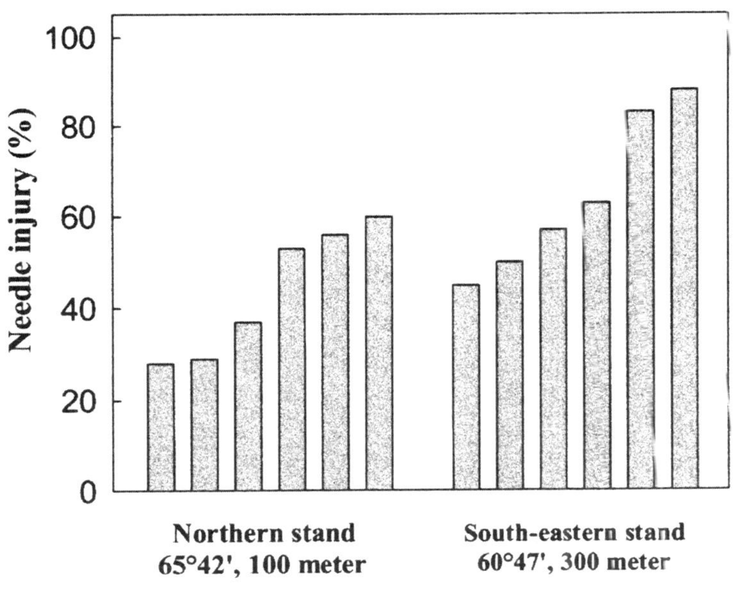

Figure 2. The differences between 6 half-sib families (open pollination) of *Picea abies* in needle injury in a northern stand and between 6 full-sib families (control crosses) in a south-eastern stand in Norway (latitudes and altitudes indicated). The differences between the least and the most hardy families are comparable to a provenance difference of 3-4° in latitudes (data from Skrøppa, 1991 and Johnsen and Østreng, 1994).

provenance. In spite of this difference among the mother trees, only minor difference between the northern Harz stand progenies and the native northern provenances were found in timing of bud-set. Large differences were however observed between the progenies from the northern located Harz stand and the original Harz provenance (Skrøppa & Kohmann, 1997; Fig. 3). Tests in the autumn of 1999 of seedlings from new collections of seeds from several other planted stands of southern imported provenances planted in Norway have confirmed the results; a rapid phenotypic change occurs in Norway spruce (Skrøppa & Breyholtz, unpublished). The authors argue that natural selection and pollen flow from the natural northern stands could not be the only explanation for the fast adaptation to local condition from one generation to the next. This is due to the fact that mortality was low in the northern Harz stand. The seed parents were 24% larger in diameter than comparable trees in the adjacent northern control stand, and a large variability among the open pollinated families from the northern Harz stand was present (Skrøppa and Kohmann, 1997).

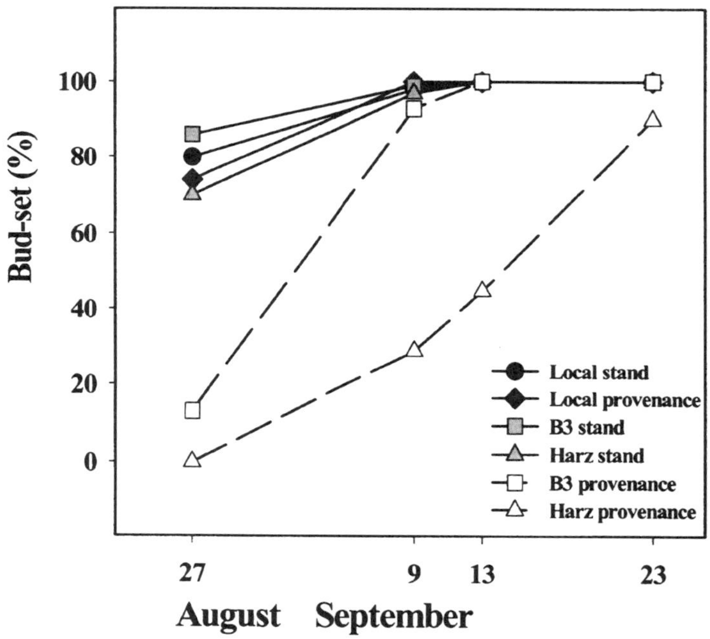

Figure 3. Percent of seedlings with a terminal bud (bud-set) the autumn of 1996 at Stiklestad nursery (63°45'N). Seeds were collected from field trial stands located in the central part of Norway (63°45'N, 400 m elevation). The planted parents originated from a local stand (63°45', 400 m), a stand with a south-eastern B3-provenance in Norway (60°30'N, 300 m elevation), and a planted stand with the German provenance Harz (51°45'N, 700 m elevation). Seeds produced and collected in the original native geographic areas were used as provenance controls. The progenies from the northern-planted stands had attained a bud phenology tuned towards the northern stand and provenance after one generation only (data from Skrøppa and Kohmann, 1997).

4. INFLUENCE OF ENVIRONMENT DURING SEXUAL REPRODUCTION

Another phenomenon may increase our understanding of the observed variability patterns and fast population differentiation in boreal conifers. Our recent findings show that climate and weather condition experienced by the parents during sexual reproduction influence the progeny in Norway spruce (Johnsen et al., 1996; see reviews by Johnsen & Skrøppa, 1996, 1997; Skrøppa & Johnsen, 2000). The phenomenon is expressed in seed orchards established by moving parent trees, propagated as grafts, from north to south, from high to low elevation, or from outdoor to indoor greenhouse conditions (translocated seed orchards). The seeds produced under warm conditions give rise to seedlings with a delayed dehardening and flushing in spring, an extended growth period and delayed growth cessation during summer, and later development of frost hardiness in the autumn than seedlings from seeds of the same parents reproduced under colder conditions. The after effect on progenies endures for at least 17 years after germination (Edvardsen et al., 1996), we suspect the changed performance to be present throughout the lifetime of the trees, although no information for this is available. Confirming result were obtained with controlled cross families, making identical crosses in contrasting environment in outside seed orchards located at latitudes 62 and 66°N in Finland (Skrøppa et al., 1994), and inside a heated greenhouse and in a nearby seed orchard in Norway (Johnsen et al., 1995). The warm conditions in the south or inside the greenhouse gave rise to seedlings with a delayed development of frost hardiness in the autumn than their full-sibs produced in the northern seed orchard or the colder outside seed orchard. It was of great interest to verify whether this was an after-effect of the male, the female or the general parental environment. Moreover, it was important to know whether the after-effect was related to the differences in environmental condition during the year when the seeds were produced only, or rather reflected an accumulated historical effect of many years growth under different temperature conditions. Experiments were conducted with potted grafts, which had received identical growth conditions until the year of sexual reproduction. There was no evidence for an altered autumn hardiness of the progenies that was related to difference in temperature and photoperiod given during male meiosis, microsporogenesis and pollen maturation. However, differences in progeny frost hardiness could be related to the female reproductive environment (Johnsen et al., 1996). It was found that an early start of the reproductive process inside the greenhouse produced progenies, which were less autumn frost hardy than their full-sibs from crosses inside the greenhouse in late spring. The most frost hardy progenies originated from the outdoor crossing environment (Fig. 4). The existence of a regulatory mechanism is suspected, which senses the changes in temperature and/or photoperiod at some stage during the reproductive process in the female flowers (Johnsen et al., 1996), and then modifies the expression of genes controlling adaptive traits in the progeny.

Practical consequences of these effects have been observed both in field trials and in plantations. They resemble our experiences of provenance transfers of Norway spruce. The progenies from translocated seed orchards have generally shown superior height growth, they have suffered in early autumn frost events and have benefited in late spring frost due to a later bud flushing compared with trees from the native provenance (Johnsen et al., 1989; Skrøppa unpublished). Seed year effects within single seed orchards have also been reported. In two seed orchards, seedlings from seeds produced in a cool seed year (1987) formed terminal buds at shorter nights both in growth chambers and in nurseries than did seedlings from a warm seed year (Kohmann & Johnsen, 1994; Fig. 5). These effects of the female reproductive environment thus represent a non-Mendelian phenomenon of which phenology and hardiness of progenies are tuned towards the characteristics typical for the genotypes native to the environment where reproduction takes place. Reproduction in cold climates under long day conditions advances bud-set as well as frost hardiness development of the progenies, both hardening in the autumn and dehardening in the spring. Reproduction in warmer climates under shorter days delays these processes even if progenies are produced from the same parental material. Thus, provenance variation is not only influenced by classical evolutionary forces, but also by directional maternal effects which deterministically influence the progeny according to climatic condition at the sites where the mothers are growing. Similar effects are also expressed to some degree in Scots pine (Lindgren & Wang, 1986; Dormling & Johnsen, 1992; Andersson, 1994; Lindgren & Wei, 1994), white spruce (Bigras & Beaulieu, 1997; Stoehr et al., 1998) and *Larix spp.* (Greenwood & Hutchison, 1996).

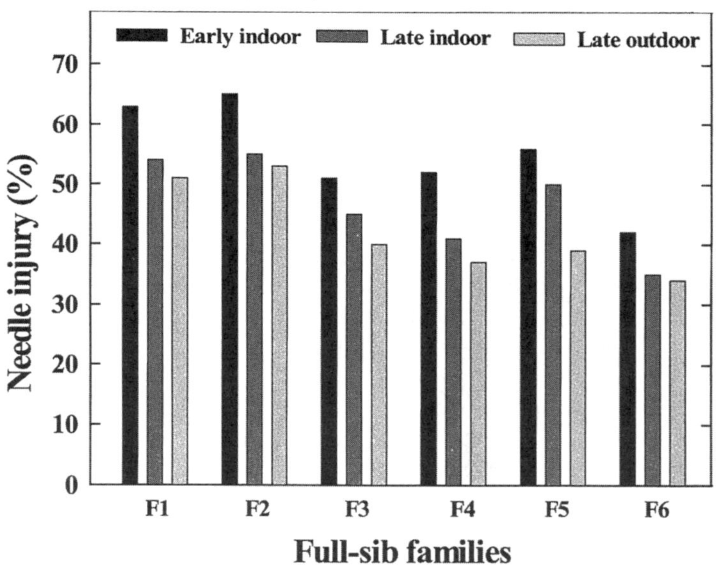

Figure 4. Freezing injury of six unrelated full-sib families as affected by the female reproductive environment. Pollination in the early indoor crosses (using a heated greenhouse) was performed in March 1993, and the late crosses in May 1993, then all the indoor-growing grafts were moved out of the greenhouse in the middle of June (data from Johnsen et al., 1996).

5. POSSIBLE MECHANISMS

When seeds are produced under non-native conditions, both temperature and photoperiod will be different from normal. An increased heat sum in the spring induces an earlier start of the reproductive process at a shorter photoperiod, and a decreased heat sum will delay

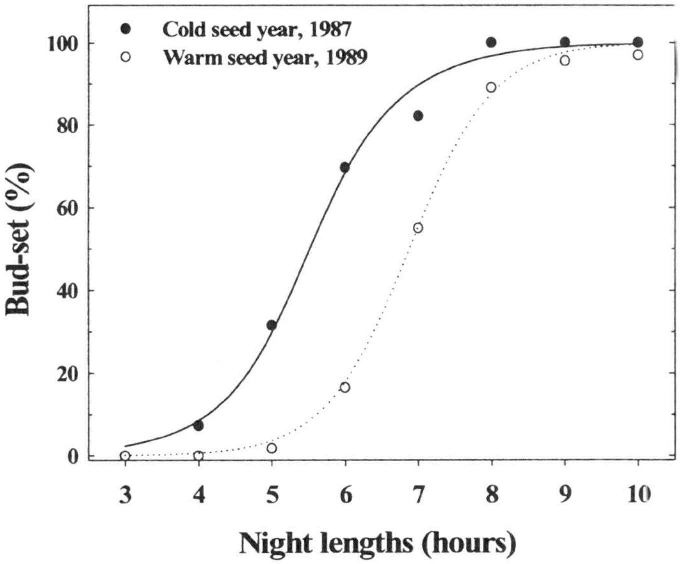

Figure 5. Photoperiodic reaction (terminal bud-set as a function of night length applied) as affected by a warm 1989 and a cold 1987 seed year in Lyngdal seed orchard (Dale compartment). The difference in response is comparable to a provenance difference of 3° latitude (data from Kohmann and Johnsen, 1994).

the reproductive processes to start at a longer photoperiod. Latitudinal transfer will always give longer or shorter photoperiod during the entire reproductive process, but a strict altitudinal transfer will alter photoperiod during stages in the reproductive cycle in an interactive manner. Recent unpublished results (Johnsen, Skrøppa & Dæhlen, 2000) indicate that the after effect of the female reproductive environment is mainly a temperature response, in which heat differences accumulated during the entire reproductive period, from dormant seed cone buds to mature seeds produced, create differences in the progeny performance. Short periods of heat treatments during female meiosis, pollen tube growth, and fertilization is apparently not enough to trigger a visible effect in the progeny. However, we also have some preliminary data indicating that photoperiod may be involved, but this result calls for further confirmations.

The reproductive cycle offers possibilities for selection both during the haploid and the diploid phase. The hypothesis of a gametophytic selection during male meiosis and pollen maturation (meiotic drive) was rejected in two independent experiments (Johnsen et al., 1996). Thus, the process with the greatest potential for selection, due to the enormous amount of pollen grains produced, does not seem to be involved. There are further possibilities for selection in the female flower during pollen tube growth, fertilisation, megaspore degeneration and post-meiotic embryo competition. However, only 3-7 pollen grains can be accommodated in the pollen chamber, and 2-4 embryos are initially present in the developing seed (Owens & Blake, 1985). The potential selection intensity of a directional selection at each of these stages is low, and it has to go in the same direction at each consecutive stage to account for the changes observed. We therefore tend to conclude that the observed results are most likely not due to a directional selection operating during the reproductive cycle.

A second mechanism may be that the influencing environmental factor at some reproductive stage in the female cones regulates the level of gene expression in the progenies and that this potential gene regulation has a long-lasting effect. Genomic imprinting is increasingly being accepted as a fundamental process determining, in ways not predicted by the laws of Mendelian inheritance, whether a particular gene should be expressed or not (Matzke & Matzke, 1993; Fagard & Vaucheret, 2000). Interestingly, Meyer at al. (1992) found that environmental factors influenced 35s promoter methylation of maize A1-gene construct in transgenetic petunia and its coloured phenotype. While blossoms on field-grown plants flowering early in the season were predominantly red, later flowers on the same plants showed weaker coloration. The reduction of the A1-speceific phenotype correlated with methylation of the 35s promoter. Moreover, they found that the stability of pigmentation correlated with the time of seed production. The A1 gene construct was rather insensitive to DNA methylation in progenies from flowers of young parental plants produced early in the season, but became susceptible to methylation within progenies from subsequent later crosses. However, so far we can only speculate that this is a possible cause of the observed influences of the maternal reproductive environment on the phenotypic expression of adaptive traits of Norway spruce progenies. If methylation is involved in the same manner as for vernalization in *Arabidopsis* (Sheldon et al., 1999), the after effect of the female reproductive environment should remain mitotically stable in the somatic cells throughout the lifetime of the trees. The changed methylation will however not be transmitted to the progenies (Michaels & Amasino, 1999; Sheldon et al., 2000), but each generation may tune their progeny performance according to the environmental condition the year the seeds are produced (e.g. Kohman & Johnsen, 1994). The primary genes and the gene products we suspect may be involved are the phytochromes, which seems to play an important role in the regulation of growth and growth cessation and frost hardiness in trees (Olsen et al., 1997; Eriksson, 2000; Clapham et al., 2001).

6. IMPLICATIONS

Regardless of the type of mechanisms involved, the observed influences of the maternal crossing environment on the phenotypic variability of the progenies may be an important component in the climatic adaptation of Norway spruce. Depending on the type of genetic control involved, this force may contribute to the maintenance of a high genetic variability within populations. Its effect on fitness may be similar to that of phenotypic plasticity (Mátyás, 1996; Eriksson, 1998a), disguising the genotypic value and maintaining a larger portion of the genotypes even under a strong selection pressure. It is, however, different in the respect that it operates through the phenotypes of the next generation. Adaptive plasticity (Donohue & Schmitt, 1998) is a proper term for the reported effects. The observed effects can then be a contributing explanation of the strong clinal variation relative to latitude and altitude found for traits that describe the annual growth cycle in *Picea abies* (Hänninen et al., 2001). They must therefore be taken into account when interpreting provenance differences in these and related traits. On the other hand, it may enhance survival of progenies leaving more genotypes to breed. Similarly, they also contribute to the understanding of why trait relationships at the provenance level and among families within populations may be so different, due to different causal factors influencing the phenotypic variability at the two genetic levels (Johnsen & Apeland, 1988; Johnsen, 1989; Skrøppa, 1991; Johnsen & Skrøppa, 2000). The influence of the crossing environment on the phenotypes of the next generation should be taken into account both in tree breeding and in the conservation of forest genetic resources. It can directly influence the phenology of seedlings from the seed orchard crop after a transfer of the parents to a warmer climate, as observed in Norway. Considerations should therefore be made of both the location of seed orchards and where controlled crosses for the generation of breeding material are made. For the conservation activities, its importance will depend on how the mechanism operates and its genetic control, which are yet not known. It should also be accounted for in the models predicting the effects of global warming on both natural and planted boreal forests.

REFERENCES

Aitken, S.N., & Adams, W.T., 1997. Spring cold hardiness under strong genetic control in Oregon populations of *Pseudotsuga menziesii* var *menziesii*. *Canadian Journal of Forest Research* **27**: 1773-1780.

Aitken, S.N., Adams, W.T., Schermann, N. & Fuchigami, L.H., 1996. Family variation for fall cold hardiness in two Washington populations of coastal Douglas-fir (*Pseudtsuga menziesii* var. *menziesii* (Mirb.) Franco). *Forest Ecology and Management* **80** 187-195.

Andersson, B., 1994. After-effect of maternal environment on autumn frost hardiness in *Pinus sylvestris* seedlings in relation to cultivation techniques. *Tree Physiology* **14**: 13-322.

Bigras, F. & Bonlieu, J., 1997. Influence of maternal environment on frost tolerance of Picea glauca progenies. In: Proceedings of the twenty-sixth meeting of the Canadian Tree Improvement Association, Part 2, 91.

Campbell, R. K., Pawuk, W.A. & Harris, A.S., 1989. Microgeographic genetic variation of Sitka spruce in southeastern Alaska. *Canadian Journal of Forest Research* **19**:1004-1013.

Clapham, D., Ekberg, I., Little, C.H.A., & Savolainen, O. 2001. Molecular biology of conifer frost tolerance and potential applications to tree breeding. In: Brigas, F.J. & Colombo, S.J., (eds.). *Conifer cold hardiness*. Kluwer Academic Publishers, Dordrecht, The Netherlands. 187-219.

Dietrichson, J., 1993. Climatic change and Norway spruce breeding. In: Rone, V., (ed.) *Norway spruce provenances and breeding*. Proceeding of the IUFRO S.2.2.-11 symposium, Latvia, Riga. 157-170.

Dobzhansky, T., 1956. What is an adaptive trait? *American Naturalist* **90**: 337-347.

Dobzhansky, T., 1968a. On some fundamental concepts of Darwinian biology. *Evolutionary Biology* **2**: 1-34.

Dobzhansky, T., 1968b. Adaptedness and fitness. In: Lewontin, R.C., (ed). *Population Biology and Evolution*. Syracuse University Press. 109-121.

Donohue, K. & Schmitt, J., 1998. Maternal environmental effects in plants. Adaptive plasticity? In: Mousseau, T.A. & Fox, C.W., (eds). *Maternal effects as adaptations*. Oxford University Press, New York, Oxford, 137-158.

Dormling, I. & Johnsen, Ø., 1992. Effects of the parental environment on full-sib families of *Pinus sylvestris*. *Canadian Journal of Forest Research* **22**: 88-100.

Dæhlen, A.G., Johnsen, Ø. & Kohmann, K., 1995. Autumn frost hardiness in young seedlings of Norway spruce from Norwegian provenances and seed orchards. (Norwegian with English summary). *Research paper of Skogforsk* **1/95**: 1-24.

Edvardsen, Ø.M., Johnsen, Ø. & Dietrichson, J., 1996. Growth rhythm and frost hardiness in northern progeny trials with plants from Lyngdal seed orchard. (Norwegian with English summary). *Research paper of Skogforsk* **9/96**: 1-9.

Ennos, R., Worrell, R. & Malcolm, D.C., 1998. The genetic management of native species in Scotland. *Forestry* **71**: 1-23.

Eriksson, G., 1998a. Evolutionary forces influencing variation among populations of *Pinus sylvestris*. *Silva Fennica* **32**: 173-184.

Eriksson, G., 1998b. Sampling for genetic resources populations in the absence of genetic knowledge. In Turok, J., Collin, E., Demesure, B., Eriksson, G., Kleinschmit, J., Rusanen, M. & Stephan R. (compilers): *Noble Hardwood Network*. Report of the second meeting, 22-25 March 1997. Lourizàn, Spain. International Plant Genetic Resources Institute, Rome, Italy. 61-75.

Eriksson, M. E. 2000. The role of phytochrome A and gibberellins in growth under long and short day conditions. Studies in hybrid aspen. Doctor's dissertation. *Acta Universitatis Agriculturae Sueciae. Silvestria* **164**. ISBN 91-576-5898-6.

Fagard, M. & Vaucheret, H. 2000. (Trans)gene silencing in plants: How many mechanisms? *Annu. Rev. Plant Physiol. Mol. Biol.* **51**: 167-194.

Greenwood, M.S. & Hutchison, K.W., 1996. Genetic aftereffetcs of increased temperature in *Larix*. In: Hom, J., Birdsey, R. & O'Brian, K., (eds.) Proceedings 1995 meeting of the Northern Global Change Program 14-16 March 1995, 56-62.

Hannerz, M., 1998. Genetic and seasonal variation in hardiness and growth rhythm in boreal and temperate conifers – a review and annotated bibliography. *SkogForsk Report* **2**: 1- 140

Heide, O.M., 1985. Physiological aspects of climatic adaptation in plants with special reference to high-latitude environments. In: Kaurin, Å., Junttila, O. & Nilsen. J., (eds.). *Plant production in the north.* Norwegian University Press, Tromsø, 1-22.

Hurme. P., Repo, T., Savolainen, O., Pääkkönen, T., 1997. Climatic adaptation of bud set and frost hardiness in Scots pine (*Pinus sylvestris*). *Canadian Journal of Forest Research* **27**: 716-723.

Hänninen, H., Beuker, E., Johnsen, Ø., Leinonen, I., Murray, M., Sheppard, L. & Skrøppa, T., 2001. Impacts of climate change on cold hardiness. In: Brigas, F.J. & Colombo, S.J., (eds.). *Conifer cold hardiness*. Kluwer Academic Publishers, Dordrecht, The Netherlands. 305-333.

Johnsen, Ø., 1989. Freeze-testing young *Picea abies* plants. A methodological study. *Scandinavian Journal of Forest Research* **4**: 351-367.

Johnsen, Ø. & Apeland, I., 1988. Screening early autumn frost hardiness among progenies from Norway spruce seed orchards. *Silva Fennica* **22**: 203-212.

Johnsen, Ø. & Østreng, G., 1994. Effects of plus tree selection and seed orchard environment on progenies of *Picea abies*. *Canadian Journal of Forest Research* **24**: 32-38.

Johnsen, Ø. & Skrøppa, T., 1996. Adaptive properties of *Picea abies* progenies are influenced by environmental signals during sexual reproduction. *Euphytica* **92**: 67-71.

Johnsen, Ø & Skrøppa, T., 1997. Parents of Norway spruce adjust the performance of their progeny according to changes in climate and weather conditions during female flowering. A review. In: Mohren, G.M.J., Kramer, K. & Sabaté, S., (eds.). *Impacts of Global Change on Tree Physiology and Forest Ecosystems*, Kluwer Academic Publisher, Dordrecht, Boston, London, 159-163.

Johnsen, Ø. & Skrøppa, T., 2000. Provenances and families show different patterns of relationship between bud-set and frost hardiness in *Picea abies*. *Can J. For. Res.* **30**: 1858-1866.

Johnsen, Ø., Dietrichson, J. & Skaret, G., 1989. Phenotypic changes in progenies of northern clones of *Picea abies* (L.) Karst. grown in a southern seed orchard. III. Climatic damage in a progeny trial. *Scandinavian Journal of Forest Research* **4**: 343-350.

Johnsen, Ø., Skrøppa, T., Haug, G., Apeland, I. & Østreng, G., 1995. Sexual reproduction in a greenhouse and reduced autumn frost hardiness of *Picea abies* progenies. *Tree Physiology* **15**: 551-555.

Johnsen, Ø., Skrøppa, T., Junttila, O. & Dæhlen, O.G., 1996. Influence of the female flowering environment on autumn frost-hardiness of *Picea abies* progenies. *Theoretical and Applied Genetics* **92**: 797-802.

Kleinschmit, J., Racz, J., Weissgerber, H., Dietze, W., Dieterich, H. & Dimpflmeier, R., 1974. Ergebnisse aus dem internationalen Douglasien-herkunftversuch von 1970 in der Bundesrepublik Deutschland. *Silvae Genetica* **23**: 167-176.

Kohmann, K., 1996. Night length reactions of Norway spruce plants of different provenances and seed orchards. . (Norwegian with English summary). *Research report of Skogforsk* **15/96**: 1-15.

Kohmann. K. & Johnsen, Ø., 1994. The timing of bud-set in seedlings of *Picea abies* from seed crops of a cool versus a warm summer. *Silvae Genetica* **43**: 328-332.

Kullmann, L., 1995. New and firm evidence for Mid-Holocene appearance of *Picea abies* in the Scandes Mountains, Sweden. *Journal of Ecology* **83**: 439-447.

Langlet, O., 1936. Studier över tallens fysiologiska variabilitet och dess samband med klimatet. *Meddelanden från Statens Skogforsöksanstalt* **29**: 219-470.

Levitt, J., 1980. *Responses of plants to environmental stresses*. Chilling, freezing, and high temperature stresses. Academic Press, New York.

Lindgren, D. & Wang, Q., 1986. Are genetic results influenced by the environment during seed maturation? In: *Proceedings of the IUFRO Conference on Breeding Theory, Progeny Testing and Seed Orchards*. Williamsburg, Virginia, USA, 192-199.

Lindgren, D. & Wei R-P., 1994. Effects of maternal environment on mortality and growth in young *Pinus sylvestris* in field trials. *Tree Physiology* **14**: 323-327.

Mátyás, C., 1996. Climatic adaptation of trees: rediscovering provenance tests. *Euphytica* **92**: 45-54.

Matzke, M. & Matzke, A. J. M., 1993. Genomic imprinting in plants: Parental effects and transinactivation phenomena. *Annual Review of Plant Physiology and Plant Molecular Biology* **44**: 53-76.

Meyer, P., Linn, F., Heidmann, I., Meyer, Z.A.H., Niedenhof, I. & Saedler H. 1992. Endogenous and environmental factors influence 35s promoter methylation of maize a1-gene contruct in transgenic petunia and its colour phenotype. *Molecular & General Genetics* **231**: 345-352.

Michaels, S.D. & Amasino, R.M. 1999. Invited Review: Memories of winter: vernalization and the competence to flower. *Plant Cell and Environment* **23**: 1145-1153.

Moe, D., 1970. The post-glacial immigration of *Picea abies* into Fennoscandia. *Botaniske Notiser* **123**: 61-66.

Nielsen, U. B., 1994. Genetic variation in Sitka Spruce (*Picea sitchensis* (Bong.) Carr.) with respect to height growth, stem form and frost hardiness - investigated on the basis of Danish provenance, progeny and clonal field experiments. (Danish with Eng. summary). *Forskingsserien*. Danish Landscape and Forest Research Institute, Hørsholm, Denmark. No **11**: 1-330.

Nielsen, U. B., 1999. Comparison of Danish first generation or later seed sources with direct imports - Examples from Sitka spruce, Nordmanns fir and Noble fir. *Aktuelt fra skogforskningen* **3/99**: 11.

Olsen, J.E., Junttila, O., Nilsen, J., Eriksson, M.E., Martiniunssen, I., Olsson, O., Sandberg, G. & Moritz, T., 1997. Ectopic expression of oat phytochrome A in hybrid aspen changes critical daylength for growth and prevents cold acclimatization. *Plant Journal* **12**: 1339-1350.

Owens, J. N., & Blake, M.D., 1985. Forest tree seed production. Information Report PI-X-53 Petawawa National Forestry Institute, Chalk River, Ontario.

Rehfeldt, G.E., 1979. Ecological adaptations in Douglas-Fir (*Pseudotsuga menziesii* var. *glauca*) populations. *Heredity* **43**: 383-397.

Rehfeldt, G. E., 1988. Ecological genetics of *Pinus contorta* from the Rocky Mountains (USA): a synthesis. *Silvae Genetica* **37**: 131-135.

Rehfeldt, G.E., 1989. Ecological adaptations in Douglas-fir (*Pseudotsuga menziesii* var *glauca*): a synthesis. *Forest Ecology and Management* **28**: 202-215.

Rehfeldt, G.E., 1995. Genetic variation, climate models and the ecological genetics *of Larix occidentalis*. *Forest Ecology and Management* **78**: 21-37.

Rehfeldt, G.E., Ying, C.C., Spittlehouse, D.L. & Hamilton D.A.Jr., 1999. Genetic responses to climate in *Pinus contorta*: niche breadth, climate change, and reforestation. *Ecological Monographs* **69**: 375-407.

Rehfeldt, G.E., 2000. Genes, climate and wood. *The Lesilie L. Schaffer Lectureship in Forest Science*, Wednesday, February 2, 2000. UBC, Vancouver B.C. Canada. 15 pp.

Savolainen, O., 1997. Pines beyond the polar circle. In Tigerstedt, P.M.A. (ed.): *Adaptation in plant breeding*. Kluwer Academic Publishers. 153-159.

Schmidt-Vogt, H., 1977. *Die Fichte*. Band I. Taxonomie - Verbreitung - Morphologie - Ökologie - Waldgensellschaften. Verlag Paul Parey, Hamburg und Berlin.

Schwartz, M., 1991. Potential effects of global climate change on the biodiversity of plants. *Forestry Chronicle* **68**: 462-471.

Sheldon, C.C., Burn, J.E., Perez, P.P., Metzger, J., Edwards, J.A., Peacock, W.J. & Dennis, E.S. 1999. The FLF MADS box gene: A repressor of flowering in Arabidopsis regulated by vernalization and methylation. *The Plant Cell* **11**: 445-458.

Sheldon, C.C., Rouse, D. T., Finnegan, E.J., Peacock, W.J. & Dennis, E.S. 2000. The molecular basis of vernalization: The central role of *FLOWERIN LOCUS C* (*FLC*). *PNAS* **97 (6)**: 3753-3758.

Skrøppa, T., 1991. Within-population variation in autumn frost hardiness and its relationships to bud-set and height growth in *Picea abies*. *Scandinavian Journal of Forest Research* 6: 353-363.

Skrøppa, T. & Kohmann, K., 1997. Adaptation to local conditions after one generation in Norway spruce. *Forest Genetics* **4**: 171-177.

Skrøppa, T. & Johnsen, Ø., 2000. Patterns of adaptive genetic variation in forest tree species; the reproductive environment as an evolutionary force in *Picea abies*. In Mátyás, C., (ed.). *Forest Genetics and Sustainability*. Kluwer Academic Publications, 49-58.

Skrøppa, T., Nikkanen, T., Routsalainen, S. & Johnsen, Ø., 1994. Effects of sexual reproduction at different latitudes on performance of the progeny of *Picea abies*. *Silvae Genetica* **43**: 297-303.

Stoehr, M.U., L'Hirondelle, S.J., Binder, W,D. & Webber, J.E., 1998. Parental environmental aftereffects on germination, growth and adaptive traits in selected white spruce families. *Canadian Journal of Forest Research* **28**: 418-426.

Turesson, G., 1922. The geographical response of the plant species to the habitat. *Hereditas* **3**: 211-350.
Turesson, G., 1925. The plant species in relation to habitat and climate. Contribution to the knowledge of genecological units. *Hereditas* **6**: 146-235.

SECTION IV

Future Aspects of Tree Physiology

TREE PHYSIOLOGY GOES GENOMICS

Perspectives related to Working Group I: Growth and Development

B. SUNDBERG
Umeå Plant Science Centre, Department of Forest Genetics and Plant Physiology, Faculty of Forestry, Swedish University of Agricultural Sciences, SE-90183 Umeå, Sweden

Forest trees accumulate an extensive amount of secondary xylem (wood), which is of significant economic value for man. In silviculture, it has always been an aim to improve biomass production and/or wood properties by breeding and by management practices. To speed up this process we need to advance our understanding of the genetics and physiology underlying growth and development of forest trees.

The development of a forest tree from seed to seed follows the same principles as in any higher plant. Knowledge created in annual plants can therefore serve as a guideline in research on tree physiology. However, certain aspects can be best approached in woody perennials, such as induction of dormancy and low temperature hardiness, transition from juvenile to mature growth habit and wood formation. Much knowledge about environmental influence on tree growth has been reached by investigating effects of light, photoperiod, temperature, water stress and nutriments. This has translated into guiding principles in forest management such as thinning and fertilisation strategies, and in large-scale production of seedlings to be used in reforestation. Seedling production also involves breeding programs based on quantitative genetics and methods for large scale cloning of elite genotypes by somatic embryogenesis.

However, the genetic and molecular mechanisms underlying growth and development have been a black box for a long time. It is evident that the information for spatiotemporal control of cell division and differentiation to tissues and organs is imprinted in the genome. Also, environmental stimuli have to be perceived at the molecular level, and the signal transduced to the proper tissue to induce gene expression and the appropriate growth response. Some of the key molecules in these events, such as plant hormones and light receptors, have been identified by biochemical approaches. Functional analysis of regulating molecules includes experiments where synthetic homologues or inhibitors have been applied to the target tissues, together with localisation and quantitative analysis of the molecule of

S. Huttunen et al. (eds.),
Trends in European Forest Tree Physiology Research, Cost Action E6: EUROSILVA, 225-227

interest. However in general, the progress obtained with these tools has been slow. This trend was dramatically changed with the introduction of molecular techniques into developmental and experimental plant biology. Genes underlying growth and development are now identified at an increased rate, and the significance of their gene products is investigated by descriptive and experimental analysis of mutant plants with up- or down-regulated levels of the corresponding protein. The challenge of the new millennium is to unravel the function of genes identified through large-scale genomics programs, and to put gene function into the context of physiological processes.

Currently, *Arabidopsis* is the best-suited system to identify and study gene function (Ausubel 2000). *Arabidopsis* research has significantly increased our understanding in classical areas of physiology such as perception and signal transduction of plant hormones, tropisms, flowering and light responses, to mention a few. The development of techniques for making transgenic trees, both conifers and broad leaves, allows for functional genomic approaches also in tree research. The induction of precocious flowering in poplar by expression of the *Arabidopsis leafy* gene was the first demonstration of a direct application of gene finding research in *Arabidopsis* in a forest tree (Weigel and Nilsson 1995). This followed by an increasing number of transgenic trees expressing both foreign and endogenous genes, with the aim to unravel molecular control of growth and development. A key target has been the secondary cell wall of woody fibres because of its great economic value, and many genes in the biosynthesis pathway of cellulose, hemicellulose and lignin have been identified (Mellerowicz et al. 2001). Other examples where growth and development of trees has been approached with molecular tools are studies on dormancy, photoperiodism and hormonal control of wood formation (Tuominen et al. 1997; Eriksson et al. 2000; Olsen et al. 2000; Rhode & Boerjan chapter in this issue)

To facilitate genomic research in trees, large-scale sequencing has been initiated in several woody species. Major projects that are currently ongoing include EST sequencing of aspen in Sweden (http://www.biochem.kth.se/PopulusDB/), birch in Finland (http://www.biocenter.helsinki.fi/bi/forest.html) and loblolly pine in USA (http://www.cbc.med.umn.edu/ResearchProjects/Pine/DOE.pine/index.html), and more initiatives are likely to follow. The databases that are emerging from these projects will facilitate identification of genes in trees. Access to a large number of gene sequences also opens a perspective to investigate patterns of global gene expression by microarray analysis (Richmond & Sommerville 2000) Microarrays reveal gene clusters induced in different developmental or environmental conditions, and may suggest a function of previously unknown genes. A functional analysis can rapidly be done by expressing a gene in *Arabidopsis,* or by identifying a knockout mutant if a corresponding orthologue is present in the *Arabidopsis* genome. In a longer perspective, the gene function can be investigated in the tree itself by a transgenic approach. Microarray analysis can also give information about the genetic regulation of complete pathways of primary and secondary metabolism under different environmental conditions, provided that the genes coding for enzymes in these pathways have been identified. At present, the microarrays have been established for poplar and pine (Hertzberg et al. 2001; Whetten et al. 2001). Although microarray analysis is still in its infancy with many shortcomings yet to be solved, it will certainly change the perspectives of experimental plant biology and a burst of data is likely to be released in the near future. In a similar way, large-scale protein analysis (proteomics) currently introduced into plant sciences will supplement the information on the gene expression and will give a more holistic and comprehensive view on physiological processes. The identification of proteins that are highly abundant in developing xylem tissue has been initiated in poplar (Vander Mijnsbrugge et al., 2000) and pine (Plomion et al., 2000).

Although much focus is presently centred around large scale genome analysis and mutant screening, it should be emphasised that only an integration of genetic, genomic, biochemical, physiological, and morphological data will result in a deeper insight into developmental and physiological processes. This calls for collaborative networks of labs with different expertise. To co-ordinate the European effort in tree genomics a new COST initiative has just been approved.

REFERENCES

Ausubel FM (2000) Arabidopsis Genome. A Milestone in Plant Biology. *Plant Physiology* **124**: 1451-1454.

Eriksson ME, Israelsson M, Olsson O, Moritz T 2000. Increased gibberellin biosynthesis in transgenic trees promotes growth, biomass production and xylem fiber length. *Nature Biotechnology* **18**: 784-788.

Hertzberg M, Sievertzon M, Aspeborg H, Nilsson P, Sandberg G, Lundeberg J (2001) cDNA microarray analysis of small plant tissue samples using a cDNA tag target amplification protocol. *Plant Journal* (in press).

Mellerowicz EJ, Baucher M, Sundberg B, Boerjan W (2001) Unravelling cell wall formation in the woody dicot stem. *Plant Molecular Biology* (in press).

Olsen JE, Junttila O, Nilsen J, Eriksson ME, Martinussen I, Olsson O, Sandberg G, Moritz T (2000) Ectopic expression of oat phytochrome A in hybrid aspen changes critical daylength for growth and prevents cold acclimatization. *Plant Journal* **12**: 1339-1350.

Plomion C, Pionneau C., Brach J, Costa P, Baillères H (2000) Compression wood-responsive proteins in developing xylem of maritime pine (*Pinus pinaster* Ait.). *Plant Physiology* **123**: 959-969.

Richmond T, Somerville S (2000) Chasing the dream: plant EST microarrays. Current Opinion *Plant Biology* **3**:108-116.

Tuominen H., Puech L, Fink S, Sundberg B 1997 A radial gradient of indole-3-acetic acid is related to secondary xylem development in hybrid aspen. *Plant Physiology* **115**: 577-585.

Vander Mijnsbrugge K, Meyermans H, Van Montagu M, Bauw G, Boerjan W (2000) Wood formation in poplar: identification, characterization, and seasonal variation of xylem proteins. *Planta* **210**: 589-598.

Weigel D, Nilsson O (1995) A developmental switch sufficient for flower initiation in diverse plants. *Nature* **377**: 495-500.

Whetten R, Ying-Hsuan S, Zhang Y, Sederoff R (2001) Functional genomics and cell wall biosynthesis in loblolly pine. *Plant Molecular Biology* (in press).

WATER, NUTRIENTS AND CARBON; ALTITUDE AND AGEING

Notes on the Activities of Working Group II: Tree nutrition and water relations

P.G. JARVIS[1], S LINDER[2] AND S. HUTTUNEN[3]

[1]*Institute of Ecology & Resource Management, University of Edinburgh, Edinburgh, EH9 3JU, UK.*

[2]*Department for Production Ecology, Swedish University of Agricultural Sciences, P.O. Box 7042, SE-750 07 Uppsala, Sweden*

[3]*Department of Botany, University of Oulu, P.O. Box 3000, FIN-90014 University of Oulu, Finland*

1. INTRODUCTION

As with the other Working Groups, Working Group II set out to generate transfer of ideas, concepts and methodologies amongst the Participants in the Action through (a) movements of young scientists from centre to centre for up to three months *via* the programmes of Short Term Missions, (b) through short training courses, and (c) through discussion meetings, held in general alongside the other two Working Groups and meetings of the Management Committee. The scientific content of these activities was quite widely spread within the overall framework of the Action, but there were certain timely issues on which the focus dwelt during the course of the Action. Initially these related to water and nutrient stress but as time went by issues surfaced that were not evident at the time of initiation of the Action. For example, in December 1997 the Kyoto Protocol emerged from the 3rd Conference of the Parties of the United Nations Framework Convention on Climate Change with an emphasis in Article 3.3 on the *change in stock of carbon* in managed forest ecosystems; the Afforestation, Reforestation and Deforestation (ARD) article (Steffen *et al.* 1998). This led during the latter part of the Action to an enhanced emphasis on processes in forest soils where much of the stock of carbon in forest ecosystems is stored. This brief review is of necessity highly selective, focussing on just a few of the issues addressed by WG II.

S. Huttunen et al. (eds.),
Trends in European Forest Tree Physiology Research, Cost Action E6: EUROSILVA, 229-240

2. WATER POTENTIAL MEASUREMENT

During the early 1990s the cohesion-tension theory of water transport came under unexpected renewed attack as the result of measurements made with the pressure probe on the xylem of twigs and branches in the crowns of transpiring trees (Balling & Zimmermann 1990, Zimmermann *et al.* 1993 a&b). The measurements purported to demonstrate that the tensions in the xylem were extremely small, much smaller than inferred from pressure 'bomb' measurements, thus also reigniting earlier controversy from the 1960s regarding the validity of pressure 'bomb' measurements of xylem tension. This controversy became intense in the early 1990s (Zimmermann *et al.* 1994) so that at the first Discussion Meeting in Thessaloniki in 1996, Tyree (1999) developed a physical analysis of the system and brought forward evidence from several sources that supported measurements made by the pressure 'bomb', and, by implication, placed the doubt on the pressure probe measurements made in the *xylem* of wood, particularly the measurements made in the unstable environment of the crowns of trees. Steudle (1999) at the same time demonstrated the satisfactory use of the pressure probe in analysing the flow of water across *living cells* of roots in relation to pressure and solute gradients. More recently, Wei, Tyree & Steudle (1999 a&b) have maintained ongoing, real time, measurements of negative xylem pressures for hours down to an all time low potential of –1 MPa, while changing environmental conditions, and this has enabled confirmation of the pressure 'bomb' method over a much wider range of pressures than previously. Thus they have experimentally demonstrated the validity of the cohesion-tension theory using the pressure probe, as well as the pressure 'bomb'. They continue to question both the concepts and methodologies of Zimmermann and his colleagues, and in a recent paper describe the essential requirements for direct pressure measurements in both xylem and living cells (Wei *et al.* 2001).

In the meantime, others have reaffirmed their confidence in the validity of pressure 'bomb' measurements for determining optimal hydraulic architecture, and water stress effects on photosynthesis and on growth and development. For example, Pereira (1999) using the pressure 'bomb' has shown that the impact of water stress on growth and development depends on their time of occurrence in the seasonal growth cycle. Water deficits occurring at the beginning of the growing season may have a major impact on shoot growth and development and, therefore, on foliage expansion and leaf area index. This, in turn, determines radiation interception and total carbon assimilation over the year.

3. NUTRIENT AVAILABILITY AND SUPPLY

The upper limit for forest biomass production on a particular site is set by the amount of incoming radiation. Actual production is, however, determined by the amount of radiation that is intercepted by the canopy during the *active period* of growth and, to a lesser degree, by the efficiency of conversion of intercepted radiation into biomass, moderated by overriding variables (Landsberg *et al.* 1996). The size, dynamics and conversion efficiency of a forest canopy are genetically determined, but in natural conditions, are strongly controlled by the climate, availability of water and mineral nutrients. If the 'natural' fertility of a site is too low to allow 'maximum' production, yield can in most cases be increased by supplying the limiting nutrients (Linder & Rook 1984, Tamm 1991). Following principles developed initially by Torsten Ingestad in the laboratory (Ingestad 1987, 1988), there have been several long-term, controlled fertilisation experiments, covering a significant part of the life cycle of plantations; some of these experiments are ongoing (e.g. Allen *et al.* 1990, Bergh *et al.* 1999). The principal aim of the experiments has been to eliminate water and mineral nutrients as growth-limiting variables, whilst at the same time avoiding leaching of nutrients into the groundwater Nutrients have been applied in very low concentrations daily throughout the growing season in spray irrigation water or in a single annual dose, more akin to forestry practice. In the ongoing experiments, target values of foliage concentrations of each nutrient element are defined, and on the basis of repeated foliar analysis and predicted growth response the proportions and amounts of nutrients applied are adjusted annually (Linder 1995).

Figure 1 summarises results from several such experiments on a range of species growing in contrasting environments in different climatic regions. The growth responses have been spectacular and production has far exceeded that predicted by conventional forest yield tables. The examples shown emphasise that most forest stands are growing at rates far below their potential level. In all the stands in which the availability of nutrients and soil water was optimised by means of combined fertilisation and irrigation, the yield of stemwood was similar to or surpassed the best yields obtained by conventional silvicultural means.

It is commonly believed that boreal tree growth is constrained by temperature (*cf* Jarvis & Linder 2000). However, these data demonstrate that inadequate nutrition is the major reason for the difference between potential and actual yield in the boreal region, although in other regions with a dry temperate or Mediterranean climate potential yield cannot be achieved without also improving soil water availability by irrigation.

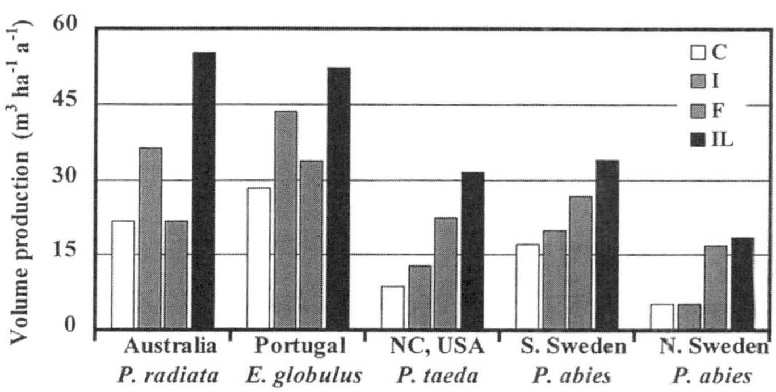

Figure 1. The current annual increment in a number of forest yield optimisation experiments in contrasting environments. The treatments were non-treated controls (C), irrigation (I), annual application of a solid fertilisers (F), and a complete liquid fertiliser combined with irrigation (IL).For more details regarding the individual experiments see Snowdon & Benson (1992), *Pinus radiata;* Pereira *et al.* (1994), *Eucalyptus globulus*; Albaugh *et al.* (1998), *Pinus taeda*; Bergh *et al.* (1999), *Picea abies.*

In natural conditions, the growth rate of trees is largely regulated by the amounts of nutrients made available for uptake through mineralisation of soil organic matter per unit time and unit area, i.e. the nutrient flux density from soil to roots (Ingestad 1988). In general, the forests that now remain after man's depredations, confined to the poorer soils over much of the world particularly in the northern temperate and boreal regions, are chronically nutrient deficient and consequently their growth is very sensitive to the small atmospheric nutrient inputs and changes in rates of mineralisation associated with global climate change (Tamm *et al.* 1999).

4. MYCORRHIZAS, MICROBES AND NUTRIENT ACQUISITION

In boreal and temperate forests, the roots of most tree species are colonised by fungi, which form a mantle over the surface of the feeding roots to produce ectomycorrhizas. Fungal hyphae extend from the mantle out into the soil where they ramify to form "the wood-wide web" (Read 1997). Fungal hyphae thus form the interface between root surfaces and soil organic matter, and play a key role in the *capture* of nutrients released through decomposition of soil organic matter and their *transport* into living cells within the trees (Read 1997). Carbon entering the soil from the roots as exudates, secretions, lysates, gases and mucilage, the total rhizodeposition, stimulates microbial community structure and activity and is a key process influencing nutrient availability, directly through the chelation of cations and indirectly through the stimulation of soil microbial activity (Grayston *et al.* 1996).

5. SOIL, NUTRITION, TEMPERATURE AND THE FOREST CARBON SINK AT HIGH LATITUDES

Forests are the interface between the atmosphere and soil and act effectively to transfer carbon from the one to the other. With just a few exceptions, the 80 or so forests so far investigated are presently annual carbon sinks in the range of 0.5 to 7.5 t C ha^{-1} (Malhi *et al.* 1999). Boreal forests occupy approximately 21% of the forested land surface – second only to tropical forests in areal extent — and contain 13% of the total carbon stored in terrestrial ecosystems. The large amount of carbon stored in the soils — over four times as much as in the trees (Dixon *et al.* 1994) — makes boreal forests extremely sensitive to climate change. The large size of the boreal forest, the large amount of carbon contained in the soil, and the anticipated climate warming make the boreal forest a key biome to understand and represent correctly in global carbon models. The experimental addition of nutrients, as illustrated above, has demonstrated the large capacity of forest ecosystems for additional tree growth and production of detritus. In the boreal forest, this accumulation of carbon seems to have been constrained since the retreat of the ice after the last glaciation by the slow concurrent accumulation of nutrients, particularly nitrogen, in the ecosystem. The present small size of the trees and their low nitrogen contents suggest that the *capacity* of the system to store carbon will not by itself limit the transfer of carbon from the atmosphere to the tree and soil pools in the immediate future, particularly as increasing atmospheric inputs and the rise in temperature may enhance the availability of nitrogen. How long the current carbon sink capacity of the global forests is likely to be maintained into the future is a matter of conjecture, and several hypotheses have been proposed as the basis for quantitative explanation.

Many short term experiments have shown that the increase in photosynthesis resulting from the rise in atmospheric CO_2 concentration diminishes at moderately high CO_2 concentrations, whereas, by contrast, autotrophic and heterotrophic respiration increase exponentially with increasing temperature. Thus, it has been hypothesised that as climate changes the overall capacity of foliage to take up additional carbon from the atmosphere in photosynthesis will progressively diminish whilst the respiration driven efflux of CO_2 from the soil will increase, so that at some time in the not too distant future losses will exceed gains and the present forest carbon sinks will become sources (Scholes *et al.* 1998, Scholes 1999, Cox *et al.* 2000). However, this argument assumes, that respiration is independent of photosynthesis, that acclimation to higher CO_2 concentrations and temperatures does not occur, and that there are no significant feedbacks involving nutrition amongst the processes dependent on CO_2 concentration and temperature. The experimental evidence coming forward does not support this simple view of the carbon cycle.

In a long-term fertilisation and irrigation experiment of Norway spruce at Flakaliden in northern Sweden (64 °N), complete fertiliser (with or without irrigation) has been applied through every growing season, since 1987. After 13 years of treatment the current annual increment on the fertilised plots has increased by a factor of over three times (Figure 1), indicating that air temperature is certainly not the major direct constraint on tree growth at these latitudes. However, temperature may be influencing tree growth in the growing season indirectly through effects on the timing of soil thaw in the spring and freeze-up in the autumn, together with accelerated decomposition of soil organic matter and mineralisation of soil nutrients. A soil warming experiment (an increase of 5 °C at 10 cm depth, through the growing season) at Flakaliden is now, after five years, showing increased concentrations of foliar nutrients and a significant, >60% increase in stem wood growth of the trees on the heated plots. Thus year to year variability in temperature may possibly be influencing stem wood growth through the availability of nutrients and length of growing season (Jarvis & Linder 2000).

After five years of warming, the measured CO_2 effluxes from the forest floor in the warmed and unwarmed fertilised plots are not significantly different. Thus, the evidence from this experiment directly contradicts the hypothesis that the projected rise in temperature is likely to lead to forests that are now C sinks becoming C sources (Linder *et al.* 2001). The processes behind the observed *acclimation* to changes in soil temperature are still not understood and are the subject of current investigation. It is likely that there is a limit to the amount of soil organic matter that is readily metabolisable in response to an increase in soil temperature and that the populations of soil microorganisms have now changed in size and composition. As well there may be acclimation in the autotrophic respiration of the root systems.

Most current models of the likely effects of climatic change on the carbon balance of boreal forest ecosystems are based on the commonly accepted dogma that boreal tree growth is constrained by temperature. Because of the known short term response of decomposition of soil organic matter to temperature, these models predict that the projected rise in temperature is likely to lead to forests that are now C sinks becoming C sources in the foreseeable future, as indicated above. However, other models that endeavour to take into account feedbacks amongst the effects of temperature, atmospheric CO_2 concentration and, particularly, nutrient availability on photosynthesis, respiration and decomposition project that the current sources will increase to a steady maximum and continue for the foreseeable future (McMurtrie *et al.* 2001). It can be concluded from the experiments in progress, both those referred to above and others elsewhere, that projections of the impact of climate change on boreal forests will remain uncertain until results from more long-term field studies of carbon gains and losses in boreal ecosystems become available.

6. HIGH ALTITUDE TREE LINES

At high elevations too growth and development of trees is a function of a constellation of environmental variables, often referred to collectively as "exposure", and there has been considerable argument as to which variables are most crucial to the tree line. Possible candidates that have been proposed include high wind speeds, low winter temperatures, frost damage, winter desiccation, low summer temperatures, high summer temperatures water stress and photo-inhibition (Tranquillini 1979). Whilst it is supposed that some of these variables cause direct physical damage, others may lead to a reduction in carbon gain or increase in carbon loss, so that the annual carbon balance declines to zero as the tree line is approached (Jones & Demmers-Derks 1999).

Analyses of monthly temperature isotherms have shown that trees do not generally grow in places where the mean temperature of the *warmest* month is less than *about* 10 °C (Grace 1989) and this has been taken as evidence that growth is limited at the tree line by a short growing season and low carbon gain. However, using recently collected microclimatic data from a number of temperate and subtropical tree lines, Bernoulli and Körner (1999) found that the frequently quoted 10 °C isotherm of the warmest month is a purely coincidental figure derived from tree lines on some temperate zone mountains and has no causal or predictive value. They showed that total biomass remained approximately constant along a transect across a Swiss alpine tree line as the trees declined in height, finally adopting the krumholz form. Their evidence suggests that tree growth at its high elevation boundary is not limited by C assimilation, but by C investment influenced by greater thermal limitation of apical growth rather than of lateral growth. Hence, they suggested sink limitation of tree growth at low temperatures as the dominant variable. This is consistent with the observations made by Grace (1989) on the energy balance of buds at the tree lines as a function of tree height. Tissues on trees close to the ground experience a microclimate within which tissue temperatures may be several degrees higher than on tall trees.

Grace (1997) has suggested that some of the apparent conflicts and inconsistencies in data and interpretation may result from substantially different primary causes of trees lines in different climatic regions. He proposed that on continental mountains the primary factors are cold winters leading to tissue water stress and winter desiccation, followed by runaway cavitation and loss of needles, coupled with warm and dry summers, leading, on the one hand, to enhanced carbon gain, but, on the other hand, to water stress, drought and again needle loss. By contrast, he proposed that on oceanic mountains the primary factors are high wind speeds, leading to physical abrasion, and cooling of buds and other meristamatic tissues, compounded by low solar radiation and low summer temperatures, leading to reduced leaf areas and low rates of carbon assimilation and poor growth. He suggested that the change in growth habit in the vicinity of the tree line results from mechanical stimulation, known as thigmomorphogenesis, leading to redirection of assimilates from height growth towards tissues conferring mechanical strength (Grace 1997).

7. AGE-RELATED DECLINE IN GROWTH AND PRODUCTION

The age-related decline in tree growth is of considerable relevance to wood production and thus is of considerable interest not only to tree physiologists but also to practicing foresters. It is generally assumed that the fall in net primary productivity (NPP) is controlled by exogenous factors, whereas the process of maturation, which strongly affects plant structure and function and initiates the diversion of resources from vegetative to reproductive growth, is thought to be triggered by a shift in hormonal patterns, possibly of endogenous genetic origin. There is current debate on the physiological and ecological determinants of tree aging, and several hypotheses have been proposed. These include increased proportion of respiration relative to photosynthesis, shifts in biomass allocation, lack of nutrients and hydraulic constraints to water transport as possible mechanisms, but so far there is no definitive evidence to support any one of these as a universal mechanism (Magnani & Jarvis 2001).

The longest standing hypothesis is that the carbon balance of a tree declines as it grows larger because the mass of respiring tissue increases relative to the mass of photosynthetic tissue (Kramer & Kozlowski 1979). A number of investigations to test this hypothesis have come to the conclusion that this is not the primary mechanism for reduction in the carbon balance of aging trees (Ryan *et al.* 2001).

Shifts in allocation of resources, particularly of assimilates to the root system, is also an attractive idea. We have pointed out above that a major shift in allocation seems to be one of the features of acclimation of trees to the climatic conditions at tree lines, and nutrient deficiency leads to a shift in allocation to root systems, for example. Information on above and below ground NPP of trees is generally scarce. However, recently both above and below ground NPP have been found decline with age to much the same extent in eucalypts in Hawaii (G. Giardina pers. comm.) and boreal forests in Canada (S.T. Gower pers. comm.), suggesting that this is not a very likely hypothesis.

Of the alternative hypotheses that have been proposed nutrient and hydraulic limitations stand out as the most likely.

In the boreal coniferous forests in Canada and Siberia and in northern hardwood forests in north America, age-related decline in NPP is strongly related to increasing sequestration of nutrients into vegetation and coarse woody debris (e.g. Schulze *et al.* 1995), and experiments suggest that the decline can be relieved to some extent by addition of nutrients (Tamm 1991, Tamm *et al.* 1999). Despite empirical correlations between nutrient limitations and decline of NPP, however, a mechanistic explanation is lacking. The most likely explanation is a general reduction in the carbon balance resulting from depressed photosynthesis, coupled with restricted meristamatic activity. However, lack of ancillary measurements prevents the exclusion of other possibilities, such as the hydraulic limitations hypothesis.

The hydraulic limitations hypothesis, is based on empirical observations that hydraulic conductances (per unit leaf area) of stem and branch decrease markedly with tree age and dimensions, closing stomata and thus reducing photosynthesis and NPP, and eventually limiting maximum tree height (Ryan *et al.* 2001). Using the air injection technique to induce xylem cavitation while simultaneously measuring canopy gas exchange under constant conditions has shown by experiment that gas exchange does indeed respond directly to changes in hydraulic conductance of stem and branches in small trees (e.g. R. Hubbard pers. comm.), but this does not necessarily explain why hydraulic conductance declines with increase in stem dimensions in mature trees. A possible mechanistic explanation for this is that xylem maturation is completed in just a few years. This could lead to the decrease in hydraulic conductance per unit leaf area as stem dimensions increase and hence to the parallel decline in photosynthesis and plant and stand productivity with age. It is, therefore, possible that the process of xylem cell maturation, largely determined by hormonal balances, could hold the key to the age-related decline in productivity.

A conclusive test of the hydraulic limitation hypothesis in the field, however, remains difficult, as pointed out by Ryan *et al.* (2001). In particular, canopy structure and canopy environments may differ with tree age and size, leading to size-dependent differences in leaf, twig and branch anatomy, thus complicating the assessment of hydraulic limitations, unless measurements are made on isolated trees.

POSTSCRIPT

We hope that these rather personal vignettes of scientific topics that are portrayed here give a taste of the depth and range of the issues that have been brought up through the Cost Action and intensively discussed. The discussions have been greatly facilitated by the opportunities provided by the Action to bring together colleagues and experts from north America as well as from across Europe. The resulting discussions around these and other topics have done much to generate a greater awareness of the issues and to stimulate new approaches and activities.

REFERENCES

Albaugh, T.J., Allen, H.L., Dougherty, P.M., Kress, L.W. and King, J.S. (1998). Leaf area and above and belowground growth responses of loblolly pine to nutrient and water additions. *Forest Science* **44**: 317-328.

Allen, H.L., Dougherty, P.M. and Campbell, R.G. (1990) Manipulation of water and nutrients - practice and opportunity in Southern U.S. pine forests. *Forest Ecology and Management* **30**: 437-453.

Balling, A. and Zimmermann, U. (1990) Comparative measurements of the xylem pressure of *Nicotiana* plants by means of the pressure bomb and pressure probe. *Planta* **182**: 325-338.

Bergh, J., Linder, S., Lundmark, T. and Elfving, B. (1999) The effect of water and nutrient availability on the productivity of Norway spruce in northern and southern Sweden. *Forest Ecology and Management* **119**: 51-62.

Bergh, J., McMurtrie, R.E. and Linder, S. (1998) *Forest Ecology and Management* **110**: 125-139.

Bernoulli, M. and Körner, Ch. (1999) Dry matter allocation in tree line trees. *Phyton* **39(4)**: 7-11.

Cox, P.M., Betts, R.A., Jones, C.D., Spall, S.A. and Totterdell, I.J. (2000) Acceleration of global warming due to carbon-cycle feedbacks in a coupled climate model. *Nature* **408**: 184-187.

Dixon, R.K., Brown, S., Houghton, R.A., Solomon, A.M., Trexler, M.C. and Wisniewski, J. (1994) Carbon pools and flux of global forest ecosystems. *Science* **263**: 185-190.

Grace, J. (1989) Tree lines. *Philosophical Transactions of the Royal Society of London* B324- 233-245.

Grace, J. (1997) The Oceanic Tree-line and the Limit for Tree Growth in Scotland. *Botanical Journal of Scotland* **49**: 223-236.

Grayston, S.J., Vaughan, D. and Jones, D. (1996) Rhizosphere carbon flows in trees, in comparison to annual plants: the importance of tree root exudation and its impact on soil microbial activity and nutrient availability. *Applied Soil* Ecology **5**: 29-56.

Ingestad, T. (1987) New concepts on soil fertility and plant nutrition as illustrated by research on forest trees and stands. *Geoderma* **40**: 237-252.

Ingestad, T. (1988) A fertilization model based on the concepts of nutrient flux density and nutrient productivity. *Scandinavian Journal of Forest Research* **3**: 157-173.

Jarvis, P.G. and Linder, S. (2000) Constraints to growth of boreal forests. *Nature* **405**: 904-905.

Jones, H.G. and Demmers-Derks, H.H.W.M. (1999) Photoinhibition as a factor in altitudinal or latitudinal limits of species. *Phyton* **39(4)**: 91-98.

Kramer, P.J. and Kozlowski, T.T. (1979) *Physiology of Woody Plants.* Academic Press, New York, San Francisco, London, 786 pp.

Landsberg, J.J., Prince, S.D., Jarvis, P.G., McMurtrie, R.E., Luxmoore, R. and Medlyn, B.E. (1996) Energy conversion and use in forests: the analysis of forest production in terms of radiation utilisation efficiency *(F,)*. In: *The Use of Remote Sensing in the Modelling of Forest Productivity at Scales from the Stand to the Globe* (eds. H.L. Gholz, K. Nakane and H. Shimoda), pp 273-298. Kluwer Academic Publishers.

Linder, S. (1995) Foliar analysis for detecting and correcting nutrient imbalances in Norway spruce. *Ecological Bulletin (Copenhagen)* **44**: 178-190.

Linder, S. and Rook, D.A. (1984) Effects of mineral nutrition on carbon dioxide exchange and partitioning of carbon in trees. In: *Nutrition of Plantation Forests* (eds G.D. Bowen and E.K.S. Nambiar), pp 211-236. Academic Press, London.

Linder, S., McMurtrie, R.E. and Landsberg, J.J. (1995) Global change impacts on managed forests. In: *Global Change and Terrestrial Ecosystems* (eds. B.H. Walker and W.L. Steffen), pp 275-290. IGBP Book Series No. 2. Cambridge University Press.

Linder, S., Strömgren, M. and Gower, S.T. (2001) Temperature acclimation of soil surface $C0_2$ flux in a boreal forest. Manuscript in preparation.

Magnani, F. and Jarvis, P.G. (2001) Development and ageing of forest trees. *Trends in Ecology and Evolution* (in press).

Malhi, Y., Baldocchi, D.D. and Jarvis, P.G. (1999) The carbon balance of tropical, temperate and boreal forests. *Plant, Cell and Environment 22: 715-740.*

McMurtrie, R., Medlyn, B. and Dewar, R. (2001) Increased understanding of nutrient immobilisation in soil organic matter is critical for predicting the carbon sink strength of forest ecosystems over the next 100 years. *Tree Physiology* (in press).

Pereira, J.S. (1999) Impacts of water stress in tree growth and development. In: COST *Action E6, EUROSILVA Proceedings of the Workshop on Resource Utilization from Cell to Canopy, October 17-19, 1996, Thessaloniki, Greece* (eds. K. Radoglou and S. Huttunen), pp 45-46. Oulu University Press.

Pereira, J.S., Madeira, M.V., Linder, S., Ericsson, T., Tomé, M. and Araújo, M.C. (1994) Biomass production with optimized nutrition in *globulus* plantations. In: *Eucalyptus for Biomass Production* (eds. J.S. Periera and H. Pereira), pp 13-30. Published by Commission of the European Communities, Brussels.

Read D.J. (1997) Mycorrhizal fungi - the ties that bind. *Nature* **388**: 517 - 518.

Ryan, M.G., Bond, B.J., Hubbard, R.M. and Williams, M. (2001) An evaluation of the hydraulic limitation hypothesis. *Tree Physiology* (in press).

Scholes, R.J. (1999) Will the terrestrial carbon sink saturate soon? *Global Change Newsletter* **37**: 2-3.

Scholes, R.J., Schulze, E.-D., Pitelka, L.F. and Hall, D.O. (1998) Biogeochemistry of terrestrial ecosystems. In: *Implications of Global Change for Natural and Managed Ecosystems: A Synthesis of GCTE and Related Research* (eds. B H Walker, W.L. Steffen, J. Canadell and J.S.I. Ingram), pp 271-303. IGBP Book Series No. 4. Cambridge University Press,

Schulze, E-D., Schulze, W., Kelliher, F.M. *et al.* (1995) Above-ground biomass and nitrogen nutrition in a chronosequence of pristine Dahurian Larix stands in Eastern Siberia. C*anadian Journal of Forest Research* 25: 943-960.

Snowdon, P. & Benson, M. L. (1992) Effects of combinations of irrigation and fertilization on the growth and above-ground biomass production of Pinus *radiata. Forest Ecology and Management* **52**: 87-116.

Steffen, W., Noble, I., Canadell, J., Apps, M., Schulze, E.-D., Jarvis, P.G. *et al.* (1998) The terrestrial carbon cycle: Implications for the Kyoto protocol. *Science* **280**: 1393-1394.

Steudle, E. (1999) Long distance transport of water in trees: role of the root. In: COST *Action E6, EUROSILVA Proceedings of the Workshop on Resource Utilization from Cell to Canopy, October 17-19, 1996, Thessaloniki, Greece (eds.* K Radoglou and S Huttunen), pp 47-52. Oulu University Press.

Tamm, C.O. (1991) *Nitrogen in Terrestrial Ecosystems. Questions of Productivity, Vegetational Changes and Ecosystem Stability.* Springer-Verlag, Berlin, 116pp.

Tamm, C.O., Aronsson, A., Popovic, B. and Flower-Ellis, J. (1999) *Optimum Nutrition and Nitrogen*

Saturation in Scots Pine Stands. Studia Forestalia Suecica, No 206, 126pp.

Tranquillini, W. (1979) *Physiological Ecology of the Alpine Timberline: Tree Existence at High Altitudes with Special Reference to the European Alps.* Springer-Verlag, Berlin, Heidelberg, New York, 131 pp

Tyree, M. (1999) The cohesion-tension theory of sap ascent: current controversies. In: *COST Action E6, EUROSILVA Proceedings of the Workshop on Resource Utilization from Cell to Canopy, October 17-19, 1996, Thessaloniki, Greece* (eds. K Radoglou and S Huttunen), pp 63-66. Oulu University Press.

Wei, C., Steudle, E. and Tyree, M.T. (1999a) Water ascent in plants: do ongoing controversies have a sound basis? *Trends in Plant Science* **4**: 372-375.

Wei, C., Tyree, M.T. and Steudle, E. (1999b) Direct measurements of xylem pressure in leaves of intact maize plants. A test of the cohesion-tension theory taking hydraulic architecture into consideration. *Plant Physiology* **121**: 1191-1205.

Wei, C., Steudle, E., Tyree, M.T. and Lintilhac, P.M. (2001) The essentials of direct xylem pressure measurement. *Plant, Cell and Environment* (in press).

Zimmermann, U., Haase, A., Langbein, D. and Meinzer, F. (1993a) Mechanism of long-distance water transport in plants: a re-examination of some paradigms in the light of new evidence. *Philosophical Transactions of the Royal Society of London B* **341**: 19-31.

Zimmermann, U., Meinzer, F.C., Benkert, R., Zhu, J.M., Schneider, H., Goldstein, G., Kichenbrod, E. and Haase, A. (1994) Xylem water transport: is the available evidence consistent with the cohesion theory? *Plant, Cell and Environment* **17**: 1169-1181.

Zimmermann, U., Benkert, R., Schneider, J., Rygol, J., Zhu, J.J., Zimmermann, G. (1993b) Xylem pressure and transport in higher plants and tall trees. In: *Water Deficits: Plant Responses from Cell to Community* (eds. J.A.C. Smith and H. Griffiths), pp 87-108. Bios Science Publishers, Oxford.

ANNEX A: KEYNOTE SPEAKERS IN WORKING GROUP II

First Discussion Meeting: Thessaloniki

Resource Utilization from Cell to Canopy: Water Use and Water Transport

Pereira, J. Portugal: Impact of water stress on tree growth and development
Tyree, M. USA: The cohesion - tension theory of sap ascent: current controversies
Scarascia-Mugnozza, G. Italy: Impacts of water stress on photosynthesis
Davies, W. & Jarvis, A. United Kingdom: A parsimonious mechanistic model for predicting stomatal conductance (g_{sw}) as a function of potential evaporation (E_p), potential assimilation (A_p) and the available water supply (V_p)

Second Discussion Meeting: Ivalo-Saariselkä

Forest nutrition

Read, D.J. United Kingdom: *The role of mycorrhizas in tree nutrition*
LeeAllen, H. USA: Comparative above and below ground responses of loblolly pine to increased water and nutrient availability

Linder, S. Sweden: Constraints on forest yield - the scope and opportunities of large scale nutrition experiments
Grayston, S.J. and Millard, P. United Kingdom: Nutrient acquisition for tree growth in sustainable systems: microbial mineralisation and internal cycling

Third Discussion Meeting: Obergurgl

Tree Growth at High Altitude and High Latitude
Körner, Chr. Switzerland: Why treelines? A synthesis of current understanding
Grace, J. United Kingdom: Energy balance and ecophysiology of trees at the tree line
Matteucci, G., Valentini, R. and Scarascia-Mugnozza, G. Italy: Linking physiology to fluxes in Appennine beech forest
Gracia, C. Spain: Scaling up growth and physiology from sample plots to region

Fourth Discussion Meeting: Gozd Martuljek

Root-Soil Interactions in Trees
Read, D.J. United Kingdom: The role of mycorrhizas in the exploitation of soil organic resources
Grayston, S.J. United Kingdom: Rhizodeposition and its impact on microbial community structure and function in trees
Trumbore, S.E. and Gaudinski, J.B. USA: Radiocarbon measurements below-ground: applications to the dynamics of fine roots and sources of soil respiration
Linder, S. Sweden: The effect of soil temperature on carbon fluxes in young stands of boreal Norway spruce
Diaci, D. *et al.* Slovenia: Natural regeneration in Altimontane spruce stands - a combination of light and soil interactions

Fifth Discussion Meeting: Firenze

Development and Ageing in Forest Trees
Mencuccini, M. United Kingdom: Physiological allometry of forest tree. The role of hydraulic constraints
Gower, S.T. USA: Towards a better understanding of forest age decline
Ryan, M.G. *et al.* USA: An evaluation of the hydraulic limitation hypothesis

TRENDS IN FOREST TREE PHYSIOLOGICAL RESEARCH

Perspectives related to Working Group III: 'Biotic and Abiotic Interactions'

RAINER MATYSSEK
Forest Botany, Department of Ecology, Technische Universität München, Am Hochanger 13, D-85354 Freising, Germany

The title of Working Group III '*Biotic and Abiotic Interactions*' highlights the ecological significance of the environmental impact on trees and, in an inverse way, of the initiated physiological responses of trees to their environment. Factors of interest in related research have been, throughout decades, variable water availability (drought and flooding), temperature (frost and heat), light regimes (including light fleck dynamics and UV-B), pollutants (elevated regimes of ozone and CO_2, high N deposition) and impacts by pathogens, herbivores and competing neighboring plants (e.g. Kozlowsky et al. 1991; Smith & Hinckley 1995a,b; Kramer & Boyer 1995). It has been the aim to unravel, for each of these factors, the tree-internal mechanisms affected or induced, as these represent the 'core' of interactions, linking the pathway of impact (i.e. the kind of stress) to the 'strain' initiated in the plant (i.e. its response to the stress). This view, reflecting the physical stress concept (Larcher 1981, 1994), has often been restricted to selected processes (e.g. biochemical responsiveness of cell metabolism, stomatal regulation) while neglecting the context of the whole-plant system. It has been increasingly recognized, however, that understanding and predicting of tree response may become ambiguous without assessments of the whole-plant performance and of further external impacts which may accompany a particular stress under consideration. It was during the two last decades of the 20th century that trends in physiological research gained in importance that developed a more 'holistic' view on trees, and approached from mono-factorial towards multi-factorial perspectives (Mooney et al. 1991). By this, the analysis of plant-site relationships acquired further qualities, i.e. physical interactions between factorial pathways during impact, and physiological interactions initiated in tree response through the impacting factors (e.g.

S. Huttunen et al. (eds.),
Trends in European Forest Tree Physiology Research, Cost Action E6: EUROSILVA, 241-249

interactions between irradiance, moisture, pollutant impact and susceptibility to biotic constraints: Lefohn 1992; Matyssek et al. 1995; Heath & Taylor 1997). Such interactions may extend to the stand level, if trees modify their given habitats in response to, for example, elevated CO_2 levels in that a changing above and belowground resource allocation may alter the carbon turnover of forest ecosystems (Ceulemans et al. 1999). The need in resolving multi-factorial interactions (as arising from the forest decline studies during the eighties and nineties, Schulze et al. 1989; Sandermann et al. 1997) inherently fostered 'holistic' approaches that 'scale' such interactions across the different levels of plant functioning, from the cellular through the organ up to the whole-plant level, and even beyond that scope to forest stands and ecosystems (Ehleringer & Field 1993; Matyssek et al. 1995).

An appropriate approach to coping with the complexity in stress response is the analysis of resource allocation in plants, as a thorough characterization of allocation is inherently linked to whole-plant functioning and performance. It has been recognized as a paramount research need and challenge in plant ecophysiology to unravel the regulatory mechanisms of resource allocation, after having gained extensive knowledge, during the past decades, on the functional behavior of the plant organs (in particular of leaves; Bazzaz & Grace 1997). Understanding of such mechanisms does integrate the interactions between the different biotic and abiotic factors (as exemplified above) in plant metabolism, including the trade-offs that exist in the internal resource use for the different physiological demands in the plant (Herms & Mattson 1992; Lerdau & Gershenzon 1997). The economic perspective of 'opportunity costs', i.e. the cost/benefit relationships of contrasting resource use, can provide conceptual models on the 'strategies' plants choose in resolving conflicts in resource allocation under multi-factorial stress (Stitt & Schulze 1994). The extent to which plants follow optimization principles in allocation (Grace 1997) appears to be of minor importance, as this aspect can and must be validated for the different spatial and temporal scales of natural habitats. It is obvious, however, that the capacity in regulating resource allocation represents the functional basis of individual plant fitness, as this regulation determines both the sequestration of resources and the ability of retaining them (Bazzaz 1997). Clearly, this latter issue deserves attention in trees, as a balanced resource allocation and 'sustainability' in the associated regulation are basic pre-requisites for the survival and overall fitness (competitiveness, stress defense, reproduction, regeneration ...) of this long-lived plant life form. It is also clear that this kind of a long-term regulation may be susceptible, in particular ways, to modifications during the prolonged tree ontogeny and to factorial interactions as represented by the environmental impact.

As a trend, the need for unraveling regulatory mechanisms as mentioned above will focus research on 'conflicts' in resource allocation. Such conflicts may arise, for example, from exposure to elevated levels of both CO_2 and ozone (Saxe et al. 1998; Grams et al. 1999), and may be complicated by the nutritional status (perhaps driven by high nitrogen deposition) and water supply as well as variable light and temperature regimes (Skärby et al. 1998; Gessler & Rennenberg 1998; Polle et al. 2000). These conflicts may be reflected in ratios of above vs. belowground allocation, organ differentiation, C/N ratio or growth patterns (Maurer & Matyssek 1997). The interactions between abiotic factors set the stage for biotic impact. A major conflict here may arise between growth demands for staying competitive with neighboring plants and efforts in defense against pathogens and phytophagous insects (Herms & Mattson 1992). Complications which will provide rewarding research aims may derive from biotic belowground interactions, the mechanisms of which have been neglected, in particular, in trees (Rogers et al. 1994; Hampp this vol.). Stress responses of trees may influence the belowground respiration, the rhizospheric microflora and processes of soil formation and decomposition (Andersen & Scagel 1997; Andersen, Rennenberg, both this vol.). It is unclear if stress on aboveground tree organs may render the sink strength of root mycorrhization a burden for the whole-plant carbon relations, however, mycorrhization has been shown to interact with the tree's resistance to biotic stress (Gehring et al. 1997; cf. Matyssek & Innes 1999). The molecular and biochemical signaling of the diverse abiotic and biotic stresses towards inducing appropriate responses in resource allocation have only been partly unraveled in trees (Sandermann 1996; Sandermann & Langebartels, Kangasjärvi, Wingsle & Karpinski, Osswald et al., all this vol.). Deficits also exist in understanding tree competitiveness, given the relatively 'static' appearance of competitive interactions between tall trees in stands on a short-term scale. Concepts need to be fostered that relate resource allocation to competitive strategies in occupying canopy and soil space, as it is space sequestration which represents the primary stage of plant competition (Schulze et al. 1986; Küppers 1994; Beyschlag this vol.). Resource investments in space occupation need to be related in terms of cost/benefit relationships to the returns from the occupied space in order to establish a mechanistic basis of competitiveness which may reach beyond the scope of descriptive assessments (Grace & Tilman 1990). In this context, the 'translation' of resource allocation into growth patterns and allometric relationships, i.e. the competitive significance of crown and root morphology, will require attention (Tremmel & Bazzaz 1995).

The conflicts and trade-offs in the resource demands for competitiveness and stress defense need to be quantified in energetic terms (Poorter & Villar 1997; Lerdau & Gershenzon 1997), as the energetic costs set the extent of tree response. By quantifying these trade-offs, predispositions to stress can be assessed on a mechanistic basis (cf. Miller et al. 1997; Dreyer this vol.). This basis decides upon the dilemma in plants of staying competitive versus maintaining stress tolerance and resistance, and may clarify the relative importance of the growth-differentiation versus the carbon-nutrient balance hypothesis in this plant-internal conflict (Herms & Mattson 1992). Do hypotheses substantiate in mechanistic terms across the internal scales of metabolism that plant vigor and high competitiveness attract parasite attack, whereas growth limitations increase resistance (Fig. 1) ? Low resource availability and primary production may favor the secondary metabolism and, as a consequence, stress resistance, while growth and competitiveness stay limited. In contrast, high productivity at non-limiting resource supply and resulting competitive vigor at the expense of metabolic stress defense may predispose to impairment by stress (Laurence et al. 1994). In what ways does nutrition determine the biochemical quality and energetic costs of stress defense ? These issues are currently being assessed through the interdisciplinary research cooperation *'Growth and Parasite Defense in Economic Plants'* (SFB 607, DFG project) as conducted by about 20 research teams of three major research institutions in Munich/Germany. It is the novelty of this initiative that scientists from forestry, agronomy and basic biological sciences compare trees with herbaceous life forms for the mechanisms of resource allocation and plant fitness across the different functional

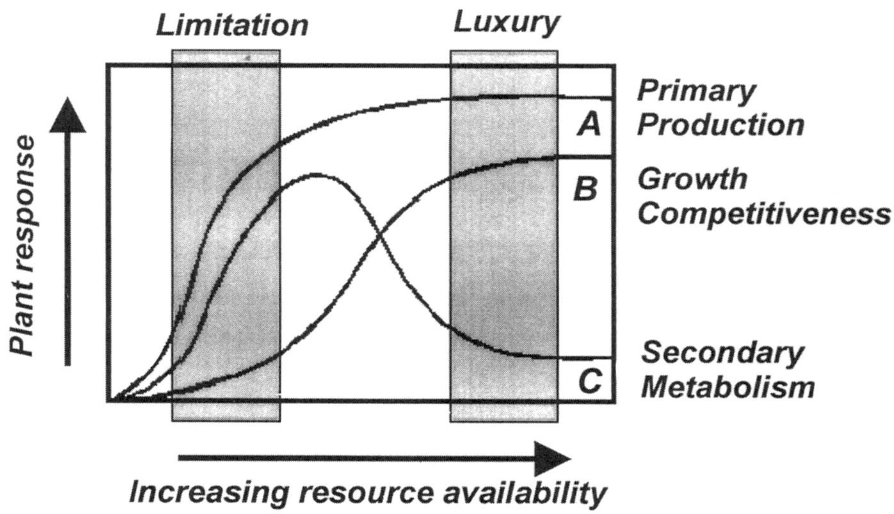

Figure 1. Plant response as expressed by the assimilatory capacity (primary production, function A), the relative growth rate (as a measure of competitiveness, function B), and the activity of the secondary metabolism (as a measure of stress defense, function C) as relating to resource availability. The hatched areas indicate contrasting scenarios where low resource availability may limit production and competitiveness but assure high defense capacity through stimulation of the secondary metabolism (left area); or, where high resource supply may induce, at the expense of the stress defense capacity (secondary metabolism), competitive vigor through enhanced production and growth (see text; adapted from Herms & Mattson 1992).

and structural scales.

As scaling of processes is crucial for obtaining an integrated view on tree performance, also the regulatory processes in resource allocation must be subject to this kind of integration. In particular, it will become a challenging task for the coming years to link the molecular level in control of tree fitness to the biochemical, physiological and structural interrelationships between cells, organs and the whole tree (Koch 1996). This kind of approach may be regarded as one 'dimension' of scaling (Fig. 2). Related to such scaling procedures, 'classical' disciplines like ecophysiology and biochemical phytopathology will develop their complementary perspectives towards closer cooperation (Manning & v. Tiedemann 1995). However, when dealing with trees, there are two further scaling schemes or 'dimensions of scaling' (Fig. 2): (1) scaling from controlled studies in the laboratory (phytotrons, glasshouses) or exposure chambers to the conditions of actual field sites and ecosystems, and (2), scaling of findings from young trees to the specificness of mature individuals. The integration of these three scaling schemes will provide new challenges to mechanistic modeling concepts (Grace 1997).

Most of the current knowledge on tree physiology (although depending on research issues) still relates to studies on young individuals as grown under controlled conditions. Scaling from the lab to the field must reach the stand level with its entire scope of abiotic

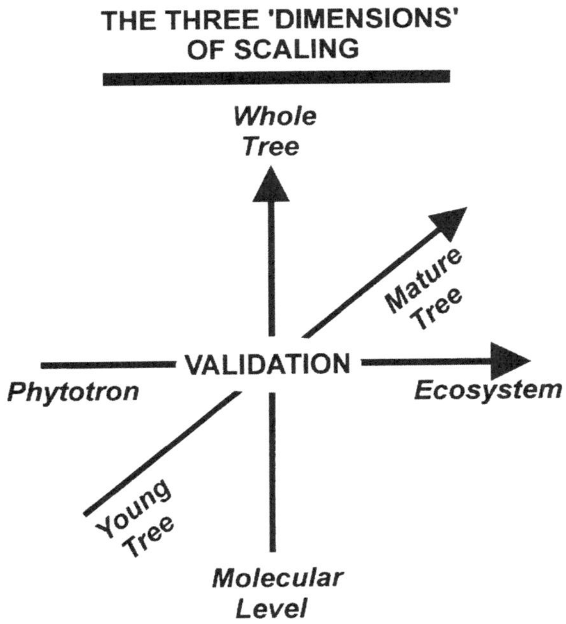

Figure 2. The three dimensions of scaling, (1) 'mechanistic scaling', from the molecular across the organ to the whole-tree level, (2) 'ontogenetic scaling', from young to mature trees, (3) 'ecological scaling', from controlled conditions (e.g. phytotrons) to actual field sites (stand, ecosystem level). Each of these scaling schemes requires validation of findings along the respective vectors (axes).

and biotic interactions, and also scaling from young to mature trees must consider the diverse interactions at given forest sites (cf. Baumgarten et al. 2000). The ontogenetic scaling requires knowledge about constraints on the carbon and water relations that are associated with the ageing process of trees and the structural dimensions acquired by this life form (Ryan et al. 1997; Ryan, Mencuccini, both this vol.). There has been growing awareness since the early nineties on this tree-specific issue and the potential, natural causes of growth limitation in mature trees (e.g. Waring & Silvester 1993; Yoder et al. 1994). Ageing processes may modify tree sensitivity to stress, e.g. to the steadily increasing CO_2 level or chronic enhancement of ozone regimes, however, the scaling abilities from young to mature trees are still too scarce to obtain reliable judgements on old-growth trees and forests (Kolb et al. 1997; Saxe et al. 1998; Matyssek & Innes 1999). Validation is required (but often still missing) of findings from young trees and controlled conditions for mature trees at forest sites (Fig. 2; Kolb, Wieser, both this vol.). However, it is the natural variability between tree individuals encountered in the field that can render validation difficult, and the often unavoidable restriction to few tall individuals can even exacerbate such a task. To overcome this difficulty, statistical concepts may be promoted that examine tree individuals for their internal consistency patterns in parameter response, regarding each individual, to some extent, as a case study. Such approaches will require interdisciplinary research on a large number of structural and functional tree parameters which must be relatable to each other on a mechanistic basis. Evolutionary trends may more readily become apparent by such approaches than by merely comparing mean values of single tree parameters (e.g. between treatments) through 'conventional' statistics. Clearly, scaling of tree-internal processes will be valuable in such analyses of consistency patterns.

A particular challenge is the validation of effects of elevated CO_2 and ozone levels, as in particular in this field the majority of findings has only limited ecological relevance with respect to actual site conditions. An ultimate tool of validation are so-called 'free-air' fumigation experiments which allow whole-tree exposure in the field to elevated regimes of CO_2 and/or ozone without bias by micro-climatic and metabolic artifacts as may be induced in chamber experiments or branch exposures in bags (Musselman & Hale 1997). The methodology of 'free-air' exposure which has been promoted during the nineties may be operated in up-growing tree plantations or in the joint canopy of tall neighboring trees in forest stands (Karnosky et al. 2001). Current examples are the Aspen FACE Project in Rhinelander/USA, the POPFACE study near Viterbo/Italy, the free-air ozone fumigation experiment in a mixed beech/spruce stand at Kranzberg Forest (Freising/Germany), or the research facility of the Kuopio University/Finland. The 'free-air' approach reaches the stand level with all its abiotic and biotic interactions (competition, parasitism, mycorrhizae) and can address aspects of ecosystem behavior. For example, belowground processes can be assessed for the given soil conditions, being of relevance for the overall carbon turnover in forest ecosystems (Ceulemans et al. 1999; Rogers et al. 1994). In this way, quantification of the carbon sink strength of forests is achievable under experimentally created 'Global-Change' conditions. Mitigation of this sink strength can be assessed through 'free-air' ozone fumigations to clarify the risks imposed by the chronic impact of enhanced ozone levels (Häberle et al. 1999), by this, taking into account that chronic ozone stress has become one component of 'Global Change'. Research on these latter issues has become important to resolve debates on forest development and its effects on global resource cycling in a changing environment (Körner 1999; IPCC reports, Kyoto protocol).

The current trends in physiological research on forest trees are reflected in the scientific progress of the EUROSILVA research cooperation. The first stage of EUROSILVA, being conducted under the auspices of EUREKA during the first half of the nineties, focused on tree-internal mechanisms at the molecular, biochemical and physiological level, but reached whole-tree integration to a minor extent only (Sandermann & Bonnet-Masimbert 1995). The aspect of scaling has been fostered during the second half of the nineties, when EURO-SILVA was staying operative as the currently ending COST E6 Action (with its outcome presented in this volume). It has been recognized during this recent EUROSILVA term that tree physiological research must seek its validation and ecological significance under given field conditions, and particular emphasis has been directed to tree ecophysiology and belowground processes. A follow-up concept for EUROSILVA, envisaged for the beginning of the 21st century and evolving from the two preceding stages, may intensify research at the stand level. Emphasis should be directed to canopy processes as well as belowground interactions and related soil processes, without neglecting, however, issues of whole-tree physiology as a determinant of ecosystem relationships.

REFERENCES

Andersen CP, Scagel CF (1997) Nutrient availability alters belowground respiration of ozone-exposed ponderosa pine. *Tree Physiology* **17**: 377–387.

Baumgarten M, Werner H, Häberle K-H, Emberson LD, Fabian P, Matyssek R (2000) Seasonal ozone response of mature beech trees *(Fagus sylvatica)* growing at high altitude in the Bavarian Forest (Germany) in comparison with young beech trees grown in the field and in phytotrons. *Environmental Pollution* **109**: 431-442.

Bazzaz FA, Grace J (1997) Plant resource allocation. Academic Press, San Diego, pp. 303.

Bazzaz FA (1997) Allocation of resources in plants: State of the science and critical questions. In: Bazzaz FA, Grace J (eds.) Plant resource allocation. Academic Press, San Diego, pp. 1-38.

Ceulemans R, Janssens IA, Jach ME (1999). Effects of CO_2 enrichment on trees and forests: Lessons to be learned in view of future ecosystem studies. *Annals of Botany* **84**: 577-590.

Ehleringer JR, Field CB (1993) Scaling of physiological processes: leaf to globe. Academic Press, San Diego, pp. 388.

Gehring CA, Cobb NS, Whitham TG (1997) Three way interactions among ectomycorrhizal mutualists, scale insects, and resistant and susceptible pi ny on pines. *American Naturalist* **149**: 824–841.

Gessler A, Rennenberg H (1998) Atmospheric ammonia: mechanisms of uptake and impacts on N metabolism of plants. In: De Kok LJ, Stulen I (eds.) Responses of plant metabolism to air pollution and global change. Backhuys Publishers, Leiden (The Netherlands), pp. 81-94.

Grace J (1997) Toward models of resource allocation in plants. In: Bazzaz FA, Grace J (eds.) Plant resource allocation. Academic Press, San Diego, pp. 279-291.

Grace JB, Tilman D (1990) Perspectives on plant competition. Academic Press, San Diego, pp. 483.

Grams TEE, Anegg S, Häberle K-H, Langebartels C, Matyssek R (1999) Interactions of chronic exposure to elevated CO_2 and O_3 levels in the photosynthetic light and dark reactions of European beech (*Fagus sylvatica*). *New Phytologist* **144**: 95-107.

Häberle K-H, Werner H, Fabian P, Pretzsch H, Reiter I, Matyssek R (1999) 'Free-air' ozone fumigation of mature forest trees: a concept for validating AOT40 under stand conditions. In: Fuhrer J, Achermann B (eds.) Critical Level for Ozone - Level II. Swiss Agency for the Environment, Forests and Landscape (SAEFL), Berne, pp. 133-137.

Heath RL, Taylor GE (1997) Physiological processes and plant responses to ozone exposure. In: Sandermann H, Wellburn AR, Heath RL (eds.) Forest decline and ozone,a comparison of controlled chamber and field experiments. Ecological Studies 127, Springer-Verlag Berlin, pp. 317-368.

Herms DA, Mattson WJ (1992) The dilemma of plants: to grow or defend. *The Quarterly Review of Biology* **67**: 283-335.

Karnosky DF, Gielen B, Ceulemans R, Schlesinger WH, Norby RJ, Oksanen E, Matyssek R, Hendrey GR (2001) FACE systems for studying the impacts of greenhouse gases on Forest Ecosystems. In: Karnosky DF, Scarascia-Mugnozza G, Ceulemans R, Innes JL (eds.) The impacts of carbon dioxide and other greenhouse gases on forest ecosystems. CABI Press (in press).

Koch KE (1996) Carbohydrate-modulated gene expression in plants. *Annual Review of Plant Physiology and Plant Molecular Biology* **47**: 509-540.

Kolb TE, Fredericksen TS, Steiner KC, Skelly, JM (1997) Issues in scaling tree size and age responses to ozone: A review. Environmental Pollution 98: 195-208.

Körner C (1999) Biologische Folgen der CO_2-Erhöhung. Biologie in unserer Zeit 29: 353-363.

Kozlowsky TT, Kramer PJ, Pallardy SG (1991) The physiological ecology of woody plants. Academic Press, San Diego, pp. 657.

Kramer PJ, Boyer JS (1995) Water relations of plants and soil. Academic Press, San Diego, pp. 495

Küppers M (1994) Canopy gaps: competitive light interception and economic space filling - a matter of whole-plant allocation. In: Caldwell MM, Pearcy RW (eds.) Exploitation of environmental heterogenity by plants - ecophysiological processes above and below-ground. Academic Press, San Diego, pp. 111-144.

Larcher W (1994) Ökophysiologie der Pflanzen. UTB Ulmer, Stuttgart, pp. 394.

Larcher W (1981) Effects of low temperature stress and forest injury on plant productivity. In: Johnson CB (ed.) Physiological processes limiting plant productivity, Butterworth, London, pp. 253-269.

Laurence JA, Amundson RG, Friend AL, Pell EJ, Temple PJ (1994) Allocation of carbon in plants under stress: An analysis of the ROPIS experiment. *Journal Environmental Quality* **23**: 412-417.

Lefohn AS (1992) Surface level ozone exposures and their effects on vegetation. Lewis Publishers, Chelsea/USA, pp. 366.

Lerdau M, Gershenzon J (1997) Allocation theory and chemical defense. In: Bazzaz FA, Grace J (eds.) Plant resource allocation. Academic Press, San Diego, pp. 265-277.

Manning WJ, v. Tiedemann A (1995) Climate change: potential effects of increase atmospheric carbon dioxide (CO_2), ozone (O_3), and ultraviolet-B (UV-B) radiation on plant diseases. *Environmental Pollution* **88**: 219-225.

Matyssek R, Reich PB, Oren R, Winner WE (1995) Response mechanisms of conifers to air pollutants. In: Smith WK, Hinckley TH (eds.) Physiological Ecology of Coniferous Forests. Physiological Ecology Series, Academic Press, New York, pp. 255-308.

Matyssek R, Innes JL (1999) Ozone - a risk factor for trees and forests in Europe? *Water, Air and Soil Pollution* **116**: 199-226.

Maurer S, Matyssek R (1997) Nutrition and the ozone sensitivity of birch (*Betula pendula*), II. Carbon balance, water-use efficiency and nutritional status of the whole plant. *Trees* **12**: 11-20.

Miller PR, Arbaugh MJ, Temple PT (1997) Ozone and its known effects on forests in Western Unites States. In: Sandermann H. Jr, Wellburn A.R., and Heath R.L., eds. Forest decline and ozone. Berlin, Heidelberg, New York: Springer-Verlag, pp. 39-67.

Mooney HA, Winner WE, Pell EJ (1991) Response of plants to multiple stresses. Academic Press, pp. 422.

Musselman RC, Hale BA (1997) Methods for controlled and field ozone exposures of forest tree species in North America. In: Sandermann H jr. Wellburn AR, Heath RL (eds.) Forest decline and ozone: a comparison of controlled chamber and field experiments. Ecological Studies 127, Springer-Verlag, Berlin, pp. 277- 315.

Polle A, Matyssek R, Günthardt-Goerg MS, Maurer S (2000) Defense strategies against ozone in trees: the role of nutrition. In: Agrawal SB, Agrawal M (eds.) Environmental pollution and plant responses. Lewis Publishers, New York, Boca Raton, pp. 223-245.

Poorter H, Villar R (1997) The fate of acquired carbon in plants: chemical composition and construction costs. In: Bazzaz FA, Grace J (eds.) Plant resource allocation. Academic Press, San Diego, pp. 39-72.

Rogers HH, Runio GB, Krupa SV (1994) Plant responses to atmospheric CO_2 enrichment with emphasis on roots and the rhizosphere. *Environmental Pollution* **83**: 155-189.

Ryan MG, Binkley D, Fownes JH (1997) Age-related decline in forest productivity: pattern and process. *Advances in Ecological Research* **27**: 213-262.

Sandermann H Jr. (1996) Ozone and plant health. In: *Annu. Rev. Phytopathol.* **34**: 347-366.

Sandermann H Jr., Bonnet-Masimbert M (1995) EUROSILVA, Contribution to forest tree physiology. Les Colloques 76, INRA, Paris/France, pp. 368.

Sandermann H JR. Wellburn AR, Heath RL (1997) Forest decline and ozone: a comparison of controlled chamber and field experiments. Ecological Studies 127, Springer-Verlag, Berlin, pp. 400.

Saxe H, Ellsworth DS, Heath J (1998) Tree and forest functioning in an enriched CO_2 atmosphere. *New Phytologist* **139**: 395-436.

Schulze E-D, Küppers M, Matyssek R (1986) The roles of carbon balance and branching pattern in the growth of woody species. In: Givnish TJ (ed.) On the economy of plant form and function. Cambridge University Press, pp. 585-602.

Schulze E-D, Lange OL, Oren R (1989) Forest decline and air pollution. Ecological Studies 77, Springer-Verlag, Berlin, pp. 475.

Skärby L, Ro-Poulsen H, Wellburn AM, Sheppard LJ (1998) Impacts of ozone forests: A European perspective. *New Phytologist* **139**: 109-122.

Smith WK, Hinckley TM (1995a) Ecophysiology of coniferous forests. Academic Press, San Diego, pp. 338.

Smith WK, Hinckley TM (1995b) Resource physiology of conifers: acquisition, allocation and utilization. Academic Press, San Diego, pp. 396.

Stitt M, Schulze E-D (1994) Plant growth, storage, and resource allocation: from flux control in a metabolic chain to the whole-plant level. In: Schulze E-D (ed.) Flux control in biological systems. Academic Press, San Diego, pp. 57-118.

Tremmel DC, Bazzaz FA (1995) Plant architecture and allocation in different neighborhoods – implications for competitive success. *Ecology* **76**: 262-271.

Waring RH, Silvester WB (1993) Variation in foliar ^{13}C values within the crowns of Pinus radiata trees. *Tree Physiology* **14**: 1203-1213.

Yoder BJ, Ryan MG, Waring RH, Schoettle AW, Kaufmann MR (1994) Evidence of reduced photosynthetic rates in old trees. *For. Sci.* **40**:513-527.

Author index

Index

Tree Physiology

1. F.J. Bigras and S.J. Colombo (eds.): *Conifer Cold Hardiness*. 2001 ISBN 0-7923-6636-0
2. S. Huttunen, H. Heikkilä, J. Bucher, B. Sundberg, P. Jarvis and R. Matyssek (eds.): *Trends in European Forest Tree Physiology Research*. Cost Action E6: EUROSILVA. 2001 ISBN 1-4020-0023-5

KLUWER ACADEMIC PUBLISHERS – DORDRECHT / BOSTON / LONDON

Nitrogen, phosphorus and sulphur utilization by fungi

Nitrogen, phosphorus and sulphur utilization by fungi

Symposium of the British Mycological Society held at the University of Birmingham April 1988

Edited by
Lynne Boddy, R. Marchant & D. J. Read

The right of the University of Cambridge to print and sell all manner of books was granted by Henry VIII in 1534. The University has printed and published continuously since 1584.

Published for The British Mycological Society by
CAMBRIDGE UNIVERSITY PRESS
Cambridge
New York New Rochelle Melbourne Sydney

Published by the Press Syndicate of the University of Cambridge
The Pitt Building, Trumpington Street, Cambridge CB2 1RP
32 East 57th Street, New York, NY10022, USA
10 Stamford Road, Oakleigh, Melbourne 3166, Australia

First published 1989

Designed and produced by **Page Design**, 48 Alan Road, Stockport SK4 4LE

Printed in Great Britain by Impact Print Services (Manchester) Ltd., Slade Lane, Manchester

British Library cataloguing in Publication Data

British Mycological Society, *(Symposium; 1988; University of Birmingham)*
Nitrogen, phosphorus and sulphur utilization by fungi
1. Fungi nutrition
I. Title II. Boddy, Lynne III. Marchant, R. (Roger) IV. Read, D. J. (David J.)
589.2'0413

Library of Congress cataloguing data available

ISBN 0 521 37405 7

Contents

Contributors

Lynne Boddy, *School of Pure and Applied Biology, University of Wales, College of Cardiff, PO Box 917, Cardiff CF2 1XH, UK.*

M. D. Coffey, *Department of Plant Pathology, University of California at Riverside, CA 92521-0122, USA.*

P. D. Crittenden, *Department of Botany, University of Nottingham, Nottingham NG7 2RD, UK.*

J. Dighton, *Institute of Terrestrial Ecology, Merlewood Research Station, Grange-over-Sands, Cumbria LA11 6JU, UK.*

S. Gianinazzi, *Laboratoire de Phytoparasitologie, Station d'Amélioration des Plantes, INRA, BV 1540, 21034 DIJON CEDEX, France.*

V. Gianinazzi-Pearson, *Laboratoire de Phytoparasitologie, Station d'Amélioration des Plantes, INRA, BV 1540, 21034 DIJON CEDEX, France.*

B. A. D. Hetrick, *Department of Plant Pathology, Kansas State University, Throckmorton Hall, Manhatten, Kansas 665016, USA.*

D. H. Jennings, *Department of Genetics and Microbiology, University of Liverpool, P.O. Box 147, Liverpool L69 3BX, UK.*

A. R. Langdale, *Department of Plant Sciences, University of Sheffield, Sheffield S10 2TN, UK.*

J. R. Leake, *Department of Plant Sciences, University of Sheffield, Sheffield S10 2TN, UK.*

R. Marchant, *Department of Biological and Biomedical Sciences, University of Ulster, Coleraine BT52 1SA, Northern Ireland, UK.*

D. G. Ouimette, *Department of Plant Pathology, University of California at Riverside, CA 92521-0122, USA.*

N. D. Paul, *Institute of Environmental and Biological Sciences, University of Lancaster, Baillrigg, Lancaster LA1 4YQ, UK.*

D. J. Read, *Department of Plant Sciences, University of Sheffield, Sheffield S10 2TN, UK.*

Anthony H. Rose, *Zymology Laboratory, School of Biological Sciences, University of Bath, Bath BA2 7AY, Avon, UK.*

J. C. Slaughter, *Department of Brewing and Biological Sciences, Heriot Watt University, Edinburgh EH1 1HX, UK.*

A. B. Tomsett, *Department of Genetics and Microbiology, The University of Liverpool, P.0. Box 147, Liverpool L69 3BX, UK.*

M. Wainwright, *Department of Microbiology, University of Sheffield, Sheffield S10 2TN, UK.*

D. R. Walters, *Department of Plant Sciences, West of Scotland College, Auchincruive, Nr Ayr KA6 5HW, UK.*

Preface

Nitrogen, phosphorus and sulphur are the major mineral nutrient elements in the biosphere and are essential for growth of both fungi and autotrophic plants. In terrestrial ecosystems fungi play a central role in the release of these elements from organic residues and therefore they indirectly influence not only plant nutrition but also primary productivity. Also, fungi which interact with autotrophs as mutualists or parasites directly influence the availability of these nutrients for the plant. Thus, understanding nutrient relationships of fungi and of fungus-plant interactions is both important and rewarding and these subjects have attracted a great deal of attention in recent years. This volume contains key papers from the 6th General Meeting of the British Mycological Society on the subject of *Nitrogen, phosphorus and sulphur metabolism*. The book is broadly based, covering four main areas: physiology and metabolism of nitrogen, phosphorus and sulphur in fungi (Chapters 1 to 6); the role of these minerals in pathogenic relationships with plants (Chapters 7 & 8); their role in mutualistic relationships with plants (Chapters 9 to 12); and the role of fungi in cycling of these elements within ecosystems (Chapter 13).

The literature concerning the physiology and biochemistry of nitrogen and phosphorus metabolism in fungi is extensive and in the first chapter Jennings has synthesized some of the major features while presenting linking hypotheses and suggesting areas for future research. The theme of nitrogen metabolism is continued in the following chapter in which Tomsett provides an account of the genetics and biochemistry of nitrate assimilation, particular emphasis being placed on ascomycetes. The utilization of inorganic sulphur is considered in Chapters 3 to 5. In the chapters by Rose and Slaughter it is apparent that much past work has been concentrated on hemiascomycetous yeasts, presumably because of their importance in the food and brewing industry, and the need to extend such studies to mycelial fungi is stressed. In contrast, inorganic sulphur oxidation has been investigated almost exclusively in mycelial fungi. Again, however, much further work is needed before such processes can be fully understood. As Wainwright points out in his chapter (4) on the utilization of reduced sulphur compounds, it is still unclear what 'benefits' the fungus obtains from the ability to metabolise such compounds - the processes do not appear to be favourable in energetic terms and it is not clear whether they are of quantitative significance in sulphur cycling within ecosystems. These three chapters therefore highlight the opportunities for further research on fungal sulphur metabolism.

The sixth chapter deals with the metabolism of phosphorus compounds, but covers the fascinating and much neglected subject of the role of phosphorus compounds as antifungal agents. Here Coffey & Ouimette examine the effect of phosphonates on Oomycetes and give some indication of how knowledge of fungal metabolism may be put to practical use.

To complement the first six chapters, which deal with the way in which fungi metabolise nitrogen, phosphorus and sulphur compounds, the following six chapters examine the processes involved in capture and transport of these nutrients by fungi growing in association with living plants, and consider the consequences of these processes for the plant. Walters (Chapter 7) and Paul (Chapter 8), in complementary chapters, consider pathogenic associations, and since little is known about necrotrophic and hemibiotrophic infections emphasis is placed on biotrophic infections which are better understood. Taken together, they provide a picture of nutrient fluxes between host plant and fungus and changes in location, content and concentrations of nitrogen, phosphorus and sulphur within whole plants and plant tissues when infected by biotrophic fungi.

The next three papers concern mutualistic relationships. It is now recognised that the roots of the majority of land plants grow in mutualistic association with fungi to form mycorrhizas. Because they receive a regular supply of carbon from their hosts the fungi involved are amongst the most active components of the soil microbial community. Their role in the nitrogen and phosphorus nutrition of the autotroph has been extensively studied but the contributions presented here (Chapters 9 to 11) emphasize recent developments in our understanding of the biology of mycorrhizas, some of which challenge conventional notions of their function. Plant ecologists increasingly recognize the overwhelming importance of nitrogen as the factor determining productivity of ecosystems, and new information presented by Read and his colleagues (Chapter 9) suggests that ericoid and some ectomycorrhizal fungi may be directly involved in the mobilization, assimilation and transport of organic nitrogen compounds. The role of mycorrhizas in the phosphorus nutrition of plants is well documented but the contributions of Hetrick (Chapter 10) and Gianinazzi-Pearson & Gianinazzi (Chapter 11) present new insights and syntheses, the latter relating in particular to biochemical and ultrastructural features of mycorrhizas in general and the former to whole plant aspects of the vesicular-arbuscular mycorrhizal symbiosis.

The other type of mutualistic association involving fungi and plants is the lichen symbiosis. As Crittenden explains in Chapter 12, knowledge of nitrogen, phosphorus and sulphur relationships in lichens is comparatively rudimentary, partly because lichenology has not benefitted from the same economic impetus as have studies of other fungal systems, but also because of the experimental difficulties imposed by lichens. Crittenden's review of current knowledge of the nitrogen relations of mat-forming lichens indicates the boundless opportunities for future research. Nonetheless, he emphasizes that a number of recent advances have been made and illustrates some experimental approaches which may help further to elucidate these relations.

In addition to pathogenic and mutualistic associations, fungi are also important as decomposers and as such play a crucial role in nutrient cycling within ecosystems. Unfortunately, although detailed nutrient budgets have been obtained for a number of ecosystems, the specific contribution of fungi to these budgets has received relatively little attention. In the final chapter Dighton & Boddy attempt to illustrate the importance of fungi in nitrogen, phosphorus and sulphur cycling in temperate forests, and to point to the many gaps which still exist in our knowledge of their role at the ecosystem level.

We are grateful to all those who contributed to the symposium, and especially to Dr Chris Caten whose skillful organization of the domestic arrangements was a major contribution to the success of the meeting. The Society wishes to thank the British Council and the Royal Society of London for donations in support of speakers visiting from outside the UK.

Lynne Boddy, University of Wales, College of Cardiff
Roger Marchant, University of Ulster
David Read, University of Sheffield

Chapter 1

Some perspectives on nitrogen and phosphorus metabolism in fungi

D. H. Jennings

Department of Genetics and Microbiology, University of Liverpool, P.O. Box 147, Liverpool L69 3BX, UK

Introduction

Nitrogen and phosphorus are two of the major elements in fungal nutrition. It is tempting to consider initially the ecological role of these two elements because they can exist in many chemical forms in natural habitats. Thus, the availability of the various forms might be expected to have very considerable selective influence on the presence of particular species within a habitat. However, I have argued (Jennings, 1987a) that minor elements may be equally if not more important in determining fungal niche. For that reason I will initially consider the relationship between fungi and the two elements from the metabolic and physiological standpoint. This consideration does then engender some thoughts which are relevant to fungal ecology.

There is a large body of literature concerning the physiology and biochemistry of nitrogen and to a lesser extent phosphorus. Beever & Burns (1980) have provided an excellent review of the uptake, storage and utilization of phosphorus in fungi, hence I have placed much more emphasis on nitrogen. When we consider the molecular details of nitrogen metabolism in fungi, the information from three species – *Aspergillus nidulans*, *Neurospora crassa* and *Saccharomyces cerevisiae* – far outweighs what we know for all other fungi. Combined biochemical and genetic analysis now allows an understanding of how the various processes of catabolism and anabolism of nitrogen in these three fungi are regulated, but extrapolation to other species must be undertaken cautiously. Nevertheless, there is sufficient information on other species to point to other aspects which need to be taken into account in any overview of the role of nitrogen in the physiology and ecology of fungi

Nitrogen metabolism in *Aspergillus nidulans*, *Neurospora crassa and Saccharomyces cerevisiae*

These fungi can use diverse sources of nitrogen including ammonium, nitrite, nitrate, numerous amino acids, acetamide, purines and proteins. Ammonium, glutamate and glutamine tend to be the favoured sources; use of the others requires the synthesis of the necessary enzymes or activation of previously existing enzymes. Maximal growth rates are obtained with these three nitrogen sources and growth on them leads to the most intense depression of a whole range of activities, including membrane transport linked to utilization of other nitrogen compounds. This depression of activities – termed nitrogen catabolite control, regulation or repression – must be lifted if *de novo* synthesis of enzymes catabolizing other nitrogen compounds is to take place. Nitrogen catabolite control might be considered to be an adaptive mechanism which gives priority to the utilization of nitrogen sources. Thus an amino acid such as arginine is only utilised effectively when ammonium, glutamate or glutamine are absent from the medium.

Nitrogen catabolite control is a transcriptional control system that governs many nitrogen-catabolic enzymes and all the evidence is that glutamine acts as the metabolic signal leading to metabolic control (Wiame, Grenson & Arst, 1985; Davis, 1986). In *S. cerevisiae* control appears to be a negative system in which glutamine represses the target enzymes, though it is possible that ammonium also has a role (Wiame *et al.*, 1985). The effect of glutamine is presumed to be mediated by an as yet unidentified protein (GDHCR) which is the product of the *gdhCR* locus (which has been named for the constitutive state of the NAD-linked glutamate dehydrogenase activity but which is not a structural gene for *either* glutamate dehydrogenase). Mutants of this locus are derepressed for many glutamine-controlled enzymes. Indeed in the presence of a *gdhCR* mutation, synthesis of all the proteins (enzymes, permeases and probably regulatory proteins) except one have been found to be derepressed under nitrogen-repress-ing conditions. The exception is glutamine synthetase, synthesis of which is repressed even in a *gdhCR* mutant.

In the two filamentous fungi it appears that glutamine opposes a positive regulatory element encoded by *areA*, in the case of *A. nidulans*, and by *nit2* in the case of *N. crassa* (Wiame *et al.*, 1985; Davis, 1986; and see Tomsett, Chapter 2). The regulator produced is required for expression of nitrogen-catabolizing enzymes including nitrate reductase (Cove, 1979; Marzluf, 1981).

Nitrogen catabolite control can be seen as one aspect of the ability of these three fungi to adapt to a changing environment with respect to the form and amount of nitrogen present. In *S. cerevisiae* there is rapid adaptation to new nitrogen sources. Thus, only 3 - 5 min elapses between the addition of arginine or urea to the growth medium and the appearance of the enzymes degrading arginine or allantoin (which is broken down to ammonia via urea); incidentally, the time required for induction is the same as that needed for the induction of β-galactosidase in *Escherichia coli* (Cooper, 1982a). It must be realised that not only is there induction of the necessary catabolic enzymes within the cytosol but also the necessary permease(s). Indeed a prominent part of nitrogen catabolite control is via inducer exclusion, i.e. the inducer fora particular metabolic pathway is unable to enter the protoplasm of the fungus. Of particular interest in this respect is the effect of ammonium (probably through the *gdhCR* gene product GDHCR) on the general amino acid permease (Wiame *et al.*, 1985). This transport system has more than 20 natural substrates including all the L-amino acids which are constituents of proteins. No activity of the permease is detectable in wild type cells grown in the presence of ammonium. The available data indicate that the permease can be activated and inactivated and that ammonium promotes the inactivation process, thus excluding inducers taken up by this particular transport system.

In growing green plants, nitrogen metabolism is envisaged as a flux of the element from inorganic form via amino acid to protein. In fungi, the ability, indeed the necessity, to use a range of nitrogen sources means that both catabolic and anabolic activity must take place simultaneously. For ordered growth to occur, there must be sophisticated regulation of metabolism particularly in those situ-

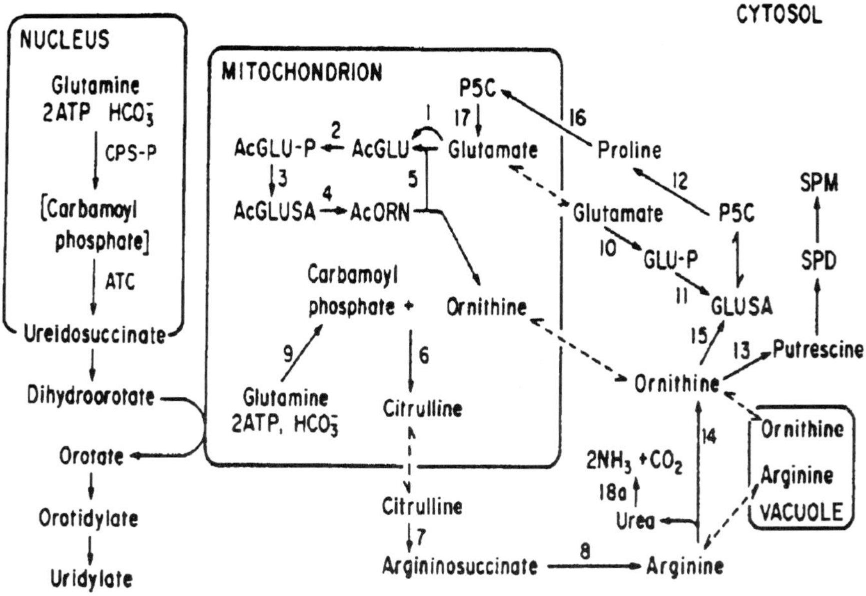

Fig. 1.1A (above): Organisation of arginine, pyrimidine, proline and polyamine metabolism in *Neurospora crassa*. The locations of reactions 16 and 17 have not been proved but are shown as in *Saccharomyces cerevisiae*. Fig. 1.1B (facing page): Organisation of selected enzymes of arginine metabolism in *S. cerevisiae*. Abbreviations: ATC, aspartate carbamoyltransferase; CPS-P, pyrimidine-specific carbamoyl-P synthetase; Ac-, acetyl; GLU, glutamate; GLUSA, glutamate semialdehyde; ORN, ornithine; P5C, pyrroline 5-carboxylate; SPD, spermidine; SPM, spermine. Identities of the enzymes: 1, acetylglutamate synthase; 2, acetylglutamate kinase; 3, acetylglutamyl-P reductase; 4, acetyl-ornithine transaminase; 5, acetylornithine-glutamate acetyltransferase; 6, ornithine carbamoyltransferase; 7, argininosuccinate synthetase; 8, argininosuccinate lyase; 9, carbamoyl-P synthetase A; 10, glutamate kinase; 11, glutamyl-P reductase; 12, pyrroline 5-carboxylate reductase; 13, ornithine decarboxylase; 14, arginase; 15, ornithine transaminase; 16, proline oxidase; 17, pyrroline 5-carboxylate dehydrogenase; l8a, urease; 18b, urea amidohydrolase. From Davis (1986).

ations when common intermediate(s) exist. Perhaps the best studied carbon source has been arginine in which the catabolic pathway consists of the hydrolysis of arginine to ornithine and urea, the breakdown of urea to ammonia and carbon dioxide and the conversion of ornithine to glutamate. The anabolic pathway involves the synthesis of ornithine and of carbamoyl phosphate, with the conversion of these two compounds to arginine. Questions are posed as to how catabolic and anabolic processes can both proceed with ornithine as a common intermediate and how both activities take place given that ornithine and carbamoyl phosphate might be diverted into other pathways. Arginine metabolism has been studied most extensively in *N. crassa* and *S. cerevisiae* and we now know that similar metabolic problems are solved in different ways in these two species (Davis, 1986).

A major difference between *N. crassa* and *S. cerevisiae* is the location of carbamoyl phosphate synthetase A (responsible for the conversion of glutamine to carbamoyl phosphate) and ornithine carbamoyl-transferase (responsible for the conversion of carbamoyl phosphate and ornithine to citrulline). In *N. crassa* these two enzymes are in the mitochondrion; in *S. cerevisiae* they are in the cytosol. Thus both these enzymes differ in kinetic characteristics as befits their different locations (Fig. 1.1). Nevertheless, while there are these very distinct differences between the two organ-

Fig. 1.1B. The organization of arginine metabolism in *S. cerevisiae*

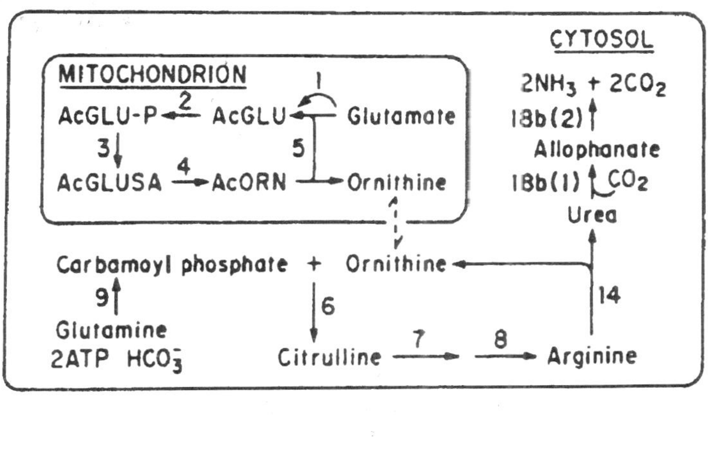

isms, this should not distract us from seeing a similar role of cell compartments in helping to modulate metabolic activity, either by storage, as is the case of the vacuole, or by separating one pathway from another, as is the case when enzymes are located in the mitochondrion. Thus for instance, when arginine is added to a culture of *N. crassa* in minimal medium, the vacuole allows the rapid onset of catabolism of arginine by discharging ornithine into the cytosol, while arginine minimises a futile ornithine cycle by preventing the ornithine in the cytosol from reaching the ornithine carbamoyl transferase in the mitochondrion (Bowman & Davis, 1977). The vacuole acts as a store for most of the amino acids which are basic; 90% of them are vacuolar (with acidic amino acids being predominantly in the cytosol) (Messenguy, 1987).

It is clear that, since ammonium, glutamate and glutamine bring about nitrogen catabolite control, the enzyme glutamate dehydrogenase must play a key role in nitrogen metabolism as well as also being a branch point between carbon and nitrogen metabolism. This enzyme activity can catalyze either the reductive amination of 2-oxoglutarate to produce glutamate or the oxidative deamination of glutamate to produce ammonia. In the three fungi considered here, there are two enzymes, one requiring NADPH and apparently serving to synthesize glutamate (NADP-GDH), the other requiring NAD^+ and leading to the production of 2-oxoglutarate and ammonium (NAD-GDH). A great deal is known about NADP-GDH from *N. crassa*. It is a hexamer of six identical subunits, each of which is a polypeptide of 452 amino acid residues; the polypeptide sequence is known (Holder *et al.*, 1975) and there is much information about allosteric behaviour (Kinsey *et al.*, 1980); the gene for the enzyme (*am*) has been sequenced (Kinnaird & Fincham, 1983). More recently, considerable attention has been given to the effects of specific amino acid replacements on enzyme properties (Fincham, Kinnaird & Burns, 1985).

However, despite the considerable molecular knowledge, little is known about the regulation of NADP-GDH or of NAD-GDH (Marzluf, 1981; Cooper, 1982a). One difficulty may be that both enzymes appear to be controlled by both nitrogen and carbon circuits. Another is that glutamine can be synthesised via the activity of glutamine synthetase (GS) and glutamate synthase (glutamine (amide) : 2-oxoglutarate amino transferase, with acronym GOGAT). Thus:

$$\text{Glutamate}+\text{ATP}+\text{NH}_3 \xrightarrow{\text{GS}} \text{Glutamine}+\text{ADP}+\text{P}_i$$

$$\text{Glutamine}+\text{2-oxoglutarate}+\text{NADPH}+\text{H}^+ \xrightarrow{\text{GOGAT}} \text{2Glutamate}+\text{NADP}^+$$

The foregoing is a somewhat superficial overview of the regulation of the metabolism of relatively simple nitrogen compounds which ensure the coordinated production of the specific amino acids necessary for protein synthesis. Much is known about these pathways; those in *A. nidulans* and *N. crassa* have been reviewed by Pateman & Kinghorn (1976) and those in *S. cerevisiae* by Jones & Fink (1982). It is not appropriate to consider here how enzyme activity of a pathway is regulated by controls specific to that pathway but an example is provided in Fig. 1.2 which gives control patterns observed for transport and catabolism of arginine, proline and the allantoin pathway intermediates in *S. cerevisiae*. However, attention needs to be drawn to the fact that many enzymes of amino acid biosynthesis are under complex general or cross-pathway control. Thus, starvation for any one of a number of amino acids will lead to derepression of the same spectrum of enzymes of several pathways. Those pathways which are now known to have an enzyme or enzymes under general control include arginine, histidine, tryptophan, lysine, leucine, isoleucine and valine. As the number of enzymes in a pathway under general control can differ, the pattern of regulation occurring when the general control mechanism is triggered can be complex. This general control is very relevant to amino acid metabolism during starvation which will be considered below. I will not, however, discuss how the relatively simple nitrogen compounds enter the three fungi as this is treated in detail elsewhere (Pateman & Kinghorn, 1976; Eddy, 1980; Iaccarino, Guardiola & De Felice, 1980; Wolfinbarger, 1980a; Cooper, 1982b; Eddy, 1982; Wiame *et al.*, 1985).

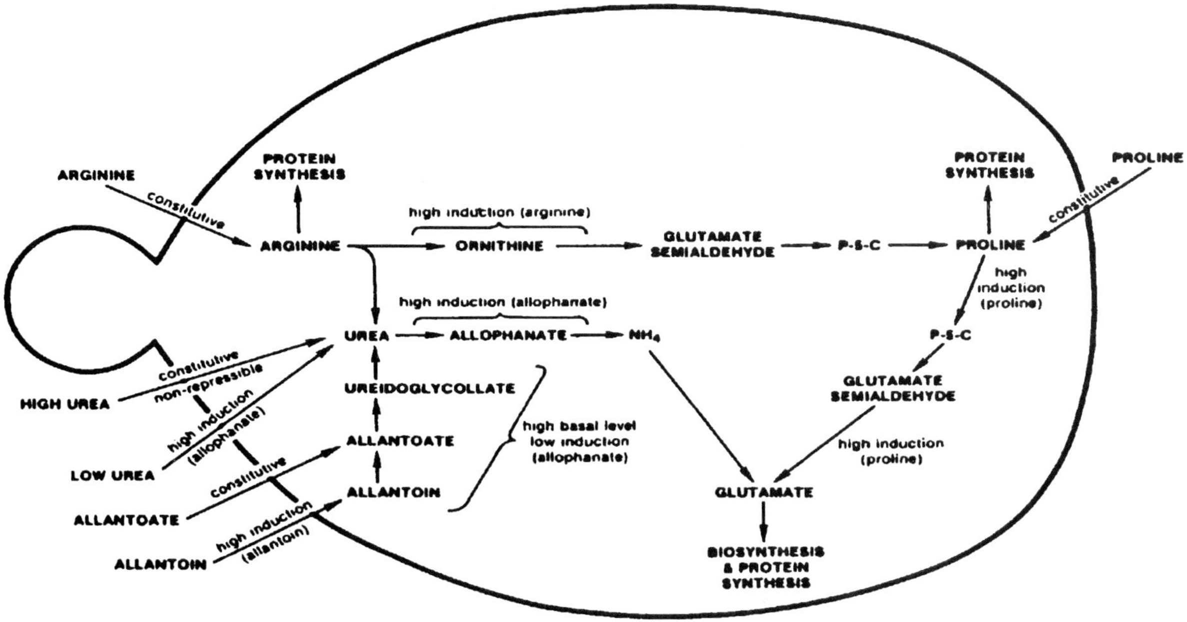

Fig. 1.2. Summary of control patterns observed from the transport and catabolism of arginine, proline and allantoin pathway intermediates. From Cooper (1982a).

If cells of a wild-type culture of *S. cerevisiae* undergoing exponential growth are transferred to distilled water, cell division continues for one or two generations, although the time interval between divisions increases. The cells are drawing upon their stores of nitrogen compounds to synthesise the necessary proteins. In particular the vacuolar pools of nitrogen are mobilised. Under the conditions in which amino acids are released from the vacuole, their metabolism is such as to be no longer under general or cross-pathway control. While information is available about the molecular basis of the general control system (Cooper, 1982a; Henry, Klig & Loewy, 1984; Messenguy, 1987), there are many uncertain aspects (Cooper, 1982a). In particular we do not know the relationship between the control system and flux of amino acids into and out of the vacuole or the relation between it and the cell cycle.

Proteins in yeast cells are constantly degraded. It is this process of degradation which allows the cell to adapt its protein composition to changes in the external environment. The rate of protein degradation increases in cells growing normally from around 0·5 - 1·0 to 2 - 3% h^{-1} under starvation conditions (Lopez & Gancedo, 1979) and proteinases are clearly involved. There is now considerable information about proteinases in *S. cerevisiae* and *N. crassa* (Wolf, 1980, 1986; Jones, 1984). The proteinases of the vacuole are very unspecific and it is this nonspecificity which, together with the properties of other vacuolar enzymes, have led to the vacuole being equated with a lysosome. Studies with mutants leave little doubt that the proteinases found in the vacuole are involved in rather nonspecific protein degradation induced by nitrogen starvation, and make both exogenous and endogenous peptides accessible for primary metabolism. Jones (1984) points out that the proteinases of the vacuoles are brought into play when the only source of amino acids for adaptation and restructuring is within the cell itself. The vacuole can therefore be viewed as the resting place of machinery to be mobilized in 'emergencies'.

It is becoming increasingly clear that proteinases play a highly significant specific role in the metabolism and development of cells of *S. cerevisiae*. Thus proteinases have been implicated in septum formation, sporulation, catabolite inactivation, carbon starvation-induced degradation of NADP-dependent glutamate dehydrogenase, nitrogen starvation-induced degradation of

NAD-dependent glutamate dehydrogenase and of glutamate synthase, enzyme secretion, localization of mitochondrial enzymes, processing of enzyme precursors, production of the killer toxin and of the pheromone α-factor, destruction of the α-factor, degradation of nonsense proteins and nonsense fragments and cell cycle-regulated protein degradation (Jones, 1984).

Both *Aspergillus* species and *N. crassa* can secrete extracellular proteinases, the production of which is regulated by carbon-, nitrogen-, sulphur- and even phosphorus-catabolite repression (Cohen, 1980, 1981; North, 1982). Starvation for any of the these elements results in the production of proteinase. In *Aspergillus*, relief of catabolite repression is sufficient for production; in *N. crassa*, protein must also be present. It should be noted that for growth to occur using the products of protein breakdown, it is not necessary for there to be the release of free amino acids. In *N. crassa*, there is an oligopeptide transport system capable of transporting peptides with up to five amino acid residues (Wolfinbarger, 1980b).

Nitrogen metabolism in other fungi

A number of themes could be followed when considering nitrogen metabolism in the less studied fungi. For example, there are the two pathways of lysine biosynthesis in fungi (Vogel, 1964): one involves α,ε-diaminopimelic acid and is associated with the Oomycetes *sensu stricto*; the other involves α-aminoadipic acid and is found in the Chytridiales, Mucorales and the so-called higher fungi. The presence of these pathways is associated with other biochemical features (Fig. 1.3; LéJohn, 1971). In particular, NADP-linked glutamate dehydrogenase is confined to the Ascomycotina, Basidiomycotina and Deuteromycotina, and within the lower fungi the NAD-linked glutamate dehydrogenases can be divided into three classes on the basis of their regulatory properties (LéJohn, 1971). This argues the need for concerted further study of the nitrogen metabolism of lower fungi.

There are differences in the specificity of nitrogen sources for growth. All fungi seem able to utilise some kind of organic nitrogen source. The great majority of fungi can use ammonium as the sole source of nitrogen. A number of fungi can use nitrate in addition to ammonium but few Basidiomycotina and Saprolegnia-

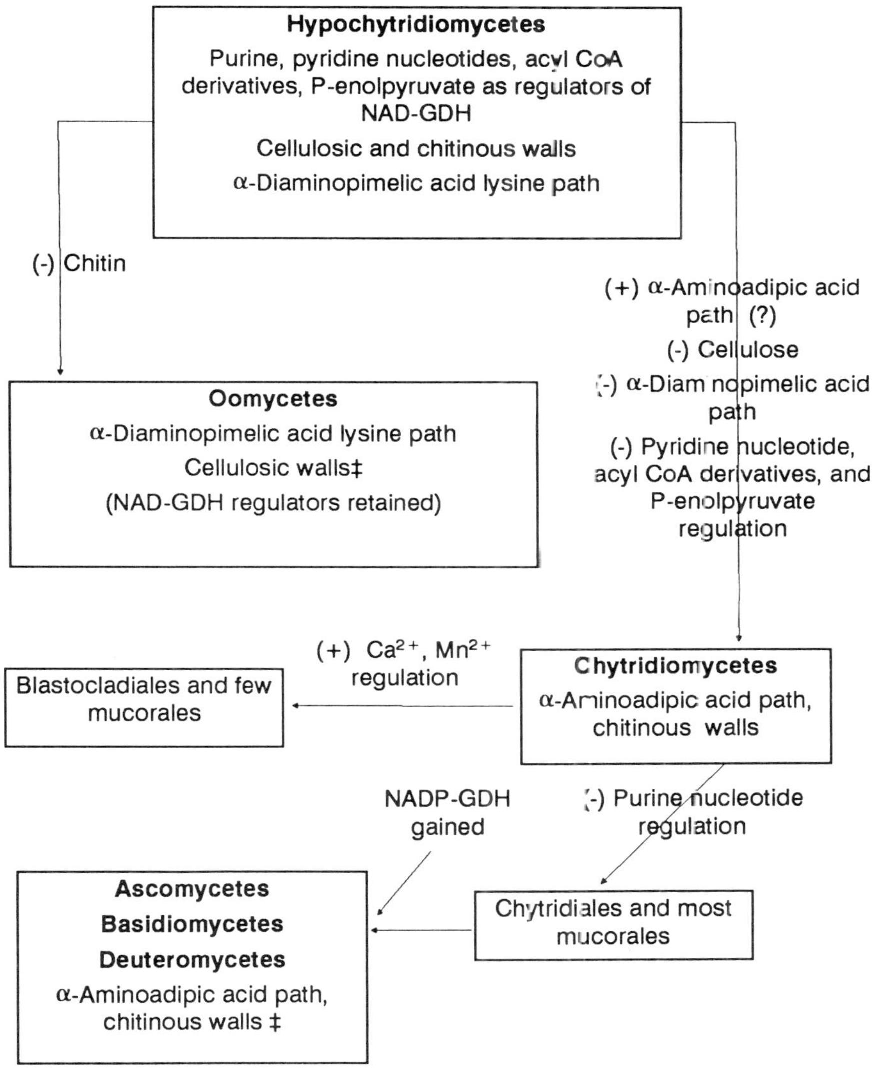

Fig. 1.3. Suggested path in the evolution of NAD-linked glutamate dehydrogenases in fungi and the relationship to cell wall structure, lysine biosynthesis and phylogeny; (+) signifies acquisition of ability; (-) loss of capacity. The suffix, ‡, means that exceptions are known. After LéJohn (1971).

ceae have this ability (Cochrane, 1963). There are indications that certain plant pathogens are apparently unable to use nitrate or ammonium as a source of nitrogen, and the situation is probably complex. Thus, in the two rust fungi whose nutrition has been critically assessed in axenic culture, it has been found that *Gymnosporangium juniperi-virginianae* can use either nitrate or ammonium as well as a wide range of amino acids and other organic nitrogen sources, whereas *Puccinia graminis* cannot use nitrate (Maclean, 1982). Further, in the presence of 8 - 80 mM ammonium, *P. graminis* and many rust fungi require a sulphur amino acid, e.g. 4 mM cysteine or 0·3 mM methionine. Likewise, while it is generally considered that species of *Phytophthora* prefer organic nitrogen sources, only very few species – and then only certain strains – have an absolute requirement (Hohl, 1983). In considering the ability of a fungus to grow on an inorganic nitrogen source, it must be remembered that absence of growth may not mean that the fungus cannot convert the nitrogen from that source into amino acids. It may simply be that the fungus has a specific additional requirement for one or a few amino acids. This appears to be the case for *Thraustochytrium* with respect to lysine; apparently the fungus cannot synthesise the amino acid (Paton & Jennings, 1988) though it seems to be able to metabolise inorganic nitrogen sources (Bahnweg, 1970).

With regard to ammonia assimilation in the higher fungi, it is clear, as indicated above, that both NAD-linked and NADP-linked glutamate dehydrogenases are present in vegetative mycelium (Fawole & Casselton, 1972). In *Coprinus cinereus*, NAD-GDH is subject to catabolite repression and urea derepression, while NADP-GDH is catabolite derepressed and repressed by urea, with the former enzyme seemingly involved in ammonia assimilation (Stewart & Moore, 1974). However when the carpophore of this fungus develops there is amplification of activity in the tricarboxylic acid and urea cycles in the cap (Moore, 1984). NADP-GDH is one of the enzymes which also increases in activity and the regulation appears to involve accumulation of acetyl CoA under conditions in which ammonia is limiting (Moore, 1981). NAD-GDH activity also increases in the cap; this enzyme appears to be required for amination of 2-oxoglutarate, the glutamate produced being converted to succinate via the glutamate oxidation loop involving glutamate decarboxylase and 4-aminobutyrate aminotransferase. These steps allow a modified tricarboxylic acid cycle to proceed in spite of the absence of detectable activity of 2-oxoglutarate dehydrogenase (Moore & Ewaze, 1976).

The increase in activity of NADP-GDH in the cap is accompanied by an increase in the activity of glutamine synthetase (Ewaze, Moore & Stewart, 1978). Moore, Horner & Liu (1987) have shown that the two enzymes are probably co-ordinately controlled and together they act as an ammonia scavenging system; ammonium apparently inhibits basidiospore formation in a manner similarto its inhibition of sporulation in *S. cerevisiae*.

Finally, with respect to primary nitrogen metabolism in *C. cinereus*, the stimulation of the urea cycle in the carpophore leads to accumulation of urea, which becomes part of the battery of low molecular weight compounds (including alanine, arginine, glutamate and trehalose) which contribute to the generation of the solute potential required for carpophore growth (Rao & Niederpruem, 1969; Moore, 1984).

It appears from nitrogen-15 nuclear magnetic resonance (^{15}N-NMR) studies that ammonia assimilation by beech (*Fagus sylvatica*) mycorrhizas occurs mainly via the glutamine synthetase/glutamate synthase (GS/GOGAT) pathway and that glutamate dehydrogenase has no significant role (Martin *et al.*, 1986). Higher plants, of course, use this pathway (Lea & Miflin, 1974) and likewise GDH plays no major role in ammonia assimilation (Miflin & Lea, 1980). On the other hand, when the ectomycorrhizal fungus *Cenococcum graniforme* is grown in a nitrogen-rich medium it has been found, again using ^{15}N-NMR, that here too ammonia assimilation occurs via glutamine synthetase (Martin, 1985). Interestingly, the same studies indicated that, as in *S. cerevisiae*, arginine may well be associated with polyphosphate. In contrast to the results of these ^{15}N-NMR studies, Genetet, Martin & Stewart (1984) found that with nitrogen-starved mycelium of *C. graniforme* labelling of amino acids from externally supplied ^{15}N-ammonium could only be explained by assimilation being mediated by NADP-GDH. This implies that when the fungus is part of the ectomycorrhizal association there is both carbon and nitrogen repression, the former probably as a consequence of the significant flux of carbohydrate from the host root into the fungal tissue (Lewis & Harley, 1965).

Finally, it is important to note that *Agaricus bisporus*, *Coprinus cinereus* and *Volvariella volvacea* can all use protein as the sole source of carbon, nitrogen and sulphur (Kalisz, Moore & Wood, 1986; Kalisz, Wood & Moore, 1987). Unlike *Aspergillus* and *N. crassa*, proteinase activity is not repressed by the presence of glucose, ammonium and sulphate in the medium. The presence of protein is required for induction of the proteinases, though in the case of *C. cinereus* proteinase activity is derepressed in protein-free media in the absence of sulphate. It is particularly interesting to find that, when these three fungi were grown with protein as sole source of carbon, very significant levels of ammonium were found in the medium. Since the amount of protein consumed contained far more nitrogen than required for the amount of biomass produced, it must be assumed that a proportion of the protein was catabolized leading to the production of non-nitrogen containing polymers or oxidised to provide the necessary energy for growth. As a result of such processes free ammonia would be produced. In the experiments referred to here one-third to one-half of the supplied protein appears to have been assimilated and metabolised in this manner (Kalisz *et al.*, 1986).

Some thoughts on nitrogen metabolism in fungi

The many studies on *A. nidulans*, *N. crassa* and *S. cerevisiae* have produced relatively clear pictures of how nitrogen metabolism in these fungi might be regulated. From these studies we appreciate the need for compartmentation of certain of the enzymes such that catabolic and anabolic processes can occur in the same cell or hyphal compartment when it is supplied with a particular organic nitrogen compound. The complexity of regulation of nitrogen metabolism is now being revealed and, even within seemingly closely related species, there are significant differences, in spite of the pathways of nitrogen metabolism being generally very similar. Nevertheless, it would seem that most catabolic enzymes are inducible with one of the degradative intermediates acting as an inducer. Overarching this induction process is nitrogen catabolite repression.

Cooper & Sumadra (1983) have pointed out that nitrogen catabolite repression tends to give a certain primacy to such compounds as ammonium, glutamate and glutamine as nitrogen sources. They referred to the relatively long standing presumption that the combination of induction and catabolite repression enables fungi to utilise good nitrogen sources, i.e. those which give high growth rates, in preference to ones which give low growth rates. A corollary would be that the provision of a good nitrogen source to cells growing on a poor one should shift metabolism such that the good source is used. These authors drew attention to the fact that this view does not address the fate of previously formed enzymes that were degrading the poor nitrogen source, or the effects of these enzymes on the metabolism of the good source. They went on to show that in functional terms nitrogen catabolite repression is long-term. Other processes such as transport of the nitrogen compounds into the cell are much more likely to be the sites of short-term regulation such that good nitrogen sources are utilised in preference to poor ones.

On the whole, *A. nidulans*, *N. crassa* and *S. cerevisiae* make poor use of organic compounds as sources of combined carbon but it is becoming increasingly clear that this is by no means true for all fungi. As indicated above, organic nitrogen compounds as complex as proteins are readily used as carbon sources by *A. bisporus*, *C. cinereus* and *V. volvacea*, and Read, Leake & Langdale (Chapter 9) have demonstrated that mycorrhizal fungi can do the same. It is possible that this ability is even more widespread. Loll & Bollag (1983), in their review of protein transformation in soil list 40 genera of fungi said to be proteolytic, though this does not necessarily mean that they can utilise protein as sole sources of carbon. Nevertheless, the ability to produce proteolytic enzymes is a necessary prerequisite of protein utilization.

The use of protein as a source of combined carbon, however, has certain consequences for the physiology of the fungus. Kalisz *et al.* (1986) point out that protein provides far more nitrogen than is necessary for the biomass of a fungus when it utilises protein as the sole source of carbon. It is not surprising therefore that ammonia is released into the medium. But when this occurs it raises the question as to how nitrogen metabolism is regulated particularly since ammonium is metabolically closely associated with nitrogen catabolite repression. While excess nitrogen can be lost as ammonia, the wall could be a sink for excess nitrogen through the polymerization of N-acetylglucosamine. This may be one reason why the wall of *A. bisporus* contains something like 45% (w/w) glucosamine, the amount of chitin and protein being 7-8 times greater than the relatively closely related *Schizophyllum commune* (Michalenko, Hohl & Rast, 1976).

The hypothesis that the wall might act as a sink for nitrogen in this way can only be tested by examining total wall composition of a range of fungi and relating the results to the physiology and ecology of the individual species concerned. Thus, I have pointed out (Jennings, 1982) that walls might be the sink for excess carbon in the face of a high C:N ratio of the substrate on which they are growing. Be that as it may, if fungi release ammonia when metabolising proteins, there must be some mechanism for preventing the compound from exerting its known inhibitory effects on developmental processes, such as sporulation, referred to above. Of course, once non-nitrogenous organic compounds as well as proteins are available for metabolism, it is likely that production of ammonia will decline or indeed cease altogether. In this respect, it is conceivable that a selective pressure towards mycorrhiza formation might be the provision of carbohydrate by the higher plant such that the fungus forming the association might be better able to cope with the utilization of protein available in the external environment.

It is pertinent here to refer to the contrasting situation where nitrogen is in short supply relative to degradable non-nitrogen organic compounds, Paustian (1985) and Paustian & Schnürer (1987a & b) have considered this matter with respect to soil fungi growing under carbon and nitrogen limitation and have demonstrated the importance of nitrogen recycling within the mycelium such that the element is preferentially allocated to hyphal extension. The importance of nitrogen recycling by fungi, especially wood decomposers, has been suggested by Merrill & Cowling (1966) and Levi & Cowling (1969), while Watkinson (1979, 1984) has emphasised the probability that in *Serpula lacrimans* mycelial development, including formation of strands, there is transport of nitrogen within the mycelium and reutilization of the element. This implies that, in view of what we know of regulation of nitrogen metabolism in fungi, there must be fairly rigid compartmentation such that breakdown of cell components can take place in one compartment producing a flux of nitrogen and other compounds to support normal metabolic activity in the next compartment. For recycling to take place with growth of mycelium over non-nutrient surfaces, there must be translocation of nitrogen. Currently the form(s) translocated is not known, though ^{14}C from exogenously supplied ^{14}C-aspartic acid is translocated very readily in mycelium of *S. lacrimans* (Brownlee & Jennings, 1982). However, Cárdenas & Hansberg (1982a & b) have presented good evidence that glutamine may act as the nitrogen carrier from the mycelial mat to the growing aerial mycelium of *Neurospora crassa*.

Watkinson (1984) has pointed out that a wood-decay fungus is able not only to extract nitrogen from wood but concentrate it, by which is meant that the C:N ratio in mycelium is much lower than in wood. It is not clear what is meant mechanistically by concentration (but see Dighton & Boddy, Chapter 13). What is certain is that a fungus must often degrade a considerable quantity of organic matter in wood not containing nitrogen in order to obtain the necessary supply of this element. The release into the surrounding medium of soluble residues from the organic matter so degraded would provide energy sources for competing micro-organisms. Thus, there is a need to remove such compounds. One way in which it might occur would be by metabolism of the compounds to carbon dioxide. The most effective ways in which this could occur would be by fermentation or by respiration in which electrons flow in the respiratory chain to a terminal oxidase which is cyanide-resistant, thus generating less ATP and being energetically less efficient than when electrons flow to cytochrome aa_3 (cytochrome oxidase). This is essentially the proposition of Lambers (1980) with respect to the role of the cyanide-resistant pathway in higher plant cells. Such a pathway has been shown to be present in fungi (Lloyd, 1974) and, interestingly, also in beech mycorrhizal roots (Harley, 1978).

An interesting situation is found in those fungi which decompose wood on the forest floor and which can spread from one resource to another by mycelial cords or rhizomorphs. We know that the structures can absorb phosphate (Clipson, Cairney & Jennings, 1987; Cairney *et al.*, 1988) and can break down organic phosphate (Jennings, 1988). If these structures were able to break down protein, the supply of carbohydrate from the wood (Jennings, 1987b) would assist in any problems there might be due to the relatively unfavourable C:N ratio of the protein. In any case there is a need to examine nitrogen metabolism of these fungi, particularly the extent to which it might be different in mycelium in the wood, which is degrading lignocellulose, and in cords or rhizomorphs growing through the soil. There are clear differences in the biochemistry and physiology of the two types of mycelium (Jennings, 1984b). In terms of what is discussed here, it is of great interest that ligninolytic activity of *Phanerochaete chrysosporium* becomes apparent when the mycelium becomes starved of carbon, nitrogen or sulphur (Jefferies, Choi & Kirk, 1981). These are conditions in which one would expect certain enzymes of nitrogen metabolism repressed in growing mycelium to become derepressed. Nitrogen metabolism of saprotrophic basidiomycetes is clearly a fertile field for investigation.

Finally, we need to be alert to the possibility that in fungi proteins may act as stores of nitrogen. Candidates for proteins which may act in this way are: (i) Woronin bodies, now known not to be made up of ergosterol (Markham & Collinge, 1987); (ii) crystalloids often associated with microbodies (Maxwell *et al.*, 1975) and which are almost certainly enzymic, an outstanding example being the identification of crystalloids in peroxisomes of *Hansenula polymorpha* growing on methanol as being alcohol oxidase (Veenhuis, van Dijken & Harder, 1983); and (iii) the gamma particles of *Blastocladiella emersonii* zoospores (Hohn, Lovett & Bracker, 1984). The concept of enzymes being stores of nitrogen is becoming familiar to green-plant physiologists with ribulose 1,5-bisphosphate carboxylase-oxygenase (RUBISCO) being the prime example of a protein with both an enzymic and storage function (Millard, 1988). In respect to the above, Blayney & Marchant (1977) showed that protein inclusions, with a crystalline substructure, are formed during stipe elongation in *Coprinus cinereus* and degraded during the later stages of development. These investigations suggested that nitrogen in the chitin synthesised during elongation may be derived from the degraded protein inclusions.

Phosphorus physiology

As indicated in the Introduction, Beever & Burns (1980) provide a detailed consideration of uptake, storage and utilization of phosphorus transformations in fungi. Here I wish to focus on the central position of orthophosphate in the physiology and biochemistry of phosphorus in fungi. In contrast to my considerations of nitrogen, it is appropriate first to make an ecological comment.

There are many ecological situations where phosphorus is not readily available to higher plants. The marked ability of most soils to sequester phosphorus is such that the amount of available phosphorus is very low. There is no effect of bulk water on phosphorus movement and the lack of availability is associated with the diffusive movement of phosphorus (as phosphate) to plant roots. Consequently, the more competitive plants appear to be those which can extend the length of their root systems, since this is the most effective way of increasing uptake (Jennings, 1986). Vesicular-arbuscular mycorrhizas are seen as another means of increasing the length of the root system (Tinker, 1975; Hetrick, Chapter 10; Gianinazzi-Pearson & Gianinazzi, Chapter 11).

With regard to fungal mycelium spreading through soil there is, however, little evidence to suggest that phosphorus is not other than accessible. First, mycelium can grow towards sources of phosphorus. Second, we know that fungi can utilise sources of phosphorus other than soluble phosphate, namely condensed phosphates, insoluble phosphates and organic phosphorus compounds, including nucleic acids (Beever & Burns, 1980; Jennings, 1988). Further, some fungi, along with some bacteria and actinomycetes, are able to solubilize phosphate through secretion of organic acids (Khan & Bhatnager, 1977; Kucey, 1987). Finally, apart from the large fruit bodies produced by many basidiomycetes and some ascomycetes, fungi unlike higher plants do not have aerial structures to supply with the element.

Thus, fungal growth in nature may not be limited by phosphorus availability in spite of the low levels of phosphorus found in many environments (Beever & Burns, 1980). A study on the marine species *Rhodotorula rubra* grown in continuous culture supports this view (Button, Dunker & Morse, 1973). The yeast was found to be capable of growth at very low phosphate concentrations; the external concentration giving half maximal growth rate was 10·8 nM. Cells grown for many days in continuous culture under phosphorus limitation showed a higher maximum specific growth rate as a function of phosphate concentration than did cells grown for less than three days. Initial phosphate uptake rates that were in agreement with the steady-state flux in continuous culture were obtained by using cells and medium directly from such cultures. With this procedure, uptake rates were found to be about five hundred times greater than for cells grown in batch culture. Thus it appears that under conditions of phosphate limitation the organism synthesises more of the relevant transport system (Robertson & Button, 1979).

Similar studies of the kind just described need to be made using other fungi. If it is assumed that what has been found for *R. rubra* is generally applicable, and given the known flexibility of phosphate transport systems in fungi with respect to the internal phosphorus status, and the capability of fungi to produce extracellular phosphatases (Beever & Burns, 1980), it would seem that fungi can react very flexibly to differing availability of phosphorus in the external medium.

The above raises the query about the role of polyphosphates, which are ubiquitous in fungi (Kulaev, 1979). There has been a tendency to assume that they could act as stores of phosphorus, possibly because polyphosphates are found in both VA and ectomycorrhizas – those fungal associations which are known to benefit their higher plant partners in terms of the phosphate nutrition of these partners, allowing them to grow better in the face of a limited phosphate supply. Yet, if the role of polyphosphate was solely as a store for phosphorus, it would seem unlikely that it would be in a polymerised form which is energetically demanding, requiring the conversion of ATP to ADP. In any case, it is frequently found that, as in beech mycorrhizal roots (Jennings, 1964) and rhizomorphs of *Armillaria mellea* (Cairney *et al.*, 1988) growing in low orthophosphate concentrations, orthophosphate levels within the fungus are as high if not higher than those of polyphosphate. Finally, if polyphosphates functioned solely as stores of phosphorus, it would be surprising to find them in locales other than the vacuole. Yet as much as 30-35% of the total polyphosphate can be found in the wall of *S. cerevisiae* and substantial amounts have been found to be associated with nuclei (Kulaev & Vagabov, 1983). In this respect it is important to realise that, though polyphosphate is certainly located in the vacuole, how much is there is uncertain since it is not clear how much contamination occurs from vesicles from the endoplasmic reticulum during isolation of vacuoles (Kulaev & Vagabov, 1983).

Whatever view is held of the role of polyphosphates in fungi, account must be taken of the wealth of information now available on the metabolism of these compounds, which is well reviewed by Kulaev & Vagabov (1983). These authors believe that polyphosphates should be viewed as part of a micro-organism's armoury of 'metabolic traps' allowing it to function in the varying conditions of its external environment. Polyphosphates can be viewed as components of the homeostatic system equivalent to the hormonal and nervous regulation of multicellular organisms. Kulaev & Vagabov (1983) point out that 'availability of sufficient amounts of such valuable endogenous pools as high molecular weight polyphosphates makes micro-organisms less dependent on external conditions and, on the other hand, capable, at any suitable moment, of initiating growth and reproduction without any considerable lag period'. Polyphosphates may thus play a role equivalent to that ascribed to polyols, namely that these compounds are part of a system of 'physiological buffering' within fungal protoplasm (Jennings, 1984a).

Such a role of polyphosphates could be played out via a number of metabolic activities. Polyphosphate may act as a high-energy phosphate alternative to ATP (Kulaev & Vagabov, 1983). There is now good evidence that growing micro-organisms, including those fungi studied, control their adenine nucleotide content and also the proportion of the total pool present as ATP, ADP and AMP (Knowles, 1977). Furthermore, micro-organisms have evolved mechanisms for protecting the adenine nucleotide pools against fluctuations in the nutrient content of the environment. Polyphosphates may therefore play a similar protective role. There is evidence from ^{31}P-NMR studies on *Candida utilis* and *Brettanomyces intermedius* that under conditions of lowered ATP concentration within the cells, polyphosphate can provide an alternative source of energy (Nicolay *et al.*, 1983).

Polyphosphates are also involved in the maintenance of the cation/anion balance within fungi. It has been pointed out already that considerable amounts of basic amino acids may be linked to the negative charges of polyphosphate, but, under certain conditions, it has been shown that bivalent cations, particularly the magnesium and manganous ions, can also be so linked (Okorokov, Lichko & Kulaev, 1980; Lichko, Okorokov & Kulaev, 1982). The fact that magnesium can be so associated with polyphosphate may be part of those processes required to regulate the concentration of this bivalent cation within the cytoplasm, where it plays a significant role in phosphorus metabolism. With respect to the more general matter of cation/anion balance within a fungus, it needs to be remembered that orthophosphate will also contribute to the total negative charge within the cell. There is value in a high proportion of that charge being inorganic, such as ortho- and polyphosphate, since there is less need to produce organic acid anions to balance incoming inorganic cations. In this way, carbon is conserved for other functions within the organism.

Finally, the breakdown and synthesis of polyphosphate are two of those several processes which help to regulate cytoplasmic orthophosphate concentration. It has been shown (den Hollander *et al.*, 1981) that the concentration of orthophosphate could be an important determinant of the rate of glycolysis in *Saccharomyces cerevisiae*, depending upon the cytoplasmic pH. From these data and from other published information about the effects the interaction of orthophosphate and pH on the *in vitro* activity of hexokinase and phosphofructokinase, it could be surmised that the regulatory action of the cytoplasmic concentration on glycolysis is via the activity of these two enzymes.

Control of phosphorus concentration in the cytoplasm must also occur at the plasmalemma and the tonoplast. With respect to the former membrane, the best studied fungi are *S. cerevisiae* and *Neurospora crassa*. In the former species, there appear to be three transport systems for the monovalent anion: (i) a constitutive system which has low affinity for the anion; (ii) an inducible proton symport; and (iii) an inducible sodium symport, both of the latter having a high affinity for phosphate. Derepression of the latter two systems occurs when cells are incubated in the presence of a suitable substrate, such as glucose, in a medium lacking phosphate

(Borst-Pauwels, 1981; Nieuwenhuis & Borst-Pauwels, 1984). The protein synthesis inhibitor cycloheximide prevents derepression. At the same time that there is derepression of the high-affinity system, the low-affinity system disappears. The sum, at any one time, of the maximal rates of uptake by the two systems (constitutive and inducible) is constant, suggesting that the constitutive system is involved in both the high-affinity and low-affinity transport of the anion. More recent evidence (Jeanjean *et al.*, 1986) indicates that the transport system is composed of a cell-wall protein and another in the plasmalemma, that in the wall being involved in the high-affinity system.

A similar situation appears to hold for phosphate transport across the plasmalemma in *N. crassa* (Lowendorf, Slayman & Slayman, 1974; Lowendorf, Bazinet & Slayman, 1975; Lowendorf & Slayman, 1975; Burns & Beever, 1977; Beever & Burns, 1977, 1978). As with *S. cerevisiae*, there is a high-affinity system which is derepressed during phosphorus starvation. However, cycloheximide appears to have no effect on derepression. While there is no direct evidence that elements of the phosphate transport system in *N. crassa* involve cotransport of protons, the kinetic characteristics of the system, particularly with respect to pH, strongly suggest that this is likely to be the case (Sanders, 1986).

There is circumstantial evidence that phosphate transport into many fungi is likely to be very similar to what has been found in *S. cerevisiae* and *N. crassa*, namely that some elements of the system involve proton symport and the kinetics of transport being dependent on the internal phosphorus status of the organism (Beever & Burns, 1980; Clipson, Cairney & Jennings, 1987; Straker & Mitchell, 1987; Cairney *et al.*, 1988).

These various transport systems form part of what Metzenberg (1979) has termed the 'phosphorus acquisition system' of a fungus. The other parts are those enzymes which are concerned with 'garnering' phosphorus for the cell when there is insufficient orthophosphate in the external medium. *N. crassa* produces, extracellularly, an alkaline and an acid phosphatase, a 5′-nucleotidase and three nucleases all of which are repressible by phosphate (Metzenberg, 1979). Presumably, these enzymes release orthophosphate from their respective substrates which is then absorbed by the high affinity phosphate transport system. Further consideration of the complex field of extracellular phosphatase production by fungi is beyond the scope of this article, but the review by Beever & Burns (1980) provides some guidance. However, there is no doubt that many fungi are able to release phosphates from organic form. Thus, as is the case for nitrogen, these organisms are likely to have within their environment a much greater source of available phosphorus than the green vascular plant.

Clearly then, as postulated at the beginning of this section, it seems likely that phosphorus does not limit growth in many environments and the best perspective from which to view the relationship between phosphorus and fungi is probably from the standpoint of the regulation of phosphate concentration within the cytoplasm. This has to be achieved in the face of (i) the various ways that the fungus can acquire phosphate and the rate of transport into the cytoplasm, and (ii) the rates of processes within the cytoplasm leading to the net utilization of phosphate. There is clearly control of transport into the cytoplasm and presumably metabolic regulation of those processes utilising phosphate. In this context the vacuole, where phosphate and polyphosphate might be stored (along with those other sites for the storage of polyphosphate mentioned above), might be visualised as providing a reservoir to cushion large changes in cytoplasmic phosphate concentration which otherwise could have untoward effects on metabolism. These ideas can now be investigated experimentally since it is now possible to determine directly cytoplasmic phosphate concentration by ^{31}P-NMR, and a suitable organism would be one which could be grown in continuous culture, yet use a variety of phosphorus sources.

References

Bahnweg, G. (1970), Studies on the physiology of Thraustochytriales. 1. Growth requirements and nitrogen nutrition of *Thraustochytrium* spp., *Schizochytrium* sp., *Japonochytrium* sp., *Ulkenia* spp. and *Labyrinthuloides* spp. *Verofentlichungen der Instituts für Meeresforschung Bremerhaven*, 17, 245-268.

Beever, R. E. & Burns, D. J. W. (1977). Adaptive changes in orthophosphate uptake by the fungus *Neurospora crassa* in response to phosphorus supply. *Journal of Bacteriology*, **132**, 520-525.

Beever, R. E. & Burns, D. J. W. (1978). Does cycloheximide-induced loss of phosphate uptake activity in *Neurospora crassa* reflect rapid turnover? *Journal of Bacteriology*, **134**, 1176-1178.

Beever, R. E., & Burns, D. J. W. (1980). Phosphorus uptake, storage and utilization by fungi. *Advances in Botanical Research*, 8, 127-219.

Blayney, G. P. & Marchant, R. (1977). Glycogen and protein inclusions in elongating stipes of *Coprinus cinereus*. *Journal of General Microbiology*, **98**, 467-476.

Borst-Pauwels, G. W. F. H. (1981). Ion transport in yeast. *Biochimica et Biophysica Acta*, **650**, 88-127.

Bowman, B. J. & Davis, R. H. (1977). Arginine catabolism in *Neurospora*: cycling of ornithine. *Journal of Bacteriology*, **130**, 285-311.

Brownlee, C. & Jennings, D. H. (1982). Long distance translocation in *Serpula lacrimans*. Velocity estimates and continuous monitoring of induced perturbations. *Transactions of the British Mycological Society*, **79**, 143-148.

Burns, D. J. W. & Beever, R. E. (1977). Kinetic characterisation of the two phosphate uptake systems in the fungus *Neurospora crassa*. *Journal of Bacteriology*, **132**, 511-519.

Button, D. K., Dunker, S. S. & Morse, M. L. (1973). Continuous culture of *Rhodotorula rubra*: kinetics of phosphate-arsenate uptake, inhibition, and phosphate-limited growth. *Journal of Bacteriology*, **113**, 599-611.

Cairney, J. W. G., Jennings, D.H., Ratcliffe, R. G. & Southon, T. E. (1988). The physiology of basidiomycete linear organs. II. Phosphate uptake by rhizomorphs of *Armillaria mellea*. *New Phytologist*, **109**, 327-334.

Cádenas, M. E. & Hansberg, W. (1984a). Glutamine requirement for aerial mycelium growth in *Neurospora crassa*. *Journal of General Microbiology*, **130**, 1723-1732.

Cádenas, M. E. & Hansberg, W. (1984b). Glutamine metabolism during aerial mycelium growth of *Neurospora crassa*. *Journal of General Microbiology*, **130**, 1733-1741.

Clipson, N. J. W., Cairney, J. W. G. & Jennings, D. H. (1987). The physiology of basidiomycete linear organs. I. Phosphate uptake by cords and mycelium in the laboratory and in the field. *New Phytologist*, **105**, 449-457.

Cochrane, V. W. (1963). *Physiology of Fungi*. John Wiley: New York.

Cohen, B. L. (1980). Transport and utilization of proteins. In *Microorganisms and Nitrogen Sources*, ed. J. W. Payne, pp. 411-430. John Wiley: Chichester.

Cohen, B. L. (1981). Regulation of protease production in *Aspergillus*. *Transactions of the British Mycological Society*, **76**, 447-450.

Cooper, T. G. (1982a). Nitrogen metabolism in *Saccharomyces cerevisiae*. In *The Molecular Biology of the Yeast Saccharomyces: Metabolism and Gene Expression*, eds. J. N. Strathern, E. W. Jones & J. R. Broach, pp. 399-461. Cold Spring Harbor Laboratory: New York.

Cooper, T. G. (1982b). Transport in *Saccharomyces cerevisiae*. In *The Molecular Biology of the Yeast Saccharomyces: Metabolism and Gene Expression*, eds. J. N. Strathern, E. W. Jones & J. R. Broach, pp. 39-99. Cold Spring Harbor Laboratory: New York.

Cooper, T. G. & Sumadra, R. A. (1983). What is the function of nitrogen catabolite repression in *Saccharomyces cerevisiae*? *Journal of Bacteriology*, **155**, 623-627.

Cove, D. J. (1979). Genetic studies of nitrate assimilation in *Aspergillus nidulans*. *Biological Reviews*, **54**, 291-327.

Davis, R. H. (1986). Compartmental and regulatory mechanisms in the arginine pathways of *Neurospora crassa* and *Saccharomyces cerevisiae*. *Microbiological Reviews*, **50**, 280-313.

den Hollander, J. A., Ugurbil, K., Brown, T. R. & Shulman, R. G. F. (1981). Phosphorus-31 nuclear magnetic resonance studies on the effect of oxygen upon glycolysis in yeast. *Biochemistry*, **20**, 5871-5880.

Eddy, A. A. (1980). Some aspects of amino acid transport in yeast. In *Microorganisms and Nitrogen Sources*, ed. J. W. Payne, pp. 35-62. John Wiley: Chichester.

Eddy, A. A. (1982). Mechanisms of solute transport in selected eukaryotic microorganisms. *Advances in Microbial Physiology*, **23**, 1-78.

Ewaze, J. 0., Moore, D. & Stewart, G. R. (1978). Co-ordinate regulation of enzymes involved in ornithine metabolism and its relation to sporophore morphogenesis in *Coprinus cinereus*. *Journal of General Microbiology*, **107**, 343-357.

Fawole, M. O. & Casselton, P. J. (1972). Observations on the regulation of glutamate dehydrogenase activity in *Coprinus lagopus*. *Journal of Experimental Botany*, **23**, 530-551.

Fincham, J. R. S., Kinnaird, J. H. & Burns, P. A. (1985). The *am* (NADP-specific glutamate dehydrogenase) gene of *Neurospora crassa*. In *Molecular Genetics of Filamentous Fungi*, UCLA Symposia on Molecular and Cellular Biology, New Series volume 34, ed. W. Timberlake, pp. 117-125. Alan R. Liss: New York.

Genetet, I., Martin, F. & Stewart, G. R. (1984). Nitrogen assimilation in mycorrhizas. Ammonium assimilation in the N-starved ectomycorrhizal fungus *Cenococcum graniforme*. *Plant Physiology*, **76**, 395-399.

Harley, J. L. (1978). Ectomycorrhizas as nutrient absorbing organs. *Proceedings of the Royal Society of London, B*, **203**, 1-21.

Henry, S. A., Klig, L. S. & Loewy, B. S. (1984). The genetic regulation and co-ordination of biosynthetic pathways in yeast: amino acid and phospholipid synthesis. *Annual Review of Genetics*, **18**, 207-231.

Hohl, H. R. (1983). Nutrition of *Phytophthora*. In *Phytophthora: its Biology, Taxonomy, Ecology and Pathology*, eds. D. C. Erwin, S. Bartnicki-Garcia & P. H. Tsao, pp. 41-54. American Phytopathological Society: St. Paul, Minnesota, USA.

Hohn, T. M., Lovett, J. S. & Bracker, C. E. (1984). Characterisation of the major proteins, gamma particles, cytoplasmic organelles in *Blastocladiella emersonii* zoospores. *Journal of Bacteriology*, **158**, 253-263.

Holder, A. A., Wootton, J. C., Baron, A. J., Chambers, G. J. K. & Fincham, J. R. S. (1975). The amino acid sequence of *Neurospora* NADP-specific glutamate dehydrogenase. *Biochemical Journal*, **149**, 757-773.

Iaccarino, M., Guardiola, J. & De Felice, M. (1980). Genetics of amino acid transport. In *Microorganisms and Nitrogen Sources*, ed. J. W. Payne, pp. 125-151. John Wiley: Chichester.

Jeanjean, R., Bédu, S., Nieuwenhuis, B. J. W. M. & Hirn, M. (1986). Immunological evidence for the involvement of cell wall proteins in phosphate uptake in the yeast *Saccharomyces cerevisiae*. *Archives of Microbiology*, **144**, 207-212.

Jefferies, T. W., Choi, S. & Kirk, T. K. (1981). Nutritional regulation of lignin degradation by *Phanerochaete chrysosporium*. *Applied & Environmental Microbiology*, **42**, 290-296.

Jennings, D. H. (1964). Changes in the size of orthophosphate pools in mycorrhizal roots of beech with reference to absorption of the ion from the external medium. *New Phytologist*, **63**, 181-193.

Jennings, D. H. (1982). The movement of *Serpula lacrimans* from substrate to substrate over nutritionally inert surfaces. In *Decomposer Basidiomycetes*, British Mycological Society Symposium volume 4, eds. J. C. Frankland, J. H. Hedger & M. J. Swift, pp. 91-108. Cambridge University Press: Cambridge, UK.

Jennings, D. H. (1984a). Polyol metabolism in fungi. *Advances in Microbial Physiology*, **25**, 149-193.

Jennings, D. H. (1984b). Water flow through mycelia. In *The Ecology and Physiology of the Fungal Mycelium*, British Mycological Society Symposium volume 8, eds. D. H. Jennings & A. D. M. Rayner, pp. 143-164. Cambridge University Press: Cambridge, UK.

Jennings, D. H. (1986). Salt relations in cells, tissues and roots. In *Plant Physiology: A Treatise*, vol. IX, *Water and Solutes in Plants*, eds. F. C. Steward, J. F. Sutcliffe & J. E. Dale, pp. 225-379. Academic Press: Orlando, Florida, USA.

Jennings, D. H. (1987a). Presidential Address: The medium is the message. *Transactions of the British Mycological Society*, **89**, 1-11.

Jennings, D. H. (1987b). Translocation of solutes in fungi. *Biological Reviews*, **62**, 215-243.

Jennings, D. H. (1988). The ability of basidiomycete mycelium to move nutrients through the soil ecosystem. In *Field Methods in Terrestrial Ecosystem Nutrient Cycling*, eds. A. F. Harrison, P. Ineson & O. W. Heal, (in press). Elsevier.

Jones, E. W. & Fink, G. R. (1982). Regulation of amino acid and nucleotide biosynthesis in yeast. In *The Molecular Biology of the Yeast Saccharomyces: Metabolism and Gene Expression*, eds. J. N. Strathern, E. W. Jones & J. R. Broach, pp. 181-297. Cold Spring Harbor Laboratory: New York.

Jones, E. W. (1984). The synthesis and function of proteases in *Saccharomyces*: genetic approaches. *Annual Review of Genetics*, **18**, 233-270.

Kalisz, H. M., Moore, D. & Wood, D. A. (1986). Protein utilization by Basidiomycete fungi. *Transactions of the British Mycological Society*, **86**, 519-525.

Kalisz, H. M., Wood, D. A. & Moore, D. (1987). Production, regulation and release of extracellular proteinase activity in Basidiomycete fungi. *Transactions of the British Mycological Society*, **88**, 221-227.

Khan, J. A. & Bhatnagar, R. M. (1977). Studies on solubilization of insoluble phosphates by microorganisms. 1. Solubilization of Indian phosphate rocks by *Aspergillus niger* and *Penicillium* sp. *Fertilizer Technology*, **14**, 329-333.

Kinnaird, J. H. & Fincham, J. R. S. (1983). The complete nucleotide sequence of the *Neurospora crassa am* (NADP-specific glutamate dehydrogenase) gene. *Gene*, **26**, 253-260.

Kinsey, J. A., Fincham, J. R. S., Siddig, M. A. & Keighren, M. (1980). New mutational variants of *Neurospora* NADP-specific glutamate dehydrogenase. *Genetics*, **95**, 305-316.

Knowles, C. J. (1977). Microbial metabolic regulation by adenine nucleotide pools. *Symposia of the Society for General Microbiology*, **27**, 241-284.

Kucey, R. M. N. (1987). Increased phosphorus uptake by wheat and field beans inoculated with a phosphorus-solubilizing *Penicillium bilaji* strain and with vesicular-arbuscular mycorrhizal fungi. *Applied & Environmental Microbiology*, **53**, 2699-2703.

Kulaev, I. S. (1979). *The Biochemistry of Inorganic Polyphosphates*. John Wiley: Chichester.

Kulaev, I. S. & Vagabov, V. M. (1983). Polyphosphate metabolism in micro-organisms. *Advances in Microbial Physiology*, **24**, 83-171.

LéJohn, H. B. (1971). Enzyme regulation, lysine pathways and cell wall structures as indicators of major lines of evolution in fungi. *Nature*, **231**, 164-168.

Lambers, H. (1980) The physiological significance of cyanide-resistant respiration in higher plants. *Plant, Cell & Environment*, **3**, 293-302.

Lea, P. J. & Miflin, B. B. (1974). An alternative route for nitrogen assimilation in higher plants. *Nature*, **251**, 614-616.

Levi, M. P. & Cowling, E. B. (1969). Role of nitrogen in wood deterioration. VII. Physiological adaptation of wood-destroying and other fungi to substrates deficient in nitrogen. *Phytopathology*, **59**, 460-468.

Lewis, D. H. & Harley, J. L. (1965). Carbohydrate physiology of mycorrhizal roots of beech. III. Movement of sugars between host and fungus. *New Phytologist*, **64**, 256-269.

Lichko, L. P., Okorokov, L. A. & Kulaev, S. (1982). Participation of vacuole in regulation of levels of K^+, Mg^{2+} and orthophosphate ions in cytoplasm of the yeast. *Archives of Microbiology*, **132**, 289-293.

Lloyd, D. (1974). *The Mitochondria of Microorganisms*. Academic Press: New York.

Loll, M. J. & Bollag, J.-M. (1983). Protein transformation in soil. *Advances in Agronomy*, **36**, 351-382.

Lopez, S. & Gancedo, J. (1979). Effect of metabolic conditions on protein turnover in yeast. *Biochemical Journal*, **178**, 769-776.

Lowendorf, H. S., Bazinet Jr., G. F. & Slayman, C. W. (1975). Phosphate transport in *Neurospora*. Derepression of a high-affinity transport system during phosphorus starvation. *Biochimica et Biophysica Acta*, **389**, 541-549.

Lowendorf, H. S., Slayman, C. L. & Slayman, C. W. (1974). Phosphate transport in *Neurospora*. Kinetic characterization of a constitutive low-affinity transport system. *Biochimica et Biophysica Acta*, **373**, 369-382.

Lowendorf, H. S. & Slayman, C. W. (1975). Genetic regulation of phosphate transport system II in *Neurospora*. *Biochimica et Biophysica Acta*, **413**, 95-103.

Maclean, D. J. (1982). Axenic culture and metabolism in rust fungi. In The Rust Fungi, eds. K. J. Scott & A. K. Chakravorty, pp. 37-84. Academic Press: London.

Markham, P. & Collinge, A. J. (1987). Woronin bodies of filamentous fungi. *FEMS Microbiology Reviews*, **46**, 1-11.

Martin, F. (1985). ^{15}N-NMR studies of nitrogen assimilation and amino acid biosynthesis in the ectomycorrhizal fungus *Cenococcum graniforme*. *FEBS Letters*, **182**, 350-354.

Martin, F., Stewart, G. R., Genetet, I. & Le Tacon, F. (1986). Assimilation of $^{15}NH_4{}^+$ by beech (*Fagus sylvatica* L.) ectomycorrhizas. *New Phytologist*, **102**, 85-94.

Marzluf, G. A. (1981). Regulation of nitrogen metabolism and gene expression in fungi. *Microbiological Reviews*, **45**, 437-461.

Maxwell, D. P., Maxwell, M. D., Hanssley, G., Armentrout, V. N., Murray, G. M. & Hoch, H. C. (1975). Microbodies and glyoxylate-cycle enzyme activities in filamentous fungi. *Planta*, **124**, 109-123.

Merrill, W. & Cowling, E. B. (1966). Role of nitrogen in wood deterioration: amounts and distribution of nitrogen in tree stems. *Canadian Journal of Botany*, **44**, 1555-1580.

Messenguy, F. (1987). Multiplicity of regulatory mechanisms controlling amino acid biosynthesis in *Saccharomyces cerevisiae*. *Microbiological Sciences*, **4**, 150-153.

Metzenberg, R. L. (1979). Implication of some genetic control mechanisms in *Neurospora*. *Microbiological Reviews*, **43**, 361-383.

Michalenko, G. O., Hohl, H. R. & Rast, D. (1976). Chemistry and architecture of the mycelial wall of *Agaricus bisporus*. *Journal of General Microbiology*, **92**, 251-263.

Miflin, B. J. & Lea, P. J. (1980). Ammonia assimilation. In *The Biochemistry of Plants: A Comprehensive Treatise*, vol. 5, *Amino Acids and Derivatives*, ed. B. B. Miflin, pp. 169-202. Academic Press: New York.

Millard, P. (1988). The accumulation and storage of nitrogen by herbaceous plants. *Plant, Cell & Environment*, **11**, 1-8.

Moore, D. (1981). Evidence that the NADP-linked glutamate dehydrogenase of *Coprinus cinereus* is regulated by acetyl-CoA and ammonium levels. *Biochimica et Biophysica Acta*, **661**, 247-254.

Moore, D. (1984). Developmental biology of the *Coprinus cinereus* carpophore: metabolic regulation in relation to cap morphogenesis. *Experimental Mycology*, **8**, 283-297.

Moore, D. & Ewaze, J. O. (1976). Activities of some enzymes involved in metabolism of carbohydrate during carpophore development in *Coprinus cinereus*. *Journal of General Microbiology*, **97**, 313-322.

Moore, D., Horner, J. & Liu, M. (1987). Co-ordinate control of ammonium-scavenging enzymes in the fruit-body cap of *Coprinus cinereus* avoids inhibition of sporulation by ammonium. *FEMS Microbiology Letters*, **44**, 239-242.

Nicolay, K., Scheffers, W. A., Briunenberg, P. M. & Kapstein, R. (1983). *In vivo* ^{31}P NMR studies on the role of the vacuole in phosphate metabolism in yeasts. *Archives of Microbiology*, **134**, 270-275.

Nieuwenhuis, B. J. W. M. & Borst-Pauwels, G. W.F. H. (1984). Derepression of the high affinity phosphate uptake system in the yeast *Saccharomyces cerevisiae*. *Biochimica et Biophysica Acta*, **770**, 40-46.

North, M. J. (1982). Comparative biochemistry of the proteinases of eukaryotic microorganisms. *Microbiological Reviews*, **46**, 308-340.

Okorokov, L. A., Lichko, L. P. & Kulaev, I. S. (1980). Vacuoles: main compartments of potassium, magnesium and phosphate ions in *Saccharomyces carlsbergensis* cells. *Journal of Bacteriology*, **144**, 661-665.

Pateman, J. A. & Kinghorn, J. R. (1976). Nitrogen metabolism. In *The Filamentous Fungi*, vol. 2, *Biosynthesis and Metabolism*, eds. J. E. Smith & D. R. Berry, pp. 159-237. Edward Arnold: London.

Paton, F. M. & Jennings, D. H. (1988). Evidence that *Thraustochytrium* is unable to synthesise lysine. *Transactions of the British Mycological Society*, in press.

Paustian, K. (1985). Influence of fungal growth pattern on decomposition and nitrogen mineralisation in a model system. In *Ecological Interactions in Soil*, ed. A. H. Fitter, pp. 159-174. Blackwell Scientific Publications: Oxford.

Paustian, K. & Schnürer, J. (1987a). Fungal growth response to carbon and nitrogen limitation: application of a model to laboratory and field data. *Soil Biology & Biochemistry*, **19**, 621-629.

Paustian, K. & Schnürer, J. (1987b). Fungal growth response to carbon and nitrogen limitation: a theoretical model. *Soil Biology & Biochemistry*, **19**, 613-620.

Rao, P. S. & Niederpruem, D. J. (1969). Carbohydrate metabolism during morphogenesis of *Coprinus lagopus* (*sensu* Buller). *Journal of Bacteriology*, **100**, 1222-1228.

Robertson, B. R. & Button, D. K. (1979). Phosphate-limited continuous culture of *Rhodotorula rubra*: kinetics of transport, leakage and growth. *Journal of Bacteriology*, **138**, 884-895.

Sanders, D. (1986). Generalised kinetic analysis of ion-driven co-transport systems. II. Random ligand binding as a simple explanation of non-Michaelian kinetics. *Journal of Membrane Biology*, **90**, 67-81.

Stewart, G. R. & Moore, D. (1974). The activities of glutamate dehydrogenase during mycelial growth and sporophore development in *Coprinus lagopus* (*sensu* Lewis). *Journal of General Microbiology*, **83**, 73-81.

Straker, C. J. & Mitchell, D. T. (1987). Kinetic characterization of a dual phosphate uptake system in the endomycorrhizal fungus of *Erica hispindula* L. *New Phytologist*, **106**, 129-137.

Tinker, P. B. H. (1975). Effects of vesicular-arbuscular mycorrhizas on higher plants. In *Symbiosis*, Society for Experimental Biology volume 29, eds. D. H. Jennings & D. L. Lee, pp. 325-349. Cambridge University Press: Cambridge, UK.

Veenhuis, M., van Dijken, J. P. & Harder, W. (1983). The significance of peroxisomes in the metabolism of one-carbon compounds in yeasts. *Advances in Microbial Physiology*, **24**, 1-82.

Vogel, H. J. (1964). Distribution of lysine pathway among fungi: evolutionary implications. *The American Naturalist*, **98**, 435-446.

Watkinson, S. C. (1979), Growth of rhizomorphs, mycelial strands, coremia and sclerotia. In *Fungal Walls and Hyphal Growth*, British Mycological Society Symposium volume 2, eds. J. H. Burnett & A. P. J. Trinci, pp. 93-113. Cambridge University Press: Cambridge, UK.

Watkinson, S. C. (1984). Morphogenesis of the *Serpula lacrimans* colony in relation to its function in nature. In *Ecology and Physiology of the Fungal Mycelium*, British Mycological Society Symposium volume 8, eds. D. H. Jennings & A. D. M. Rayner, pp. 165-184. Cambridge University Press: Cambridge, UK.

Wiame, J. M., Grenson, M. & Arst Jr, H. N. (1985). Nitrogen catabolite repression in yeasts and filamentous fungi. *Advances in Microbial Physiology*, **26**, 1-87.

Wolf, D. H. (1980). Control of metabolism in yeast and other lower eukaryotes through action of proteinases. *Advances in Microbial Physiology*, **21**, 267-338.

Wolf, D. H. (1986). Cellular control in the eukaryotic cell through action of proteinases: the yeast *Saccharomyces cerevisiae* as a model organism. *Microbiological Sciences*, **3**, 107-114.

Wolfinbarger Jr., L. (1980a). Transport and utilization of amino acids by fungi. In *Microorganisms and Nitrogen Sources*, ed. J. W. Payne, pp. 63-87. John Wiley: Chichester.

Wolfinbarger Jr., L. (1980b). Transport and utilization of peptides by fungi. In *Microorganisms and Nitrogen Sources*, ed. J. W. Payne, pp. 281-300. John Wiley: Chichester.

Chapter 2

The genetics and biochemistry of nitrate assimilation in ascomycete fungi

A. B. Tomsett

Department of Genetics and Microbiology, The University of Liverpool, P.0. Box 147, Liverpool, L69 3BX, UK

Introduction

All organisms require reduced forms of nitrogen to synthesize proteins and nucleic acids. Inorganic nitrogen is assimilated to the reduced forms by two major pathways, nitrate assimilation and nitrogen fixation. Guerrero, Vega & Losada (1981) have calculated the relative contribution of these two pathways to reduced nitrogen in the biosphere: nitrate assimilation is the major pathway being 100-fold greater than the total contribution of nitrogen fixation. The study of nitrate assimilation therefore, has obvious importance for both ecology and agriculture. Through the combination of genetical, biochemical and more recently, molecular biological techniques, we have developed an excellent understanding of the enzymology and genetic regulation of this pathway.

In this paper, I have been asked to present a general overview of the subject in a limited space. Our knowledge of nitrate assimilation is now sufficiently detailed to fill this entire volume. Even a moderately comprehensive list of the literature would exceed the space available for this article. Inevitably, this review is highly selective in its content: it examines the outstanding contribution made by classical genetic analysis to our understanding of nitrate assimilation, and how biochemistry and molecular biology have built upon its conclusions and hypotheses. Most of this article concerns the study of the two ascomycete fungi *Aspergillus nidulans* and *Neurospora crassa*, although I refer to other organisms where the investigations are more advanced. I have also attempted to

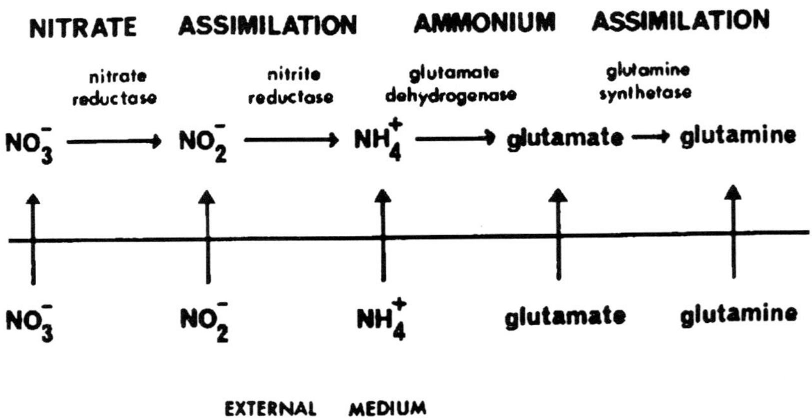

Fig. 2.1. The pathways of nitrate and ammonium assimilation.

write this for the non-specialist. For those wishing to pursue this field, there are a number of excellent more extensive reviews (Garrett & Amy, 1978; Cove, 1979; Guerrero *et al.*, 1981; Marzluf, 1981; Dunn-Coleman, Smarrelli & Garrett, 1984; Wiame, Grenson & Arst, 1985).

The pathway of nitrate assimilation

The pathway of nitrate assimilation in the filamentous fungi *A. nidulans* and *N. crassa* involves the complete reduction of nitrate to ammonium, via nitrite (Fig. 2.1). This process occurs in two steps: the reduction of nitrate to nitrite by nitrate reductase (NADPH: nitrate oxidoreductase, E.C. 1.6.6.3.), a reaction involving a two-electron transfer; and the reduction of nitrite to ammonium by nitrite reductase (NAD(P)H: nitrite oxidoreductase, E.C. 1.6.6.4.), a six-electron transfer. The ammonium thus formed is then assimilated into glutamate and glutamine by NADP-linked glutamate dehydrogenase and glutamine synthetase.

Both *A. nidulans* and *N. crassa* can utilise a wide range of nitrogen-containing compounds when supplied as their sole source of nitrogen in the external medium. In broad terms, we can consider three classes of nitrogen-source: inducers, repressors and neutral N-sources. A number of pathways of N-source acquisition in fungi are subject to two genetic control mechanisms, i.e. they are substrate-inducible and product-repressible. For nitrate assimilation, nitrate and/or nitrite will induce enzyme synthesis but only in the absence of the reduced forms of nitrogen, ammonium and glutamine, which act as repressors; compounds such as urea and uric acid are neutral, i.e. they neither induce nor repress nitrate assimilation. For each pathway, substrate-inducibility is pathway-specific but nitrogen metabolite repression controls all the pathways for nitrogen catabolism.

Enzymology

Nitrate reductase is a homodimer of molecular weight 228,000 in *N. crassa* (Garrett & Nason, 1969) and 180,000 in *A. nidulans* (Minagawa & Yoshimoto, 1982). For *N. crassa*, Horner (1983) has reported a subunit size of 145 kD, while the *A. nidulans* subunit is 91 kD (Cooley & Tomsett, 1985). In addition to the two identical polypeptides, nitrate reductase possesses a molybdenum-containing cofactor (see Garrett & Amy, 1978). This cofactor is found in a wide variety of enzymes from diverse phylogenetic sources. In these fungi, the cofactor is also used by xanthine dehydrogenase (E.C.1.2.1.37). Nitrate reductase catalyses an electron-transfer reaction from NADPH to nitrate (Fig. 2.2a). In addition to the molybdenum cofactor, the enzyme contains flavin adenine dinucleotide (FAD) and cytochrome-*b557* (see Dunn-Coleman *et al.*, 1984). Although the physiological activity utilizes the entire electron-transfer sequence, the enzyme has a number of non-physiological partial activities which can be assayed *in vitro*. The initial part of the sequence can be measured by the reduction of certain electron acceptors, e.g. cytochrome-*c* (NADPH-CCR), the latter part by measurement of nitrite production when electrons are passed to the enzyme by reduced methyl viologen (MVH-NaR). These partial activities afford the opportunity of assessing the presence of mutant enzyme molecules.

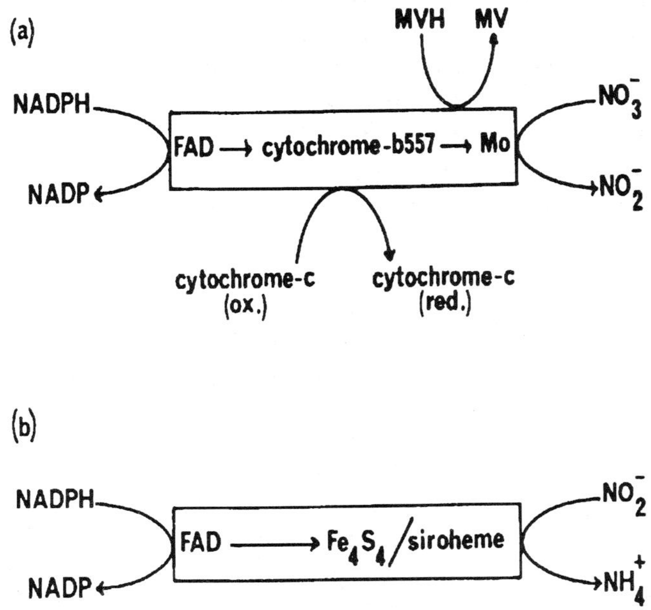

Fig. 2.2. The electron-transfer reactions of (a) nitrate reductase and (b) nitrite reductase.

Nitrite reductase has a molecular weight of 290,000, a homodimer of identical 140,000 molecular weight protomers. The enzyme catalyses a six-electron transfer from NADPH to nitrite, via FAD, an iron-sulphur centre and a siroheme cofactor (Fig. 2.2b; see Dunn-Coleman *et al.*, 1984).

Mutant isolation and characterization

The genetic analysis of nitrate assimilation has been greatly facilitated by having a positive selection method for nitrate-non-utilising mutants, based on the finding that cells lacking nitrate reductase are resistant to the toxic analogue, chlorate (Aberg, 1947). The exact mechanism of chlorate toxicity is unknown. The simplest hypothesis is that nitrate reductase catalyses the conversion of chlorate to chlorite, a highly reactive substance (Aberg, 1947). This mechanism is almost certainly correct, but is insufficient to explain all the data (Cove, 1976a & b, 1979). Nevertheless, resistance to chlorate has been used successfully to select mutants defective in nitrate assimilation in a variety of organisms, including *A. nidulans* and *N. crassa* (Cove, 1976a; Tomsett & Garrett, 1980).

Since the correlation between chlorate resistance and lack of nitrate reductase activity is not absolute, it has been equally important to isolate nitrate non-utilising mutants by other criteria. Several such mutants were obtained in *N. crassa* by a variety of workers (see Garrett & Amy, 1978). In *A. nidulans*, in addition to those isolated without positive selection, the technique of putrescine starvation yields useful numbers of nitrate non-utilising mutants (Cove, 1976a).

A third technique can be used to recognise mutants which have normal or higher levels of nitrate reductase activity. Nitrite in the presence of sulphanilic acid and N-1-naphthylethylenediamine dihydrochloride (NED) produces a pink coloration. When growing on solid media, colonies which can convert nitrate to nitrite will stain pink; colonies which excrete nitrite have a pink halo (see Tomsett & Garrett, 1980). These methods have been used to isolate a wide variety of mutants. Such mutants can be classified on the basis of their ability to utilise certain nitrogen sources, their resistance to chlorate and whether they excrete nitrite (Table 2.1). Using complementation tests, these groups of mutants can be shown to define individual genetic loci as shown in Table 2.1.

Nitrate reductase apoprotein

The nitrate reductase apoprotein is encoded by a single gene, *nia*D in *A. nidulans* and *nit*-3 in *N. crassa* (Cove, 1979; Garrett & Amy, 1978). Mutants in these genes have a unique phenotype, an inability to utilise nitrate whilst exhibiting normal growth on all other nitrogen sources. This phenotype results from a loss of NADPH-nitrate reductase activity which has been demonstrated in a large number of individual mutants (see Cove, 1979; Tomsett & Garrett, 1981). In *A. nidulans*, the genetic evidence that *nia*D encodes the apoprotein comes from temperature-sensitive *nia*D mutants which retain nitrate reductase activity when the organism

Table 2.1. Growth test classification of mutant strains.

Function	*A. nidulans* locus	*N. crassa* locus	Growth media: NH_4^+	NO_3^-	NO_2^-	Hypo xanthine	Uric acid	Uric acid +100 mM ClO_3^-	Plate test for nitrite production
	wild type		+	+	+	+	+	VS	+
NAR structural gene encoding the apoprotein	*nia*D	*nit*-3	+	–	+	+	+	most R, few S	–
NIR structural gene encoding the apoprotein	*nii*A	*nit*-6	+	–	–	+	+	VS	++
Formation of the molybdenum-containing cofactor required by both NAR and XDH	*cnx*ABC, E, F, G, H	*nit*-1, 7, 8, 9ABC	+	–	+	–	+	most R, few S	–
Regulator gene necessary for nitrate induction of nitrate assimilation	*nir*A*	*nit*-4	+	–	–	+	+	SR	–
Regulator gene responsible for nitrogen metabolite repression of nitrate assimilation and other pathways of nitrogen acquisition	*are*A*	*nit*-2	+	–	–	–	–	–	–
Nitrate permease gene	*crn*A	?	+	+	+	+	+	SR	+

NAR = nitrate reductase; NIR = nitrite reductase; XDH = xanthine dehydrogenase. For nitrogen source utilization in growth media, + indicates growth on that medium, – indicates no growth (nitrogen-starved morphology). For chlorate resistance, R indicates resistance, SR an intermediate resistance characteristic of *nir*A⁻ and *crn*A, VS indicates extreme sensitivity, and – that the strain cannot grow. For the nitrite production plate test, + indicates colony stains pink, ++ indicates the colony has a pink halo from excreted nitrite, and – that the colony does not stain. * The *nir*A and *are*A genes can mutate to give a variety of mutant phenotypes, these examples are representative of null alleles. Data compiled from Cove (1976a & b), Tomsett & Garrett (1980) and see Wiame *et al.* (1985).

is grown at 25°C but lack the activity when grown at 37°C. The enzyme from these mutants is more thermolabile than that of the wild-type and hence, has a much shorter half-life *in vitro* (MacDonald & Cove, 1974).

Despite overwhelming genetic evidence that nitrate reductase has only one specific core polypeptide, it is only recently that this has been characterized biochemically. Nitrate reductase has been extensively purified from both *N. crassa* and *A. nidulans*. In *N. crassa*, nitrate reductase was reported to have subunit sizes of 115,000 and 130,000 daltons, but one was presumed to be a degradation product of the other (Pan & Nason, 1978). In *A. nidulans*, the subunit structure was difficult to define. Downey & Focht (1974) isolated a protomer of molecular weight 49,000. Downey & Steiner (1979) later identified proteins which had subunit molecular weights of 49,000, 50,000 and 75,000 daltons. In another study, Steiner & Downey (1982) demonstrated a single polypeptide of molecular weight 54,000, suggesting a homotetrameric enzyme structure for nitrate reductase, whereas Minagawa & Yoshimoto (1982) described the *A. nidulans* nitrate reductase to be 180,000 daltons, consisting of two each of two subunits of 59,000 and 38,000 daltons. These studies of *N. crassa* and *A.nidulans* all involved extensive purification procedures, during which the enzyme was presumably exposed to proteolysis producing the variation in size reported. In the most recent studies, Horner (1983) and Cooley & Tomsett (1985) both employed rapid purification in the presence of higher concentrations of protease inhibitors. For *N. crassa* the subunit size is apparently 145,000 daltons (Horner, 1983), while in *A. nidulans*, a 91,000 dalton peptide is the *nia*D gene product (Cooley & Tomsett, 1985). The fact that proteolysis of nitrate reductase gives the discrete sizes reported above indicates that there may be one or more sites in the core polypeptide which are susceptible to proteolysis. Native enzyme can be purified because the cleaved products remain bound and hence, the enzyme retains full enzyme activity. Upon dissociation by SDS, one or more separate protein fragments are observed. Minagawa & Yoshimoto (1982) report that the 38 kD peptide subunit can be dissociated from the native enzyme by heat treatment, resulting in the loss of overall NADPH-NaR activity and of the NADPH-CCR partial activity but not MVH-NaR. This strongly suggests functional domains for *A.nidulans* nitrate reductase separated by a protease-sensitive site.

Solomonson and his coworkers have examined functional domains in the NADPH-nitrate reductase from *Chlorella vulgaris* (Solomonson *et al.*, 1986). Their model (Fig. 2.3) includes two functional domains connected by a protease-sensitive hinge region. The *Chlorella* enzyme is believed to be a homotetramer, native molecular weight 360 - 380 kD with four 90 - 100 kD subunits.

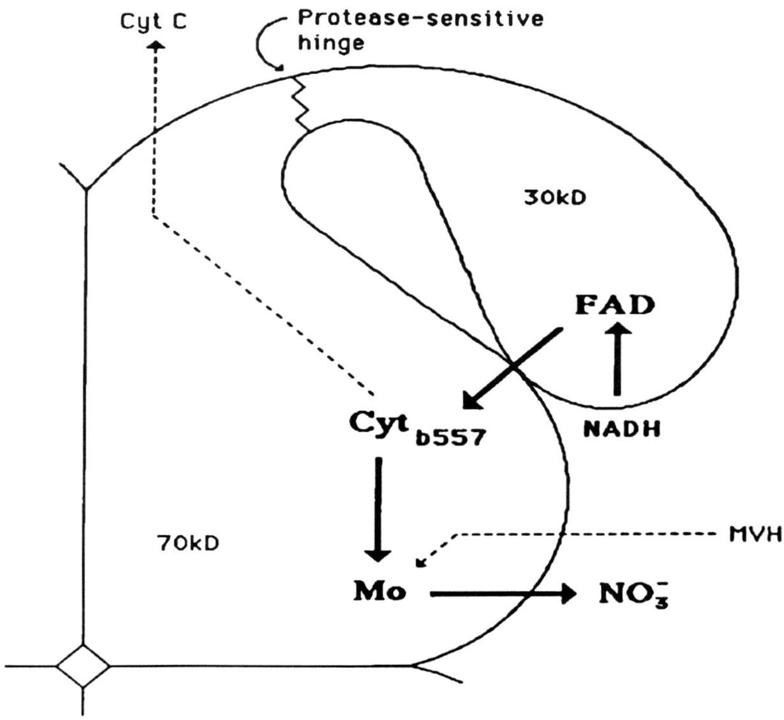

Fig. 2.3. A model for the structure of nitrate reductase from *Chlorella vulgaris* (after Solomonson *et al.*, 1986).

Each 100 kD subunit has a 70 kD domain which is at the core of the molecule and contains the *b5*-type cytochrome and the molybdenum cofactor. The second domain, about 30 kD, is exposed on the surface of the enzyme and contains the FAD/NADH-binding activities. The quaternary structure of the enzyme is maintained by interaction between the 70 kD structures: the 30 kD moieties are linked to the 70 kD molecules by the protease-sensitive hinge and have no role in maintaining the overall structure. Proteolysis releases the 30 kD peptide with consequent loss of the NADH-CCR activity. The 30 kD peptide is thought to be physically closely associated with the 70 kD core.

This model would elegantly explain the data of Minagawa & Yoshimoto (1982): native enzyme cleaved at the hinge *in vivo* or during extensive purification, retains activity by the close physical association between the two domains, but upon heat treatment the 38 kD FAD/NADPH domain is lost leaving the core enzyme (estimated at 126,000 daltons) composed of a dimer of core peptides each 59 kD. Pan & Nason (1978) observed a major 115 kD peptide (and a minor 130 kD peptide) derived from the 145 kD *nit*-3 gene product: this also can be explained by loss of a 30 kD moiety from the native subunit polypeptide. Since the 115 kD and 130 kD polypeptides have the same N-terminal amino acid (Pan & Nason, 1978), it seems likely that the lost 30 kD peptide is from the C-terminal end.

Le & Lederer (1983) have shown, by amino acid sequencing, that the heme-binding domain of *N. crassa* nitrate reductase is a member of the cytochrome-*b5* superfamily. Analysis of this superfamily reveals 13 conserved amino acid residues. Recently, Caboche and his coworkers (Calza *et al.*, 1987) have isolated cDNA clones complementary to tobacco nitrate reductase mRNA. The amino acid sequence, deduced from the DNA sequence, shows significant homology with different members of the superfamily, including all 13 of the conserved residues. This clearly demonstrates that the cytochrome-*b557* is encoded within the nitrate reductase structural gene. From the data available, the tobacco and *Neurospora* sequences show only 33 per cent homology in this region.

Calza *et al.* (1987) have also compared their sequence with others in sequence data banks. Downstream (3′ on the mRNA) of the cytochrome-*b5* homology, there is a significant homology with human flavoprotein cytochrome-*b5* reductase. This protein catalyses the reduction of cytochrome-*b5*, using NADH as an electron donor and FAD as the redox intermediate; this is analogous to the first partial activity of nitrate reductase. This NADH/FAD domain is separated by 35 amino acids from the cytochrome-*b557* domain and lies toward the C-terminal end of the polypeptide. Unfortunately, the complete sequence for the tobacco nitrate reductase is not yet available. The cDNA contains only about 60% (60 kD polypeptide) of the native polypeptide (100 kD). However, two genomic DNA clones have now been isolated from *Nicotiana tabacum*, each of which can be attributed to one or the other of the ancestors of tobacco, *Nicotiana sylvestris* and *Nicotiana tomentosiformis* (M. Caboche *et al.*, unpublished data). These reveal that each contains two introns which differ in size, but not position, between the two clones. The intron positions map to divide the 1·6 kb cDNA clone into three approximately equal sections. The native DNA sequence appears to start 800 bp (approx.) upstream (5′) of the cDNA clone and extend 1·1 kb (approx.) downstream (3′) of it (M. Caboche *et al.*, unpublished data). However, their data seem to include both the cytochrome-*b557* and the NADH/FAD domains within the same exon, i.e. the third exon downstream. Comparing this with the model of Solomonson *et al.* (1986), and the loss of the C-terminal fragment from *N. crassa* nitrate reductase (Pan & Nason, 1978), we might predict that the molybdenum-cofactor binding site should be located within exon 1 and/or exon 2 (if there is an overall structural analogy in addition to the enzymological (catalytic) similarities). If the model of Solomonson *et al.* (1986) is applicable, one might also expect to find the protease-sensitive hinge region within, or close to, the 35 amino acids separating the cytochrome-*b557* and NADH/FAD domains. This would yield a 39 kD (approx.) C-terminal fragment.

If the protease-sensitive site is a common feature of all NAD(P)H-nitrate reductases, it is tempting to suggest this has an important physiological role. Proteases have been implicated in the turnover of nitrate reductase (see Dunn-Coleman *et al.*, 1984). In *N. crassa*, two nitrate reductase inactivators, both proteolytic, have been described (Sorger, Premakumar & Gooden, 1978;

Walls *et al.*, 1978) and a role for their control of nitrate-reductase has been proposed.

As yet, many of the details of molecular studies of the *nia*D gene of *A. nidulans* and the *nit*-3 gene of *N. crassa* have not been published. The *nit*-3 gene is transcribed to give a 3·4 kb mRNA but as yet the gene has not been sequenced (Fu & Marzluf, unpublished data). The *nia*D gene transcript is about 2·8 - 3 kb; the gene contains 6 introns (J. R. Kinghorn, unpublished data). The *nia*D gene clones have been sequenced but the data are not available at the time of writing this review.

The molybdenum cofactor

Among mutants isolated for their inability to utilise nitrate are a common class which cannot grow on hypoxanthine as sole nitrogen source (see Table 2.1). This phenotype results from a loss of both NADPH-nitrate reductase activity and xanthine dehydrogenase activity (see Cove, 1979; Tomsett & Garrett, 1981). These mutants retain the NADPH-cytochrome-*c* reductase activity of nitrate reductase and lack its MVH-NaR activity which requires the molybdenum-containing cofactor. Pairwise complementation tests between such mutants reveal that they are encoded by several genes. In *A. nidulans*, the initial studies described seven complementation groups (see Cove, 1979): *cnx*E, F, G and H mutants each formed a discrete group and hence these define four separate genes; *cnx*A, B and C mutants show an overlapping complementation pattern, *cnx*B mutants fail to complement mutants in *cnx*A, *cnx*C or of course *cnx*B; *cnx*A and *cnx*C each form a discrete group which complement each other. Since *cnx*A, B and C mutants are all closely linked, it is uncertain whether this locus defines one gene, in which A and C exhibit intragenic complementation, or two genes, *cnx*B being mutant in both (for discussion see Cove, 1979, and Scazzocchio, 1980).

More recently, another *cnx* locus has been described in *A. nidulans* (Arst, Tollervey & Sealy-Lewis, 1982). Unlike the other *cnx* mutants, *cnx*J mutants are unaffected for their utilization of nitrate and hypoxanthine unless tungstate or methylammonium is present. Tungstate is a molybdate analogue, which presumably lowers the concentration of active molybdenum cofactor by competition with molybdate at the level of uptake and/or incorporation into the cofactor (see Arst *et al.*, 1982). Methylammonium, a toxic analogue of ammonium, may also lower cofactor concentrations by triggering nitrogen metabolite repression (see below) of cofactor synthesis (for further discussion, see Arst *et al.*, 1982). Because of this unusual phenotype, Arst *et al.* (1982) have tentatively suggested *cnx*J could regulate cofactor biosynthesis, although no clear evidence for this exists.

The role of the *cnx*ABC, E, F, G and H genes is also unclear. Because mutation at any of these loci results in loss of enzyme activity, all are assumed to have an obligatory function in cofactor synthesis (or its regulation). Mutants in *cnx*E, when grown on high concentrations of molybdate, have some nitrate reductase and xanthine dehydrogenase activity restored (Arst, MacDonald & Cove, 1970). This molybdenum-repairable phenotype led to their suggestion that the *cnx*E gene product is responsible for insertion of molybdenum into the cofactor. The only other evidence for the role of these genes comes from study of temperature-sensitive mutations. MacDonald & Cove (1974) measured the thermostability of nitrate reductase *in vitro*. Temperature-sensitive alleles of *cnx*E and *cnx*F synthesized nitrate reductase with the same half-life as that from wild-type strains. Nitrate reductase from *cnx*H strains was, however, more thermolabile than from wild-type strains. This evidence suggests that the *cnx*H gene product is a polypeptide which is a structural component of the nitrate reductase molecule; whereas, the *cnx*E and *cnx*F gene products are presumably enzymes responsible for the synthesis of cofactor, but are not themselves part of its structure. Arst *et al.* (1982) have also described a cryosensitive *cnx*C allele, which grows normally at 37°C but not at 25°C. This strain has an extremely labile nitrate reductase and hence it is possible that the *cnx*C gene product is also a structural component of the enzyme. A model has been proposed to accommodate the genetic evidence for cofactor biosynthesis and nitrate reductase assembly (Cove, 1979).

In *N. crassa*, four loci have been identified which are responsible for molybdenum cofactor biosynthesis, using the same criteria as for *A. nidulans* (Tomsett & Garrett, 1980). The first gene to be identified, *nit*-1, was isolated as a nitrate/hypoxanthine nonutiliser and has played a major role in the biochemical characterization of the cofactor (see Coddington, 1976, and see below).

Mutation in the three other loci, *nit*-7, *nit*-8 and *nit*-9ABC, were selected among chlorate-resistant mutants. Little is known of the role of these *N. crassa* genes. *N. crassa nit*-9ABC shows a similar complementation pattern to *cnx*ABC of *A. nidulans* (Tomsett & Garrett, 1980) but interestingly, unlike *cnx*ABC mutants, alleles of all 3 complementation groups at this locus are molybdenum-repairable (Dunn-Coleman, 1984a). Dunn-Coleman (1984b) has isolated a cloned *N. crassa* DNA fragment which, when transformed into *Escherichia coli*, can restore *E. coli* nitrate reductase activity to *chl*D mutants. *E. coli chl*D mutants produce a defective molybdenum cofactor. This same DNA fragment also transforms *N. crassa nit*-9A and *nit*-9B to enable growth on nitrate, but not *nit*-9C. As yet, however, no further details of this work have emerged. The discrepancy between the number of loci in *A. nidulans* and *N. crassa* almost certainly arises because insufficient nitrate non-utilizing mutants have been analyzed in *N. crassa*.

Biochemical analysis of the molybdenum cofactor has been pioneered by Rajagopalan and his co-workers. All molybdoenzymes except nitrogenase contain a common cofactor, which is a complex of a novel pterin, called molybdopterin (MPT),

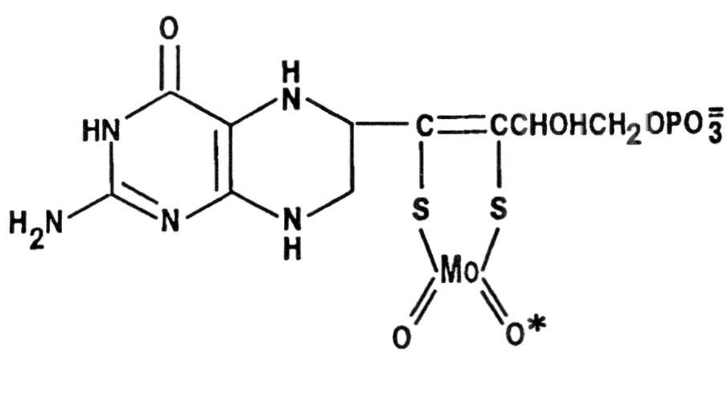

Fig. 2.4. The structure of the molybdenum cofactor (after Johnson & Rajagopalan, 1982). *Sulphur replaces oxygen at this position in xanthine dehydrogenase.

with the molybdenum atom (Johnson, 1980; Johnson *et al.*, 1984). The proposed structure for the molybdopterin with bound molybdenum is shown in Fig. 2.4 (Johnson & Rajagopalan, 1982; Rajagopalan, personal communication).

Until the recent description of the *E. coli chl* M and N loci, the *N. crassa nit*-1 mutant was unique in that it lacks molybdopterin: the other mutants are defective for cofactor biosynthesis but retain some molybdopterin (Kramer, Hageman & Rajagopalan, 1984; Johnson & Rajagopalan, 1987). This has been of great significance in these studies, because molybdopterin from a wide variety of sources can be assayed *in vitro* by its ability to reconstitute nitrate reductase activity in *nit*-1 cell extracts. Thus molybdenum cofactor can be released by denaturation of purified molybdoenzymes, e.g. by guanidine hydrochloride or sodium dodecyl sulphate, and purified by testing fractions with the *nit*-1 reconstitution assay. This has led to the description of the structure shown in Fig. 2.4. Despite the genetic evidence, attempts to identify a polypeptide component of the active cofactor (equivalent to the *cnx*H or *cnx*C gene products of *A. nidulans*) have given negative results. Hageman & Rajagopalan (1985) found no amino acids associated with the cofactor. This evidence is good, but a polypeptide component cannot be totally ruled out since we do not know the nature of the *nit*-1 lesion. It is not inconceivable that *nit*-1 extracts contain such a polypeptide which acts in the reconstitution assay. The purification procedures employed may only measure molybdopterin and hence a polypeptide could be lost at an early stage.

Nitrate and nitrite uptake

The transport of nitrate and nitrite into fungal cells has received relatively little attention in comparison to the enzymology of this pathway. Schloemer & Garrett (1974a) have made a biochemical study of nitrate transport in *N. crassa*. It appears that nitrate uptake is inducible by nitrate or nitrite, even in the presence of ammonium or glutamine; when grown on ammonium as sole nitrogen source, the nitrate uptake system was absent. Curiously, casamino acids repress nitrate uptake, and with the exception of methionine, glutamine and alanine, all individual amino acids prevent nitrate accumulation in the mycelia. These results do not conform to the normal patterns of nitrogen metabolite repression (see below). There is evidence that this is an active transport system which requires transcription and translation. Wild-type and three strains which lack NADPH-nitrate reductase (*nit*-1, *nit*-2, *nit*-3) all displayed similar characteristics for nitrate uptake.

In a parallel study, Schloemer & Garrett (1974b) demonstrated that nitrite uptake in *N. crassa* is induced by nitrate or nitrite, requires transcription and translation, and is an active transport system inhibited by metabolic poisons. They found no evidence, however, for repression or inhibition of this process by ammonium or casamino acids. Coddington (1976) measured nitrite uptake in *nit* mutant strains. He found that *nit*-1 and *nit*-3 strains which can grow on nitrite as sole nitrogen source, did not accumulate nitrite in the mycelia but depleted it from the medium. The *nit*-2 mutant, defective in nitrate and nitrite reductases, totally lacked nitrite uptake, even though it retained nitrate uptake. Mutants of *nit*-4, which lack nitrate and nitrite reductase, accumulate nitrite in the mycelia.

In *A. nidulans*, initial experiments to measure nitrate and nitrite uptake in wild-type, *nia*D, *nii*A and *crn*A mutations used mature mycelia (Tomsett, 1977). These studies demonstrated no difference between wild-type and *crn*A strains but strains lacking nitrate reductase (*nia*D) did not show any nitrate uptake activity, and *nii*A mutants, lacking nitrite reductase, did not show any nitrite uptake activity. This implies that nitrate and nitrite passively diffuse into the mycelium, with diffusion being driven by reduction. Subsequently, Brownlee & Arst (1983) have shown that the *crn*A1 mutation reduces nitrate uptake several-fold in conidia and young mycelia, but it has no effect in older mycelia. They also found that uptake was distinct from, but dependent upon, nitrate reductase activity. Expression of the *crn*A gene is not regulated by *nir*A, which mediates nitrate induction, but is subject to control by the *are*A gene mediating nitrogen metabolite repression.

Although no nitrite uptake gene has formally been described in *A. nidulans*, two mutations have been reported which result in a nitrite hypersensitive phenotype. The *nih*A mutation prevents growth of strains on elevated concentrations of nitrite (Pombeiro, Martinez-Rossi & Rossi, 1983). Since this strain grows normally on normal concentrations of nitrite and can utilize a wide range of nitrogen sources, these authors have suggested that this mutation leads to increased efficiency of nitrite uptake. An alternative explanation, which cannot be eliminated, is that nitrite efflux is reduced or prevented in these strains. The *nih*A locus maps to chromosome I, close to the *pro*A gene. The second nitrite hypersensitive mutation, *nii*C628, also maps to chromosome I, but it is distal to the *bi*A gene (at least 31 centimorgans from *nih*A). The *nii*C628 strains grow with a nitrogen-starved morphology on nitrate or nitrite, but have wild-type levels of both nitrate and nitrite reductases (Cove, 1979). These strains grow normally on all other nitrogen sources but are unable to grown on urea and nitrite. The fact that *nii*C628 has the enzymes to utilise nitrate and nitrite but that nitrite appears to inhibit growth of this mutant again suggests an altered permease function, either increased uptake or decreased efflux.

Nitrite reduction

One of the best biochemical characterizations of nitrite reductase has been undertaken by Garrett and his coworkers in *N. crassa* (Garrett, 1972, 1978; Lafferty & Garrett, 1974). The enzyme has been purified to homogeneity (Greenbaum, Prodouz & Garrett, 1978; Prodouz & Garrett, 1981). It is a homodimer of 140,000 daltons subunits, and contains a siroheme prosthetic group which is thought to serve as the site of binding and reduction of nitrite. Siroheme is also found in assimilatory sulphite reductase. The enzymology of nitrite reductase has been extensively reviewed recently by Dunn-Coleman *et al.*, (1984). Only one structural gene has been identified for nitrite reductase; *nii*A mutants in *A. nidulans* and *nit*-6 mutants of *N. crassa* have identical phenotypes being unable to utilise either nitrate or nitrite as sole nitrogen source (see Cove, 1979; Tomsett & Garrett, 1980). These mutants lack NADPH-nitrite reductase but have normal levels of NADPH-nitrate reductase activity: when nitrate is supplied, they excrete nitrite (see Cove, 1979; Tomsett & Garrett, 1980). There have been no detailed characterizations of mutations affecting siroheme biosynthesis, although a mutant which required both reduced sulphur and reduced nitrogen compounds has been isolated (see Cove, 1979).

The nitrate assimilation gene cluster of *A. nidulans*

The *crn*A, *nii*A and *nia*D genes of *A. nidulans* are tightly linked (Tomsett & Cove, 1979). The genetic organization of this region was determined by deletion mapping. Deletions were selected as spontaneous chlorate-resistant mutants which were unable to utilise either nitrate or nitrite (Cove, 1976a; Tomsett & Cove, 1979). These were easily recognised because *nii*A single mutants are chlorate sensitive and *nir*A⁻ mutants have a distinctive intermediate pattern of resistance. The frequency at which they arose was too high for them to be two independent point mutations. Mutations of *crn*A were initially mapped using heterozygous diploids. However, preliminary studies of bromate toxicity showed that *crn*A mutant strains were more resistant than the wild-type to 1mM bromate when uric acid was present as the nitrogen source. Point mutations involving either the *nii*A or the *nia*D genes do not lead to similar bromate resistance. Analysis of *nii*A *nia*D deletions revealed two groups based upon their bromate resistance. Bromate resistant deletions also lacked part or all of the *crn*A gene, whereas bromate sensitive deletions terminated within the *nii*A gene. The gene order within the cluster was thus defined: *crn*A - *nii*A - *nia*D (Tomsett & Cove, 1979).

Rand & Arst (1977) isolated a mutant, *nis*5, which was tightly linked to the gene cluster but only affected nitrite reductase activity. The *nis*-5 strains have 40% of the wild-type activity when induced, and a low constitutive activity which was insensitive to nitrogen metabolite repression. The mutation results from a translocation which apparently links a new, but inefficient promoter to the *nii*A gene. Deletion mapping showed that *nis*-5 mapped to the central region between the *nii*A and *nia*D genes (Tomsett, 1977) and this was later confirmed as a large portion of the right arm of linkage group II (Arst, Rand & Bailey, 1979). The fact that the *nii*A and *nia*D genes could be separated, but that the *nii*A retained some expression from its normal promoter, is a strong indication that the *nii*A and *nia*D genes are expressed on separate transcripts. Unfortunately, the details of the *nii*A and *nia*D gene sequences were not available to me at the time of writing this review, although the region between *nii*A and *nia*D does contain separate divergent promoters for these genes (J. R. Kinghorn, personal communication).

Nitrate induction

The synthesis of the two enzymes of nitrate assimilation is subject to two control mechanisms: induction by nitrate and/or nitrite and nitrogen metabolite repression. Both mechanisms are mediated by the products of positive regulatory genes; that is, proteins bind to DNA and switch on structural gene expression.

Nitrate/nitrite induction is a pathway-specific regulatory system. In the absence of ammonium or glutamine (the effector of nitrogen metabolite repression), the pathway of nitrate assimilation is expressed only in the presence of nitrate and/or nitrite. The *nir*A gene in *A. nidulans* and the *nit*-4 gene in *N. crassa* encode the regulatory protein mediating this process (see Cove, 1979; Tomsett & Garrett, 1980). The most common class of mutations in any gene are those which result in the synthesis of an inactive gene product or no gene product. In *nir*A and *nit*-4, most mutations result in an inability to grow on nitrate or nitrite (Cove, 1979; Tomsett & Garrett, 1980). This can be shown to result from an inability to synthesize normal wild-type levels of both nitrate reductase and nitrite reductase when supplied nitrate or nitrite. Most retain a low constitutive level of both enzymes on a neutral nitrogen source such as urea whether nitrate/nitrite are present or not. In the presence of ammonium or glutamine, however, even this low activity is repressed (Cove, 1979; Tomsett & Garrett, 1981).

In *A. nidulans*, other rare classes of *nir*A mutants have been isolated (Rand & Arst, 1978; Tollervey & Arst, 1981). The first of these, *nir*A^{c}1, was recognised by the plate test for the production of nitrite, because colonies stain pink much faster than either wild-type or *nii*A strains (see Cove, 1979). The reason for this is that *nir*A^{c}1 has constitutive high levels of nitrate-reductase and nitrite reductase when grown on a neutral nitrogen source even when inducer is absent. However, the synthesis of these enzymes in this strain is still subject to nitrogen metabolite repression. Rand & Arst (1978) identified a third class of mutation at this locus. The *nir*A$^{c/d}$101 mutant has the constitutive synthesis characteristic of *nir*A^{c}1 but also leads to strongly derepressed synthesis of nitrate and nitrite reductase in the presence of ammonium or glutamine. The *nir*A$^{c/d}$101 strain was derived from *nir*A^{c}1 by mutagenesis and hence carries two mutations within this gene. Tollervey & Arst (1981) isolated the *nir*A^{d}106 mutation which appears to be altered for the derepression site in the *nir*A gene. Crosses between *nir*A^{c}1 and *nir*A^{d}106 in the appropriate genetic background produce recombinant *nir*A^{c}1 *nir*A^{d}106 progeny, whose phenotype resembles the *nir*A$^{c/d}$101 mutant. The *nir*A gene, therefore, contains at least two separable domains: one in which mutations can lead to constitutivity, presumably representing a co-inducer binding domain; another where mutations can result in nitrogen metabolite derepression, which is either a binding site for the *are*A gene product (see below) or the site which interacts with the initiator/promoter regions adjacent to *are*A/*nir*A regulated structural genes.

The *nir*A gene is not, however, the only gene affecting nitrate induction. Mutations affecting nitrate reductase can also result in altered regulation of nitrate assimilation. Many *nia*D and *cnx* mutants in *A. nidulans* have constitutive synthesis for the nitrite reductase and the mutant nitrate reductase (Pateman *et al.*, 1964; see Cove, 1979). This is also true for the equivalent mutants of *N. crassa* (Tomsett & Garrett, 1981). Mutants of *nia*D (and *nit*-3) fall basically into two classes: those which are regulated exactly as the wild-type, and those that synthesize the enzymes constitutively (i.e. in the presence and absence of the inducer). This altered regulation is detected by comparison of enzyme levels of both nitrate reductase partial activities and NADPH-nitrite reductase on non-inducing and inducing medium. A few strains remain normally inducible like the wild-type, while others become completely constitutive. Most mutants are found to be at least partly altered with respect to regulation by induction. These observations also apply to the mutants altered for molybdenum cofactor synthesis. Although nitrate/nitrite induction is altered, all these mutants are still subject to nitrogen metabolite repression.

These data demonstrate the dual role played by nitrate reductase in nitrate assimilation. In addition to its catalytic function, the enzyme plays a key role in the regulation of this pathway. Cove (1970) proposed a model to explain this conclusion. The *nir*A gene product is thought to switch on transcription of the structural genes of both nitrate and nitrite reductase. After translation, the native nitrate reductase is assembled with the molybdenum cofactor into active enzyme. In the absence of nitrate, this nitrate reductase molecule converts the active nirA gene product to an inactive state in which it can no longer stimulate transcription. Once the low level of nitrate reductase had been degraded the cycle would start again. When nitrate was present, nitrate reductase would complex with it and no longer be able to inactivate the *nir*A gene product, leading to further transcription of the structural genes and the induced levels of these enzymes. This model explains all the mutant phenotypes observed; including *nia*D and *cnx* mutants which drastically alter the conformation of native nitrate reductase or prevent its formation, prevent *nir*A gene product inactivation and hence are constitutive, while *nia*D and *cnx* mutants which only affect catalytic activity and not conformation, remain normally regulated. The *nir*A^c mutations prevent nitrate reductase inactivating the *nir*A gene product and hence are constitutive. The *nir*A^- mutants lack the active *nir*A gene product, cannot stimulate high level expression, and hence appear non-inducible. The confirmation of this model awaits complex molecular biology experiments. As the first step towards this, Y. H. Fu & G. A. Marzluf (unpublished data) have examined *nit*-3 mRNA levels; expression requires both nitrogen derepression and nitrate induction. In *nit*-3 mutants, *nit*-3 mRNA is constitutive.

Nitrogen metabolite repression

It is not within the scope of this review to cover all the aspects of nitrogen metabolite repression. These have been eloquently explained in another recent review (Wiame *et al.*, 1985). Instead, I shall give a brief overview of the facts and discuss these in the light of recent data.

The *nit*-2 gene of *N. crassa* and the *are*A gene of *A. nidulans* encode positive-acting regulatory proteins which switch on the expression of a variety of pathways of nitrogen metabolism, including nitrate assimilation. These pathways are only switched on when glutamine is absent; when present, glutamine prevents this switch. Ammonium was formerly thought to be the active effector, but mutations preventing the assimilation of ammonium to glutamine also relieve ammonium repression (Dunn-Coleman, Tomsett & Garrett, 1979; MacDonald, 1982). The most common class of mutations in *nit*-2 and *are*A result in a repressed phenotype: an inability to utilise nitrogen sources other than ammonium or glutamine. A rarer class of mutation, *are*A^d leads to derepressed expression of *are*A regulated pathways: enzymes and permeases in these pathways are synthesized whether glutamine is present or not (but if they require a specific inducer, such as nitrate, this must also be present for synthesis).

The simplest explanation is that the *are*A and *nit*-2 genes encode regulatory proteins which, in the absence of glutamine, bind the promoter/initiator regions of the structural genes which they regulate and stimulate transcription. Glutamine could bind to the regulatory protein preventing the activation of transcription. Grove & Marzluf (1981) isolated non-histone DNA-binding proteins from *N. crassa* nuclei and identified a 22,000 dalton protein which was eluted with glutamine. This protein was present in greatly reduced quantities, or had a different electrophoretic mobility, in nuclei of *nit*-2 mutants. Recently the *nit*-2 gene has been cloned (Stewart & Vollmer, 1986). This gene is transcribed as a 3·5 kb transcript, which is present at considerably higher levels in cells which are limited for nitrogen than those growing under nitrogen-repressed conditions (Fu & Marzluf, 1987). This suggests that *nit*-2 is regulated by other control genes. The *are*A gene has also been cloned and appears to be similarly regulated (Caddick *et al.*, 1986; M. X. Caddick, personal communication).

Conclusions

The genetics and biochemistry of nitrate assimilation in fungi have made outstanding contributions to our knowledge of enzymology, gene expression and gene regulation. By disrupting the normal metabolism of these organisms, detailed information has emerged on the structure of enzymes, their catalytic activity, and the regulation of their synthesis. This has led to models of the interaction of regulatory proteins with DNA and with each other. It is only relatively recently that sophisticated molecular biology has been possible in these fungi. A number of genes have now been cloned and these are currently being sequenced. In the next few years, cloned sequences can be used directly to test the proposals made for the various regulatory interactions. The examination of sequences upstream from structural genes should reveal sequences to which the *are*A/*nit*-2 protein will bind to mediate nitrogen metabolite repression and those to which the *nir*A/*nit*-4 protein binds to mediate nitrate induction. Site-directed mutagenesis should reveal those sites in the *areA*/*nit*-2 protein responsible for binding to DNA and those which bind glutamine. Similar experiments with *nir*A/*nit*-4 could identify whether this protein can bind and be inactivated by nitrate reductase. Certainly in the next few years, we will see these proposals confirmed or refuted by molecular biology.

References

Aberg, B. (1947). On the mechanism of the toxic action of chlorates and some related substances upon young wheat plants. *Kungliga Lantbrukshogskolans Annaler*, **15**, 37-107.

Arst, H. N., MacDonald, D. W. & Cove, D. J. (1970). Molybdate metabolism in *Aspergillus nidulans*. I Mutations affecting nitrate reductase and/or xanthine dehydrogenase. *Molecular and General Genetics*, **108**, 129-145.

Arst, H. N., Rand, K. N. & Bailey, C. R. (1979). Do the tightly linked structural genes for nitrate and nitrite reductases in *Aspergillus nidulans* form an operon? Evidence from an insertional translocation which separates them. *Molecular and General Genetics*, **17**, 89-100.

Arst, H. N., Tollervey, D. W. & Sealy-Lewis, H. M. (1982). A possible regulatory gene for the molybdenum-containing cofactor in *Aspergillus nidulans*. *Journal of General Microbiology*, **128**, 1083-1093.

Brownlee, A. G. & Arst, H. N. (1983). Nitrate uptake in *Aspergillus nidulans* and involvement of the third gene of the nitrate assimilation gene cluster. *Journal of Bacteriology*, **155**, 1138-1146.

Caddick, M. X., Arst, H. N., Taylor, L. H., Johnson, R. I. & Brownlee, A. G. (1986). Cloning of the regulatory gene *are*A mediating nitrogen metabolite repression in *Aspergillus nidulans*. *EMBO Journal*, **5**, 1087-1090.

Calza, R., Huttner, E., Vincentz, M., Rouze, P., Galangau, F., Vaucheret, H., Cherel, I., Meyer, C., Kronenberger, J. & Caboche, M. (1987). Cloning of DNA fragments complementary to tobacco nitrate reductase mRNA and encoding epitopes common to the nitrate reductases from higher plants. *Molecular and General Genetics*, **209**, 552-562.

Coddington, A. (1976). Biochemical studies on the *nit* mutants of *Neurospora crassa*. *Molecular and General Genetics*, **145**, 195-206.

Cooley, R. N. & Tomsett, A. B. (1985). Determination of the subunit size of NADPH-nitrate reductase from *Aspergillus nidulans*. *Biochimica et Biophysica Acta*, **831**, 89-93.

Cove, D. J. (1970). Control of gene action in *Aspergillus nidulans*. *Proceedings of the Royal Society, London, Series B*, **176**, 267-275.

Cove, D. J. (1976a). Chlorate toxicity in *Aspergillus nidulans*: the selection and characterization of chlorate resistant mutants. *Heredity*, **36**, 191-203.

Cove, D. J. (1976b). Chlorate toxicity in *Aspergillus nidulans*: studies of mutants altered in nitrate assimilation. *Molecular and General Genetics*, **146**, 147-159.

Cove, D. J. (1979). Genetic studies of nitrate assimilation in *Aspergillus nidulans*. *Biological Reviews*, **54**, 291-327.

Downey, R. J. & Focht, W. J. (1974). Subunit character of the NADPH-nitrate reductase from *Aspergillus nidulans*. *Microbios*, **11A**, 61-70.

Downey, R. J. & Steiner, F. X. (1979). Further characterization of the reduced NADP: nitrate oxidoreductase in *Aspergillus nidulans*. *Journal of Bacteriology*, **137**, 105-114.

Dunn-Coleman, N. S. (1984a). Biochemical characterization of the molybdenum cofactor mutants of *Neurospora crassa*: *in vivo* and *in vitro* reconstitution of NADPH-nitrate reductase activity. *Current Genetics*, **8**, 581-588.

Dunn-Coleman, N. S. (1984b). Cloning and preliminary characterization of a molybdenum cofactor gene of *Neurospora crassa*. *Current Genetics*, **8**, 589-595.

Dunn-Coleman, N. S., Smarrelli, J. Jr. & Garrett, R. H. (1984). Nitrate assimilation in eukaryotic cells. *International Review of Cytology*, **92**, 1-50.

Dunn-Coleman, N. S., Tomsett, A. B. & Garrett, R. H. (1979). Nitrogen metabolite repression of nitrate reductase in *Neurospora crassa*: effect of the *gln*-1a locus. *Journal of Bacteriology*, **139**, 697-700.

Fu, Y. H. & Marzluf, G. A. (1987). Characterization of *nit*-2, the major nitrogen regulatory gene of *Neurospora crassa*. *Molecular and Cellular Biology*, **7**, 1691-1696.

Garrett, R. H. (1972). The induction of nitrite reductase in *Neurospora crassa*. *Biochimica et Biophysica Acta*, **264**, 481-489.

Garrett, R. H. (1978). Nitrite reduction in fungi. In *Microbiology 1978*, ed. D. Schlessinger, pp. 324-329. American Society of Microbiology Publication, Washington, USA.

Garrett, R. H. & Amy, N. K. (1978). Nitrate assimilation in fungi. *Advances in Microbial Physiology*, **18**, 1-65.

Garrett, R. H. & Nason, A. (1969). Further purification and properties of *Neurospora* nitrate reductase. *Journal of Biological Chemistry*, **244**, 2870-2882.

Greenbaum, P., Prodouz, K. N. & Garrett, R. H. (1978). Preparation and some properties of homogeneous *Neurospora crassa* assimilatory NADPH-nitrite reductase. *Biochimica et Biophysica Acta*, **526**, 52-64.

Grove, G. & Marzluf, G. A. (1981). Identification of the product of the major regulatory gene of the nitrogen control circuit of *Neurospora crassa* as a nuclear DNA binding protein. *Journal of Biological Chemistry*, **256**, 463-470.

Guerrero, M. G., Vega, J. M. & Losada, M. (1981). The assimilatory nitrate-reducing system and its regulation. *Annual Review of Plant Physiology*, **32**, 169-204.

Hageman, R. V. & Rajagopalan, K. V. (1985). Characterization of molybdopterin, the organic portion of the molybdenum cofactor. In *Nitrogen Fixation and C02 Metabolism*, eds. P. W. Ludden & J. E. Burris, pp. 133-141. Elsevier, New York.

Horner, R. D. (1983). Purification and comparison of *nit*-1 and wild-type NADPH: nitrate reductases of *Neurospora crassa*. *Biochimica et Biophysica Acta*, **744**, 7-15.

Johnson, J. L. (1980). The molybdenum cofactor common to nitrate reductase, xanthine dehydrogenase and sulphite oxidase. In *Molybdenum and Molybdenum-Containing Enzymes*, ed. M. P. Coughlan, pp. 345-383. Pergamon Press, New York.

Johnson, J. L., Hainline, B. E., Rajagopalan, K. V. & Arison, B. H. (1984) The pterin component of the molybdenum cofactor: structural characterization of two fluorescent derivatives. *Journal of Biological Chemistry*, **259**, 5414-5422.

Johnson, J. L. & Rajagopalan, K. V. (1982). Structural and metabolic relationship between the molybdenum cofactor and urothione. *Proceedings of the National Academy of Sciences, U.S.A*, **79**, 6856-6860.

Johnson, M. E. & Rajagopalan, K. V. (1987). Involvement of *chlA*, *E*, *M*, and *N* loci in *Escherichia coli* molybdopterin biosynthesis. *Journal of Bacteriology*, **169**, 117-125.

Kramer, S., Hageman, R. V. & Rajagopalan, K. V. (1984). *In vitro* reconstitution of nitrate reductase activity of the *Neurospora crassa* mutant *nit*-1: specific incorporation of molybdopterin. *Archives of Biochemistry and Biophysics*, **233**, 821-829.

Lafferty, M. A. & Garrett, R. H. (1974). Purification and properties of the *Neurospora crassa* assimilatory nitrite reductase. *Journal of Biological Chemistry*, **249**, 7555-7567.

Le, K. H. D. & Lederer, F. (1983). On the presence of a heme-binding domain homologous to cytochrome *b5* in *Neurospora crassa* assimilatory nitrate reductase. *EMBO Journal*, **2**, 1909-1914.

MacDonald, D. W. (1982). A single mutation leads to loss of glutamine synthetase and relief of ammonium repression in *Aspergillus*. *Current Genetics*, **6**, 203-208.

MacDonald, D. W. & Cove, D. J. (1974). Studies on temperature-sensitive mutants affecting the assimilatory nitrate reductase of *Aspergillus nidulans*. *European Journal of Biochemistry*, **47**, 107-110.

Marzluf, G. A. (1981). Regulation of nitrogen metabolism and gene expression in fungi. *Microbiological Reviews*, **45**, 437-461.

Minagawa, N. & Yoshimoto, A. (1982). Purification and characterization of the assimilatory NADPH-nitrate reductase of *Aspergillus nidulans*. *Journal of Biochemistry*, **91**, 761-774.

Pan, S-S. & Nason, A. (1978). Purification and characterization of homogeneous assimilatory NADPH-nitrate reductase from *Neurospora crassa*. *Biochimica et Biophysica Acta*, **523**, 297-313.

Pateman, J. A., Cove, D. J., Rever, B. M. & Roberts, D. B. (1964). A common cofactor for nitrate reductase and xanthine dehydrogenase which also regulates the synthesis of nitrate reductase. *Nature*, **201**, 58-60.

Pombeiro, S. R. C., Martinez-Rossi, N. M. & Rossi, A. (1983). Nitrate toxicity in *Aspergillus nidulans*: a new locus in a *proA1 pabaA6 yA2* strain. *Genetical Research*, **41**, 203-207.

Prodouz, K. N. & Garrett, R. H. (1981). *Neurospora crassa* NAD(P)H-nitrite reductase: studies on its composition and structure. *Journal of Biological Chemistry*, **256**, 9711-9717.

Rand, K. N. & Arst, H. N. (1977). A mutation in *Aspergillus nidulans* which affects the regulation of nitrite reductase and is tightly linked to its structural gene. *Molecular and General Genetics*, **155**, 67-75.

Rand, K. N. & Arst, H. N. (1978). Mutations in *nir*A gene of *Aspergillus nidulans* and nitrogen metabolism. *Nature*, **272**, 732-734.

Scazzocchio, C. (1980). The genetics of the molybdenum-containing enzymes. In *Molybdenum and Molybdenum-Containing Enzymes*, ed. M. P. Coughlan, pp. 487 - 515. Pergamon Press, Oxford.

Schloemer, R. H. & Garrett, R. H. (1974a). Nitrate transport system in *Neurospora crassa*. *Journal of Bacteriology*, **118**, 259-269.

Schloemer, R. H. & Garrett, R. H. (1974b). Uptake of nitrite by *Neurospora crassa*. *Journal of Bacteriology*, **118**, 270-274.

Solomonson, L. P., Barber, M. J., Robbins, A. P. & Oaks, A. (1986). Functional domains of assimilatory NADH: nitrate reductase from *Chlorella*. *Journal of Biological Chemistry*, **261**, 11290-11294.

Sorger, G. J., Premakumar, R. & Gooden, D. (1978). Demonstration *in vitro* of two intracellular inactivators of nitrate reductase from *Neurospora*. *Biochimica et Biophysica Acta*, **706**, 33-47.

Steiner, F. X. & Downey, R. J. (1982). Isoelectric focusing and two dimensional analysis of purified nitrate reductase from *Aspergillus nidulans*. *Biochimica et Biophysica Acta*, **706**, 203-211.

Stewart, V. & Vollmer, S. J. (1986). Molecular cloning of *nit-2*, a regulatory gene required for nitrogen metabolite repression in *Neurospora crassa*. Gene, **46**, 291-295.

Tollervey, D. W. & Arst, H. N. (1981). Mutations to constitutivity and derepression are separate and separable in a regulatory gene of *Aspergillus nidulans*. *Current Genetics*, **4**, 63-68.

Tomsett, A. B. (1977). Structure and function in the *nii*A *nia*D gene region of *Aspergillus nidulans*. Ph.D. Thesis, University of Cambridge, U.K.

Tomsett, A. B. & Cove, D. J. (1979). Deletion mapping of the *nii*A *nia*D gene region of *Aspergillus nidulans*. *Genetical Research*, **34**, 19-32.

Tomsett, A. B. & Garrett, R. H. (1980). The isolation and characterization of mutants defective in nitrate assimilation in *Neurospora crassa*. *Genetics*, **95**, 649-660.

Tomsett, A. B. & Garrett, R. H. (1981). Biochemical analysis of mutants defective in nitrate assimilation in *Neurospora crassa*: evidence for autogenous control by nitrate reductase. *Molecular and General Genetics*, **184**, 183-190.

Walls, S., Sorger, G. J., Gooden, D. & Klein, V. (1978). The regulation of the decay of nitrate reductase: evidence for the existence of at least two mechanisms of decay. *Biochimica et Biophysica Acta*, **540**, 24-32.

Wiame, J. M., Grenson, M. & Arst, H. J. (1985). Nitrogen catabolite repression in yeasts and filamentous fungi. *Advances in Microbial Physiology*, **26**, 1-88.

Chapter 3

Transport and metabolism of sulphur dioxide in yeasts and filamentous fungi

Anthony H. Rose

Zymology Laboratory, School of Biological Sciences, University of Bath, Bath BA2 7AY, Avon, UK

Introduction

From a knowledge of the chemistry of the element sulphur and of its oxidised forms, it can be stated with confidence that yeasts and fungi have been exposed to sulphur dioxide (SO_2) from time immemorial. However, a deeper scientific interest in the reaction of these micro-organisms with SO_2 has recently been evident, largely for two reasons.

First, for centuries SO_2 has been used to preserve beverages and more recently foods. Although it is often stated that the ancient Greeks and Romans disinfected their amphorae with burning sulphur, there is curiously little firm evidence for this practice (Henderson, 1824). We do know, however, that sulphur was included in a mixture of pitch and resin which was used to patch cracks in fermenting jars (Allen, 1961). Sulphur is hardly ever burned nowadays to disinfect, although a visit to a traditional oast house can still provide an experience of this pungent breathtaking vapour (Lloyd-Hind, 1948). The great strides made in inorganic chemistry in the 19th Century led to the availability of a variety of sulphur salts which give rise to SO_2 when dissolved in water. Today, the principal compound used to generate sulphur dioxide and the related anions in the preservation of foods and beverages is sodium metabisulphite ($Na_2S_2O_5$), designated additive E223 in Directives of the European Economic Community (Hanssen & Marsden, 1984). Other compounds frequently employed as sulphiting agents include SO_2, potassium and sodium salts of sulphite and bisulphite and potassium metabisulphite, the common characteristic being their ability to liberate free molecular SO_2 which appears to correlate with their preservative activity.

At the present time, there is, however, deep concern over the use of SO_2 as a beverage and food preservative. This arises in part from the increasing disapproval of adding chemical compounds of any description to beverages and foods. Specific criticism of the use of SO_2 has been prompted largely by the pungent nature of this additive. The merest whiff of gaseous SO_2, when opening a bottle of wine or a packet of food, can bring on an asthmatic attack, which has in one or two instances proved fatal for persons with a history of respiratory problems. The general desire to lower concentrations of SO_2 included in beverages and foods explains in large part the recent upsurge of interest into the physiological basis of the interaction of this food additive with micro-organisms in general. Knowledge of the molecular mechanisms of this interaction, it is argued, could indicate ways in which the lethal effect of SO_2 on microorganisms might be maximised, thereby permitting use of lower concentrations.

The second reason for an intensification of interest in the reaction of yeasts and filamentous fungi with SO_2 stems from the problem of acid rain. Most fossil fuels contain various proportions of sulphur-containing minerals which, when combusted, generate SO_2. When deposited with precipitation, this constitutes acid rain which, especially in parts of Northern Europe and North America, has, partly at least by killing micro-organisms, led to complete or partial sterilization of lakes and rivers.

Effect of pH value

This article is concerned with SO_2 and related ions which accompany this compound in aqueous solution. Throughout, sulphur dioxide is referred to as SO_2, the bisulphite ion as HSO_3^- and the sulphite ion as SO_3^{2-}. It should be stressed, however, that this nomenclature, although one which is very widely used, is at variance with that recommended by I.U.P.A.C. which recommends that the oxoacid $H_2S_2O_5$ be named disulphurous acid, and its salts as disulphites rather than metabisulphites. I.U.P.A.C. also recommend that partially dissociated acids be referred to with the prefix 'hydrogen' replacing 'bi-'. Further information on the I.U.P.A.C. recommendations can be obtained from Fernelius, Leoning &

Adams (1973). An excellent account of the chemistry of the compounds and ions referred to above can be found in the text written by Wedzicha (1984).

While in solution, metabisulphite generates SO_2, HSO_3^- and SO_3^{2-}; the proportion of these species present depends on the pH value of the solution. The equilibria are:

$$SO_2.H_2O \leftrightarrow HSO_3^- \leftrightarrow SO_3^{2-}$$

Although molecular sulphur dioxide will, as already indicated, be referred to in this chapter as SO_2, Raman and infrared spectra of dissolved SO_2 and gaseous or liquid SO_2 have led workers to conclude that dissolved SO_2 exists as $SO_2.H_2O$. Dissociation constants for each of the two equilibria have been determined at low sulphite concentrations of the order of those normally encountered by yeasts and fungi. The reaction leading to ionization of SO_2 has a pK_a value of 1·86 at 25°C and zero ionic strength, while the value for the reaction leading to production of the sulphite ion under the same conditions is 7·18 (Wedzicha, 1984). Using these pK_a values, calculations have been made of the proportions of each species present in solution as a function of pH value and tabulated data have been published by King *et al.* (1981. This effect of pH value is vitally important in relation to the mechanism of action of SO_2 with yeasts and filamentous fungi.

Transport of SO_2

Passage of sulphite across the microbial plasma membrane appears to be by different mechanisms depending on the species of yeast or mycelial fungus. The currently accepted mechanism for transport of sulphite into *Saccharomyces cerevisiae*, a microbe on which the majority of studies of SO_2 transport have been carried out, is by free diffusion of the molecular form of SO_2 (Stratford & Rose, 1986) which conflicts with the active transport system previously proposed by Macris & Markakis (1974). Stratford & Rose (1986) presented strong evidence in favour of a protein not being involved in SO_2 transport by reporting near vertical Eadie-Woolf plots (Hofstee, 1959) for SO_2 transport at pH 3·0 and 4·0. At these pH values, a high proportion of free SO_2 is present in solution. Values for K_T calculated from kinetic plots of v against v/s (v is the initial velocity of accumulation of sulphite equivalents, and s the extracellular SO_2 concentration) were 3·2 mM and 0·1 mM at pH 3·0 and 4·0 respectively. These K_T values are far in excess of the concentration of SO_2 required to kill *Saccharomyces cerevisiae.* At such concentrations, all active transport systems would break down and yet the initial rate of uptake is theoretically still only half its maximum level, strongly suggesting that a passive transport system predominates under these conditions. This evidence is corroborated by the inability of carbonylcyanide-*m*-chlorophenylhydrazone (CCCP) and dinitrophenol (DNP) (Borst-Pauwels, 1981) to affect initial velocities of SO_2 accumulation. These protonophores are known to dissipate the transmembrane proton gradient and to inhibit protein-mediated transport systems. Further evidence for a lack of active transport of SO_2 came from the finding that exclusion of glucose from reaction mixtures had no effect on initial velocities of accumulation. Similarly, the inability of the glycolytic inhibitor, 2-deoxy-D-glucose, to affect SO_2 accumulation adds fuel to the theory that energy is not required for SO_2 uptake. Yet additional evidence is provided by the absence of an effect of pH value on the process, an effect characteristic of protein-mediated transport. Macris & Markakis (1974) studied the kinetics of radiolabelled SO_2 uptake by *Saccharomyces cerevisiae* var. *ellipsoideus* making some valuable observations on SO_2 toxicity and pH dependence. They found a close correlation between accumulation of radiolabelled sulphite, over the pH range 3·0 - 5·0, and concentration of molecular SO_2 in solution, which was corroborated by Stratford & Rose (1986) and Hinze & Holzer (1985). Evidence strongly suggests that, over this pH range, only the molecular form of SO_2 passes into organisms and, by inference, that *Saccharomyces cerevisiae* does not transport HSO_3^-. In this yeast, the plasma membrane merely acts as a selective barrier to free diffusion of molecular SO_2. For this reason the structure and fluidity of the plasma membrane are likely to influence transport and further investigation in this area is obviously required. A slow transport system for HSO_3^- in *Saccharomyces cerevisiae* has been tentatively suggested which is apparent in the presence of low concentrations of the ion (Stratford & Rose, 1986). As sulphite concentrations are increased, this system rapidly becomes saturated and is masked by diffusion of higher concentrations of molecular SO_2. *Saccharomyces cerevisiae* accumulates SO_2 initially very rapidly, reaching a plateau concentration after about 5 min. Intracellular sulphite concentrations at equilibrium are many times greater than those in the suspension. This can be explained by the dynamic equilibrium between the three forms of SO_2 in solution and the presence of sulphite-binding compounds inside cells (Burroughs & Sparks 1964). A third factor is the difference in pH value between the cell suspension and that inside organisms. Intracellular pH values in *Saccharomyces cerevisiae* are around 6·5 where only 0·0015% of free sulphite exists as molecular SO_2 (King *et al.*, 1981). As the extracellular pH value is below 5·5 in most yeast cultures, molecular SO_2 will diffuse into cells, dissociate to HSO_3^- and H^+, thereby allowing further diffusion into organisms until theoretically the concentration of SO_2 is equal on both sides of the plasma membrane.

Recent reports from Stratford, Morgan & Rose (1987) and Pilkington & Rose (1988) have shown that sulphite transport by *Saccharomycodes ludwigii* and accumulation by *Zygosaccharomyces bailii*, respectively, were very similar to these processes in *Saccharomyces cerevisiae*. Evidence was presented that diffusion alone of molecular SO_2 occurred in both of these yeasts, while both again were shown to be able to accumulate large quantities of sulphite intracellularly.

The few reports that have appeared on sulphite transport in other eukaryotic micro-organisms show that this process differs from that in the yeasts already referred to. Benitez *et al.* (1983) and Garcia *et al.* (1983) directed their studies towards sulphate transport in *Candida utilis* and postulated a common transport system for sulphate, sulphite and thiosulphate. Data presented in these publications yield strong evidence in favour of sulphate transport being an active, energy-dependent process and that sulphate analogues, such as molybdate, selenate and chromate, competitively inhibit sulphate incorporation. In addition, the sulphur oxyanions sulphite, thiosulphate and dithionate cause substantial inhibition. With the exception of dithionate, which acts as a mixed-type inhibitor, all of the other compounds cause competitive inhibition. Uptake of sulphate is also under endogenous control. If sulphite, sulphate or thiosulphate is allowed to accumulate, subsequent uptake of sulphate is specifically inhibited. Thiosulphate and sulphite have also been shown to be extracellular inhibitors of sulphate transport in *Penicillium chrysogenum* (Yamamoto & Segel, 1966). Selenate-resistant mutants of *C. utilis* have been isolated (Garcia *et al.*, 1983). They have a common transport defect showing inability to grow in media with either sulphite, sulphate or thiosulphate as the sole source of sulphur. The wild type grows with any one of these sources. These results imply that sulphite is transported via the common transport system. This contrasts with the situation in *Saccharomyces cerevisiae, Saccharomycodes ludwigii* and *Zygosaccharomyces bailii.* A possible explanation is that transport of the bisulphite ion is biphasic in behaviour and that the common active transport system is the same as that intimated by Stratford & Rose (1986). The experiments of Benitez *et al.* (1983) were carried out at pH 6·1 where molecular SO_2 concentrations are negligible and active transport of bisulphite predominates. This might also explain the absence of sulphite toxicity since only molecular SO_2 is believed to have antimicrobial activity.

Other references to sulphite transport include work on filamentous fungi. Tweedie & Segel (1970) recognized a common active transport system for the ions SO_4^{2-}, $S_2O_3^{2-}$, SeO_4^{2-} and MnO_4^{2-} entering mycelia of *Penicillium* and *Aspergillus* species. In addition, mycelia appear to possess distinct permeases for sulphite or tetrathionate. This is borne out by the ability of sulphate permease-deficient mutants to accumulate either sulphite or tetrathionate. Anion transport in mycelia, as in *C. utilis* (Alonso, Benitez & Diaz, 1984), is regulated by both extracellular and intracellular anion concentrations. Mycelia grown in methionine-rich media accumulate few sulphur anions whereas those grown under sulphur-deficient conditions show a much greater rate of accumulation of substrate.

Evidence for a sulphite-specific permease is still questionable. Uptake of sulphite by sulphate permease-deficient mycelia recorded by Tweedie & Segel (1970) could be equally explained by simple leakage. All transport studies using multi-anionic systems are fraught with problems due to oxidation and cross reaction of anions. Tweedie & Segel (1970) clearly recognized these disadvantages but did not consider the equilibrium position of sulphite. Wherever HSO_3^- exists in solution, some proportion must be present as molecular SO_2 depending upon the pH value. It is possible that molecular SO_2 could diffuse into filamentous fungi as readily as into *Saccharomyces cerevisiae* and other yeasts. Accumulation of sulphite may be misleading in this case and, in fact, merely reflect molecular SO_2 accumulation.

The variety of possible transport mechanisms described does not readily help us to understand the mode of action of sulphite, but only to reveal that its transport is likely to be specific for each organism. Certainly in those yeasts that are a major cause of food spoilage, it seems likely that diffusion of molecular SO_2 is common and is possibly dependent on membrane composition.

Metabolism of SO_2

Sulphite can have a damaging or a beneficial effect upon yeasts and filamentous fungi, depending on the concentration to which the micro-organism is exposed. The remainder of this review deals with the ways in which sulphite, having made contact with a yeast or filamentous fungus, is chemically modified. The first section summarizes the manner in which sulphite can deleteriously interact with biologically important molecules. This is followed by a brief summary of the reactions by which SO_2 acts as a sulphur-containing nutrient for yeasts and filamentous fungi.

Reaction with biologically important molecules

All three molecular species that are found in solutions of sulphite, especially the bisulphite ion, are chemically very reactive. Not surprisingly, a voluminous literature exists on the chemistry of the interaction of these species, most of it dealing with model and low molecular-weight compounds. Molecular mechanisms for the interactions have, in many instances, been the subject of detailed studies, which are comprehensively reviewed in Wedzicha (1984). These studies have revealed that biologically important molecules, both molecularly complex and otherwise, contain chemical groupings that are known to react with sulphite particularly when in the form of the bisulphite ion. What is much more problematical, however, is to infer that, because a biological macromolecule (e.g. a protein or nucleic acid) possesses a grouping that interacts with sulphite, an interaction takes place that leads to inactivation or denaturation of that molecule.

Several enzymes are known to be inhibited by sulphite. Sulphite ions have the ability to react with disulphide bonds in the following reaction:

$$R\text{-}S\text{-}S\text{-}R + SO_3^{2-} \leftrightarrow R\text{-}S\text{-}S\text{-}O_3^- + RS^-$$

The catalytic activity of an enzyme depends on the tertiary conformation of the protein molecule. This conformation is dictated by the primary sequence of amino acid residues in the molecule which in turn governs the manner in which the molecule becomes folded into its tertiary structure. Disulphide bonds forged between juxtaposed cysteine residues help to stabilize the tertiary structure of proteins. A potential point of attack for sulphite on enzyme molecules must be disulphide bridges. However, it cannot automatically be assumed that, because an enzyme molecule contains disulphide bridges, interaction with sulphite takes place leading to inactivation of the enzyme. It is likely, nevertheless, that this reaction is responsible for inactivation of many enzymes.

Sulphite ions also react with a range of coenzymes which are essential for enzyme activity and with prosthetic groups conjugated with biologically active proteins. Inhibition of enzyme activity by sulphite may, therefore, be attributable on occasions to inactivation of a coenzyme or prosthetic group brought about by the oxoanion. Pyridine reacts with sulphite to form an addition compound. An important biological molecule containing a pyridine residue is nicotinamide adenine dinucleotide (NAD^+) and the related $NADP^+$. The reaction with NAD^+ can be represented as follows (Pfleiderer, Sann & Stock, 1960):

$$NAD^+ + SO_3^{2-} \leftrightarrow NADSO_3^-$$

Sulphite can also react with NAD^+ and $NADH^+$ when these coenzymes are bound to dehydrogenase enzymes (Parker, Lodola & Holbrook, 1978).

In addition to NAD^+ and $NADP^+$, a variety of other organic compounds have been shown to be essential for enzyme activity. Unlike NAD^+ and $NADH^+$ those compounds known as prosthetic groups remain attached to the enzyme molecule and are essential for the biological activity of the enzyme. Several of them are inactivated by sulphite. Two prosthetic groups concerned in oxidation-reduction reactions, namely flavins and haems, are both inactivated by sulphite. Other prosthetic groups that are susceptible to sulphite inactivation are the vitamins folic acid, pyridoxal and thiamin. Little is known of the chemical nature of the reaction between folic acid and sulphur oxoanions, although the reaction is thought to involve preferentially the bisulphite ion (Wedzicha, 1984). The reaction between pyridoxal and sulphite almost certainly involves the carbonyl group on the vitamin molecule. A discussion of the reaction between carbonyl compounds and sulphite appears later in this section of the article. Thiamin is cleaved by sulphite to produce pyrimidine sulphonate and a thiazole residue-containing product.

Bisulphite ions can inactivate nucleic acids by undergoing an addition reaction with pyridine residues. The reaction of cytosine and uracil with bisulphite has been closely studied. Residues of the first of these pyrimidines are found in DNA while uracil residues occur in RNA. Reaction of sulphite with DNA presumably terminates polymerization of new DNA on a DNA template, as well as transcription of DNA into RNA. Similarly, interaction of uracil residues in a mRNA molecule would abort the process of translation into polypeptide.

One of the most important and most extensively studied reactions involving sulphite is that with carbonyl groups in organic compounds. The reaction for formation of an adduct with acetaldehyde is:

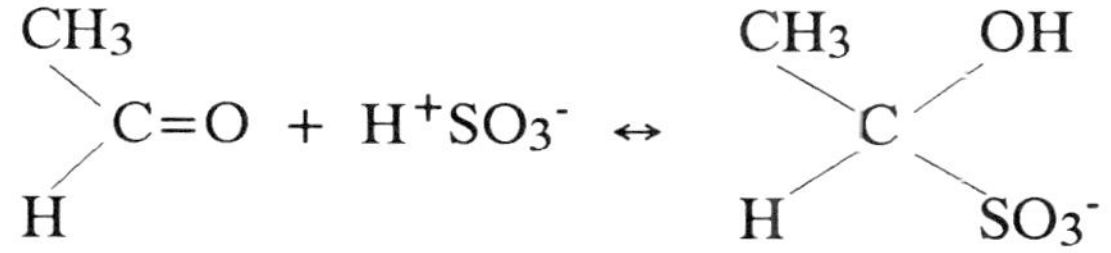

The rate constant for dissociation of the adduct differs for each carbonyl-containing compound, and also with pH value. The data in Table 3.1 illustrate these points, and show the extremely high affinity which acetaldehyde has for sulphite. The table does not include information on the ability of the more common sugars to bind sulphite, although some interesting data in this connection were reported by Ingram & Vas (1950).

Table 3.1. Apparent equilibrium constants of some α-hydroxysulphonates.

Carbonyl compound	Concentration (mM) of		Equilibrium constant at	
	Carbonyl compound	Total SO_2	pH 3·0	pH 4·0
acetaldehyde	6·0	4·0	$1{\cdot}5\times10^{-6}$	$1{\cdot}4\times10^{-6}$
2,5-diketogluconic acid	2·0	0·6-7·2	$4{\cdot}5\times10^{-4}$	$4{\cdot}3\times10^{-4}$
galacturonic acid	6·0	4·0	$1{\cdot}6\times10^{-2}$	$2{\cdot}1\times10^{-2}$
2-ketoglutaric acid	2·0	2·0-10·0	$4{\cdot}9\times10^{-4}$	$7{\cdot}0\times10^{-4}$
D-threo-2,5-hexodiulose	2·0	2·0-10·0	$3{\cdot}4\times10^{-4}$	$3{\cdot}3\times10^{-4}$
pyruvic acid	2·0	0·8-5·0	$1{\cdot}4\times10^{-4}$	$2{\cdot}2\times10^{-4}$
L-xylosone	2·0	2·0-10·0	$1{\cdot}4\times10^{-3}$	$1{\cdot}4\times10^{-3}$

Adapted from Burroughs & Sparks (1973).

Finally, there is the interaction of sulphite with molecular oxygen, the stoicheiometric equation for which is:

$$2SO_3^{2-} + O_2 \rightarrow 2SO_4^{2-}$$

This is a much studied reaction, data on which are thoroughly reviewed by Wedzicha (1984). In deducing kinetic data for this equation, the concentration of sulphite is that of SO_3^{2-}since the HSO_3^- ion is much less easily oxidized by oxygen. The reaction can take place spontaneously or through a reaction catalyzed by a transition metal ion. Postgate (1963) cautioned workers who store solutions of sulphite for experimental work against the possibility of oxidation. He recommended storing solutions at -20°C in the presence of 5 mM EDTA to sequester metal ions.

The future

Research into transport and metabolism of SO_2 by yeasts and mycelial fungi will continue in the foreseeable future, spurred on for the reasons adduced in the Introduction to this article. Pressure to lower the concentrations of sulphite used to preserve beverages and foods is, if anything, increasing year by year, and brings with it an ever increasing curiosity into the molecular mechanisms of the interaction of yeasts and mycelial fungi to this compound. No less interest is shown in production of acid rain, particularly in countries in the northern hemisphere. When precipitated, vapourised sulphur derivatives are in the form of sulphate. While potentially providing micro-organisms with a source of the element sulphur, the response all too often is to lower the pH value of an environment - lake or soil - to one which will not permit microbial growth. Politically, a solution to this problem has to be found.

References

Allen, H. W. (1961). *A History of Wine*, pp. 17-42. Faber & Faber: London.

Alonso, A., Benitez, J. & Diaz, M. A. (1984). A sulfate, sulfite and thiosulfate incorporating system in *Candida utilis*. *Folia Microbiologica*, **29**, 8-13.

Benitez, J., Alonso, A., Delgado, J. & Kotyk, A. (1983). Sulphate transport in *Candida utilis*. *Folia Microbiologica*, **28**, 6-11.

Borst-Pauwels, G. W. F. H. (1981). Ion transport in yeast. *Biochimica et Biophysica Acta*, **650**, 88-127.

Burroughs, L. F. & Sparks, A. H. (1964). The identification of sulphur dioxide-binding compounds in apple juices and ciders. *Journal of the Science of Food and Agriculture*, **15**, 176-185.

Fernelius, W. C., Leoning, K. & Adams, R. M. (1973). Notes on nomenclature. *Journal of Chemical Education*, **50**, 341-342.

Garcia, A., Benitez, J., Delgado, J. & Kotyk, A. (1983). Isolation of sulphate transport defective mutants of *Candida utilis*; further evidence for common transport system for sulphate, sulphite and thiosulphate. *Folia Microbiologica*, **28**, 1-5.

Hanssen, M. & Marsden, J. (1984). *E for Additives*, pp. 60-65. Thorsons Publishers Ltd.: Wellingborough, Northamptonshire, UK.

Henderson, A. (1824). *The History of Ancient and Modern Wines*, pp. 47-59. Baldwin, Cradock & Joy: London.

Hinze, H. & Holzer, H. (1985). Effect of sulfite or nitrate on the ATP content and the carbohydrate metabolism in yeast. *Zeitschrift für Lebensmitteluntersuchung und -Forschung*, **181**, 87-91.

Hofstee, B. H. J. (1959). Non-inverted versus inverted plots in enzyme kinetics. *Nature*, **184**, 1296-1298.

Ingram, M. & Vas, K. (1950). Combination of sulphur dioxide with concentrated orange juice. I. Equilibrium states. *Journal of the Science of Food and Agriculture*, **1**, 21-27.

King, A. D. Jr., Ponting, J. D., Sanschuck, D. W., Jackson, R. & Mihara, K. (1981). Factors affecting death of yeast by sulphur dioxide. *Journal of Food Protection*, **44**, 92-97.

Lloyd-Hind, H. (1948). *Brewing Science and Practice*, vol. 1, *Brewing Materials*. Chapman & Hall: London.

Macris, B. J. & Markakis, P. (1974). Transport and toxicity of sulfur dioxide in *Saccharomyces cerevisiae* var. *ellipsoideus*. *Journal of the Science of Food and Agriculture*, **25**, 21-29.

Parker, D. M., Lodola, A. & Holbrook, J. J. (1978). Use of the sulphite adduct of nicotinamide adenine dinucleotide to study ionisations and the kinetics of lactate dehydrogenase and malate dehydrogenase. *Biochemical Journal*, **173**, 959-967.

Pfleiderer, G., Sann, E. & Stock, A. (1960). The mechanism of action of dehydrogenases. The reactivity of pyridine nucleotides (PN) and PN-models with sulphite as the nucleophilic reagent. *Berichte der Deutschen Chemischen Gesellschaft*, **93**, 3083-3099.

Pilkington, B. J. & Rose, A. H. (1988). Reactions of *Saccharomyces cerevisiae* and *Zygosaccharomyces bailii* to sulphite. *Journal of General Microbiology*, **134**, 2823-2830.

Postgate, J. R. (1963). The examination of sulphur auxotrophs: a warning. *Journal of General Microbiology*, **30**, 481-484.

Stratford, M. & Rose, A. H. (1986). Transport of sulphur dioxide by *Saccharomyces cerevisiae*. *Journal of General Microbiology*, **132**, 1-6.

Stratford, M., Morgan, P. & Rose, A. H. (1987). Sulphur dioxide resistance in *Saccharomyces cerevisiae* and *Saccharomycodes ludwigii*. *Journal of General Microbiology*, **133**, 2173-2179.

Tweedie, J. W. & Segel, I. W. (1970). Specificity of transport processes for sulfur, selenium and molybdenum anions by filamentous fungi. *Biochimica et Biophysica Acta*, **196**, 95-106.

Wedzicha, B. L. C. (1984). *Chemistry of Sulphur Dioxide in Foods*. Elsevier Applied Science Publishers: London and New York.

Yamamoto, L. A. & Segel, I. W. (1966). The inorganic sulphate transport system of *Penicillium chrysogenum*. *Archives of Biochemistry and Biophysics*, **114**, 523-531.

Chapter 4

Inorganic sulphur oxidation by fungi

M. Wainwright

Department of Microbiology, University of Sheffield, Sheffield S10 2TN, UK

Introduction

The oxidation of reduced sulphur involves three main groups of microorganisms: (a) colourless sulphur bacteria, including members of the genus *Thiobacillus*, (b) photosynthetic sulphur bacteria, and (c) heterotrophs (Wainwright, 1978a). Although the involvement of organisms of groups (a) and (b) in sulphur oxidation in the environment has been generally emphasised, a wide range of heterotrophs, including bacteria and fungi, have also been implicated in the process, particularly in soils (Wainwright, 1984a).

The ability of fungi to oxidize sulphur *in vitro* was recognised during the early part of this century. For example, Waksman (1918), in a review on nutrient cycling by fungi made a passing reference to the ability of *Fusarium* species to oxidize elemental sulphur, while thiosulphate oxidation by fungi was demonstrated by Armstrong (1921). Abbott (1923) also showed that fungi could oxidize sulphur both *in vitro* and when grown in autoclaved soils. Surprisingly however, Starkey (1934) concluded that fungi were incapable of oxidizing thiosulphate. Fungi were also observed contaminating sulphur-enrichment cultures used in the original isolation of *Thiobacillus thiooxidans* (Lipman, Waksman & Joffe, 1922), and Joffe (1922) concluded that fungi, including species of *Aspergillus*, *Fusarium* and *Penicillium* participate in the early stages of sulphur oxidation, prior to the involvement of sulphur bacteria.

Interest in sulphur utilization by heterotrophs, including fungi, eventually declined as the role of the chemolithotrophic sulphur oxidizing bacteria was emphasised (Waksman & Joffe, 1922) since the rates of sulphur oxidation achieved by these bacteria *in vitro* suggested that heterotrophs would play only a minor role in the process in soils. Recent evidence suggests, however, that fungi and heterotrophic bacteria play a more important role in environmental sulphur oxidation than was previously thought, most notably in soils lacking significant populations of chemolithotrophic sulphur oxidizers (Wainwright, 1984a).

Modern studies on fungal sulphur oxidation began with the isolation by Wainwright (1978b & c) of sulphur oxidizing species from vegetation exposed to heavy atmospheric pollution. More recently a number of studies have appeared on the physiology of sulphur oxidation by both yeasts (Kurek, 1979, 1983, 1985) and filamentous fungi (Grayston, Nevell & Wainwright, 1986), and research has also been directed towards determining the role which fungi play in the process in soils (Wainwright, 1984a; Grayston *et al.*, 1986; Lawrence & Germida, 1988a & b).

My aim here is to review the literature on fungal sulphur oxidation, paying particular attention to recent studies which show that fungi may play a more important role in the process in soils than is generally recognized.

Isolation and *in vitro* study of sulphur oxidizing fungi

Fungi capable of oxidizing sulphur to thiosulphate and tetrathionate can be readily isolated from most soils. The isolates can be obtained using conventional media and then screened for their ability to oxidize sulphur using spot-tests for the formation of oxyanions, such as thiosulphate and tetrathionate. Specific media have also been reported for the isolation of sulphur-oxidizing fungi. The first of these media to be described consists of Czapek-Dox medium in which sulphur is precipitated by the reaction between polysulphide and hydrochloric acid (Wainwright, 1978a & b); a modification of a carbon-free medium devised by Wieringa (1966) to isolate chemolithotrophic sulphur-oxidizers. Sulphur oxidizing fungi growing on the surface of this medium oxidize the finely precipitated sulphur and produce a clear halo in the milky-white medium (Fig. 4.1). Colonies producing these haloes are considered presumptive sulphur oxidizers and can be removed and sub-cultured for further study. This medium has been described erroneously as a polysulphide medium and, therefore, has been criticized for containing a sulphur source not commonly added to soils as a fertilizer. This is not the case however, since the reaction between polysulphide and acid forms a fine precipitate of sulphur. Germida (1985) later used a medium for isolating sulphur oxidizers, including fungi, which contains colloidal sulphur as the sulphur source; sulphur oxidation being confirmed by the use of indicators to detect acid production. Lawrence & Germida (1988b) have also recently reported the use of a 'most probable

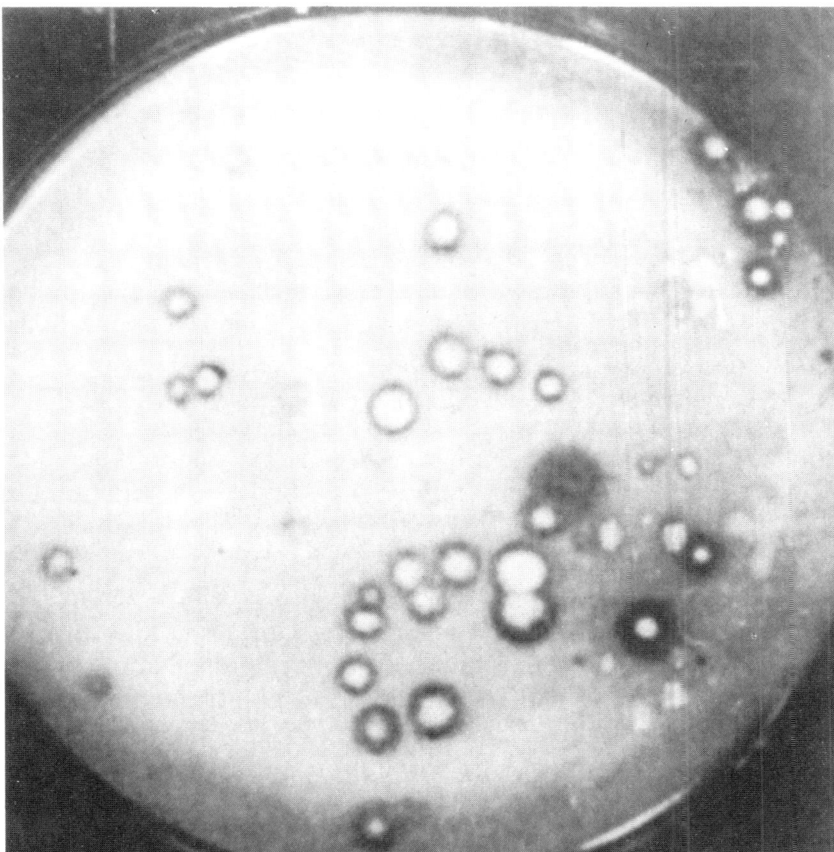

Fig. 4.1. Fungi, isolated from soil, clearing precipitated sulphur when growing on Wainwright's medium (Wainwright, 1978b).

number' method to enumerate heterotrophic thiosulphate sulphur oxidizers, including fungi, in soils.

Once presumptive sulphur oxidizing fungi have been isolated their ability to oxidize elemental and reduced forms of sulphur can be readily demonstrated in shake-cultures, using media such as liquid Czapek-Dox, and the production of thiosulphate and tetrathionate can be measured colorimetrically after cyanolysis. Sulphate is generally measured turbidimetrically. Fungal sulphur oxidation can also be detected by measurements of oxygen uptake; Grayston (1987), for example, showed that oxygen uptake over a 6 hour period was doubled when *Aspergillus niger* oxidized elemental sulphur.

Range of fungi capable of oxidizing sulphur

The ability of fungi to oxidize elemental and reduced forms of sulphur is not restricted to one specialised group, but occurs widely over all classes. However, to date most of the reported studies on the process have been devoted to mesophilic Deuteromycotina, although the list of sulphur oxidizers also includes thermophilous species (Table 4.1), and at least one marine fungus (*Asteriomyces crucicatus*). Free-living soil fungi appear to be particularly active sulphur oxidizers, and some ectotrophic mycorrhizal species and wood rotting basidiomycetes can also oxidize the element (Grayston & Wainwright, 1988). Fungi capable of oxidizing sulphur have been isolated from soils (Wainwright, 1984a; Germida, 1985; Krol, 1983), vegetation (Wainwright, 1978c), H_2S-impregnated filter beds (Langenhove, Wuyts & Schamp, 1986) and even from sulphur fungicides (Adams, 1932). The widespread occurrence of the sulphur oxidative ability among fungi isolated from diverse habitats, suggests that these organisms play an important role in sulphur oxidation in nature. Nor does this ability need to be acquired by long term exposure to reduced sulphur, since isolates from culture collections, even without prior exposure to sulphur, can immediately oxidize sulphur when grown *in vitro*. The widespread occurrence among fungi of the ability to oxidize sulphur suggests that it is either a directly useful process, or a fortuitous product of metabolic reactions which are not themselves associated with the ability to oxidize sulphur (see below). Resistance to the fungicidal properties of elemental sulphur is an obvious pre-

Table 4.1. Fungi reported to oxidize sulphur *in vitro*

(a) Deuteromycotina and Zygomycotina	
Absidia glauca	*Acremonium*
Alternaria tenuis	*Aspergillus flavus*
Aspergillus fumigatus	*Aspergillus niger*
Aureobasidium pullulans	*Epicoccum nigrum*
Fusarium episphaeria	*Fusarium tricinctum*
Monilia sp.	*Mortierella isabellina*
Mucor flavus	*Penicillium pinetorum*
Trichoderma hamatum	*Trichoderma harzianum*
Trichoderma viride	*Trichoderma sp.*
Zygorhynchus molleri	*Zygorhynchus vuilleminii*
(b) Mycorrhizal fungi	
Amanita muscaria	*Hymenoscyphus ericae*
Paxillus involutus	*Pisolithus tinctorius*
Rhizopogon roseolus	*Suillus bovinus*
(c) Cord-forming Basidiomycotina	
Hypholoma fasciculare	*Phanerochaete velutina*
(d) Yeasts	
Rhodotorula sp.	
(e) Thermophilous fungi	
Geosmithia argillacea	*Geosmithia emersonii*
Myceliophthora thermophila	

Sources: Germida (1985); Grayston & Wainwright (1988); Wainwright (1984a).

requisite if a fungus is to oxidize the element, so that those phytopathogenic fungi which are very sensitive to the fungicidal effect of sulphur cannot, *a priori* participate in the process.

Sulphur oxidation by fungi *in vitro*

The majority of reports on sulphur oxidation by fungi have involved shake culture studies using carbon-rich media e.g. liquid Czapek-Dox. The products of elemental sulphur oxidation by fungi (e.g. *Fusarium solani*) are usually thiosulphate, tetrathionate, and sulphate – the terminal oxidation product (Wainwright & Killham, 1980). It is generally assumed that the following direct polythionate pathway is involved in these transformations, although polythionates may result from side reactions, or even from non-biological oxidation:

S^0	→	$S_2O_3^{2-}$	→	$S_4O_6^{2-}$	→	SO_4^{2-}
Elemental-S		thiosulphate		tetrathionate		sulphate

Wainwright & Killham (1980) showed that elemental sulphur-oxidation by *F. solani* was optimal at pH 7·0, increased with increasing amounts of added sulphur and doubled as the C:N ratio was increased from 5:1 to 10:1. The process caused a decrease in the pH of the medium (from 6·5 to 2·4 over 10 days) as sulphuric acid was formed. However, most fungi, like *Aspergillus niger*, generally produce only thiosulphate and sulphate when oxidizing elemental sulphur *in vitro* (Fig. 4.2), with oxidation leading to a reduction in the pH of the culture medium to pH 2·0. Thiosulphate oxidation, on the other hand, usually results in a less dramatic effect on medium pH and to the production of tetrathionate as well as sulphate (Fig. 4.3). Fungi achieve differing rates of sulphur oxidation *in vitro*. For example, after 21 days growth with elemental sulphur in Czapek-Dox medium *Penicillium nigricans* produced 430 $\mu g\ SO_4^{2-}\ ml^{-1}$ compared to values for *Aspergillus niger* of 280; *Trichoderma harzianum*, 250; *Mucor flavus*, 28 and *Phanerochaete velutina*, 21 (Grayston, 1987). However, these amounts of sulphate cannot be used directly to list fungi in order of their sulphur oxidative ability, since the values represent net sulphate levels and do not account for differences in the rate of sulphate assimilation between species (Raistrick & Vincent, 1948). It should again be emphasised that these rates of oxidation achieved by fungi are in no way comparable to those seen when thiobacilli oxidize sulphur *in vitro*, although they do compare favourably with rates of oxidation achieved by other heterotrophs (Pepper & Miller, 1978).

When fungi are grown *in vitro* with elemental sulphur they invariably adsorb the sulphur particles onto their mycelial surface, thereby removing them from solution (Fig. 4.4). Particle adsorption by fungi is a widespread phenomenon and since particulate materials like clays are also adsorbed onto hyphae (Wainwright,

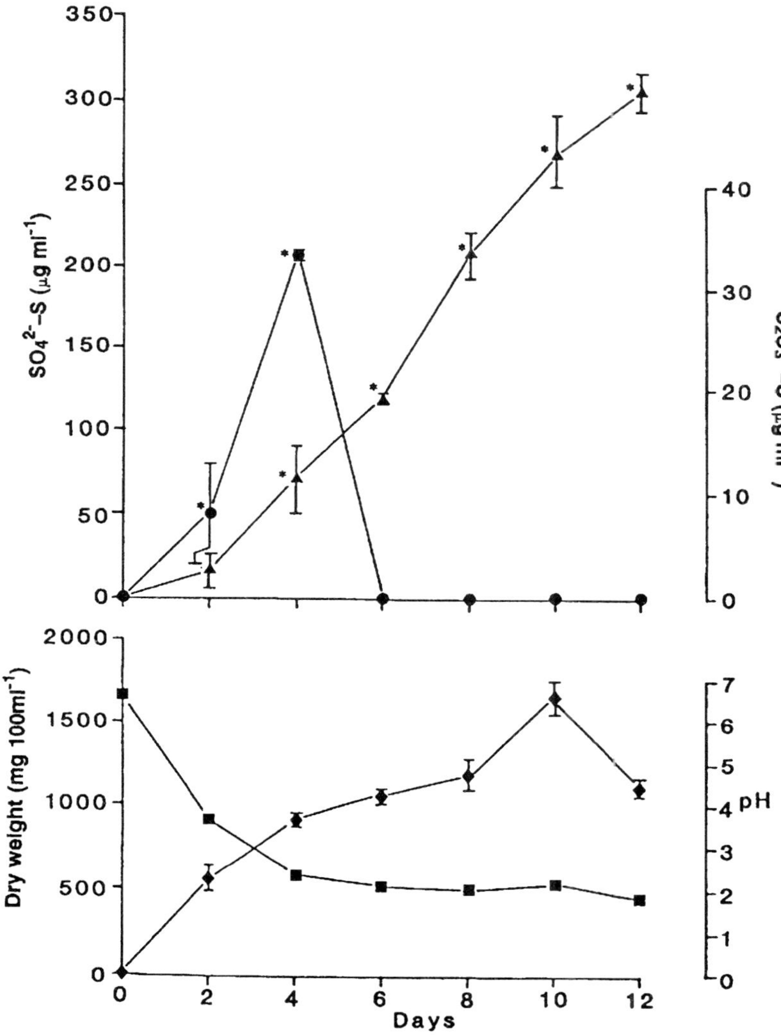

Fig. 4.2. Oxidation of elemental sulphur (1% w/v) to thiosulphate (closed circles) and sulphate (closed triangles) by *Aspergillus niger in vitro*; dry weight is shown as diamonds, and medium pH as closed squares. Values are excess over uninoculated controls, and are the means of triplicates. Asterisks indicate values which are significantly increased over the controls at $P = 0{\cdot}05$. Error bars are ±1 SD.

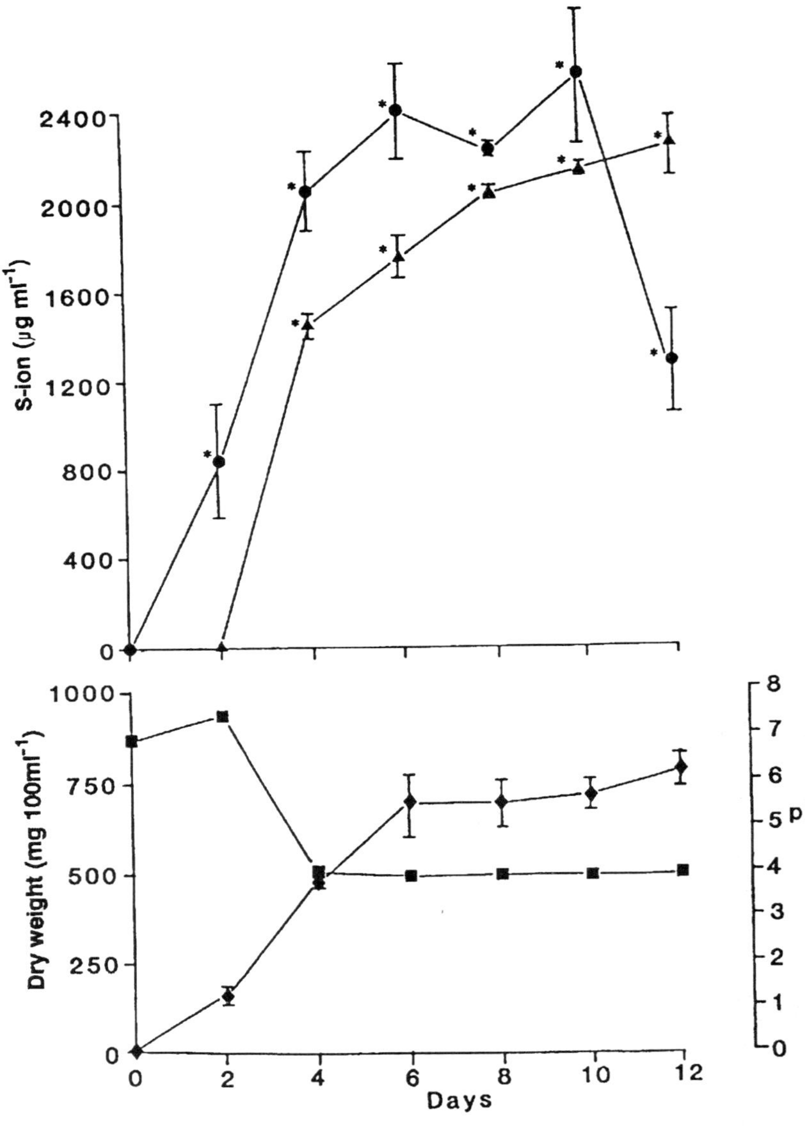

Fig. 4.3. Oxidation of thiosulphate to tetrathionate (closed circles) and sulphate (triangles) by *Aspergillus niger*. Dry weight is shown as diamonds, pH as squares. Asterisks indicate values which are significantly increased over the controls at $P = 0{\cdot}05$. Error bars are ± 1 SD.

Grayston & DeJong, 1986) the ability of fungi to adsorb sulphur may be unrelated to the fact that they oxidize the element.

Although fungal sulphur oxidation is generally studied using carbon-rich media, low, but measurable, rates of oxidation can occur in media containing relatively small amounts of carbon (10 to 100 μg C ml^{-1}) (Grayston *et al.*, 1986; Wainwright & Grayston, 1988), and some active species such as *A. niger* can clear carbon-free polysulphide medium solidified with colloidal silica. The ability of fungi to oxidize sulphur in media containing low concentrations of carbon is obviously relevant to the question of whether they participate in the process in soils and other environments where supplies of carbon are likely to be limited.

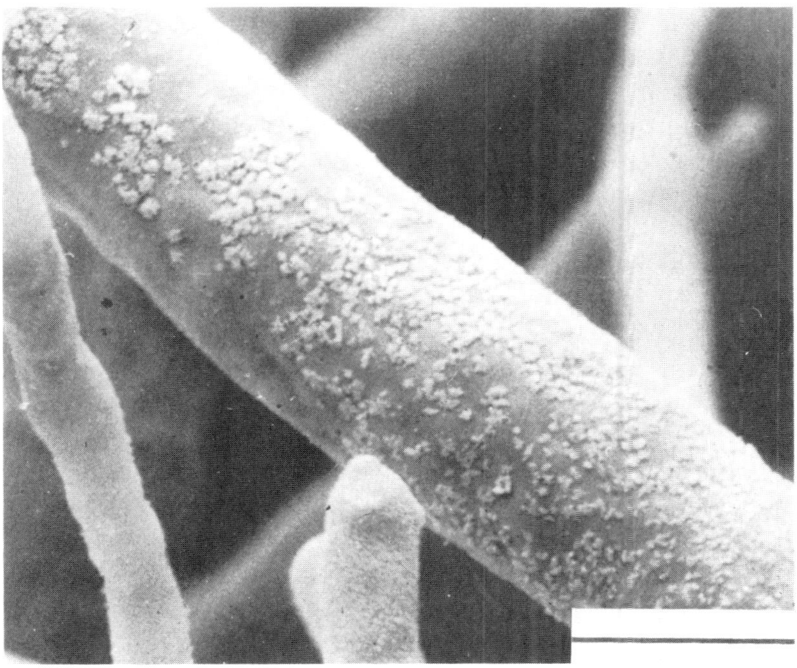

Fig. 4.4. Scanning electron-micrograph (scale bar = 20 μm) of hyphae of *Mucor flavus* adsorbing elemental sulphur (1% w/w) after 7 days growth at 25°C.

A wide range of carbon substrates can support sulphur oxidation by fungi *in vitro* (Grayston, 1987), with rates of oxidation depending upon the carbon source employed. For example, larger amounts of sulphate were produced by both *A. niger* and *T. harzianum* when glucose or sucrose acted as energy source than when amino acids provided the carbon source (Grayston & Wainwright, 1987). In contrast, Krol (1983) reported that the fungi used in her studies achieved higher rates of sulphur oxidation when amino acids, rather than sugars were provided as the carbon source. Cereal straw and acetic acid, the latter being an anaerobic breakdown product of straw, can also support fungal sulphur oxidation, so plant residues should be capable of supplying carbon to support sulphur oxidation by fungi in agricultural soils (Grayston *et al.*, 1986).

Most *in vitro* studies on fungal sulphur oxidation have involved the use of elemental sulphur, thiosulphate or tetrathionate as sulphur source, although Skerman, Dermetjev & Skyring (1957) also showed that some species can oxidize H_2S. Wainwright & Grayston (1986) reported that *A. niger* oxidizes the sulphides of copper, lead and zinc, but not cadmium. Since metal ions were not found free in solution as a result of these oxidation reactions it was concluded that any metal released during sulphide oxidation was rapidly adsorbed by the fungus.

Studies by Germida (1985) suggest that fungi may interact synergistically when oxidizing sulphur on solid media. Similarly, Grayston & Wainwright (1988) showed synergism between *A. niger*, *T. harzianum* and *M. flavus* in liquid culture, with larger quantities of sulphate being initially produced in mixed, compared to single culture. Interactions between sulphur oxidizing fungi and other sulphur oxidizing micro-organisms will probably occur in nature, and maximum rates of oxidation are likely to be achieved by mixed communities of sulphur oxidizers.

Biochemistry of fungal sulphur oxidation

Little is known about the biochemical pathways involved in fungal sulphur oxidation. The process has, however, been shown to be enzymic in *Aureobasidium pullulans* (Killham, Lindley & Wainwright, 1981) and in a species of the yeast *Rhodotorula* (Kurek, 1985). The latter was shown to possess both thiosulphate and sulphite oxidases, with the activity of both enzymes being coupled with ferricyanide and native, but not mammalian, cytochrome-*c*. The thiosulphate enzyme catalyzed the following reaction:

$$2S_2O_3^{2-} + 2Fe(CN)_6^{3-} \rightarrow S_4O_6^{2-} + 2Fe(CN)_6^{4-}$$

While the sulphite oxidase catalyzed the following:

$$2SO_3^{2-} + \text{cytochrome-c } Fe^{3+} + H_2O \rightarrow SO_4^{2-} + 2 \text{ cytochrome-c } Fe^{2+} + 2H^+$$

Both enzymes were inhibited by thiol-inhibiting, but not chelating agents, and they were both denatured by proteolytic enzymes and sodium deoxycholate. It appears, therefore, that not only proteins, but also phospholipids are important in the activity of this sulphur-oxidizing enzyme complex, and that the sulphite oxidase of *Rhodotorula* is similar to that found in thiobacilli (Kurek, 1985).

The fact that H_2S is often liberated when fungi are grown in culture with elemental sulphur and thiosulphate suggests an alternative means by which sulphate may be produced. Sciarini & Nord (1943) suggested that elemental sulphur serves mainly as a hydrogen acceptor in dehydrogenations and alcoholic fermentations of hexoses and pentoses in *Fusarium lini* thus:

$$S^0 + 2H \rightarrow H_2S$$

$$2H_2S + O_2 \rightarrow 2S + 2H_2O$$

$$2H_20 + H_2S + 4O \rightarrow SO_4^{2-} + 2H_3O^+$$

To date no clear explanation has been put forward to account for why fungi oxidize sulphur. The following possibilities have been suggested, however, and will be considered in turn below:

(1) The process may be fortuitous or incidental to the normal heterotrophic metabolism of the organisms, and may be merely an *in vitro* phenomenon.

(2) Conversely, the process may yield energy, which the organism can presumably use largely chemolithoheterotrophically.

(3) Fungi may oxidize inorganic sulphur to sulphate to meet their assimilatory requirements for sulphur.

(4) Inorganic sulphur oxidation may involve the same metabolic reactions which function in sulphur mineralization.

(5) Sulphur oxidation may provide a means by which the toxic effects of elemental sulphur and heavy metals can be avoided.

(6) Finally, the process may involve free radicals and as a result, may protect fungi from the damaging effects of, for example, superoxide.

The first of these possibilities, namely that the ability of fungi to oxidize sulphur is fortuitous or incidental to the organism's normal metabolism, is a widely held belief and it is generally assumed that no energy is obtained from the process, an assumption which has, however, largely gone unchallenged. Although fungi are generally regarded as being strict heterotrophs there is no *a priori* reason why they should not obtain energy from processes such as sulphur oxidation and nitrification. An inorganic, energy yielding pathway, has been implicated by Aleem, Lees & Lyric (1964) to account for nitrification by fungi, a view which was challenged although not necessarily disproved by Doxtader & Alexander (1966). The usual argument proposed against the view that fungi gain energy from nitrification and sulphur oxidation is that maximum rates of these processes occur towards the end of the fungus growth cycle. However, where this has been observed, carbon rich media were employed, so that the fungus could meet all its energy and carbon requirements by metabolizing sugars without the need to gain energy from inorganic oxidation reactions. The use of such media would tend to mask the relatively small energy gains obtained from the oxidation of reduced nitrogen and sulphur compounds.

The second possibility listed above (i.e. chemolithoheterotrophy) was strengthened when Grayston & Wainwright (1987) and Wainwright & Grayston (1988) showed that thiosulphate stimulated fungal growth at low carbon concentrations (Fig. 4.5), suggesting that they might grow chemolithoheterotrophically gaining energy from thiosulphate oxidation while growing hetero-

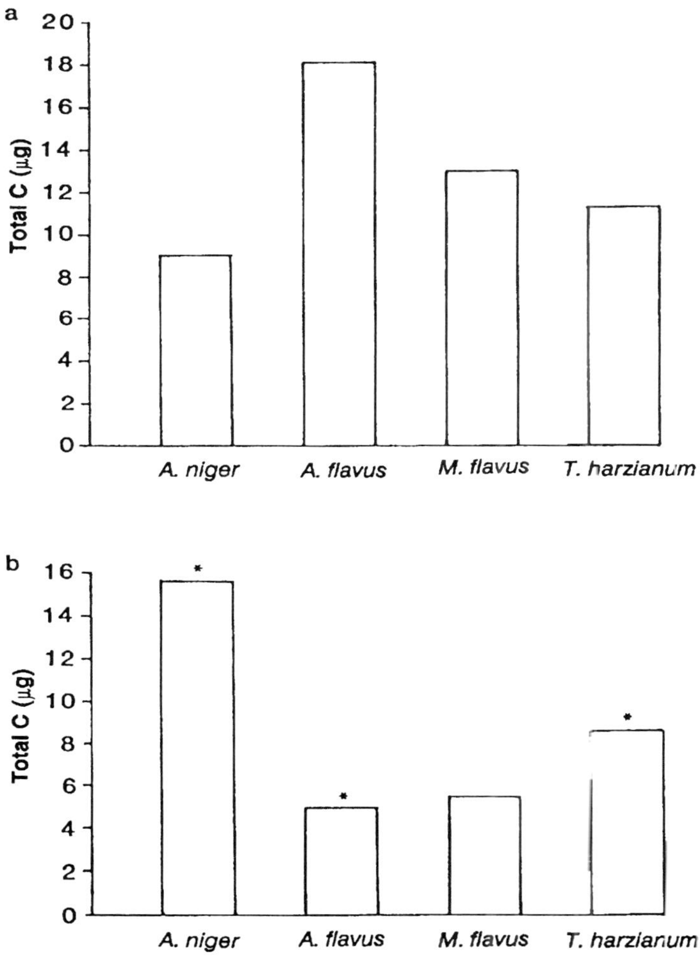

Fig. 4.5. Growth of fungi in media lacking added carbon (a), and (b) increase in fungal growth (7 days at 25°C) in this medium due to thiosulphate (1 mg S ml^{-1}). Asterisks indicate bars which show a significant increase above growth in the absence of thiosulphate at $P = 0{\cdot}05$. From Wainwright & Grayston, 1988.

trophically on the carbon provided. The result of such a strategy under carbon-limited conditions would be an increase in biomass. Similar thiosulphate induced growth stimulation has also been reported for the mycorrhizal fungus *Glomus caledonium* (Hepper, 1984), and heterotrophic bacteria (Mason & Kelly, 1988). Unfortunately, however, chemolithoheterotrophy is difficult to confirm, particularly by using batch cultures, and proof that fungi are able to grow by oxidizing thiosulphate in this way must await further studies involving the use of chemostats. An aid to such studies would be the recognition that fungi are metabolically more versatile than they are usually assumed to be. For example, they participate in many oxidation-reduction reactions which were formally thought to be mediated solely by bacteria. Evidence has been provided to show that they can fix CO_2 autotrophically and heterotrophically, as well as growing oligotrophically by scavenging traces of both carbon and nitrogen (Wainwright, 1988). It is probable, therefore, that our obsession with growing these organisms on carbon rich media has masked many of the more interesting aspects of their physiology, particularly in relation to transformations of nitrogen and sulphur (Wainwright, 1988).

The fact that fungi can oxidize reduced forms of sulphur in the presence of large amounts of sulphur would appear to dismiss quickly the view, given as point (3) above, that fungi oxidize sulphur solely to meet their assimilatory needs for this element. The possibility (point 4) that sulphur oxidation by fungi involves reactions common to sulphur mineralization processes, while possibly valid, has yet to be tested.

In support of suggestion (5), that the ability of fungi to oxidize sulphur may confer other benefits unrelated to energy gain, Wainwright & Grayston (1983) showed that the thiosulphate and tetrathionate produced during elemental sulphur oxidation can protect fungi from the toxic effects of metals such as mercury. Whether the production of polythionate and thiols, and the associated protective effects of the formation of metal complexes occur in nature is not known. In a similar way, sulphur oxidation may enable fungi to avoid the toxic effect of elemental sulphur and since hydrogen sulphide production appears to play a role here it would clearly be beneficial for a fungus to be able to oxidize both elemental sulphur and H_2S to innocuous sulphate. This would imply that plant-pathogenic fungi which are susceptible to the fungicidal effects of elemental sulphur, are incapable of oxidizing elemental sulphur, a view which is unfortunately difficult to substantiate experimentally.

Finally, Wood (1988) has suggested that nitrification by fungi may involve free radical mechanisms largely because nitrification has been linked with fungal cell lysis. The involvement of such reactions is even more likely to account for fungal sulphur oxidation since the redox potentials for sulphate/reduced forms of sulphur couples are much lower than for nitrate/nitrite or nitrite/ammonium couples (P. M. Wood pers. comm.). An example of the autooxidation of sulphite has now been well characterized (McCord & Fridovich, 1969).

The role of fungi in sulphur oxidation in soil

Our interest in the ability of fungi to oxidize sulphur began with a desire to know if these organisms make a major contribution to sulphur oxidation in soils - a process which is often considered to be mediated solely by the thiobacilli. There is, in fact, an increasing awareness of the role which heterotrophs play in sulphur oxidation in soil, particularly where thiobacilli are lacking; and recent evidence suggests that a wide range of fungi and bacteria (Pepper & Miller, 1978), including actinomycetes (Wainwright, Skiba & Betts, 1984), might all play a role in the process.

The general paucity of carbon is usually cited as a major limitation on heterotrophic sulphur oxidation in soils, although it could occur in the rhizosphere or in close proximity to decomposing plant residues where carbon should be more readily available. However, since fungi can grow and oxidize sulphur under oligocarbotrophic conditions, low levels of carbon in soils may not be such an important limitation as it first appears. Further, actinomycetes can also oxidize sulphur under oligotrophic conditions (Wainwright *et al.*, 1984), so it is clear that carbon deficiency in soils need not necessarily limit the ability of heterotrophs to oxidize sulphur. Also, recent studies by Lawrence & Germida (1988b) have confirmed that heterotrophs play an important role in sulphur oxidation in soil. Using glucose amendment they showed that microbial biomass was produced in direct proportion to an increase in the rate of sulphur oxidation, and with the aid of metabolic inhibitors demonstrated the role of fungi and other heterotrophs in the process.

Although fungi and other heterotrophs readily oxidize sulphur when inoculated into autoclaved soils (Wainwright & Killham, 1980; Grayston *et al.*, 1986; Pepper & Miller, 1978), from which are liberated large amounts of available carbon during sterilization, the process is difficult to demonstrate in non-sterilized soils. Evidence for the ability of fungi to oxidize sulphur while growing in non-autoclaved soils was, however, provided by Wainwright & Killham (1980), who inserted into soil glass capillary pedoscopes containing modified polysulphide medium, a fungal spore germination inhibitor and antibacterial antibiotics. After a period of incubation in the soil, the capillaries were removed and examined for evidence of clearing of the opaque milk-white medium which would indicate sulphur oxidation. As clearing was observed around fungal hyphae, it was concluded that fungi can grow from the soil (the germination inhibitor prevented spores from germinating at the soil-agar interface) and oxidize the sulphur, the inclusion of antibacterial antibiotics avoiding the possibility that sulphur oxidizing bacteria growing on the hyphal surface were responsible for the observed sulphur oxidation. More recent work from my laboratory has shown that ectotrophic mycorrhizal fungi such as *Suillus bovinus* and *Amanita muscaria* are also able to oxidize sulphur *in vitro* (Table 4.1), and limited evidence has also been provided to indicate that these fungi can oxidize the element when growing as mycelial fans in soil (Grayston & Wainwright, 1988). Wood decomposing basidiomycetes e.g. *Hypholoma fasciculare* are also capable of slow rates of oxidation *in vitro*, but no evidence was found by Grayston & Wainwright (1988) to suggest that these fungi actively oxidize the element when growing from inoculated wood blocks into the soil. These fungi receive carbon from a host or food base, but nevertheless they seem to be incapable of oxidizing sulphur at rates achieved by free-living species, at least *in vitro*. This could indicate that the latter group are adapted for growth under the rigours of carbon limitation in soil, and that free-living fungi may have evolved the ability to gain energy by oxidizing sulphur in order to supplement the limited energy supplies normally available to them.

Conclusion

Although the role of the colourless sulphur bacteria in sulphur oxidation has been generally emphasised it is becoming increasingly clear that heterotrophs, including fungi, play a major role in the process. As yet, however, insufficient is known about the biochemical and physiological processes involved in fungal sulphur oxidation to allow us to assess accurately the advantages which these organisms derive from participating in these oxidative reactions. Similarly, while evidence increasingly shows that fungi contribute to the oxidation of sulphur in soils, it is unclear what proportion of the sulphate produced can be attributed to the chemolithotrophs and what proportion derives from sulphur oxidation by fungi and other heterotrophs. Future research will hopefully provide a clearer picture of the part played by fungi in sulphur oxidation and their overall role in the sulphur cycle in nature.

Acknowledgements. The contributions of the following postgraduates to this work are gratefully acknowledged, Susan J. Grayston, Wendy Nevell, K. Killham, Ute Skiba and I. Singleton. Funding from the NERC and SERC is also acknowledged.

References

Adams, J. M. (1932). Bacterial and fungus flora in certain fungicides. *Phytopathology*, **22**, 785-786.

Abbot, E. V. (1923). The occurrence and action of fungi in soils. *Soil Science*, **16**, 207-216.

Aleem, M. I. H., Lees, H. & Lyric, R. (1964). Ammonium oxidation by cell-free extracts of *Aspergillus wentii*. *Canadian Journal of Biochemistry*, **42**, 989-999.

Armstrong, G. M. (1921). Studies in the physiology of the fungi – sulphur nutrition, the use of thiosulphate as influenced by hydrogen ion concentration. *Annals of the Missouri Botanical Garden*, **8**, 237-248.

Doxtader, K. G. & Alexander, M. (1966). Nitrification by growing and replacement cultures of *Aspergillus*. *Canadian Journal of Microbiology*, **12**, 807-815.

Germida, J. J. (1985). Modified sulphur-containing media for studying sulphur-oxidizing micro-organisms. In *Planetary Ecology*, eds. D. E. Caldwell, J. A. Brierley & C. L. Brierley, pp. 333-344. Nostrand Reinhold, New York.

Grayston, S. J. (1987). Sulphur oxidation and nitrification by fungi *in vitro* and in soils. PhD thesis, University of Sheffield, England.

Grayston, S. J., Nevell, W. & Wainwright, M. (1986). Sulphur oxidation by fungi. *Transactions of the British Mycological Society*, **87**, 193-198.

Grayston, S. J. & Wainwright, M. (1987). Fungal sulphur-oxidation: effect of carbon source and growth stimulation by thiosulphate. *Transactions of the British Mycological Society*, **88**, 213-219.

Grayston, S. J. & Wainwright, M. (1988). Sulphur oxidation by soil fungi including some species of mycorrhizae and wood-rotting basidiomycetes. *FEMS Microbiology Ecology*, **53**, 1-8.

Hepper, C. M. (1984). Inorganic sulphur nutrition of the vesicular-arbuscular mycorrhizal fungus *Glomus caledonium*. *Soil Biology and Biochemistry*, **16**, 669-671.

Joffe, J. S. (1922). Preliminary studies on the isolation of sulfur oxidizing bacteria from sulfur floats - soil composts. *Soil Science*, **13**, 161-172.

Killham, K., Lindley, N. D. & Wainwright, M. (1981). Inorganic sulphur oxidation by *Aureobasidium pullulans*. *Applied and Environmental Microbiology*, **42**, 629-631.

Krol, M. (1983). Occurrence in soils and activity of sulphur oxidizing micro-organisms. *Pamietnik Pulawski Prace Iung Zeszyt*, **79**, 45-61.

Kurek, E. (1979). Oxidation of inorganic sulphur compounds by yeasts. *Acta Microbiologica Polonica*, **28**, 169-172.

Kurek, E. (1983). An enzymatic complex active in sulphite and thiosulphate oxidation by *Rhodotorula* sp. *Archives of Microbiology*, **134**, 143-147.

Kurek, E. (1985). Properties of an enzymatic complex active in sulphite and thiosulphate oxidation by *Rhodotorula* sp. *Archives of Microbiology*, **143**, 277-282.

Langenhove, H. van, Wuyts, E. & Schamp, N. (1986). Elimination of hydrogen sulphide from odorous air by a wood bark filter. *Water Research*, **20**, 1471-1476.

Lawrence, J. R. & Germida, J. J. (1988a). A most probable number method (MPN) for the enumeration of heterotrophic thiosulphate producing sulfur oxidizers in soil. *Soil Biology and Biochemistry*, **20**, 577-578.

Lawrence, J. R. & Germida J. J. (1988b). Relationship between microbial biomass and elemental sulfur oxidation in agricultural soils. *Soil Science Society of America Journal*, **52**, 662-667.

Lipman, J. G., Waksman, S. A. & Joffe, J. S. (1921). The oxidation of sulfur by soil micro-organisms. *Soil Science*, **12**, 475-489.

Mason, J. & Kelly, D. P. (1988). Thiosulphate oxidation by obligately heterotrophic bacteria. *Microbial Ecology*, **15**, 123-134.

McCord, J. M. & Fridovich, I. (1969). The utility of superoxide dismutase in studying free radical reactions. *Journal of Biological Chemistry*, **244**, 6056-6063.

Pepper, I. L. & Miller, R. H. (1978). Comparison of the oxidation of thiosulphate and elemental sulphur by two heterotrophic bacteria and *Thiobacillus thiooxidans*. *Soil Science*, **126**, 9-14.

Raistrick, H. & Vincent, J. M. (1948). Studies in the biochemistry of micro-organisms. A survey of fungal metabolism of inorganic sulphates. *Biochemical Journal*, **43**, 90-99.

Sciarini, L. J. & Nord, F. F. (1943). Elementary sulphur as a hydrogen acceptor in dehydrogenations by living Fusaria. *Archives of Biochemisty and Biophysics*, **3**, 261-267.

Skerman, V. B. D., Dermentjev, S. & Skyring, G. W. (1957). Deposition of sulphur from hydrogen sulphide by bacteria and yeast. *Nature*, **179**, 742.

Starkey, R. L. (1934). The production of polythionates from thiosulphate by micro-organisms. *Journal of Bacteriology*, **28**, 387-400.

Wainwright, M. (1978a). Microbial sulphur oxidation in soil. *Science Progress, Oxford*, **65**, 459-475.

Wainwright, M. (1978b). A modified sulphur medium for the isolation of sulphur oxidizing fungi. *Plant & Soil*, **49**, 191-193.

Wainwright, M. (1978c). Sulphur oxidizing micro-organisms on vegetation and in soils exposed to atmospheric pollution. *Environmental Pollution*, **17**, 167-174.

Wainwright, M, (1984a). Sulphur oxidation in soils. *Advances in Agronomy*, **37**, 349-396.

Wainwright, M. (1984b). Sulphur oxidation by some thermophilous fungi. *Transactions of the British Mycological Society*, **83**, 721-724.

Wainwright, M. (1988). Metabolic diversity of fungi in relation to growth and mineral cycling in soil. *Transactions of the British Mycological Society*, **90**, 159-170.

Wainwright, M. & Grayston, S. J. (1983). Reduction in heavy metal toxicity towards fungi by addition to media of sodium thiosulphate and sodium tetrathionate. *Transactions of the British Mycological Society*, **81**, 541-546.

Wainwright, M. & Grayston, S. J. (1986). Oxidation of heavy metal sulphides by *Aspergillus niger* and *Trichoderma harzianum*. *Transactions of the British Mycological Society*, **86**, 269-272.

Wainwright, M. & Grayston, S. J. (1988). Fungal growth and stimulation by thiosulphate under oligocarbotrophic conditions. *Transactions of the British Mycological Society*, in press.

Wainwright, M. & Killham, K. (1980). Sulphur oxidation by *Fusarium solani*. *Soil Biology and Biochemistry*, **12**, 555-558.

Wainwright, M., Grayston, S. J. & De Jong, P. (1986). Adsorption of insoluble compounds by mycelium of the fungus *Mucor flavus*. *Enzyme and Microbial Technology*, **8**, 597-600.

Wainwright, M., Skiba, U. & Betts, R. P. (1984). Sulphur oxidation by *Streptomyces* sp. growing in a carbon deficient medium and autoclaved soil. *Archives of Microbiology*, **139**, 272-276.

Waksman, S. A. (1918). The importance of mould action in soil. *Soil Science*, **6**, 137-155.

Waksman, S. A. & Joffe, J. S. (1922). Micro-organisms concerned in the oxidation of sulfur in the soil. II. *Thiobacillus thiooxidans* a new sulfur oxidizing organism isolated from the soil. *Journal of Bacteriology*, **11**, 239-256.

Wieringa, R. T. (1966). Solid media with elemental sulphur for detection of sulphur oxidizing microbes. *Antonie van Leeuwenhoek*, **32**, 183-186.

Wood, P. M. (1988). Monoxygenase and free radical mechanisms for biological ammonia oxidation. In *The Nitrogen and Sulphur Cycles*, Society for General Microbiology Symposium 42, eds. J. A. Cole & S. Ferguson, pp. 65-98. Cambridge University Press, Cambridge, UK.

Chapter 5

Sulphur compounds in fungi

J. C. Slaughter

Department of Brewing and Biological Sciences, Heriot Watt University, Edinburgh EH1 1HX, UK

Introduction

This chapter is concerned with compounds containing divalent sulphur at its lowest oxidation state of -2. In its sulphide, thioester or thioether forms, sulphur plays a significant part in the biochemical function of a wide range of compounds. The high reactivity of the S-X bond compared to the equivalent O-X bond is mainly responsible for this and is itself due to the fact that as a larger atom, sulphur has a lower charge density and so forms longer, weaker and less polarized bonds than does oxygen. An idea of the effect of substituting sulphur for oxygen can easily be gained by comparing the chemical and physical properties of water and hydrogen sulphide. H_2S has a boiling point of -61°C, is toxic to living organisms at trace levels, reacts readily with metals and is explosive in air at concentrations between 4 and 46%. A notable exception to this reactivity rule is that disulphides (-S-S-) are much more stable than peroxides (-O-O-).

Biosynthesis of sulphide

The fungi can normally utilize inorganic sulphur for all their requirements. The most common source is sulphate which is the most highly oxidized and stable form of the element. Sulphate is actively taken up (Bradfield *et al.*, 1970; McCready & Din, 1974) and reduced to sulphide by a pathway which appears to be quite well understood and occurs in all organisms investigated which can utilize sulphate (Fig. 5.1). The reactions absorb a considerable amount of energy as the free energy difference (ΔG^0) between sulphate and sulphide is about 800 kJ mol^{-1} (Huxtable, 1986). The first two steps of the pathway can be seen as an 'activation' of sulphate. First, ATP sulphurylase (sulphate adenylyl transferase) catalyses reaction between sulphate and ATP to form adenosine-5'-phosphosulphate (APS) and inorganic pyrophosphate. The chemical equilibrium of the reaction is very much in favour of sulphate and ATP. A biosynthetic flux is only possible because pyrophosphate is removed through the action of pyrophosphatase, and the K_m value for APS of the next enzyme in the sequence, APS kinase, is very low. APS kinase catalyses reaction between APS and a further molecule of ATP to form adenosine-3'-phospho-5'-phosphosulphate (PAPS). The characteristics of APS sulphurylase from *Penicillium chrysogenum* have been described (Tweedie & Segal, 1970) but little seems to be known about APS kinase.

The first reduction of the sulphur atom is catalyzed by sulphotransferases using either APS or PAPS as substrate. Variation exists from one organism to another but it appears likely that, in most fungi, PAPS is in the main stream of sulphur reduction. The

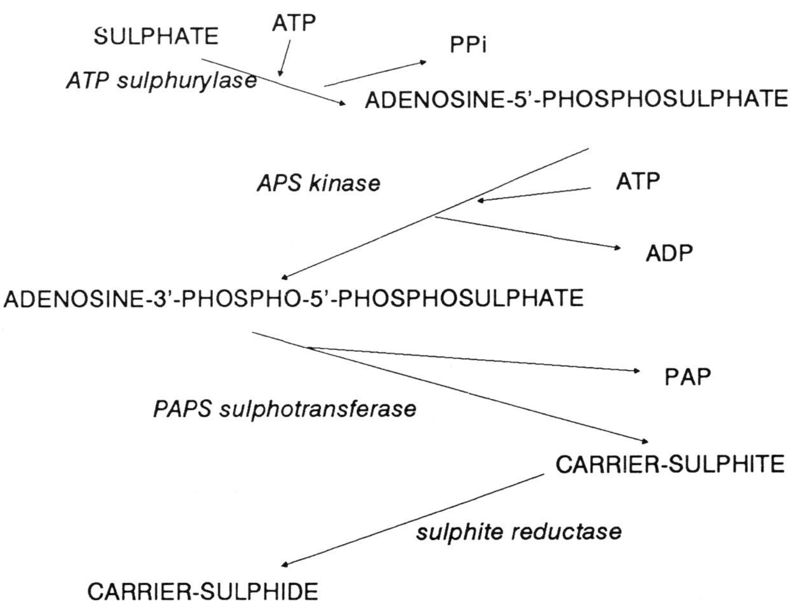

Fig. 5.1. Biosynthesis of sulphide. PAP = adenosine 3', 5'-diphosphate.

sulphate group is transferred to a sulphydryl group on a carrier molecule to form a sulphite ion attached to the carrier through a disulphide bond (a thiosulphonate). The reaction releases either AMP or PAP (adenosine-3'-phospho-5'-phosphate) depending on whether APS or PAPS was the substrate. NADPH is probably required in this step. The carrier is most likely to be a protein but can be a simple thiol and in some organisms glutathione seems to fulfill the function. In the final step the sulphite ion, still bound to the carrier, is reduced to sulphide through the action of sulphite reductase. The source of reducing power is not always clear but in yeast ferredoxin is required in addition to NADPH, and lipoic acid is a cofactor (Cooper, 1983). It is likely that thioredoxin is also involved, as in bacteria and unicellular algae. Furthermore, it may be that in fungi, certainly in *Saccharomyces cerevisiae*, there are two distinct sulphite reductases: one acting on carrier bound sulphite and the other using free sulphite (Umbarger 1978).

Biosynthesis of cysteine and methionine

Reduced sulphur is incorporated into carbon compounds through the reaction of sulphide with various precursors to form the amino acids cysteine and homocysteine (Fig. 5.2). Cysteine is formed by reaction between serine or O-acetylserine and sulphide, in free or bound form, catalyzed by cysteine synthase. O-acetylserine is the preferred substrate (de Robichon-Szulmajster & Surdin-Kerjan, 1971). In an analogous way homocysteine synthase catalyses formation of homocysteine from O-acetylhomoserine and sulphide. Homocysteine can also be formed from cysteine through reaction with O-acetylhomoserine to form cystathionine (cystathionine-γ-synthase) followed by breakdown to yield homocysteine, pyruvate and ammonia (cystathionine-β-lyase). In the reverse reactions from homocysteine to cysteine through cystathionine, homocysteine reacts with serine rather than pyruvate, and α-ketobutyrate is the other product rather than O-acetylhomoserine. The reactions in this direction are catalyzed by two quite separate enzymes, cystathionine-β-synthase and cystathionine-γ-lyase. Methionine is synthesised from homocysteine by methylation of the sulphur atom with N^5-methyltetrahydrofolate as the methyl donor.

Evidence from studies with mutant organisms suggests that though all the pathways mentioned above may be present, they are

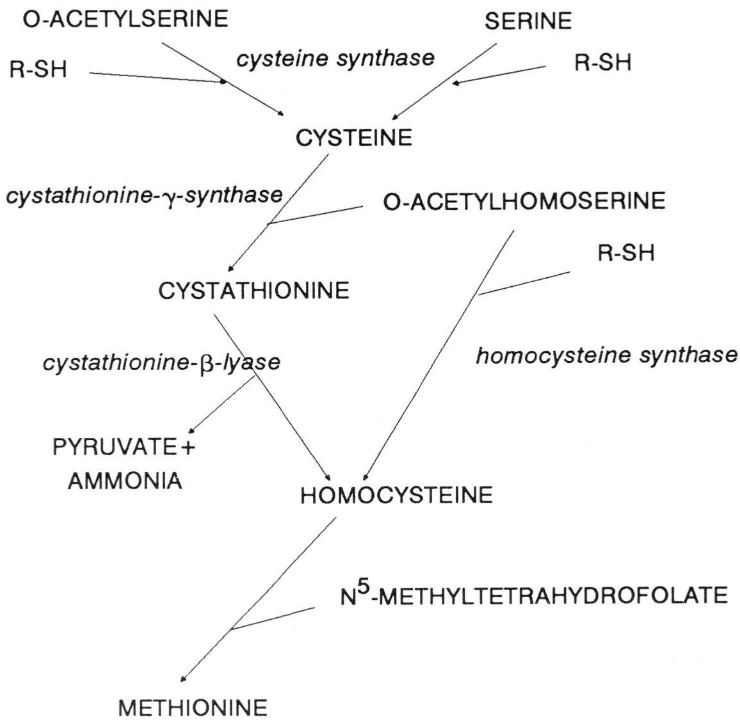

Fig. 5.2. Biosynthesis of cysteine, homocysteine and methionine.

put to different uses in different species. For example, methionine auxotrophs of *Neurospora crassa* can grow when supplied with cystathionine but not when homoserine and sulphide are the substrates, despite the presence of homocysteine synthase (Huxtable, 1986). In *N. crassa* the latter enzyme is possibly involved in sulphur salvage as it functions well with thiomethane (methyl sulphide), a product of cysteine breakdown.

In hemiascomycetous yeasts, it seems likely that the direct sulphydration reaction to homocysteine is the important biosynthetic route (Cherest, Eicher & de Robichon-Szulmajster, 1969). Homocysteine can still be formed from cystathionine but the activity of cystathionine-γ-synthase is very low (Kerr & Flavin, 1970). Savin

& Flavin (1972) have suggested that in yeast the O-acetylhomoserine sulphydrase (homocysteine synthase) and cystathionine-γ-synthase activities may be properties of a single protein.

Further metabolism of cysteine and methionine

Both cysteine and methionine have several biochemical functions. These are well described in standard textbooks of biochemistry and so only a brief outline will be given here to illustrate the highly significant functions of the reduced divalent sulphur atom.

Proteins

Both cysteine and methionine are incorporated into proteins. Cysteine is particularly important as the sulphydryl group is involved in the chemistry of the catalytic centre of many enzymes and in the binding of metal ions. Two highly developed examples where the function of a protein depends heavily on sulphydryl groups are the thioredoxins, which participate in many redox reactions (Holmgren, 1985) and the metallothioneins which are responsible for binding considerable amounts of metals, such as zinc, within the cell (Hamer, 1986). In addition, formation of the disulphide amino acid, cystine, from two cysteine residues is an important factor in the tertiary structure of proteins. Apart from the peptide bond, the disulphide bond is the only type of covalent linkage between amino acids in proteins.

Peptides

In common with other organisms fungi appear to synthesise the ubiquitous tripeptide, glutathione (γ-glutamylcysteinylglycine). The β-lactam antibiotics, i.e. the penicillins and cephalosporins, are produced by several fungi from the tripeptide δ-(L-aminoadipyl)-L-cysteinyl-D-valine. The toxic principles of some fungi are peptides, e.g. *Amanita phalloides* produces phalloidin, a cyclic peptide containing a thioether bond (Wieland & Faulstich, 1978).

S-adenosyl methionine

S-adenosyl methionine is formed from methionine and ATP with release of pyrophosphate and inorganic phosphate. It is an important methylating agent and is also the source of the propylamino groups used in the synthesis of polyamines.

Vitamins and growth factors

Several vitamins and growth factors contain divalent sulphur. The functional group of Coenzyme A is the sulphydryl group of cysteamine, the decarboxylation product of cysteine. The vitamins biotin and thiamine both contain thioether linkages. The exact source of the sulphur is still unclear but methionine appears to be the most likely donor. Lipoic acid contains a disulphide grouping which is reversibly reduced and oxidised in the multienzyme dehydrogenation of pyruvic acid and α-ketoglutaric acid. As mentioned above, lipoic acid is also involved in the biological reduction of sulphur.

Volatile sulphur compounds

In addition to the major sulphur compounds mentioned in the foregoing sections, fungi also produce a limited range of volatile sulphur compounds usually at low concentrations. These appear to represent a more or less accidental loss of sulphur from the biological system and in most cases at least, do not seem to have any function in the intermediary metabolism of the fungus. However, some may have a significance in terms of the life cycle of certain fungi, e.g. they may act as animal attractants for fungi lacking the ability to distribute their own spores (Bellina-Agostinone, D'Antonio & Pacioni, 1987) and they do have significance for man. Sulphur compounds of this type tend to have very low odour thresholds, i.e. they can be detected as smells very easily. For example, in aqueous solution threshold values are 5-10 $\mu g\ l^{-1}$ for hydrogen sulphide, 30 $\mu g\ l^{-1}$ for dimethyl sulphide and 2 $\mu g\ l^{-1}$ for methyl mercaptan. In addition to the low odour thresholds, these sulphur compounds also tend to have very characteristic and easily recognizable smells and tastes and our attitude to a particular fungus may be determined by, amongst other things, its ability to produce volatile sulphur compounds. The topic does not seem to have been investigated systematically but there is considerable information available concerning certain situations where there is an obvious economic significance and these are dealt with in the following paragraphs.

Production by Saccharomyces cerevisiae

The flavour of fermented beverages is a very important characteristic and beer is known to contain a range of sulphur volatiles which can affect the flavour (Garza-Ulloa, 1980). It is not clear whether all these arise from yeast metabolites and some of the mercaptans can certainly be formed chemically during processing, particularly hop boiling. From the point of view of yeast involvement, attention has focussed mainly on hydrogen sulphide and dimethyl sulphide (CH_3-S-CH_3).

Hydrogen sulphide

Most beers contain hydrogen sulphide and, at concentrations just below the flavour threshold, it may be a positive contributor to the 'yeasty' flavour. However, above the flavour threshold, it is always undesirable. In commercial practice brewery fermentations occasionally produce considerable quantities of hydrogen sulphide and Wainwright (1970) demonstrated that the gas was produced abundantly by pantothenate-requiring yeasts in the absence of the vitamin. Pantothenate is an essential part of the Coenzyme A molecule and it was suggested that deficiency of the vitamin in the medium leads to deficiency of Coenzyme A within the cell. As Coenzyme A is the major acylating agent in the cell, one effect of this would be lowered levels of O-acetylhomoserine. This compound is an intermediate in the biosynthetic route to methionine between cysteine and homocysteine (Fig. 5.2). Thus, lack of pantothenate could result in accumulation of cysteine. It was then envisaged that the excess cysteine, a compound which is toxic to cells, would be broken down to pyruvate, ammonia and hydrogen sulphide by the enzyme cysteine desulphydrase (Lawrence & Cole, 1968). In support of these ideas, Tokuyama *et al.* (1973) found that the activity of cysteine desulphydrase was nearly doubled in cells grown in pantothenate-deficient media. They further suggested that some of the external hydrogen sulphide could come directly from the biosynthetic pool of sulphide although they felt this would be a minor source of the compound.

In 1971, de Robichon-Szulmajster & Surdin-Kerjan reported that cysteine synthase used O-acetylserine as substrate in preference to serine. This should mean, using the logic of Wainwright (1970), that pantothenate deficiency would cause a blockage at

Table 5.1. Effect of addition of amino acids to the medium on the production of hydrogen sulphide.

Amino acid	μg Hydrogen sulphide produced 1·25 mg l^{-1} pantothenate	0·125 mg l^{-1} pantothenate
none	0	6·0
Met	0	0
Cys	17·5	19·0
His	2·4	24
Asp	2·5	23
Glu	11·0	50
Hser	11·0	2·5
Tyr	2·4	50
Ala	2·0	30

All amino acids were present at 5 mM except cysteine where the concentrations were 1 mM in the pantothenate-sufficient medium and 0·5 mM in the pantothenate-deficient medium. Production of hydrogen sulphide from cysteine was strongly concentration dependent and the quoted concentrations are those which led to the highest values. For experimental details see Jordan & Slaughter (1986).

the level of cysteine synthesis as well as at homocysteine synthesis. In that case it is unlikely that the intracellular concentration of cysteine would rise and, rather than hydrogen sulphide being produced by breakdown of cysteine, it seems likely that hydrogen sulphide would arise almost entirely from the biosynthetic pool.

This latter hypothesis has been supported by a number of experiments carried out by Jordan & Slaughter (1986) using a single polyploid strain of *Saccharomyces cerevisiae* (NCYC 1108) and defined glucose/salts media. Under the conditions used, hydrogen sulphide was produced only at pantothenate concentrations less than 0·2 mg l^{-1} and then in inverse proportion to the concentration of the vitamin. The bulk of the gas was formed towards the end of the growth phase of the cells. Table 5.1 shows the effect of adding single amino acids at 5 mM to media with sufficient pantothenate (1·2 mg l^{-1}) or just limiting pantothenate (0·125 mg l^{-1}). Methionine prevented production of hydrogen sulphide in the pantothenate limiting medium and this is in line with earlier reports that methionine controls sulphate uptake and sulphide pro-

Table 5.2. Effect of amino acids on cysteine desulphydration activity of cell extracts

Amino acid	Cysteine desulphydration activity (% of the control)		
	Addition to medium		
	1·25 mg l^{-1} pantothenate	0·0125 mg l^{-1} pantothenate	Addition to assay
Asp	82	133	91
Glu	97	117	91
His	63	118	96
Hser	72	99	85
Tyr	91	85	81
Ala	96	101	90

The amino acids were present at 5 mM in the medium or in the assay as appropriate (Jordan & Slaughter, 1986).

duction (Marzluff, 1972; Breton & Surdin-Kerjan, 1977). In contrast, all the other amino acids stimulated hydrogen sulphide production regardless of whether or not they could act as sources of sulphur themselves. Measurement of the cysteine desulphydrase activity in extracts of yeast cells grown with the addition of non-sulphur amino acids and measurement of the effect of addition of amino acids to the enzyme assay (Table 5.2) failed to show any correlation between the influence of the amino acids on enzyme activity and hydrogen sulphide production. These observations support the idea that hydrogen sulphide results more from metabolic disturbances leading to excess sulphate uptake rather than simple desulphydration of cysteine.

In an extension of this work, the cysteine pool size, hydrogen sulphide production, and cysteine desulphydrase activity were compared in yeast grown in three batches of standard medium varying as follows: 1·25 mg l^{-1} pantothenate, 0·0125 mg l^{-1} pantothenate and 0·0125 mg l^{-1} pantothenate with 5 mM methionine (Table 5.3). It is clear that the cysteine pool, whether measured per mg of yeast or per culture, is inversely proportional to the hydrogen sulphide production. This would be expected if the metabolic blockage came after sulphide formation but before cysteine synthesis. Tokuyama *et al.* (1973) also noticed that the concentration

Table 5.3. Influence of medium composition on cysteine pool size, hydrogen sulphide production and cysteine desulphydration activity

Pantothenate (mg l^{-1})	Yeast crop (mg dry wt per culture)	Cysteine pool nmole per mg dry yeast	Cysteine pool μmole per culture	Hydrogen sulphide production per culture (μmole)	Cysteine desulphydration activity nmole h^{-1} mg^{-1} protein	Cysteine desulphydration activity μmole h^{-1} per culture
1·25	128	38·0	4·74	0	67	3·43
0·0125	42	8·8	0·37	7·27	140	2·45
0·0125*	47	21·4	1·00	0·30	63	1·18

of cysteine fell under pantothenate deficiency but attributed this to excessive activity of cysteine desulphydrase. However, although the specific activity of cysteine desulphydrase rises about two fold in pantothenate deficiency, the lower amount of growth which occurs under these conditions in comparison to a pantothenate sufficient medium leads to an overall lower catalytic capability in the vitamin deficient culture. The capacity to desulphydrate cysteine is highest in the cultures which do not produce hydrogen sulphide and intermediate in those which produce the most.

This evidence, taken along with the theoretical likelihood of a metabolic blockage in the synthesis of cysteine from O-acetylserine, strongly suggests that biosynthetic sulphide rather than cysteine is the precursor of hydrogen sulphide produced by yeast under these conditions.

Dimethyl sulphide

Dimethyl sulphide is a common flavour active compound in a variety of foodstuffs (Anness & Bamforth, 1982) and the main source is thought to be thermal decomposition of the plant amino acid, S-methyl methionine. Dimethyl sulphide was first noticed in beer by Kepner, Strating & Weurman (1963) but began to excite particular interest when it was found to be associated with lager beer rather than ales (Sinclair, Hall & Thorburn-Burns, 1969). Subsequent work has indicated that decomposition of S-methyl

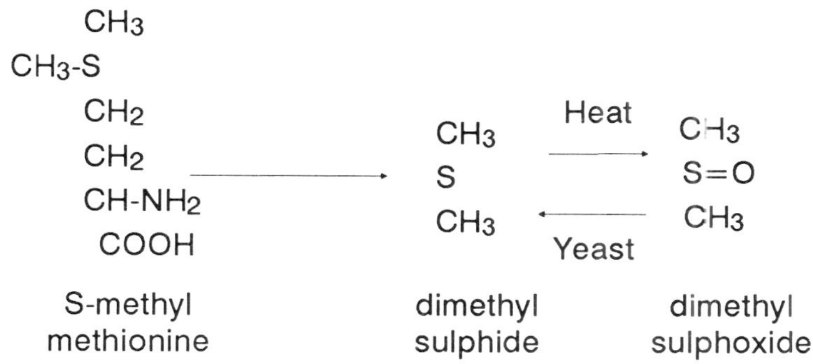

Fig. 5.3. Synthesis of dimethyl sulphide.

methionine during the various heating stages of the brewing process accounts for most of the dimethyl sulphide found in beer (Dickenson, 1983) but in some cases yeast can be responsible for a proportion of the compound (Anness & Bamforth, 1982). Yeast cannot form dimethyl sulphide directly from S-methyl methionine, but Anness & Bamforth (1982) showed that the dimethyl sulphide precursor in fermenting worts is dimethyl sulphoxide which is itself produced from S-methyl methionine during kilning of the malt (Fig. 5.3). Extracts of yeast can reduce dimethyl sulphoxide to dimethyl sulphide using NADPH as a cofactor. It seems unlikely, however, that there is a special dimethyl sulphoxide reductase in yeast and studies have indicated that the reaction is catalyzed by methionine sulphoxide reductase. The affinity of this enzyme is much higher for the methionine derivative than for dimethyl sulphoxide and the catalytic rate is over 600 times faster. The functional enzyme has three components: a methionine sulphoxide reducing protein, thioredoxin and thioredoxin reductase (Bamforth, 1980; Bamforth & Anness, 1981). The normal physiological function of the enzyme is unclear but, presumably, could form part of a sulphur salvage system or possibly be acting to protect the methionine pool against oxidation.

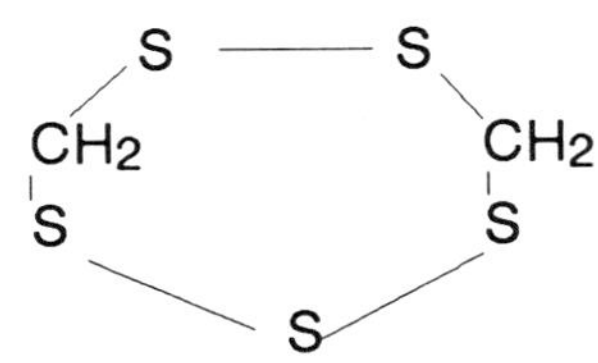

Fig. 5.4. Structure of lenthionine.

Production by other fungi

There do not seem to be many reports on the mechanism of production of volatile sulphur compounds by fungi other than *S. cerevisiae*. However, it is clear that compounds of this type are important in some cases. Over 40 years ago, methyl mercaptan was identified as a product of the wood-rotting fungus *Schizophyllum commune* (Birkinshaw, Findlay & Webb, 1942). This organism can also produce dimethyl sulphide and dimethyl disulphide when grown on a complex medium, and *Scopulariopsis brevicaulis* forms volatile thiols when grown on a bread medium (Challenger & Charlton, 1947). A compound reported as *bis*-methylthiomethane (presumably identical to dimethyl sulphide) was recognised as a product of the white truffle, *Tuber magnatum* (Fiecchi *et al.*, 1967) and thiobismethane (also identical to dimethyl sulphide) has been identified as a component of the odour of the truffle *Tuber aestivum* (Bellina-Agostinone *et al.*, 1987). The unusual compound lenthionine (Fig. 5.4) has been isolated from the edible shiitake mushroom *Lentinula edodes* (Morita & Kobayashi, 1966). The compound appears to be responsible for much of the flavour of the mushroom but is formed during soaking of the dried tissue, which is part of the food preparation process, rather than occurring as a normal metabolite. The natural precursor and the enzyme involved in the process are both unknown but this sort of situation occurs widely in food processing, e.g. the formation of garlic flavour in the crushed bulb due to the formation of allicin from alliin by the action of allinase. No doubt the nose of mycologists has detected volatile sulphur compounds from a wide range of fungi but the individual compounds have not been identified or their function investigated.

Conclusions

There seems to be no feeling that important compounds containing reduced sulphur atoms are waiting to be discovered and there is a clear appreciation of the role and importance of several compounds of this type, e.g. cysteine and methionine, both as intermediary metabolites and combined in proteins. However, the metabolic significance of others, e.g. glutathione, is as uncertain in the fungi as it is in other organisms. Understanding of biosynthesis varies from a good knowledge of the intermediates and enzymes involved, with information on possible control mechanisms, as in sulphide biosynthesis from sulphate, to virtually complete ignorance, as in biotin and thiamine synthesis. Because of the tight, reversible interconversions among the central reduced sulphur compounds the exact metabolic routes followed in real situations are often not clear, even when considerable information is available on the enzymes and intermediates involved. In contrast to the biosynthetic situation, there appears to be a great lack of information on the breakdown of sulphur compounds in fungi, both in terms of general turnover and with regard to the fate of specific compounds, such as methylthioadenosine, formed as products during the normal functioning of metabolism, in this case formation of polyamines from S-adenosyl methionine. The degree to which sulphur is salvaged and the effort necessary to maintain the reduced state in an oxidising world are two other areas where there is little knowledge at present.

References

Anness, B. J. & Bamforth, C. W. (1982). Dimethyl sulphide – a review. *Journal of the Institute of Brewing*, **88**, 244-252.

Bamforth, C. W. (1980). Dimethyl sulphoxide reductase of *Saccharomyces* spp. *FEMS Microbiology Letters*, **7**, 55-59.

Bamforth, C. W. & Anness, B. J. (1981). The role of dimethyl sulphoxide reductase in the formation of dimethyl sulphide during fermentation. *Journal of the Institute of Brewing*, **87**, 30-34.

Bellina-Agostinone, C., D'Antonio, M. & Pacioni, G. (1987). Odour composition of the summer truffle, *Tuber aestivum*. *Transactions of the British Mycological Society*, **88**, 568-569.

Birkinshaw, J. H., Findlay, W. P. K. & Webb, R. A. (1942). Biochemistry of the wood-rotting fungi. 3. The production of methyl mercaptan by *Schizophyllum commune* Fr. *Biochemical Journal*, **36**, 526-529,

Bradfield, G., Sommerfield, P., Meyen, T., Holly, M., Babcock, D., Bradley, D, Segel, I. H. (1970). Regulation of sulphate transport in filamentous fungi. *Plant Physiology*, **46**, 720-727.

Breton, A. & Surdin-Kerjan, Y. (1977). Sulfate uptake in *Saccharomyces cerevisiae* biochemical and genetic study. *Journal of Bacteriology*, **132**, 224-232.

Challenger, F. & Charlton, P. T. (1947). Studies on biological methylation. Part X. The fission of the mono- and di-sulphide links by moulds. *Journal of the Chemical Society*, 424-429.

Cherest, H., Eicher, F. & de Robichon-Szulmajster, H. (1969). Genetic and regulatory aspects of methionine biosynthesis in *Saccharomyces cerevisiae*. *Journal of Bacteriology*, **97**, 328-336.

Cooper, A. J. L. (1983). Biochemistry of sulphur-containing amino acids. *Annual Review of Biochemistry*, **53**, 187-222.

de Robichon-Szulmajster, H. & Surdin-Kerjan, Y. (1971). Nucleic acid and protein synthesis in yeasts: regulation of synthesis and activity. In *The Yeasts*, eds. A. H. Rose & A. H. Harrison, pp. 335-418. Academic Press: London & New York.

Dickenson, C. J. (1983). Dimethyl sulphide – Its origin and control in brewing. *Journal of the Institute of Brewing*, **89**, 41-46.

Fiecchi, A., Galli-Kienle, M., Scala, A. & Cabella, P. (1967). *Bis*-methylmethane, an odorous substance from white truffle, *Tuber magnatum* Pico. *Tetrahedron Letters*, **18**, 1681-1682.

Garza-Ulloa, H. (1980). Analytical control of sulphur compounds in beer – a review. *Brewers' Digest*, January, 20-25.

Hamer, D. H. (1986). Metallothionein. *Annual Review of Biochemistry*, **55**, 913-951.

Holmgren, A. (1985). Thioredoxin. *Annual Review of Biochemistry*, **54**, 237-271.

Huxtable, R. J. (1986). *Biochemistry of Sulfur*. Plenum Press: New York.

Jordan, B. & Slaughter, J. C. (1986). Sulphate availability and cysteine desulphydration activity as influences on the production of hydrogen sulphide by *Saccharomyces cerevisiae* during growth in a defined glucose-salts medium. *Transactions of the British Mycological Society*, **87**, 525-531.

Kepner, R. E,, Strating, J. & Weurman, C. (1963). Quantitative estimation of esters in beer by gas chromatographic analysis of head space volume. *Journal of the Institute of Brewing*, **69**, 399-405.

Kerr, D. S. & Flavin, M. (1970). The regulation of methionine synthesis and the nature of cystathionine γ-synthase in *Neurospora*. *Journal of Biological Chemistry*, **245**, 1842-1855.

Lawrence, W. C. & Cole, E. R. (1968). Yeast sulphur metabolism and the formation of hydrogen sulphide in brewery fermentations. *Wallerstein Laboratory Communications*, **31**, 95-115.

Marzluff, G. A. (1972). Control of synthesis, activity and turnover of enzymes of sulfur metabolism in *Neurospora crassa*. *Archives of Biochemistry and Biophysics*, **150**, 714-724.

McCready, R. G. L. & Din, G. A. (1974). Active sulphate transport in *Saccharomyces cerevisiae*. *FEBS Letters*, **38**, 361-363.

Morita, K. & Kobyashi, S. (1966). Isolation and synthesis of lenthionine, an odorous substance of Shiitake, an edible mushroom. *Tetrahedron Letters*, **6**, 573-577.

Savin, M. A. & Flavin, M. (1972). Cystathionine synthesis in yeast: an alternative pathway for homocysteine synthesis. *Journal of Bacteriology*, **122**, 299-303.

Sinclair, R., Hall, R. D. & Thorburn-Burns, D. (1969). The analysis of volatile sulphur compounds in beer. *Proceedings of the 12th European Brewery Convention*, 427-444.

Tokuyama, T., Kuraishi, H., Aida, K. & Vemura, T. (1973). Hydrogen sulphide evolution due to pantothenic acid deficiency in the yeast requiring this vitamin, with special reference to the effect of adenosine triphosphate on yeast cysteine desulphydrase. *Journal of General and Applied Microbiology*, **19**, 439-466.

Tweedie, J. W. & Segal, I. H. (1970). Specificity for transport processes for sulfur, selenium and molybdenum anions by filamentous fungi. *Biochimica et Biophysica Acta*, **196**, 95-106.

Umbarger, H. E. (1978). Amino acid synthesis and its regulation. *Annual Reviews of Biochemistry*, **47**, 533-606.

Wieland, T. & Faulstich, H. (1978). Amatoxins, phallotoxins, phallolysin and antamamide - biologically active components of poisonous *Amanita* mushrooms. *Critical Reviews in Biochemistry*, **5**, 185-260.

Wainwright, T. (1970). Hydrogen sulphide production by yeast under conditions of methionine, pantothenate or vitamin B_6 deficiency. *Journal of General Microbiology*, **61**, 107-119.

Chapter 6

Phosphonates: antifungal compounds against Oomycetes

M. D. Coffey & D. G. Ouimette

Department of Plant Pathology, University of California at Riverside, CA 92521-0122, USA

Introduction

While phosphorus is normally considered as an essential element necessary for the growth of microorganisms and microbial communities in nature (see Jennings, Chapter 1), recent studies have demonstrated that some simple phosphorus compounds have powerful and selective antifungal properties. Historically, one of the more frustrating elements in the long search for anti-fungal compounds has been the lack of discovery of molecules with good selective activity against Oomycete plant pathogens in higher plants. Necrotrophic pathogens in this class include many *Pythium* and *Phytophthora* species which cause damping off, root rot, crown and foot rot diseases as well as postharvest rots of various fruits and vegetables. In addition, the downy mildew fungi represent an important group of biotrophic Oomycete parasites causing serious diseases of many valuable food plants. The discovery, in the 1970s, of the phenylamide fungicides, and particularly metalaxyl, marked an important milestone in the improved chemical control of many of these diseases. At about the same time, another class of anti-Oomycete compounds was also discovered in France by researchers at Rhône-Poulenc Agrochimie. These compounds are phosphonates and they include salts of inorganic phosphonic (phosphorous) acid and of short chain organic derivatives, notably aluminium tris-*O*-ethyl phosphonate (fosetyl aluminium) (Fig. 6.1). This latter compound possessed some unique properties apparently unlike those of other known antifungal agents. The systemic activity of fosetyl aluminium proved to be remarkable since foliar applications frequently led to control of soil-borne diseases caused by Oomycetes. Further, fosetyl aluminium displayed very low activity *in vitro* against the majority of culturable Oomycete fungi, including different *Pythium* and *Phytophthora* species. Finally, the compound, despite its very simple structure, displayed antifungal properties which were unusually selective, possibly suggestive of a site-specific mode of action. For instance, whilst it provides effective disease control against *Plasmopora viticola* on grapevines and *Pseudoperonospora humuli* on hops, activity against *Peronospora hyoscyami* on tobacco is poor. Perhaps even more remarkable is the behaviour of this compound against diseases caused by some *Phytophthora* species. Control of many diseases caused by *Phytophthora citricola*, *P. cinnamomi*, *P. citrophthora*, *P. fragariae*, *P. cactorum*, and *P. palmivora* is generally excellent but, in contrast, little or no control of diseases caused by *P. megasperma* f.sp. *glycinea* or *P. infestans* has been possible.

The low activity of fosetyl aluminium *in vitro* against culturable Oomycetes has led some scientists to the viewpoint that this compound triggers a resistance reaction in the host plant (Bompeix, Fettouche & Saindrenan, 1981; Guest, 1984a). However, a careful review of some of the more pertinent patent literature led us to an alternative and still controversial conclusion, namely that phosphonate fungicides have important direct antifungal activity against their target pathogens. Crucial to this opinion was our realization that the first phosphonate fungicides with anti-Oomycete activity to be discovered by the Rhône-Poulenc scientists, were not organic derivatives such as fosetyl aluminium, but in fact simple salts of phosphonic (phosphorous) acid (Thizy *et al.* 1978). Subsequently, our own studies with the potassium salt of phosphonic acid led to the conclusion that this compound has significant antifungal activity against many *Phytophthora* species, but has little or no activity against the majority of other fungi tested (Coffey & Bower, 1984; Fenn & Coffey, 1984).

The phosphonate story, which is slowly emerging, is novel, unexpected and challenges currently held concepts. The purpose of this review is to present some of the new information now available on phosphonate chemistry and biology as it relates to specific target effects on the growth and development of *Phytophthora* species, its behaviour in higher plants and probable fate in the soil environment.

Phosphonates

The simplest phosphonate compounds are phosphorous acid [$P(OH)_3$] and phosphonic acid [$H\text{-}PO(OH)_2$]. These two inorganic oxy acids are in tautomeric equilibrium (Guthrie, 1979), but due to the strength of the phosphoryl group (P=O) this equilibrium overwhelmingly favours phosphonic acid (Fig. 6.1). Previous literature has tended to use the term phosphorous acid

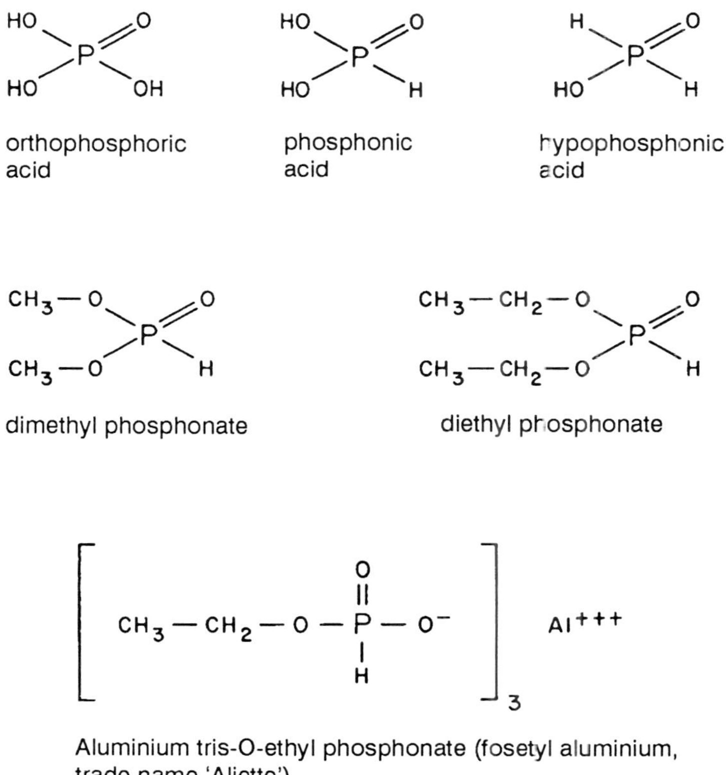

Fig. 6.1. Structures of phosphorus oxy acids, including phosphonic acid, and alkyl-substituted phosphonates.

indiscriminately when referring to a tautomeric mixture consisting predominantly of phosphonic acid. Phosphonic acid is dibasic (pK_1 = 1·3, pK_2 = 6·7) and has one hydrogen bound to phosphorus rather than oxygen. This hydrogen exchanges rapidly with the deuterium of heavy water and this rate can be increased further by the addition of HCl (Martin, 1959). In marked contrast, the sodium and potassium salts of phosphonic acid exhibit negligible or very slow deuterium exchange, respectively. This property has been used to advantage in preparing relatively stable isotopes of phosphonate salts where tritium is bound exclusively to the phosphorus atom (Fenn & Coffey, 1985, 1988). Phosphonate salts such as sodium or potassium phosphonate have strong buffering action in the pH 5·5 to 7·5 range, and in solution are stable to oxidation over a pH range of 1·5 to 7·6 for many months (Robertson & Boyer, 1956a).

Alkyl esters of phosphorous and phosphonic acid are termed phosphites and phosphonates, respectively (Cohen & Coffey, 1986). Again, the tautomeric equilibrium favours the phosphonate esters. A striking feature of phosphonic acid, inorganic phosphonate salts and simple alkyl esters such as dimethyl phosphonate (Fig. 6.1) is their high solubility in water. Aluminium tris-*O*-ethyl phosphonate (fosetyl aluminium) is currently the best known example of an alkyl phosphonate which is a systemic fungicide with good activity against some serious diseases caused by Oomycetes (Bruin & Edgington, 1983; Cohen & Coffey, 1986). It has a solubility in water of 12% and is routinely used as a 10% formulation for injection into tree trunks for disease control (Darvas, Toerien & Milne, 1984).

Most early studies on phosphonate activity against pathogenic fungi concentrated on fosetyl aluminium since this was the commercial product. Little or no direct antifungal activity was detected until concentrations exceeded 200 $\mu g\ ml^{-1}$ even with Oomycete species known to be affected *in planta* (Bompeix *et al.*, 1981; Farih, Tsao & Menge, 1981; Guest, 1984b). Consequently, research had tended to focus on the possibility of indirect effects of the compound, perhaps through stimulation of the host defence system (Bompeix *et al.*, 1980; Guest, 1984a). However, two important facts were largely overlooked. First, Rhône-Poulenc Agrochimie had already provided technical data with the release of their experimental compound, indicating that fosetyl aluminium was degraded to phosphonic acid. Second, the patent literature indicated that phosphonic acid had strong antifungal properties specifically against Oomycetes (Thizy *et al.*, 1978). Interestingly, Bompeix *et al.* (1980) did suggest that phosphonic acid was the active product of degradation of fosetyl aluminium in plant tissues. However, the possibility that phosphonic acid was directly antifungal was not investigated until much later (Bompeix & Saindrenan, 1984; Fenn & Coffey, 1984).

Direct antifungal activity

Mycelial growth

Potassium phosphonate was investigated for its antifungal activities against a range of fungi grown on liquid and solid media (Coffey & Bower, 1984; Fenn & Coffey, 1984; Dolan & Coffey, 1988). The compound was active against a number of *Phytophthora* species and isolates in the range of 2 to 90 $\mu g\ ml^{-1}$ based on the concentration needed to cause 50% inhibition (EC_{50}) of mycelial growth (Table 6.1). The most sensitive species were *Phytophthora cinnamomi*, *P. citricola*, *P. citrophthora*, and *P. palmivora* with EC_{50} values of 5 to 10 $\mu g\ ml^{-1}$. At the other extreme were several with EC_{50} values of about 90 $\mu g\ ml^{-1}$, especially isolates of *P. megasperma* from lucerne (Table 6.1). However, by far the most insensitive species was *P. infestans*, where inhibition of mycelial growth at 200 $\mu g\ ml^{-1}$ ranged from 71·2 to only 30·4% for various isolates (Coffey & Bower, 1984). Similar observations with different *Phytophthora* species were also made by Bompeix & Saindrenan (1984).

The direct activity of phosphonate salts against different *Pythium* species has not been investigated in the same detail as *Phytophthora*. Of four species examined, an isolate of *Pythium aphanidermatum* was the most sensitive to potassium phosphonate, with an EC_{50} value of about 140 $\mu g\ ml^{-1}$ on 0·5% cornmeal agar (Fenn & Coffey, 1984). Other fungi, including Deuteromycotina, Ascomycotina, Zygomycotina and Basidiomycotina, were not affected by potassium phosphonate.

Thus phosphonate demonstrates high selectivity against some culturable Oomycete pathogens, notably certain *Phytophthora*

Table 6.1. EC_{50} values for inhibition of mycelial growth of representative *Phytophthora* species and isolates.

Species	Isolate	Host	EC_{50}[a] (μg ml^{-1})
P. citrophthora	M143	citrus	5·2
P. palmivora	P376	cocoa	6·6[b]
P. citricola	P1287	avocado	6·8
P. citricola	P1273	avocado	7·0
P. cinnamomi (A1)	Pc97	camellia	9·0
P. cinnamomi (A2)	Pc356	avocado	9·9
P. palmivora	P736	cocoa	≈10·0[b]
P. citrophthora	P1163	citrus	10·4
P. cinnamomi (A2)	Pc402	avocado	11·9
P. capsici	P1091	capsicum	18·5
P. megasperma f.sp. *glycinea*	P405	soybean	22·3
P. capsici	P1314	capsicum	30·6
P. parasitica	M134	citrus	30·9
P. capsici	P1319	capsicum	34·7
P. megasperma	P1253	chickpea	91·2
P. megasperma	P1316	lucerne	88·9
P. infestans	P1300	potato	224·4[c]

Data from Coffey & Bower, 1984; Dolan & Coffey, 1988. [a]The majority of isolates were grown on a defined medium containing 0·84 mM potassium phosphate; [b]*P. palmivora* was grown on 0·5% cornmeal agar and [c]*P. infestans* was grown on rye-seed medium.

species within the Peronosporales, while no such activity can be detected among a wide range of non-Oomycete fungi.

Sporulation

The life cycle of a pathogenic Oomycete fungus such as *Phytophthora* involves not only mycelial growth in its host, but also the formation of various spore types. Sporulation is a vital component of such life cycles and disease progress depends critically on the quantity and quality of such processes. Interference with the life cycle, by inhibition of germination and/or sporulation processes provides a window of opportunity for an antifungal compound. Production of zoosporangia is especially sensitive to potassium phosphonate. The EC_{50} values for inhibition of sporangial development range from 0·1 μg ml^{-1} with an isolate of *Phytophthora palmivora* (Dolan & Coffey, 1988) to 1·4 and 1·8 μg ml^{-1}, respectively, for *P. citricola* and *P. cinnamomi* (Coffey & Joseph, 1985). EC_{90} values for the same isolates of *P. citricola* and *P. cinnamomi* were 5·5 and 3·8 $\mu g\ ml^{-1}$ respectively. Oospore production by the same isolate of *P. citricola* was also sensitive, being inhibited by as little as 1 $\mu g\ ml^{-1}$ of phosphonate. Oospore and especially chlamydospore formation in *P. cinnamomi* were not as sensitive to phosphonate; around 50 $\mu g\ ml^{-1}$ caused significant inhibition of their production (Coffey & Joseph, 1985).

In contrast to spore production, germination of fungal propagules was much less affected. The one exception was indirect germination of zoosporangia by cytoplasmic cleavage to form motile zoospores. Such zoospore release with *Phytophthora cinnamomi* and *P. citricola* was inhibited by 50% in the presence of 6 $\mu g\ ml^{-1}$ phosphonate (Coffey & Joseph, 1985).

Production of resistant strains

The development of laboratory strains of *Phytophthora* species with high resistance to phosphonate provides important evidence supporting the hypothesis of direct and selective inhibition of these fungi by phosphonate. By use of the chemical mutagen *N*-methyl-*N*'-nitro-*N*-nitrosoguanidine (MNNG), Bower & Coffey (1985) succeeded in producing strains of *Phytophthora capsici* with high resistance to phosphonate. One strain of *P. capsici* (P1361), derived from a single-zoospore wild type isolate (P1319), possessed an EC_{50} on 0·5% cornmeal agar of 415 $\mu g\ ml^{-1}$ phosphonate compared to a value of 37 $\mu g\ ml^{-1}$ for P1319 (Table 6.2). Studies by Fenn & Coffey (1988) led to the production of further mutant strains of both *P. capsici* and *P. parasitica* var. *nicotianae* with resistance to phosphonate (Table 6.2). The choice of *P. capsici* and *P. parasitica* var. *nicotianae* was based on the fact that these species had been used by Bompeix *et al.* (1981) and Guest (1984a), in their studies supporting an indirect role for phosphonate and/or fosetyl aluminium via stimulation of host defence mechanisms. In tomato leaflets, the phosphonate-resistant mutant of *P. capsici* (P1361) demonstrated an EC_{50} of >348 $\mu g\ ml^{-1}$ for lesion development (Fenn & Coffey, 1985). Likewise with tobacco seedlings, a mutant of *P. parasitica* var. *nicotianae*, with an EC_{50} *in vitro* on 0·5% cornmeal agar of 665 $\mu g\ ml^{-1}$ phosphonate, was only partially controlled when the phosphonate level in the plant

Table 6.2. EC_{50} values ($\mu g\ ml^{-1}$) for resistance to phosphonate on 0·5% cornmeal agar of mutant strains of *Phytophthora palmivora*, *P. capsici* and *P. parasitica* var. *nicotianae*.

Phytophthora species and isolate	Wild type	Mutant strains
P. palmivora (P376)	7	118
		130
P. capsici (P1319)	37	415
P. capsici (P1503)	57	311
		462
		815
P. parasitica (P1495)	<50	270
		301
P. parasitica (P1352)	47	514
		535
		650
		665

Data taken from Fenn & Coffey (1988).

tissue was 753 $\mu g\ g^{-1}$ fr. wt. In contrast, the wild type isolate was almost completely controlled by a phosphonate concentration of 279 $\mu g\ g^{-1}$ fr. wt. (Fenn & Coffey, 1988).

One possible criticism of much of the work on the mode of action of the phosphonate fungicides, potassium phosphonate and fosetyl aluminium, is that neither *P. capsici* nor *P. parasitica* var. *nicotianae* are very sensitive to these compounds (Table 6.1). To overcome this criticism we chose to obtain mutants of *P. palmivora* (P376), an isolate selected because of its high sensitivity to phosphonate *in vitro* (Table 6.2, Fig. 6.2). A range of mutants was obtained *in vitro* using MNNG with EC_{50} values varying from 37 to 130 $\mu g\ ml^{-1}$ phosphonate on 0·5% cornmeal agar (Dolan & Coffey, 1988). Using tomato as a host plant, it was possible to demonstrate that the range of responses obtained *in vivo* paralleled those found *in vitro* (Fig. 6.2).

Finally, there has been a single report of resistance to fosetyl aluminium occurring in the field with *P. cinnamomi* (Vegh *et al.*, 1985). This finding has yet to be corroborated by other workers,

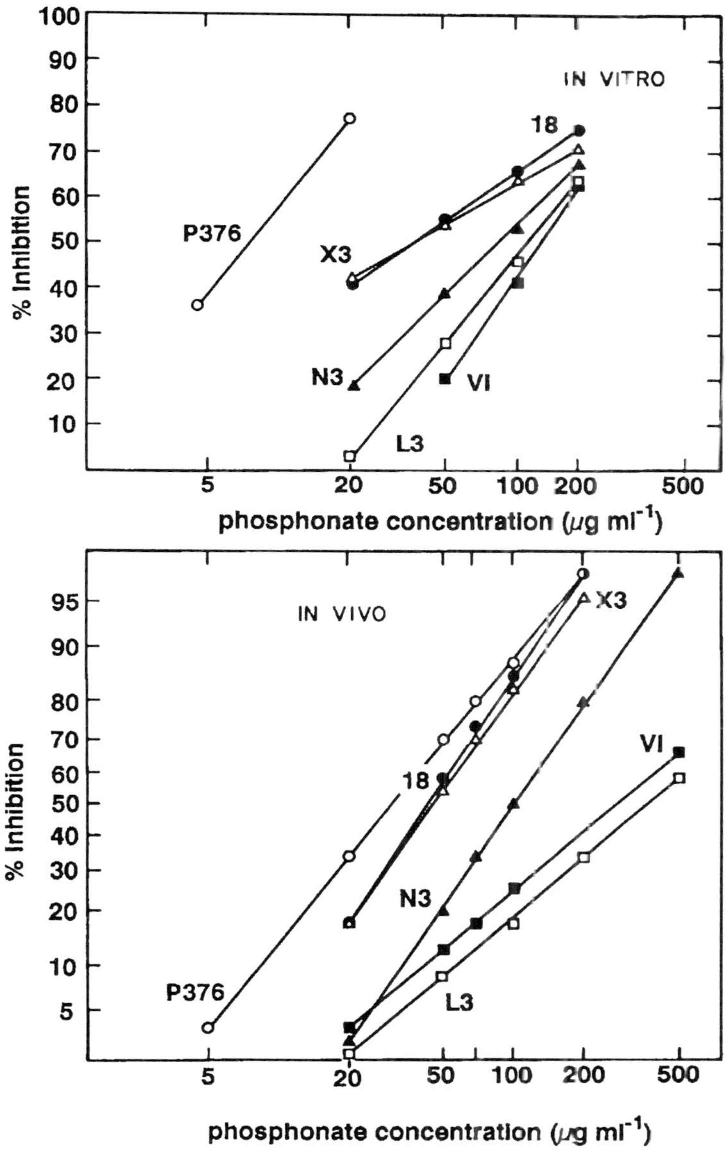

Fig. 6.2. Comparison of (a) *in vitro* and (b) *in vivo* responses to potassium phosphonate of a wild type (P376) and five resistant mutants of *Phytophthora palmivora* (from Dolan & Coffey, 1988).

but in view of the relative ease with which resistant strains of some *Phytophthora* species can be produced in the laboratory, it is not entirely unexpected.

Phosphonate behaviour in higher plants

Metabolism and uptake of organic phosphonates

Although it is known that fosetyl aluminium is degraded to phosphonate in plant tissues (Piedallu & Jamet, 1985; Saindrenan, Darakis & Bompeix, 1985), good quantitative data on levels and persistence in plant tissues has not been available. Recently, Fenn & Coffey (1988) found that tomato leaflets treated with 400 μg ml^{-1} fosetyl aluminium for 48 h contained only 14 μg g^{-1} fr. wt. ethyl phosphonate and 358 μg g^{-1} fr. wt. phosphonate (Table 6.3). Results with whole tomato plants provide a similar data base (Luttringer & de Cormis, 1985). Ethyl phosphonate was only detectable for a few hours following a soil application of 8 mg ml^{-1} and subsequent uptake by the plant roots. Phosphonate accumulated over 7 days to a final concentration of about 150 μg g^{-1} fr. wt. in the leaves, but only about 50 μg g^{-1} fr. wt. in stems.

The levels of ethyl phosphonate and phosphonate were determined in avocado seedlings following either soil or foliar applications of fosetyl aluminium (Ouimette & Coffey, 1989). With soil treatment, ethyl phosphonate was detectable at low levels (2-3 μg g^{-1} fr. wt.) in roots and stems after 1 week. The phosphonate le-

Table 6.3. Phosphonate and ethyl phosphonate concentrations in detached tomato leaflets floating on solutions of fosetyl aluminium for 48 hours

Fosetyl aluminium in floating solution (μg ml^{-1})	Phosphonate concentration μg g^{-1} fr. wt.	Ethyl phosphonate concentration μg g^{-1} fr. wt.
0	0	0
80	49	0·5
160	88	3·0
240	181	4·0
320	222	10·0
400	358	14·0

Data taken from Fenn & Coffey (1988).

Table 6.4. Concentration of phosphonate (HPO_3^{2-}) in tissues of avocado seedlings up to 8 weeks after soil or foliar application with either fosetyl aluminium or potassium phosphonate at equivalent HPO_3^{2-} concentrations

		$\mu g\ HPO_3^{2-}\ g^{-1}$ fr. wt. tissue					
		Soil application[a]			Foliar application[b]		
Fungicide	Week	Roots	Stems	Leaves	Roots	Stem	Leaves
Fosetyl	1	208	191	57	8	43	140
aluminium	6	706	1084	86	15	121	73
	8	488	543	105	12	95	49
Potassium	1	356	221	221	15	21	42
phosphonate	6	512	784	72	11	130	74
	8	213	382	47	14	75	19

[a]Potassium phosphonate and fosetyl aluminium were applied at a volume of 500 ml per seedling at 2·1 and 3·0 mg ml^{-1} respectively. [b]Potassium phosphonate and fosetyl aluminium were applied twice, with a 24 hour interval between applications, at 2·1 and 3·0 mg ml^{-1}, respectively. Data from Ouimette & Coffey, 1989.

vels in roots and stems were 208 and 191 $\mu g\ g^{-1}$ fr. wt., respectively. After 8 weeks the level in roots was 488 $\mu g\ g^{-1}$ fr. wt. phosphonate (Table 6.4). With foliar application, no ethyl phosphonate residues were detected in avocado tissues even at 1 week. Concentrations of phosphonate in foliage were 140 μg and 49 $\mu g\ g^{-1}$ fr. wt. at 1 and 8 weeks, respectively (Table 6.4). Levels in root tissues remained fairly constant at 8 to 15 $\mu g\ g^{-1}$ fr. wt. over the experimental period. Interestingly, much more phosphonate was present in root tissues following soil, as compared to foliar, application of fosetyl aluminium (Table 6.4). The levels of phosphonate in avocado tissues was found to be quite similar when potassium phosphonate was used at equivalent rates to fosetyl aluminium (Table 6.4; Ouimette & Coffey, 1989). The data are consistent with the hypothesis that fosetyl aluminium is rapidly degraded in higher plant tissues to phosphonate and that it persists in this form for at least several months.

Other short chain alkyl-substituted phosphonates such as dimethyl and diethyl phosphonate (Fig. 6.1) have been shown to provide disease control in plants against various Oomycetes (Abblard, Gaulliard & Lacroix, 1983; Ouimette & Coffey, 1985). Pepper plants were treated with 5 mM solutions of either potassium phosphonate, monoethyl, diethyl or dimethyl phosphonate as root drenches for 24 h and then analyzed for their phosphonate content (Ouimette & Coffey, 1988). In addition to residues of the respective parent compound, the organic phosphonate treatments contained high levels of inorganic phosphonate, indicating that partial metabolism had occurred within 24 h. Plants treated with dimethyl phosphonate contained the highest levels of phosphonate, 397 $\mu g\ g^{-1}$ fr. wt. in roots, which was slightly higher than the amounts found in roots of plants treated with potassium phosphonate.

Factors affecting the degradation of different organic phosphonates to inorganic phosphonate are unknown. In addition, there are no data available on the influence of different alkyl substitutions on their uptake and subsequent distribution in the plant. Most likely they enter the plant by diffusion, since it is difficult to envisage selective uptake processes being involved. Once in the plant tissue, rapid degradation to phosphonate near the site of entry could generate a chemical gradient, perhaps facilitating loading into the symplast and eventually transport in the phloem. The possibility also exists that phosphonate might be taken up actively into plant tissues as is orthophosphate, since the two molecules possess a very similar spatial arrangement of atoms (Fig. 6.1). In mammalian and yeast systems, however, phosphonate has been shown to exert little effect on phosphate-mediated metabolism such as cell respiration (Robertson & Boyer, 1956b). It was concluded that phosphonate and phosphonate esters were essentially inert, even at relatively high ionic strengths (0·067 M–0·2 M). This fundamental biological inertness relative to phosphate might be explained by differences in electrostatic charge distribution (Robertson & Boyer, 1956b). The persistence of phosphonate in higher plants is also consistent with this hypothesis. The ability of phosphonate to mimic phosphate in certain plant metabolic processes, particularly active uptake by cells, needs to be investigated.

Influence of phosphate on phosphonate activity

Bompeix *et al.* (1980) were the first to study the influence of phosphate concentrations on the efficacy of fosetyl aluminium using tomato leaflets and *Phytophthora capsici* as a pathogen. They observed that 250 μg ml^{-1} fosetyl aluminium no longer controlled *P. capsici* on detached leaflets floated on a solution in which the sodium phosphate level had been increased from 10 to 67 mM. However, neither pre- nor post-application of the fungicide-treated leaflets with 67 mM phosphate influenced fungicide efficacy. In contrast, increasing the potassium phosphate concen-

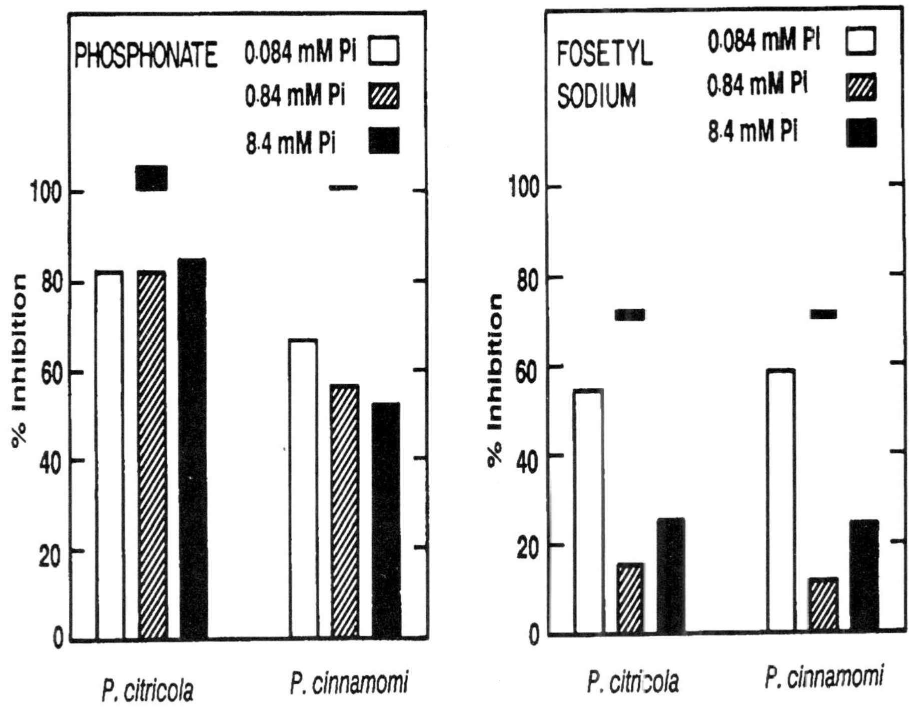

Fig. 6.3. Influence of different potassium phosphate levels (Pi) on the inhibition in defined liquid medium of *Phytophthora citricola* (P1273) and *P. cinnamomi* (Pc356) by (a) 47 μg ml^{-1} potassium phosphonate and (b) 132 μg ml^{-1} fosetyl sodium (from Fenn & Coffey, 1984).

tration of the external medium from 0 to 10 mM with *P. palmivora* on tomato seedlings actually enhanced the antifungal effects of both fosetyl sodium and potassium phosphonate (Dolan & Coffey, 1988). Fenn & Coffey (1984) examined the effect of potassium phosphate levels on the efficacy of both fosetyl sodium and potassium phosphonate directly against *P. cinnamomi* and *P. citricola* in liquid culture. Fosetyl sodium was chosen in preference to fosetyl aluminium since the latter compound formed a precipitate, probably of aluminium hydroxide, when the pH of the solution was adjusted to a more suitable physiological level, typically from 3·1 to 6·2-6·5. It was found that an increase in phosphate from 0·084 to 0·84 mM reduced the efficacy of fosetyl sodium (132 μg ml^{-1}) towards both pathogens in a dramatic fashion (Fig. 6.3). In con-

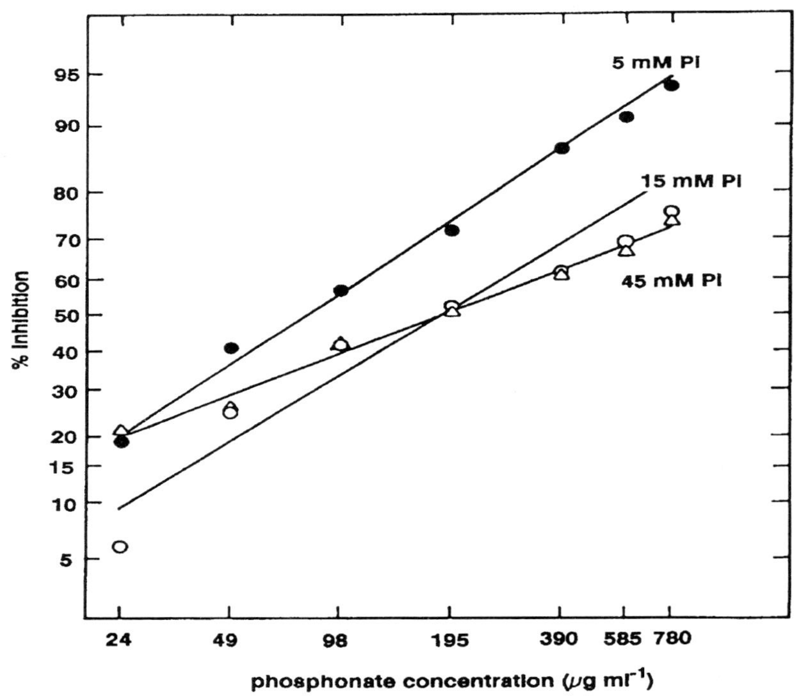

Fig. 6.4. Influence of 5, 15 and 45 mM potassium phosphate (Pi) on the *in vitro* response to potassium phosphonate of *Phytophthora capsici*, isolate P1319 (from Fenn & Coffey, 1988).

trast, there was absolutely no response to an increase in phosphate level to 8·4 mM, with potassium phosphonate (47 μg ml^{-1}) against *P. citricola*. The effect on efficacy of phosphonate against *P. cinnamomi* was slight, reducing inhibition of fungal growth from 67% at 0·084 mM to 52% at 8·4 mM phosphate concentration (Fig. 6.3). Recently, the effect of phosphate concentration on *P. capsici* (P1319)(Fig. 6.4) and *P. parasitica* var. *nicotianae* (P1352) was examined with respect to its influence on phosphonate efficacy (Fenn & Coffey, 1988). The EC_{50} values for inhibition of mycelial growth of *P. capsici* with 5, 15 and 45 mM potassium phosphate concentrations were 77, 184 and 186 μg ml^{-1} phosphonate (Fig. 6.4). Corresponding EC_{50} values for *P. parasitica* var. *nicotianae* at 5, 15 and 45 mM phosphate levels were 55, 123 and 78 μg ml^{-1}. Summarizing these effects, it can be seen that increasing the level of phosphate can have an effect on the fungicidal efficacy of fosetyl aluminium and fosetyl sodium (Bompeix *et al.*, 1980; Fenn & Coffey, 1984), though it may change efficacy in either a positive or negative direction, or alternatively its influence may be negligible.

A key issue in all these considerations of the phosphate:phosphonate interaction is whether phosphate levels in plant tissues are sufficient to reduce the direct antifungal activity of phosphonate against the target pathogen. Proponents of an indirect mode of action for both fosetyl aluminium and phosphonate argue that phosphate levels in plant tissues are too high to account for a direct mode of action of these compounds against the pathogen (Bompeix *et al.*, 1980; Saindrenan *et al.*, 1988). Fenn & Coffey (1988) have produced data indicating that the level of inhibition by phosphonate of two *Phytophthora* species, *P. capsici* and *P. parasitica* var. *nicotianae*, is similar both *in vitro* and *in vivo* at comparable phosphate concentrations (Table 6.5). It was also shown that the responses of resistant strains of these same fungi to phosphonate was quite similar in culture and in plant tissues. In addition, values for phosphonate levels in plant tissues may not accurately reflect the concentrations to be found in fungal tissues. Barchietto, Saindrenan & Bompeix (1988) demonstrated that *P. citrophthora* grown in a synthetic liquid medium could take up phosphonate against a concentration gradient and that this process was inhibited by 10 or 25 mM sodium azide. Consequently,

Table 6.5. Comparison of *in vivo* and *in vitro* antifungal activity of potassium phosphonate (HPO_3^{2-}) towards growth of HPO_3^{2-}-sensitive and resistant isolates of *Phytophthora capsici* and *P. parasitica* var. *nicotianae*.

Isolate[a]	Growth medium[b]	HPO_3^{2-} concentration[c]	HPO_4^{2} concentration (mM)	% inhibition
P. capsici	Tomato leaflet	78-110	8·9	61
(P1319)	*In vitro* on SM	77	5·0	50
P. capsici	Tomato leaflet	484-642	8·9	37
(P1361)	*In vitro* on SM	344	5·0	50
P. parasitica	Tobacco stem	279	3·2	96
(P1352)	*In vitro* on CMA	279	$<1{\cdot}0$	86
P. parasitica	Tobacco stem	753	3·2	62
(P1755)	*In vitro* on CMA	665	$<1{\cdot}0$	50

[a]P1319 and P1352 are HPO_3^{2-}-sensitive parental isolates. P1361 and P1755 are HPO_3^{2-}-resistant mutants obtained by treating zoospores of the parental isolates with the chemical mutagen MNNG. [b]The *in vitro* antifungal activity of HPO_3^{2-} was determined by measuring radial growth on either a synthetic medium (SM) or on 0·5% cornmeal agar (CMA) at various concentrations of HPO_3^{2-}. [c]Concentrations of HPO_3^{2-} are expressed as $\mu g\ g^{-1}$ fr. wt. of plant tissue or $\mu g\ ml^{-1}$ of culture medium. Data taken from Fenn & Coffey (1988).

phosphonate levels determined in plant tissues may not accurately reflect those present in fungal tissues.

Influence of amino-oxy acetic acid on phosphonate efficacy

Bompeix *et al.* (1981) showed that 300 μM 2-amino-oxy acetic acid (AOA) could negate the effect of 250 $\mu g\ m^{-1}$ fosetyl aluminium in inhibiting lesion development of *P. capsici* on detached tomato leaflets. Since AOA is known to be an inhibitor of phenylalanine ammonia-lyase (PAL), it was hypothesized that the mode of action of fosetyl aluminium might involve a stimulation of host defence mechanisms. This was based on the assumption that AOA, by interference with PAL activity, blocked phenylpropanoid biosynthesis and that the latter was essential for host defence. Contradictory evidence was obtained in this study, however, since L-2-amino-oxy-3-phenylpropionic acid (AOPP), did not interfere with the antifungal activity of fosetyl aluminium (Bompeix *et al.*, 1981). AOPP is known to be a potent inhibitor of PAL activity *in vitro*, much more so than AOA. One additional factor that was overlooked by Bompeix *et al.* (1981) was the need for a control consisting only of AOA. Fenn & Coffey (1985) attempted to repeat these results concentrating on the effect of AOA on fosetyl aluminium and potassium phosphonate activity against *P. capsici*. They found that 80 μM AOA alone caused a 28% increase in lesion length of *P. capsici* on tomato leaflets. At higher concentrations of fosetyl aluminium (260 μg ml^{-1}) or potassium phosphonate (180 μg ml^{-1}), AOA had no influence on lesion development. At lower concentrations of these two fungicides (130 μg and 90 μg ml^{-1} respectively), 80 μM AOA did increase lesion size by 62% and 48%. Since leaflets treated with AOA alone were

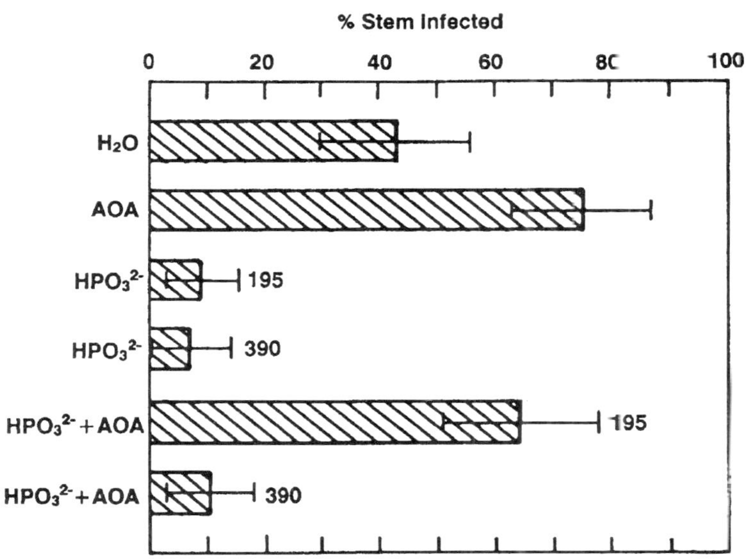

Fig. 6.5. Effect of 0·08 mM 2-aminooxy acetic acid (AOA) and/or potassium phosphonate (HPO_3^{2-}) with respect to control of *Phytophthora parasitica* var. *nicotianae* (P1352) on tobacco seedlings. Concentrations of HPO_3^{2-} used were 195 and 390 μg ml^{-1} (From Fenn & Coffey, 1988).

made more susceptible to *P. capsici*, this result might be explained by assuming that more fungicide was required to achieve a similar level of control. In addition, 80 μM AOA was shown to reduce the *in vitro* uptake of phosphonate into *P. capsici* by 62% after 3 h. Consequently, the primary effect of AOA against phosphonate is more likely to be through its effect on uptake by the fungus (Fenn & Coffey, 1985).

These results have been repeated with *Phytophthora parasitica* var. *nicotianae* on tobacco seedlings (Fenn & Coffey, 1988). AOA used alone markedly increased the susceptibility of tobacco to *P. parasitica* var. *nicotianae* (Fig. 6.5). Higher levels of potassium phosphonate (390 μg ml^{-1}) controlled *P. parasitica* var. *nicotianae* and were not affected by the presence of 80 μM AOA. Lower levels of the fungicide (195 μg ml^{-1}) were affected by AOA, however (Fig. 6.5). AOA inhibited *in vitro* uptake of phosphonate by *P. parasitica* var. *nicotianae* by 80% over 4 h. The combination of enhanced host susceptibility and reduced uptake of phosphonate by the fungus offer a feasible explanation of the results obtained with AOA in these two studies (Fenn & Coffey, 1985; 1988).

Behaviour of phosphonates in the field

Fosetyl aluminium provides effective control of some important diseases caused by both necrotrophic and biotrophic members of the Peronosporales (Table 6.6). An interesting feature of its behaviour is its more marginal or poor activity against closely-related Oomycete species causing similar diseases. Such high selectivity within a single taxonomic group might suggest it has a biochemical mode of action which is very site-specific. Compared with some other selective fungicides with anti-Oomycete properties such as metalaxyl, fosetyl aluminium is much less active (Cohen & Coffey, 1986); rates used for foliar application range from 100 to 400 g hl^{-1} and 2-10 kg ha^{-1}. Conversely, the number of applications necessary are low and the persistence of efficacy can be exceptional. With pineapples, control of *Phytophthora cinnamomi* and *P. parasitica* was achieved with a single pre-planting dip which was effective for 18 months (Rohrbach & Schenck, 1985). In the case of avocado root rot caused by *P. cinnamomi*, as little as two applications, as a trunk injection, resulted in some circumstances in a reversal of the symptoms of tree decline and restoration of good

Table 6.6. Examples of the relative efficacy of fosetyl aluminium against selected fungal pathogens within the Peronosporales (Oomycetes).

Pathogens	Efficacy rating High	 Moderate	 Low
Phytophthora spp.	*P. cactorum*	*P. capsici*	*P. infestans*
	P. cinnamomi	*P. palmivora*	*P. megasperma* f.sp. *glycinea*
	P. citricola	*P. parasitica*	*P. megasperma* f.sp. *medicaginis*
	P. citrophthora		
	P. fragariae		
Pythium spp.	*Py. splendens*	*Py. aphanidermatum*	Many *Pythium* spp.
Downy mildews and *Albugo* spp.	*Peronospora parasitica*	*Bremia lactucae*	*Peronospora hyoscyami*
	Pseudoperonospora humuli		
	Albugo candida		

root and shoot growth in the following season (Darvas *et al.*, 1984; Pegg *et al.*, 1985). Practical exploitation of the unique phloem-mobile properties of phosphonate fungicides has been both rapid and innovative. A good example is the use of foliar sprays or trunk injections to control root rots caused by various *Phytophthora* species. Currently, however, basic information on the mechanisms which regulate phloem transport of phosphonates is lacking.

Application of fosetyl aluminium as a drench to soils for control of *Pythium* or *Phytophthora* diseases has also been utilized, sometimes with good results. Fosetyl aluminium is rapidly degraded to phosphonate in soils. In a study with three different soil types the half-lives of the fungicide were 0·32, 0·73 and 0·77 days at 20°C (Piedallu & Jamet, 1985). Ouimette & Coffey (1989) found no trace of fosetyl aluminium present in soil 2 weeks after application of 3 g l^{-1}. Phosphonate was detectable in this same soil for 4-6 weeks after application of fosetyl aluminium, though it was not detectable after 8 weeks. Adams & Conrad (1953) demonstrated that phosphonate was oxidised by soil microorganisms to phosphate. Many different soil microorganisms including bacteria, fungi and actinomycetes can apparently utilize phosphonate as a phosphorus source (Casida, 1960; Malacinski & Konetzka, 1966). However, the rate of conversion of phosphonate to phosphate in soils would appear to be quite slow (MacIntire *et al.*, 1950), with a half-life of about 16 weeks recorded for one clay loam soil (Adams & Conrad, 1953). Since phosphate reduction can occur naturally under anaerobic conditions (Tsubota, 1959), it has been suggested that phosphonate is part of a natural phosphorus cycle, with general features similar to that of the nitrogen and sulphur cycles (Alexander, 1977). Since many bacteria have retained the genetic ability to oxidise phosphonate, even though they preferentially utilize phosphate, it has been suggested that they must be frequently exposed to it in nature (Malacinski & Konetzka, 1967).

Conclusions

The active ingredient of fosetyl aluminium, a unique phloem-mobile xenobiotic possessing remarkable antifungal activity against some important Oomycete plant pathogens, is phosphonate. This simple anion is believed to be an important component of the phosphorus cycle in soils (Alexander, 1977). Many microorganisms have retained the enzyme systems necessary to oxidise it to phosphate. What is different about the biology or metabolism of some Oomycetes that causes phosphonate to be inhibitory to their growth? A simple inability by Oomycetes to utilize phosphonate as a phosphorus source (Fothergill & Child, 1964) would appear unlikely to provide an explanation of this selective inhibition. Neither higher plants, nor mammals (Robertson & Boyer, 1956b), can utilize phosphonate, and yet it exhibits no toxic effects. We believe that the answer to the ability of phosphonate to inhibit selectively certain Oomycetes, may lie in unique features of their phosphorus metabolism. Wang & Bartnicki-Garcia (1973) have demonstrated that novel phosphoglucans are abundant in the storage vacuoles of some *Phytophthora* species at certain stages in their life cycle, notably during sporangium development. Polyphosphates are common in filamentous fungi and yeasts, but have not been detected in Oomycetes (Dietrich, 1976). Could it be that Oomycetes store their phosphate in a unique fashion, quite unlike that of other fungi, or higher organisms? Are specific metabolic steps in this storage process highly sensitive to phosphonate? Certainly, the ability of extremely low concentrations of phosphonate (1 to 2 $\mu g\ ml^{-1}$) to inhibit zoosporangium production by *Phytophthora* (Coffey & Joseph, 1985; Dolan & Coffey, 1988) may indicate selective interference with key biosynthetic events. It is our belief that phosphonate will prove not only to be a valuable tool in the control of important plant diseases, but may also be used experimentally to unmask new and important properties concerning both phloem translocation in plants and unique features of phosphorus metabolism in Oomycetes.

Acknowledgements. This research was supported in part by a grant from the California Avocado Commission. The authors are very grateful to BettyAnn Merrill who typed the text and to Linda Bobbitt and John Kitasako for preparing the figures.

References

Alexander, M. (1977). Microbial transformations of phosphorus. In *Introduction to Soil Microbiology*, pp. 347-348. John Wiley & Sons, New York.

Abblard, J., Gaulliard, J.-M. & Lacroix, G. (1983). Fungicidal compositions. *United States Patent Number 4,382,928.*

Adams, F. & Conrad, J. P. (1953). Transition of phosphite to phosphate in soils. *Soil Science*, **75**, 361-371.

Barchietto, T., Saindrenan, P. & Bompeix, G. (1988). Uptake and utilisation of phosphonate ions by *Phytophthora citrophthora* and *Nectria haematococca* in relation to their selective toxicity. *Pesticide Science*, **22**, 159-167.

Bompeix, G., Fettouche, F. & Saindrenan, P. (1981). Mode d'action du phoséthyl Al. *Phytiatrie-Phytopharmacie*, **30**, 257-272.

Bompeix, G., Ravisé, H., Raynal, G., Fettouche, F. & Durand, M. C. (1980). Modalités de l'obtention des nécroses bloquantes sur fueilles détachées de tomate par l'action du tris-*O*-éthyl phosphonate d'aluminium (phoséthyl d'aluminium), hypothèses sur son mode d'action *in vivo*. *Annales de Phytopathologie*, **12**, 337-351.

Bompeix, G. & Saindrenan, P. (1984). *In vitro* antifungal activity of fosetyl Al and phosphorous acid on *Phytophthora* species. *Fruits*, **39**, 777-786.

Bower, L. A. & Coffey, M. D. (1985). Development of laboratory tolerance to phosphorous acid, fosetyl-Al, and metalaxyl in *Phytophthora capsici*. *Canadian Journal of Plant Pathology*, **7**, 1-6.

Bruin, G. C. A. & Edgington, L. V. (1983). The chemical control of diseases caused by zoosporic fungi. In *Zoosporic Plant Pathogens*, ed. S. T. Buczacki, pp. 193-232. Academic Press, London & New York.

Casida, L. E. (1960). Microbial oxidation and utilization of orthophosphite during growth. *Journal of Bacteriology*, **80**, 237-241.

Coffey, M. D. & Bower, L. A. (1984). *In vitro* variability among isolates of eight *Phytophthora* species in response to phosphorous acid. *Phytopathology*, **74**, 738-742.

Coffey, M. D. & Joseph, M. C. (1985). Effects of phosphorous acid and fosetyl-Al on the life cycle of *Phytophthora cinnamomi* and *P. citricola*. *Phytopathology*, **75**, 1042-1046.

Cohen, Y. & Coffey, M. D. (1986). Systemic fungicides and the control of oomycetes. *Annual Review of Phytopathology*, **24**, 311-338.

Darvas, J. M., Toerien, J. C. & Milne, D. L. (1984). Control of avocado root rot by trunk injection with phosethyl-Al. *Plant Disease*, **68**, 691-693.

Dietrich, S. M. C. (1976). Presence of polyphosphate of low molecular weight in Zygomycetes. *Journal of Bacteriology*, **127**, 1408-1413.

Dolan, T. E. & Coffey, M. D. (1988). Correlative *in vitro* and *in vivo* behavior of mutant strains of *Phytophthora palmivora* expressing different resistances to phosphorous acid and fosetyl-Na. *Phytopathology*, **78**, 974-978.

Farih, A., Tsao, P. H. & Menge, J. A. (1981). Fungitoxic activity of efosite aluminum on growth, sporulation, and germination of *Phytophthora parasitica* and *P. citrophthora*. *Phytopathology*, **71**, 934-936.

Fenn, M. E. & Coffey, M. D. (1984). Studies on the *in vitro* and *in vivo* antifungal activity of fosetyl-Al and phosphorous acid. *Phytopathology*, **74**, 606-611.

Fenn, M. E. & Coffey, M. D. (1985). Further evidence for the direct mode of action of fosetyl-Al and phosphorous acid. *Phytopathology*, **75**, 1064-1068.

Fenn, M. E. & Coffey, M. D. (1988). Quantification of phosphonate and ethyl phosphonate in tobacco and tomato tissues and significance for the mode of action of two phosphonate fungicides. *Phytopathology*, **78**, in press.

Fothergill, P. G. & Child, J. H. (1964). Comparative studies of the mineral nutrition of three species of *Phytophthora*. *Journal of General Microbiology*, **36**, 67-78.

Guest, D. I. (1984a). Modification of defence responses in tobacco and capsicum following treatment with fosetyl-Al [aluminium tris (*O*-ethyl phosphonate)]. *Physiological Plant Pathology*, **25**, 125-134.

Guest, D. I. (1984b). The influence of cultural factors on the direct antifungal activities of fosetyl-Al, propamocarb, metalaxyl, SN 75196 and Dowco 444. *Phytopathologische Zeitschrift*, **111**, 155-164.

Guthrie, J. P. (1979). Tautomerization equilibria for phosphorous acid and its ethyl esters, free energies of formation of phosphorous and phosphonic acid and their ethyl esters, and p*Ka* values for ionization of the P-H bond in phosphonic acid and phosphonic esters. *Canadian Journal of Chemistry*, **57**, 236-239.

Luttringer, M. & de Cormis, L. (1985). Absorption, dégradation et transport du phoséthyl-Al et de son métabolite chez la tomate (*Lycopersicon esculentum* Mill.). *Agronomie*, **5**, 423-430.

MacIntire, W. H., Winterberg, S. H, Hardin, L. J., Sterges, A. J. & Clements, L. B. (1950). Fertilizer evaluation of certain phosphorus, phosphorous, and phosphoric materials by means of pot cultures. *Agronomy Journal*, **42**, 543-549.

Malacinski, G. & Konetzka, W. A. (1966). Bacterial oxidation of orthophosphite. *Journal of Bacteriology*, **91**, 578-582.

Malacinski, G. M. & Konetzka, W. A. (1967). Orthophosphite-nicotinamide adenine dinucleotide oxidoreductase from *Pseudomonas fluorescens*. *Journal of Bacteriology*, **93**, 1906-1910.

Martin, R. B. (1959). The rate of exchange of the phosphorus bonded hydrogen in phosphorous acid. *Journal of the American Chemistry Society*, **81**, 1574-1576.

Ouimette, D. G. & Coffey, M. D. (1985). *In vivo* efficacy of five phosphite compounds against *Phytophthora capsici* on pepper plants. *Phytopathology*, **75**, 1330 (abstract).

Ouimette, D. G. & Coffey, M. D. (1988). Quantitative analysis of organic phosphonates, phosphonate, and other inorganic anions in plants and soil using high performance ion chromatography. *Phytopathology*, **78**, 1150-1155.

Ouimette, D. G. & Coffey, M. D. (1989). Phosphonate levels in avocado (*Persea americana* Mill.) seedlings and soil following treatment with fosetyl-Al or potassium phosphonate. *Plant Disease*, **72**, in press.

Pegg, K. G., Whiley, A. W., Saranah, J. B. & Glass, R. J. (1985). Control of *Phytophthora* root rot of avocado with phosphorus acid. *Australian Journal of Plant Pathology*, **14**, 25-29.

Piedallu, M.-A. & Jamet, P. (1985). Cinétique de disparition et dégradation du phoséthyl-Al dans le sol en conditions contrôlées. In *Fungicides for Crop Protection: 100 Years of Progress*, ed. I. M. Smith, pp. 297-300. British Crop Protection Council, Croydon, UK.

Robertson, H. E. & Boyer, P. D. (1956a). Orthophosphite as a buffer for biological studies. *Archives of Biochemistry and Biophysics*, **62**, 396-401.

Robertson, H. E. & Boyer, P. D. (1956b). The biological inactivity of glucose 6-phosphite, inorganic phosphites and other phosphites. *Archives of Biochemistry and Biophysics*, **62**, 380-395.

Rohrbach, K. G. & Schenck, S. (1985). Control of pineapple heart rot, caused by *Phytophthora parasitica* and *P. cinnamomi*, with metalaxyl, fosetyl Al, and phosphorous acid. *Plant Disease*, **69**, 320-323.

Saindrenan, P., Barchietto, T., Avelino, J. & Bompeix, G. (1988). Effects of phosphite on phytoalexin accumulation in leaves of cowpea infected with *Phytophthora cryptogea*. *Physiological and Molecular Plant Pathology*, **32**, 425-435.

Saindrenan, P., Darakis, G. & Bompeix, G. (1985). Determination of ethyl phosphite, phosphite and phosphate in plant tissues by anion-exchange high-performance liquid chromatography and gas chromatography. *Journal of Chromatography*, **347**, 267-273.

Thizy, A., Pillon, D., Debourge, J.-C. & Lacroix, G. (1978). Fungicidal compositions containing phosphorous acid and derivatives thereof. *United States Patent Number 4,075,324*.

Tsubota, G. (1959). Phosphite reduction in the paddy field. *Soil Plant Food*, **5**, 10-15.

Vegh, I., Leroux, P., LeBerre, A. & Lanen, C. (1985). Détection sur *Chamaecyparis lawsoniana* 'Ellwoodii' d'une souche de *Phytophthora cinnamomi* Rands résistante au phoséthyl-Al. *Pépiniériste, Horticulteurs, Maraîchers. Revue Horticol*, **262**, 19-21.

Wang, M. C. & Bartnicki-Garcia, S. (1973). Novel phosphoglucans from the cytoplasm of *Phytophthora palmivora* and their selective occurrence in certain life cycle stages. *Journal of Biological Chemistry*, **248**, 4112-4118.

Chapter 7

Phosphorus and nitrogen fluxes between plant and fungus in parasitic associations

D. R. Walters

Department of Plant Sciences, West of Scotland College, Auchincruive, Nr Ayr KA6 5HW, UK.

Introduction

Infection with pathogenic fungi results in considerable alterations in host plant physiology and metabolism, including profound effects on the uptake and translocation of nutrients (see Paul, Chapter 8). In certain host/pathogen interactions these effects are accompanied by changes in the utilization and metabolism of some nutrients (eg. nitrogen-containing compounds and phosphorus), which may in turn affect their uptake from the rooting medium and subsequent transport throughout the plant (see Walters, 1985). Although in many plant/parasite systems, altered nutrient uptake and translocation may be the indirect result of other changes in host physiology (eg. transpiration), such changes, if they result in transport of nutrients to the fungal domain, will prove of benefit to the pathogen. Thus, altered rates of xylem and phloem transport within the diseased plant may lead to nutrient and metabolite accumulation at sites of fungal infection (Scott, 1972) and the fungus may be able to draw upon these compounds to support its growth and sporulation. In this chapter the phenomenon of nutrient accumulation at infection sites is first examined. The movement of nutrients, mainly nitrogen and phosphorus, from the host to the fungus is then considered. Because most of the available information relates to biotrophic fungal pathogens, major emphasis will be placed on plants infected with rusts or powdery mildews.

Phosphorus and nitrogen in infected leaves

Changes in whole leaf concentrations

Leaves infected with rust or powdery mildew fungi show considerable changes in nutrient concentrations (Paul, Chapter 8). Elevated phosphate concentrations have often been reported. For example, rust-infected leeks showed increases of 21% (Roberts & Walters, 1988a), barley infected with brown rust showed increases of 20% (Ahmad *et al.*, 1982) and in various powdery mildew infections increased accumulations of 20-25% have been reported (Comhaire, 1963; Fric, 1978; Walters & Ayres, 1981). However, changes in nitrogen concentrations in infected tissues do not always take the form of an increase. Raised total nitrogen concentrations have been reported in rusted groundsel (Paul & Ayres, 1988) and rusted leek (Roberts & Walters, 1988a), whereas Ahmad *et al.* (1982) found a reduced total nitrogen concentration in rusted barley. Study of alterations in total nitrogen will, however, mask the changes occurring in concentrations of individual nitrogenous compounds. Information on individual nitrogenous compounds is important since changes will reflect responses of the host to infection and may in turn provide information concerning the nutrient requirements of the pathogen.

Recent work on rust-infected leeks has indicated that most of the increase in total nitrogen is probably due to increases in amino acid and nitrate concentrations, rather than ammonium, especially since ammonium concentration was reduced in infected leaves (Roberts & Walters, 1988b). Piening (1972) also found an increased nitrate concentration in rust-infected rye and although no change was observed in nitrate in powdery mildew infected barley leaves, amino acid and amide concentrations increased (Sadler & Scott, 1974).

Changes in discrete regions of the infected leaf: nutrient accumulation at infection sites

A common feature of infection with biotrophic fungi is the accumulation of metabolites at infection sites. Gottlieb & Garner (1946), who were the first clearly to demonstrate this phenomenon, showed that radioactive phosphorus accumulated in wheat leaves infected with the stem rust fungus (*Puccinia graminis* f.sp.

Table 7.1. Concentrations of free amino acids (μmoles g^{-1} f.wt.) in pustules of *Uromyces phaseoli* var *typica* on Pinto bean leaves, in regions surrounding pustules (3 and 5 mm)* at primary sporulation, and in healthy control Pinto bean leaves.

Amino acids	Pustule	First ring	Second ring	Control
Cysteic acid	0·3	0·2	0·1	0·1
Aspartic acid	4·1	2·3	1·1	1·0
Threonine	1·1	0·3	0·2	0·3
Asparagine	4·4	10·9	8·7	1·6
Glutamic acid	5·2	2·4	1·7	1·8
Glutamine	11·0	3·1	1·8	0·5
Cystine	2·3	2·1	1·1	0·5
Leucine	0·4	0·1	0·1	0·2
Tyrosine	0·6	0·3	0·2	0·1
Lysine	1·2	0·9	0·3	0·4
Arginine	2·0	2·6	0·8	0·3

* Discrete regions were sampled as follows: pustules – discs of 1 mm diameter containing mainly stroma and host cells invaded by parasite; first ring – discs of 3 mm diameter containing the 'green-island' zone; second ring – discs of 5 mm diameter, from beyond the 'green island' zone. (Adapted from Raggi, 1974).

tritici), the label being concentrated in pustule regions. Since that time several workers have demonstrated nutrient accumulation at infection sites. Thus, Shaw & Samborski (1956) found substantial accumulations of labelled phosphorus in pustule regions of stem rust on wheat and of powdery mildew on barley. Phosphorus accumulation was also observed at *Uromyces* infection sites on bean leaves (Yarwood & Jacobsen, 1955). Later work by Raggi (1974, 1976) showed that amino acids and amides also accumulated in pustule areas on bean leaves infected with *Uromyces phaseoli* (Table 7.1) and on peach leaves infected with the powdery mildew *Sphaerotheca pannosa persicae*.

Two factors appear to be important in the accumulation of nutrients at infection sites: (1) an increased rate of nutrient transport towards the pustule area, and (2) a reduction in nutrient movement away from these regions. Thus, rusted primary leaves of bean were shown to import as much as 40 times more assimilate than controls (Livne & Daly, 1966), while substantial reductions in assimilate movement out of infected leaves has been demonstrated in wheat infected with yellow rust (Doodson, Manners & Myers, 1965) and in rusted bean (Zaki & Durbin, 1965; Livne & Daly, 1966). It should be noted however, that evidence for increased phloem import into infected monocot leaves is lacking. Thus, although nutrient accumulation at infection sites on monocot leaves might reflect a redistribution of material within the infected leaf or import of nutrients in the xylem, no data exists to support the view that phloem import into infected leaves of say, grasses or cereals, takes place. Work on brown rust-infected barley has shown that when $^{14}CO_2$ is fed to a non-infected leaf, very little label appears in infected leaves (Farrar, 1984). Similarly in leeks infected with rust, the construction of a carbon budget for individual leaves has shown that there is no assimilate movement into infected leaves (Roberts, 1987).

The mechanisms underlying the movement of nutrients to infection sites have been the subject of much research. Some workers have suggested that the vigorously growing fungal mycelium represents a powerful sink for the diversion of nutrients from other parts of the plant (Thrower, 1965). Indeed, partly because of their ability to 'create' sink effects, plant growth substances like auxins and cytokinins have been implicated in the mobilization of nutrients towards infection sites (Shaw & Hawkins, 1958; Pozsar & Kiraly, 1966; Brian, 1967). Furthermore, nutrient accumulation at infection sites of biotrophic pathogens, together with retention of chlorophyll and of the capacity for protein synthesis, are probably indicative of the maintenance of a 'juvenile' condition at and around infection sites. Evidence of such juvenility takes the form of the so-called 'green-islands', which are typical of rust and powdery mildew infections of many hosts. Interest in the possible involvement of cytokinins in green-island formation on infected leaves has been partly due to their ability to induce 'green-island' formation on detached leaves. The increased cytokinin activity, at least in rusted bean, was suggested to be of host origin (Dekhuijzen & Staples, 1968), but rust and powdery mildew spores are known to contain high levels of cytokinins (Kiraly, El-Hammady

& Pozsar, 1967). Some authors have suggested that localized delay of senescence and other phenomena associated with green-island formation on infected leaves may be the result of release of cytokinins by the parasite (Kiraly, Pozsar & El-Hammady, 1966). If this is the case, then it would represent the flux of an important nitrogen-containing compound from the fungus to the host. Although all of the research on cytokinins in fungally-infected plants has involved the use of bioassays, and as such is open to some criticism, it is interesting to note that using monoclonal antibodies raised against zeatin riboside and dehydrozeatin riboside, increases in cytokinins have been detected in rusted leek leaves, the largest increases occurring in and around pustules (Roberts, 1987).

Polyamine accumulation at infection sites

Cytokinins are not the only substances capable of delaying senescence and causing green-island formation. Among the substances that possess this property, is a group of simple aliphatic amines called polyamines. The most common of these compounds are putrescine (a diamine), spermidine (a triamine) and spermine (a tetra-amine). These compounds are known to be essential for growth in most organisms and among their many proposed functions are included the *in vivo* regulation of senescence (Galston, 1983; Slocum, Kaur-Sawhney & Galston, 1984; Walters, 1987). As nitrogenous compounds, the polyamines represent major pools of tissue nitrogen, and their formation and possible mobilization have large potential effects on fluxes of the element in host-biotroph relations. Greenland & Lewis (1984) were the first to examine the possibility that polyamines may be involved in the delay in senescence and green-island formation in a plant infected with a biotrophic fungus. They showed that barley leaves infected with brown rust accumulated spermidine and, to a lesser extent, spermine, with peak accumulation coinciding with fungal sporulation (Fig. 7.1). Soon thereafter, putrescine, spermidine and spermine were found to accumulate in mildewed barley leaves (Walters, Wilson & Shuttleton, 1985), the largest increases occurring in and around fungal pustules (Walters & Wylie, 1986) (Fig. 7.2).

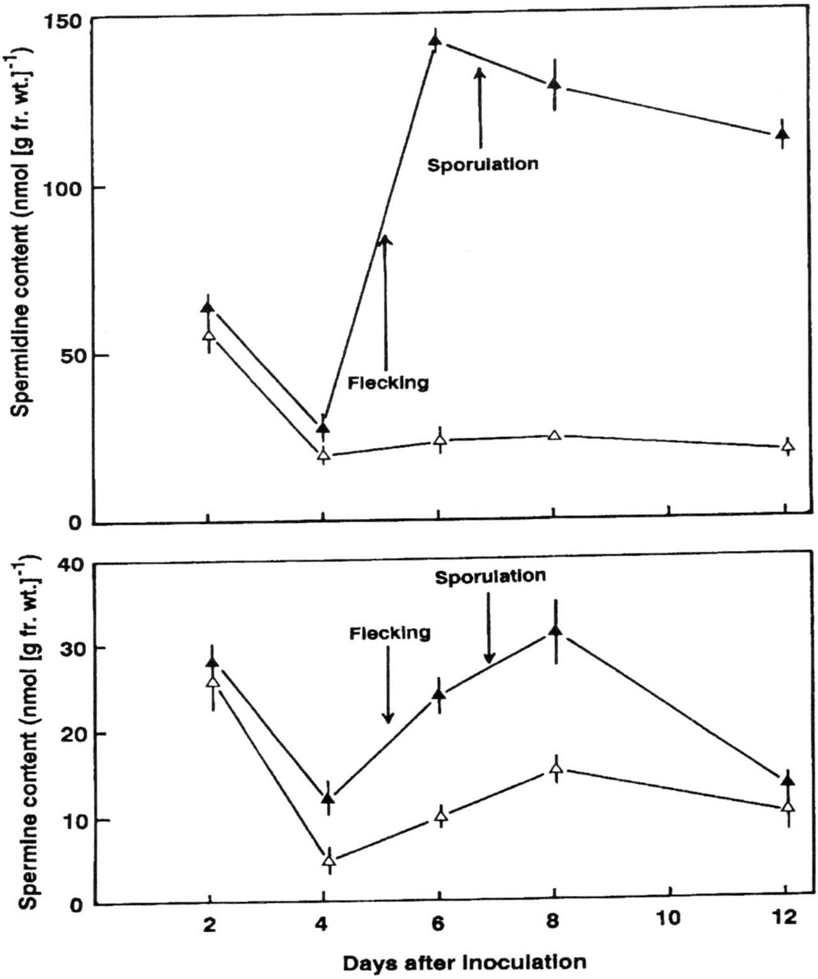

Fig. 7.1. Changes in the concentrations of (a) spermidine and (b) spermine plus an unidentified amine, in healthy (open triangles) primary leaves of barley and comparable leaves infected with *Puccinia hordei* (closed triangles). Vertical bars represent twice the standard errors of the means; where none are shown the limits are within the size of the symbols. (Adapted from Greenland & Lewis, 1984).

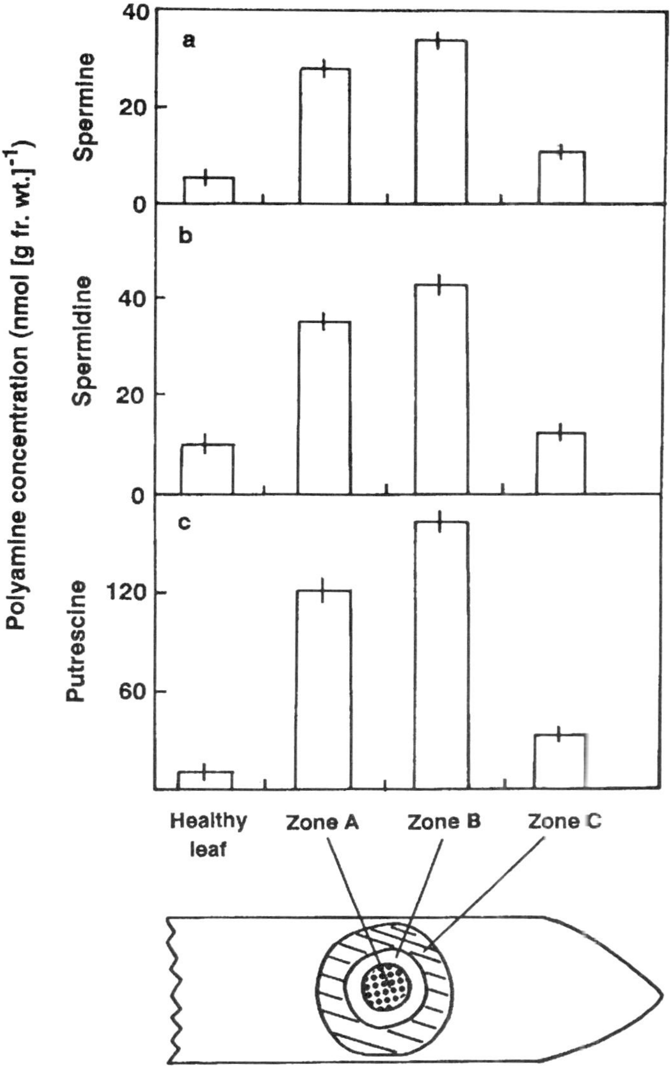

Fig. 7.2. The concentrations of spermine (a), spermidine (b) and putrescine (c) in healthy leaves and in discrete regions of mildewed barley leaves. Zone A, pustule; Zone B, region up to 5 mm from pustule boundary; Zone C, region up to 5 mm from boundary to Zone B. Vertical bars represent SEM of four replicates. (From Walters & Wylie, 1986).

However, on mildewed barley, green-islands are not visible until late on in infection at which stage pustules and surrounding regions remain green while the remainder of the leaf senesces. Green-islands can be induced to form more rapidly if senescence is hastened by detaching the leaf. In this way, recent work has detected substantial increases in the concentrations of the three polyamines in green-island tissues on mildewed barley leaves (S. E. Coghlan & D. R. Walters, unpublished). Because the peak in polyamine accumulation occurs at about the time of fungal sporulation, it is possible that much of the increase in amines could be located in the fungus. Indeed, recent work using autoradiographic localization of ornithine decarboxylase (ODC), the rate limiting

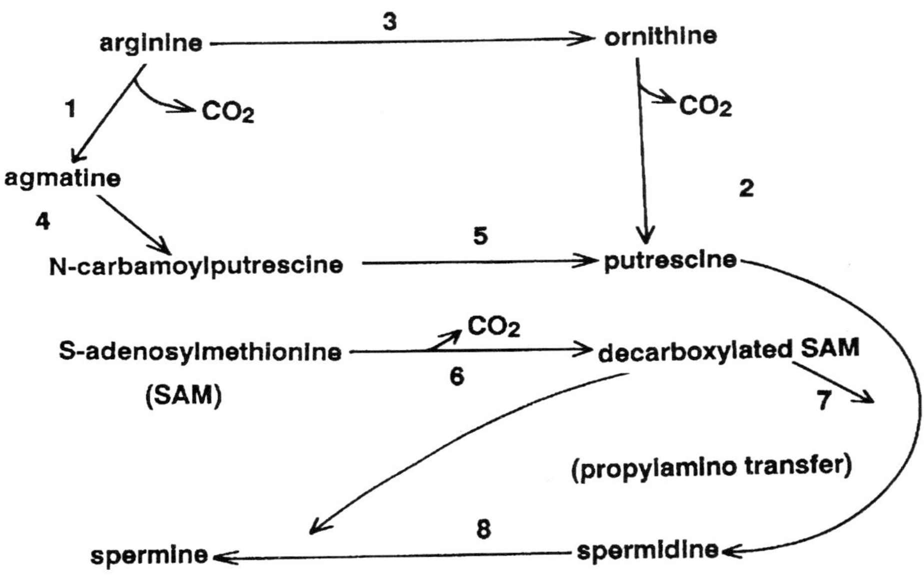

Fig. 7.3. Biosynthetic pathways of putrescine, spermidine and spermine. Putrescine can be formed either from ornithine in a reaction catalyzed by ornithine decarboxylase or from arginine in a reaction catalyzed initially by arginine decarboxylase. Spermidine is formed by the addition of a propylamino group to putrescine; the addition of another propylamino moiety to spermidine leads to spermine. 1 = arginine decarboxylase; 2 = ornithine decarboxylase; 3 = arginase; 4 = agmatine iminohydrolase; 5 = N-carbamoylputrescine aminohydrolase; 6 = S-adenosylmethionine decarboxylase; 7 = spermidine synthase; 8 = spermine synthase.

enzyme in fungal polyamine biosynthesis and one of two enzymes (ODC and arginine decarboxylase, ADC) involved in plant amine biosynthesis (Fig. 7.3), has shown that most of the increased ODC in brown rust infected barley appears to be located in the fungal material of the pustules, much less being found elsewhere (Bailey, Bower & Lewis, 1987). The situation in mildewed barley is rather different. Here, increases in both ODC and ADC were found in mildewed leaves, especially in green-islands (S. E. Coghlan & D. R. Walters, unpublished). Moreover, removal of surface fungal growth, leaving haustoria within epidermal cells, reduced, but did not eliminate, the increased polyamine concentrations and enzyme activities; it would thus appear that in mildewed barley, increased polyamine accumulation is of both host and fungal origin.

In the course of the above studies of green-islands in mildewed barley, it became apparent that substantial changes were occurring in the concentration of another polyamine, cadaverine. Cadaverine, like putrescine, is a simple diamine and can be synthesized in plants from lysine in a reaction catalyzed by lysine decarboxylase. Cadaverine concentration was found to decrease in mildewed leaves, including green-islands, but it virtually disappeared from senescing areas well away from pustules (S. E. Coghlan & D. R. Walters, unpublished). A reduction in cadaverine biosynthesis or an increased cadaverine breakdown should result in a build-up of its precursor lysine, which could then be used for other purposes. It is known that some rusts take up lysine from their hosts (see Mendgen, 1981; p. 147) and it is possible that any lysine accumulating as a result of decreased cadaverine synthesis or enhanced breakdown, could be transported toward mildew pustules and be taken up by the fungus.

Inhibition of fungal polyamine biosynthesis

Since fluxes of nitrogen, in the form of polyamines, between host and fungus, may be key processes in the nutrient relations of the association, much interest surrounds the possibility that disease control could be achieved by inhibition of their synthesis. Most fungi appear to be dependent upon ODC for polyamine biosynthesis, thus, providing the fungus cannot obtain all of its polyamine requirement from its host, specific inhibition of fungal

Table 7.2. Mean number of lesions induced by uredospores of *Uromyces phaseoli* on unifoliate leaves of bean plants exposed to α-difluoromethylornithine (DFMO) before or after inoculation.

DFMO concentration (mM)	Lesions cm^{-2} DFMO treatment Preinoculation	DFMO treatment Postinoculation
None	59 ± 4	61 ± 2
0·025	34 ± 1**	28 ± 2**
0·050	29 ± 4**	17 ± 2**
0·10	14 ± 2**	5 ± 1**
0·25	2 ± 1**	2 ± 1**
0·50	0**	0**

Each value is the mean ± SEM, based on six replicates (one leaf per plant); ** denotes significant differences from controls at the 1% level. (Adapted from Rajam, Weinstein & Galston, 1985).

ODC might appear to offer a unique means of inhibiting fungal growth. Although we have no information concerning either the ability of biotrophic fungi to take up polyamines from plant cells or of the fungal requirements for host polyamines, it would seem likely that necrotrophic fungi might take up polyamines from the contents of host cells killed in advance of colonization. Disease control via inhibition of polyamine biosynthesis is possible because of the existence of several ODC specific inhibitors; e.g. α-difluoromethylornithine (DFMO) — a specific, irreversible inhibitor of ODC. It has been shown that excellent control of rust infection of *Phaseolus vulgaris* (Rajam, Weinstein & Galston, 1985) and *Vicia faba* (Walters, 1986) can be achieved using pre-inoculatory or post-inoculatory sprays of DFMO (Table 7.2; Fig. 7.4). Weinstein *et al.* (1987) confirmed this effect of DFMO on several rust diseases of wheat, while work at Auchincruive (West & Walters, 1988) has shown that powdery mildew infection of barley can be controlled using several different inhibitors of ODC or other inhibitors of polyamine biosynthesis.

It is interesting to note that work so far on hemibiotrophs and necrotrophic fungi has highlighted differences in the responses to

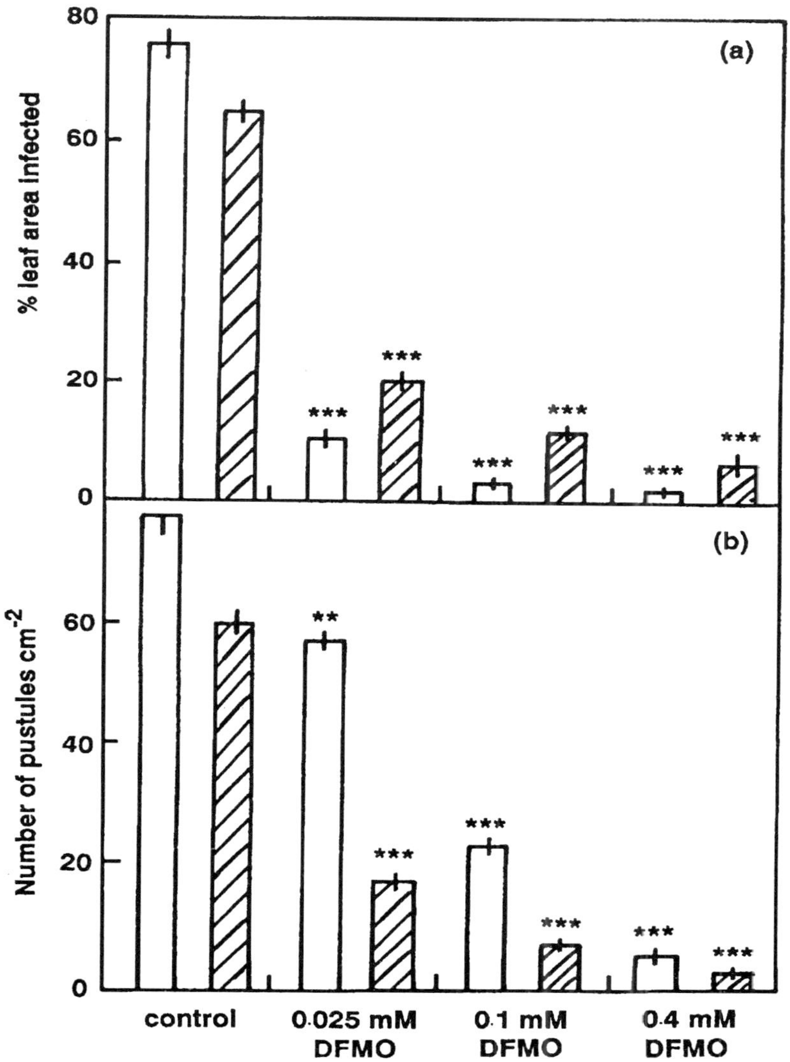

Fig. 7.4. Effect of post-inoculatory (open bars) or pre-inoculatory (hatched bars) treatment with α-difluoromethylornithine (DFMO) on (a) the percentage leaf area infected and (b) pustules per cm^2 on leaves of broad bean inoculated with rust, *Uromyces viciae-fabae*. Vertical bars represent the standard error of the means and significant differences are shown at $P \leq 0{\cdot}05$ *; $P \leq 0{\cdot}01$ **; $P \leq 0{\cdot}001$ ***. (From Walters, 1986).

inhibitors of polyamine biosynthesis in different types of fungi. Thus, although very good control of rust and powdery mildew fungi can be achieved with this approach, some fungi, e.g. *Pyrenophora teres*, require much higher inhibitor concentrations for effective control of growth, while polyamine biosynthesis inhibitors appear to have no effect on the growth of other fungi, e.g. *Fusarium culmorum* (H. M. West & D. R. Walters, unpublished). These differences in fungal response to inhibitors may reflect differences in sensitivity of, for example, fungal ODCs to DFMO and/or they may reflect differences in the abilities of different fungi to take up DFMO (or other inhibitors) from host tissues.

Role of intercellular hyphae and haustoria in N, P and S uptake in biotrophic fungi

For many powdery mildew fungi, the only contact with host cells and therefore the only structure through which nutrient uptake can occur is the haustorium. Indeed, Gay and his colleagues (see Manners & Gay, 1983; Gay, 1984) have convincingly shown that haustoria are the main means by which assimilates are transported to powdery mildew fungi. According to Manners & Gay (1983), haustoria offer several advantages over intercellular hyphae for nutrient uptake. In particular, stimulation of solute efflux over an area surrounding the haustorium is provided by the establishment of an isolated apoplast and by modification of membrane function at localized areas of the host cell. It would appear that by virtue of structural and functional modification of the host plasmalemma in the haustorial region, the powdery mildew controls efflux from the host to its own advantage. Indeed, it has been shown, using wheat inoculated with powdery mildew, that ^{32}P and ^{35}S presented to host tissues did not enter the fungus until haustoria had developed (Mount & Ellingboe, 1969; Slesinski & Ellingboe, 1971).

In plants infected with rusts or Oomycetes little information is available concerning the relative contributions to uptake of haustoria and intercellular hyphae. It has been shown that certain solutes supplied to host tissues are incorporated into fungal material before haustoria are formed whereas other compounds are not incorporated until haustorial establishment has occurred (Onoe, Tani & Haito, 1973; Andrews, 1975). In a microautoradiographic study of uptake of tritiated lysine during the early stages of rust infection on bean, Mendgen (1979) showed that maximum uptake occurred via mature haustoria and very little occurred via intercellular hyphae. Interestingly, only the plasmalemma of pea powdery mildew haustoria possess ATPase activity. The plasma membranes of intercellular hyphae of bean rust possess no detectable ATPase activity (Spencer-Phillips & Gay, 1981), and very little activity is found in intercellular hyphae of *Albugo candida* (Woods & Gay, 1983). Gay (1984) has suggested that intercellular hyphae of the bean rust may thus be unable to take up nutrients at concentrations lower than those obtained internally. However, according to Farrar (1984), there will be movement of nutrients to fungal surfaces in the apoplast, in which case it seems important to know how much fungal growth can be supported in this manner. Hancock & Huisman (1981) have calculated that the concentration of nutrients in intercellular spaces is sufficient to sustain fungal growth. More recently, by calculating the surface areas of haustoria and intercellular hyphae, Kneale & Farrar (1985) were able to show that all of the hexose needed to support growth of brown rust on barley could be taken up through either fungal structure. However, this does not rule out the possibility that haustoria may be necessary for the uptake of specific substances e.g. amino acids like lysine.

Uptake of phosphorus by biotrophic fungi

While there are several reports of phosphorus accumulation at infection sites of biotrophically infected leaves (see above), it is important to know whether the fungus takes up this accumulated phosphorus. Martin & Ellingboe (1978) showed that *Erysiphe graminis tritici* took up labelled phosphorus from wheat leaves as early as 10 h after inoculation, and this continued as long as fungal growth occurred. Martin (1974, cited in Ellingboe, 1978) found that transfer of radioactive phosphorus to the fungus was ten times greater in an incompatible host/parasite interaction than in a compatible interaction. Ellingboe (1978) has suggested that because an incompatible interaction between the plant and the fungus places great stress on the host, it might cause a substantial movement of phosphorus to sites of interaction, especially since phosphorus is important in cell energy metabolism. Earlier work by Hsu

& Ellingboe (1972) had shown that uptake of ^{35}S by barley mildew depended upon successful primary infections and genotypes with high infection efficiency (i.e. those which produced more haustoria and elongating secondary hyphae) had high uptake rates. Conversely, mildew genotypes with lower infection efficiencies had lower uptake rates.

Similar experiments with rust infections do not appear to have been carried out, probably because of the difficulty in separating fungus from host. In fact, few data exist on the transfer of phosphorus from host to parasite. Ahmad, Farrar & Whitbread (1984) made a detailed study of phosphorus fluxes from barley to the brown rust fungus on the day of sporulation. They found that phosphorus retranslocation from first leaves was greatly reduced by infection and as a result, phosphorus concentration in infected leaves increased substantially. It was considered that accumulation was largely in the fungus. By calculating a phosphorus budget for healthy and infected leaves, these authors were able to show that phosphorus fluxes from the host to the rust could be sustained largely because the leaf was supplied by way of the xylem. Interestingly, very little change was observed in the concentrations of phosphorus in the xylem sap from healthy and rusted barley plants. Because the flux of phosphorus from the host symplast to the apoplast was greater than phosphorus transfer to the fungus, there appeared to be no need to invoke pathogen-induced membrane leakiness to provide intercellular rust hyphae with phosphorus (Ahmad *et al.*, 1984). In fact, later work showed that brown rust infection of first leaves of barley did not result in gross impairment of host membrane integrity (Ahmad, Farrar & Whitbread, 1985).

Ahmad *et al.* (1984) found that brown rust stored most of its phosphorus as polyphosphate. Earlier work by Bennett & Scott (1971) had shown that stem rust infected primary leaves of wheat accumulated eight times more polyphosphate than controls and most of the polyphosphate appeared to be located in the stem rust uredospores (Table 7.3). Apparently, production of brown rust spores on barley leaves used phosphorus of xylem origin and the fungus did not mobilize its own polyphosphate store (Ahmad *et al.*, 1984). These authors have speculated that polyphosphate may represent a store of phosphorus which the fungus can 'leak' to the host once it begins to senesce. In view of the importance of inor-

Table 7.3. Amounts of orthophosphate and 'soluble' polyphosphate in non-infected and rusted wheat leaves and in uredospores and saprophytic mycelium of *Puccinia graminis tritici*.

Source of material	Orthophosphate	'Soluble' polyphosphate
	(mg P g^{-1} f.wt.)	
Non-infected wheat leaves[b]	0·08	0·03
Rusted wheat leaves[b,c]	0·19	0·29
Uredospores	0·22	2·44
Infected leaves after removal of uredospores from sori	0·17	0·11
Saprotrophic mycelium	0·14	0·50

[a]Cold acid-soluble, not adsorbed to charcoal, acid labile, shown by paper chromatography to be almost entirely polyphosphate with a trace of pyrophosphate.
[b]Primary leaves of wheat 17 days from sowing.
[c]Ten days after inoculation, uredospores constituted 7·5% of the fresh weight of the infected leaf. (From Bennett & Scott, 1971).

ganic phosphate in controlling the rate of photosynthesis and perhaps also in controlling sink activity, the fungal polyphosphate store may represent a means of maintaining host function in a disease lesion while the rest of the leaf senesces (Ahmad *et al.*, 1984).

Uptake of nitrogen by the fungus

As previously described, infection can cause large changes in nitrogen in host tissues. Thus, twelve days after inoculation of barley with mildew, total nitrogen in mildewed areas of the leaf increased considerably (Bushnell & Gay, 1978) and most of this increase appeared to be in the mildew fungus (Fig. 7.5). Various workers have shown that soluble nitrogen compounds e.g. aspartic acid, glutamic acid, and their corresponding amides, asparagine and glutamine, can accumulate in powdery mildew infected tissues (Sadler & Scott, 1974). Bushnell & Gay (1978) have postulated that powdery mildews take up and utilize these host amino acids and amides. Recent work by Murray & Ayres (1986) supports this premise. They found that the mildew accumulated organic nitrogen at a time when there was no nitrate reduction in infected shoots. Prior to sporulation, mildew infection resulted in only a minor alteration in the rate of nitrogen recirculation to the roots and in the flux of nitrogen to the shoots via the xylem. They thus suggested that early in infection the mildew draws on normal supplies of nitrogen available in the shoot. However, the recycling of nitrogen to the roots stops after fungal sporulation, and nitrogen which is normally recycled becomes available for the mildew. It

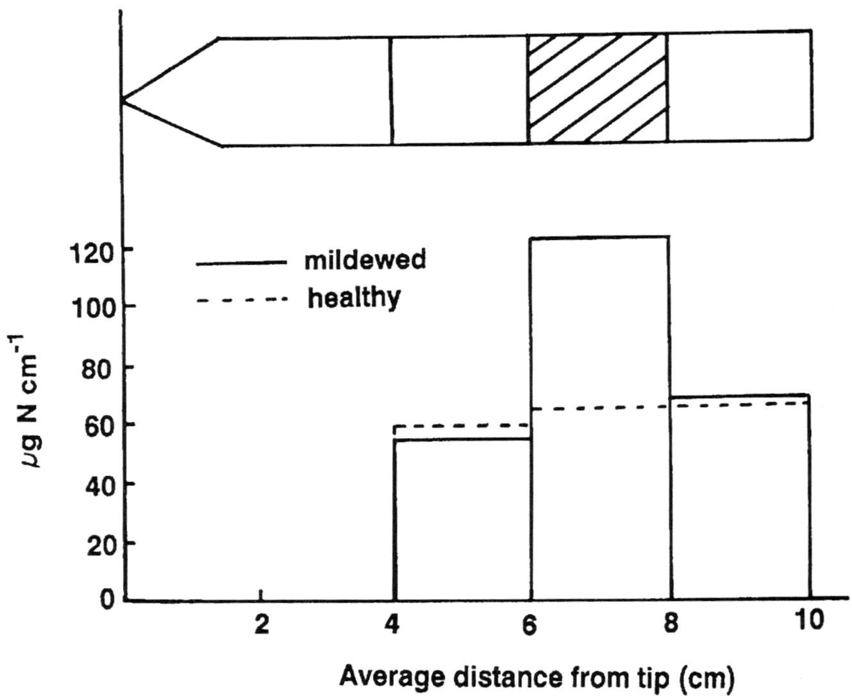

Fig. 7.5. Accumulation of nitrogen in mildewed tissues as shown by the total nitrogen content of mildewed and adjacent non-mildewed segments of attached barley leaves twelve days after inoculation with *Erysiphe graminis* f.sp. *hordei*. All spores were included in the mildewed sample. Nitrogen content of the host tissues in mildewed segments is relatively unchanged. Hatched area represents powdery mildew infection. (From Bushnell & Gay, 1978).

would appear that the fungus cannot obtain all of its nitrogen in this way and it must draw upon a supply of reduced nitrogen either from the xylem (as suggested for phosphorus uptake by brown rust, see p. 144) or by redirecting the flow of nitrogen within the shoot. Unfortunately, we have very little data on transfer of nitrogen-containing compounds from hosts to mildew and, because mildew fungi cannot be grown axenically, we know little of their nutritional requirements.

Interestingly, it has been suggested that alterations in purine metabolism are important in facilitating the successful establishment of powdery mildew infection on barley (Butters, Burrell & Hollomon, 1985). The powdery mildew fungus appears unable to synthesize purines *de novo* (Hollomon, 1979) and so it must obtain these compounds from the host plant. Butters *et al.* (1985) found that although purines were taken up from the leaf surface and metabolized by the powdery mildew fungus, it was difficult to determine what purines were transported to the mildew mycelium through the haustoria. They were able to show that [^{3}H]-adenosine supplied to the host, was taken up by the fungus and extensively deaminated to inosine. Apparently, the substantial activity of the enzyme adenosine deaminase, present in the fungus and in infected leaves but not in healthy leaves and which is responsible for the conversion of adenosine to inosine, would help to maintain an adequate supply of purines for fungal growth and development (Butters *et al.*, 1985).

Some data are available on the uptake of nitrogenous compounds, mainly of amino acids and amides, by rust fungi from their hosts. Jager & Reisener (1969) found that lysine and arginine were probably transported intact into the stem rust fungus from wheat. Amino acids such as alanine and glycine appeared to be metabolized prior to uptake and incorporation into spores. Autoradiographic analysis of [^{3}H]-lysine uptake by resistant and susceptible bean/rust interactions showed that although lysine uptake by rust haustoria occurred in both types of interaction, only in the compatible combination did uptake occur in the very young developing haustorium (Mendgen, 1979). In an incompatible interaction between the cultivar Golden Gate Wax and rust, labelled lysine uptake was much less than that in the compatible interaction, although haustoria were still labelled when cell contents of both partners had become disorganised and broken down. Mendgen (1979) considered that the label had simply diffused from host cell to haustorium via damaged membranes. However, not all incompatible interactions are the same. In another incompatible interaction, lysine uptake was constant and similar to that observed in the compatible combination, suggesting that here, nutrient uptake by the haustorium remains normal until other metabolic processes result in death of the fungus and the host cell (Mendgen, 1979). Experiments by Burrell & Lewis (1977) on infection of coltsfoot (*Tussilago farfara* L.) with the rust *Puccinia poarum*, indicated that serine and alanine were probably taken up directly by the rust, whereas aspartic acid, glutamic acid and glutamine were not readily taken up by the parasite. Indeed, these authors considered serine and alanine to be important in the successful establishment and growth of the rust on its host. Bushnell (1984), on the other hand, considered glutamine to be the most important of the amino acids or amides to be utilized by rusts. Thus, it is known to accumulate in rust infections, is readily translocated in plants and together with ammonia and asparagine, can be an important product of proteolysis (Lea & Miflin, 1980). Glutamine is also a favoured source of nitrogen for axenically grown rusts and it is a precursor in fungal chitin synthesis (e.g. Raggi, 1974).

Conclusions

In their 1946 paper describing the effects of rust infection on phosphorus distribution in wheat leaves, Gottlieb & Garner stated that '... Despite the many years which have elapsed since the early work of Marshall Ward (1901; 1902) on the physiology of obligate parasitism, this complex is still in many respects an enigma'. Today, some 42 years later, this statement is still true. In spite of much excellent work in certain areas, our knowledge of nutrient movement from host to parasite is still fragmentary and many questions remain unanswered. Moreover, very little detailed information exists on the movement of solutes into hemibiotrophic and necrotrophic fungi. This is surprising in view of the obvious importance of nutrient fluxes in infected tissues (and indeed, in whole infected plants) in controlling the growth potential of pathogenic fungi. However, because of the difficulty in separating and isolating host and fungus in infected tissues, studying the movement of nutrients and metabolites from host to pathogen is not an easy task. Nevertheless, it is clear that more detailed experiments on nutrient fluxes in infected plants are required (viz the work of Ahmad *et al.*, 1984 and Murray & Ayres, 1986) and especially on fluxes between host and parasite. Given the importance of polyamines in the growth and development of various organisms, including fungi, and the potential for controlling fungal infections by inhibiting polyamine biosynthesis, it would be useful to know more about fungal polyamine requirements in host tissue. Further, it is desirable to know whether fungal pathogens are able to take up polyamines from their hosts or whether precursors are absorbed and used in polyamine biosynthesis in fungal tissue. Thus, a study of solute fluxes between host and fungus in parasitic associations is not solely of academic interest, since a better understanding of such phenomena may be useful in the search for much needed novel crop protection agents.

References

Ahmad, I., Owera, S. A. P., Farrar, J. F. & Whitbread, R. (1982). The distribution of five major nutrients in barley plants infected with brown rust. *Physiological Plant Pathology*, **21**, 335-346.

Ahmad, I., Farrar, J. F. & Whitbread, R. (1984). Fluxes of phosphorus in leaves of barley infected with brown rust. *New Phytologist*, **98**, 361-375.

Ahmad, I., Farrar, J. F. & Whitbread, R. (1985). Membrane integrity in leaves of barley infected by brown rust: an examination using tracer efflux and *in vivo* chlorophyll fluorescence. *New Phytologist*, **99**, 107-115.

Andrews, J. H. (1975). A freeze-etch study of membranes of *Plasmodiophora*-infected and non-infected cabbage root hairs. *Canadian Journal of Botany*, **52**, 1441-1449.

Bailey, J. P., Bower, A. J. & Lewis, D. H. (1987). Localization of ornithine decarboxylase in barley leaves infected by brown rust (*Puccinia hordei*). *Transactions of the British Mycological Society*, **89**, 83-87.

Bennett, J. & Scott, K. J. (1971). Inorganic polyphosphates in the wheat stem rust fungus and in rust-infected wheat leaves. *Physiological Plant Pathology*, **1**, 185-198.

Brian, P. W. (1967). Obligate parasitism in fungi. *Proceedings of the Royal Society of London, Series B*, **168**, 101-118.

Burrell, M. M. & Lewis, D. H. (1977). Amino acid movement from leaves of *Tussilago farfar* L. to the rust, *Puccinia poarum* Neils. *New Phytologist*, **79**, 327-333.

Bushnell, W. R. (1984). Structural and physiological alterations in susceptible host tissue. In *The Cereal Rusts*, vol. 1, eds. W. R. Bushnell & A. P. Roelfs, pp. 477-507. Academic Press, Orlando.

Bushnell, W. R. & Gay, J. (1978). Accumulations of solutes in relation to the structure and function of haustoria in powdery mildews. In *The Powdery Mildews*, ed. D. M. Spencer, pp. 183-235. Academic Press, London.

Butters, J. A., Burrell, M. M. & Hollomon, D. W. (1985). Purine metabolism in barley powdery mildew and its host. *Physiological Plant Pathology*, **27**, 65-74.

Comhaire, F. (1963). Study of the absorption and repartition of phosphorus in *Triticum vulgare* Link. infected by *Erysiphe graminis*. *Leujeunia*, N. S. **15**, 1-87.

Dekhuijzen, H. M. & Staples, R. C. (1968). Mobilization factors in uredospores and bean leaves infected with bean rust fungus. *Contributions from Boyce Thompson Institute*, **24**, 39-52.

Doodson, J. K., Manners, J. G. & Myers, A. (1965). Some effects of yellow rust (*Puccinia striiformis*) on 14 carbon assimilation and translocation in wheat. *Journal of Experimental Botany*, **16**, 304-317.

Ellingboe, A. H. (1978). A genetic analysis of host-parasite interactions. In *The Powdery Mildews*, ed. D. M. Spencer, pp. 159-181. Academic Press, London.

Farrar, J. F. (1984). Effects of pathogens on plant transport systems. In *Plant Diseases: Infection, Damage and Loss*, eds. R. K. S. Wood & J. G. Jellis, pp. 87-104. Blackwell Scientific Publications, Oxford.

Fric, F. (1978). Absorption and translocation of phosphate in barley plants infected with powdery mildew. *Phytopathologische Zeitschrift*, **91**, 23-32.

Galston, A. W. (1983). Polyamines as modulators of plant development. *Bioscience*, **33**, 382-388.

Gay, J. L. (1984). Mechanisms of biotrophy in fungal pathogens. In *Plant Diseases: Infection, Damage and Loss*, eds. R. K. S. Wood & J. G. Jellis, pp. 49-59. Blackwell Scientific Publications, Oxford.

Gottlieb, D. & Garner, J. M. (1946). Rust and phosphorus distribution in wheat leaves. *Phytopathology*, *36*, 557-564.

Greenland, A. J. & Lewis, D. H. (1984). Amines in barley leaves infected by brown rust and their possible relevance to formation of 'green islands'. *New Phytologist*, **96**, 283-291.

Hancock, J. G. & Huisman, O. C. (1981). Nutrient movement in host-pathogen systems. *Annual Review of Phytopathology*, **19**, 309-331.

Hollomon, D. W. (1979). Evidence that ethirimol may interfere with adenine metabolism during primary infection of barley powdery mildew. *Pesticide Biochemistry and Physiology*, **10**, 181-189.

Hsu, S. C. & Ellingboe, A. H. (1972). Elongation of secondary hyphae and transfer of 35S from barley to *Erysiphe graminis* f.sp. *hordei* during primary infection. *Phytopathology*, **62**, 876-882.

Jäger, K. & Reisener, H. J. (1969). Untersuchungen über Stoffwechselbeziehungen zwischen Wirt und Parasit am Beispiel von *Puccinia graminis* var *tritici* auf Weizen. I. Aufnahme von Aminosäuren aus dem Wirtsgewebe. *Planta*, **85**, 57-72.

Kiraly, Z., Pozsar, B. I. & El-Hammady, M. (1966). Cytokinin activity in rust-infected plants: Juvenility and senescence in diseased leaf tissues. *Acta Phytopathologica Academiae Scientiarum Hungaricae*, **1**, 29-38.

Kiraly, Z., El-Hammady, M. & Pozsar, B. I. (1967). Increased cytokinin activity of rust infected bean and broad bean leaves. *Phytopathology*, **57**, 93-94.

Kneale, J. & Farrar, J. F. (1985). The localization and frequency of haustoria in colonies of brown rust on barley leaves. *New Phytologist*, **101**, 495-505.

Lea, P. J. & Miflin, B. J. (1980). Transport and metabolism of asparagine and other nitrogen compounds within the plant. In *The Biochemistry of Plants*, vol. 5, ed. B. J. Miflin, pp. 569-607. Academic Press, New York.

Livne, A. & Daly, J. M. (1966). Translocation in healthy and rust affected beans. *Phytopathology*, **56**, 170-175.

Manners, J. M. & Gay, J. L. (1983). The host-parasite interface and nutrient transfer in biotrophic parasitism. In *Biochemical Plant Pathology*, ed. J. A. Callow, pp. 163-195. John Wiley & Sons, Chichester.

Martin, T. J. & Ellingboe, A. H. (1978). Genetic control of ^{32}P transfer from wheat to *Erysiphe graminis* f.sp. *tritici* during primary infection. *Physiological Plant Pathology*, **13**, 1-11.

Mendgen, K. (1979). Microautoradiographic studies on host-parasite interactions. II. The exchange of ^{3}H-lysine between *Uromyces phaseoli* and *Phaseolus vulgaris*. *Archives of Microbiology*, **123**, 129-135.

Mendgen, K. (1981). Nutrient uptake in rust fungi. *Phytopathology*, **71**, 983-989.

Mount, M. S. & Ellingboe, A. H. (1969). ^{32}P and ^{35}S transfer from susceptible wheat to *Erysiphe graminis* f.sp. *tritici* during primary infection. *Phytopathology*, **59**, 235.

Murray, A. J. S. & Ayres, P. G. (1986). Uptake and translocation of nitrogen by mildewed barley seedlings. *New Phytologist*, **104**, 355-365.

Onoe, T., Tani, T. & Haito, N. (1973). The uptake of labelled nucleosides by *Puccinia coronata* grown in susceptible oat leaves. *Report of the Tottori Mycological Institute (Japan)*, **10**, 303-312.

Paul, N. D. & Ayres, P. G. (1988). Nutrient relations of groundsel (*Senecio vulgaris* L.) infected by rust (*Puccinia lagenophorae* Cooke.) at a range of nutrient concentrations. I. Concentrations, contents and distribution of N, P and K. *Annals of Botany*, **61**, 489-498.

Piening, L. J. (1972). Effects of leaf rust on nitrate in rye. *Canadian Journal of Plant Science*, **52**, 842-843.

Pozsar, B. I. & Kiraly, Z. (1966). Phloem transport in rust-infected plants and cytokinin directed long distance movement of nutrients. *Phytopathologische Zeitschrift*, **56**, 297-309.

Raggi, V. (1974). Free and protein amino acids in the pustules and surrounding tissues of rusted bean. *Phytopathologische Zeitschrift*, **81**, 289-300.

Raggi, V. (1976). Amino acids in mycelium of *Sphaerotheca pannosa* var. *persicae* and in the infected and surrounding tissues of peach leaves. *Phytopathologia Mediterranea*, **15**, 110-114.

Rajam, M. V., Weinstein, L. H. & Galston, A. W. (1985). Prevention of a plant disease by specific inhibition of fungal polyamine biosynthesis. *Proceedings of the National Academy of Science, USA*, **82**, 6874-6878.

Roberts, A. M. (1987). Photosynthesis, nitrogen metabolism and shoot: root equilibria in leeks infected with the rust, *Puccinia allii* Rud. Ph.D. Thesis, University of Glasgow.

Roberts, A. M. & Walters, D. R. (1988a). Shoot: root interrelationships in leeks infected with the rust, *Puccinia allii* Rud.: growth and nutrient relations. *New Phytologist*, (in press).

Roberts, A. M. & Walters, D. R. (1988b). Nitrogen assimilation and metabolism in rusted leek leaves. *Physiological Plant Pathology*, **32**, 229-235.

Sadler, R. & Scott, K. J. (1974). Nitrogen assimilation and metabolism in barley leaves infected with the powdery mildew fungus. *Physiological Plant Pathology*, **4**, 235-247.

Scott, K. J. (1972). Obligate parasitism by phytopathogenic fungi. *Biological Reviews*, **47**, 537-572.

Shaw, M. & Samborski, D. J. (1956). The physiology of host-parasite relations. I. The accumulation of radioactive substances at infections of facultative and obligate parasites including tobacco mosaic virus. *Canadian Journal of Botany*, **34**, 389-405.

Shaw, M. & Hawkins, A. R. (1958). The physiology of host-parasite relations. V. A preliminary examination of the level of free endogenous indoleacetic acid in rusted and mildewed cereal leaves and their ability to decarboxylate exogenously supplied radioactive indoleacetic acid. *Canadian Journal of Botany*, **36**, 1-16.

Slesinski, R. S. & Ellingboe, A. H. (1971). Transfer of ^{35}S from wheat to the powdery mildew fungus with compatible and incompatible parasite/host genotypes. *Canadian Journal of Botany*, **49**, 303-310.

Slocum, R. D., Kaur-Sawhney, R. & Galston, A. W. (1984). The physiology and biochemistry of polyamines in plants. *Archives of Biochemistry and Biophysics*, **235**, 283-303.

Spencer-Phillips, P. T. N. & Gay, J. L. (1981). Plasma membrane ATP-ase domains and transport through infected cells. *New Phytologist*, **89**, 393-400.

Thrower, L. B. (1965). Host physiology and obligate fungal parasites. *Phytopathologische Zeitschrift*, **52**, 319-334.

Walters, D. R. (1985). Shoot: root interrelationships: the effects of obligately biotrophic fungal pathogens. *Biological Reviews*, **60**, 47-79.

Walters, D. R. (1986). The effects of a polyamine biosynthesis inhibitor on infection of *Vicia faba* L. by the rust fungus, *Uromyces viciae-fabae* (Pers.) Schroet. *New Phytologist*, **104**, 613-619.

Walters, D. R. (1987). Polyamines: the Cinderellas of Cell Biology. *Biologist*, **33**, 73-76.

Walters, D. R. & Ayres, P. G. (1981). Phosphate uptake and transport by roots of powdery mildew infected barley. *Physiological Plant Pathology*, **18**, 195-205.

Walters, D. R., Wilson, P. W. F. & Shuttleton, M. A. (1985). Relative changes in levels of polyamines and activities of their biosynthetic enzymes in barley infected with the powdery mildew fungus, *Erysiphe graminis* D.C. ex. Merat f.sp. *hordei* Marshal. *New Phytologist*, **101**, 695-705.

Walters, D. R. & Wylie, M. A. (1986). Polyamines in discrete regions of barley leaves infected with the powdery mildew fungus, *Erysiphe graminis*. *Physiologia Plantarum*, **67**, 630-633.

Ward, H. M. (1901). The bromes and their rust fungus (*Puccinia dispersa*). *Annals of Botany*, **15**, 560-562.

Ward, H. M. (1902). Experiments on the effect of mineral starvation on the parasitism of the uredine fungus, *Puccinia dispersa*, on species of *Bromus*. *Proceedings of the Royal Society of London*, **71**, 138-151.

Weinstein, L. H., Osmeloski, J. F., Wettlaufer, S. H. & Galston, A. W. (1987). Protection of wheat against leaf and stem rust and powdery mildew diseases by inhibition of polyamine metabolism. *Plant Science*, **51**, 311-316.

West, H. M. M. & Walters, D. R. (1988). Novel control of fungal plant diseases using inhibitors of polyamine biosynthesis. *Crop Research*, **28**, 97-108.

Woods, A. M. & Gay, J. L. (1983). Evidence for a neckband delimiting structural and physiological regions of the host plasmalemma associated with haustoria of *Albugo candida*. *Physiological Plant Pathology*, **23**, 73-88.

Yarwood, C. E. & Jacobson, L. (1955). Accumulation of chemicals in diseased areas of leaves. *Phytopathology*, **45**, 43-48.

Zaki, A. I. & Durbin, R. D. (1965). The effect of bean rust on the translocation of photosynthate from diseased leaves. *Phytopathology*, **55**, 528-529.

Chapter 8

Effects of fungal pathogens on nitrogen, phosphorus and sulphur relations of individual plants and populations

N. D. Paul

Institute of Environmental and Biological Sciences, University of Lancaster, Bailrigg, Lancaster LA1 4YQ, UK

Introduction

Despite the crucial importance of nitrogen, phosphorus and sulphur to all aspects of plant growth, development and reproduction, our knowledge of the interactions of pathogenic fungi and these nutrients remains extremely limited. Indeed, until recently almost all studies had been concerned with pathogens of the major temperate cereals, wheat and barley. Furthermore, with the notable exception of a few detailed studies of the root pathogens of wheat (e.g. Fitt & Hornby, 1978) data on fungal pathogens other than powdery mildews and rusts were, and still are, virtually absent. Despite this comparative paucity of data, it is clear that fungal infections may have profound consequences for the nutrient relations of their hosts. It is also increasingly evident that many of the observed changes in whole plant nutrient relations following infection, by biotrophic pathogens at least, form part of an integrated response which is, in many ways, no different from responses to other environmental stresses. In considering the effects of pathogens on nitrogen, phosphorus and sulphur, I will make use of our understanding of the responses of uninfected plants to abiotic stress. The use of such data is valuable both in allowing an insight into mechanisms operating in infected plants and in considering the many aspects of nutrient relations where the effects of pathogens remain poorly understood.

In this chapter I will consider the effects of pathogens on the nutrient contents and concentrations of the host (whole plant and both infected and uninfected organs) and on the uptake of nutrients by the roots of infected plants, interactions between populations of hosts and pathogens will also be discussed. The lack of information on necrotrophic and hemibiotrophic infections makes it inevitable that greater emphasis will be placed on biotrophic infections, principally powdery mildews and rusts, and, equally inevitably, I will discuss nitrogen and phosphorus in greater detail than sulphur, which has rarely been studied even in the better worked systems.

Nutrient relations of plants infected with biotrophic pathogens

Changes in contents and concentrations

There are few reports of the contents (i.e. quantity per plant or per organ) and concentrations (i.e. quantity per unit weight) of sulphur in plants infected with powdery mildews. Yarwood & Jacobson (1954) found that exogenously applied ^{35}S accumulated at the site of infection in some mildew infections but not in others. No change in sulphur-containing amino acids was found in peach (*Prunus persicae*) infected with *Sphaerotheca pannosa* (Raggi, 1976). Changes in phosphorus relations are better understood, at least in young barley (*Hordeum vulgare*) plants infected with *Erysiphe graminis* f. sp. *hordei* where phosphorus concentrations have invariably been found to increase in infected leaves (Comhaire, 1963; Fric, 1978; Walters & Ayres, 1981a) and root tissues (Walters & Ayres, 1981a). Powdery mildews appear to have little effect on total nitrogen concentrations in infected barley leaves (Fric, 1978; Walters & Ayres, 1980). However, nitrate concentrations have generally been found to decrease in mildewed leaves (Walters & Ayres, 1980) and reduced nitrogen (free- and protein-amino acids and free ammonium) to increase (Raggi, 1976; Walters & Ayres, 1980; Murray & Ayres, 1986). There is no consistent evidence that free ammonium contributes significantly to elevated concentrations of reduced-nitrogen (see Murray & Ayres, 1986) and it is amino acids which show the most consistent and pronounced increase following infection, although the responses of individual amino acids exhibit substantial variation (see Walters, Chapter 7).

The nutrient relations of rust-infected plants have been investigated in detail in only three systems: brown rust (*Puccinia hordei*) of barley seedlings up to the six leaf stage (Ahmad *et al.*, 1982), *P. lagenophorae* infected groundsel (*Senecio vulgaris*) throughout its development (Paul & Ayres, 1988a & b) and *P. allii* on young leek (*Allium porrum*) plants (Roberts & Walters, 1988a & b). Sulphur was not studied in any of these systems, but in a survey of a number of rusts, Yarwood & Jacobson (1954) found accumulation of exogenously applied ^{35}S in the pustule up to 19 times that in uninfected tissues. Raggi (1974) found accumulation of sulphur-containing amino acids in rust pustules on bean (*Phaseolus vulgaris*) but increases were confined to free amino acids, those incorporated in proteins showed some decline. Tissues around sporulating rust pustules were often found to be depleted of sulphur (Yarwood & Jacobson, 1954), perhaps because the quantity of sulphur-containing amino acids was reduced (Raggi, 1974).

In leek plants, rust infection resulted in increases in the total content of nitrogen and phosphorus, despite the lower dry weight of infected plants (Roberts & Walters, 1988b). In groundsel and barley the total amount of phosphorus per plant was depressed by rust infection and in the latter nitrogen content was also reduced (Ahmad *et al.*, 1982); in groundsel there was no significant change in nitrogen (Paul & Ayres, 1988a). However, reductions in plant dry weight were greater than any decline in nutrient content, with the result that concentrations were increased. In leek and groundsel, elevated phosphate concentration was most pronounced in, but not confined to, infected tissues, while in barley, phosphorus concentration increased more than two-fold in infected leaves but was reduced in uninfected tissues (Ahmad *et al.*, 1982). Indeed, phosphorus accumulation at sites of rust infection is known to occur in a number of rusts, partly as organic phosphate but primarily as polyphosphate. The relative importance of these compounds is discussed by Walters (Chapter 7).

In contrast to phosphorus concentration, the concentration of Kjeldahl-nitrogen in rusted barley leaves was reduced (Ahmad *et al.*, 1982) but in groundsel (Paul & Ayres, 1988a) and leek (Roberts & Walters, 1988b) the concentration of Kjeldahl-nitrogen was increased in infected leaves. More detailed analysis of the form of nitrogen in rusted leek suggested that, since free ammonium-nitrogen decreased, much of the observed increase in reduced nitrogen was due to amino acids (Roberts & Walters, 1988a). In leek, and in the leaves of rye infected with *Puccinia recondita* (Peining, 1972) the concentration of nitrate also increased. Nitrogen concentrations in the uninfected tissues of rusted plants were found to increase in all three species (Ahmad *et al.*, 1982; Paul & Ayres, 1988a; Roberts & Walters, 1988a). The form of nitrogen accumulated in uninfected tissues is not known, and indeed, the expression of nutrient concentrations on the basis of dry weight may be misleading as they may reflect reductions in the percentage dry weight of tissues rather than any change of nutrient concentration in the aqueous phase. It is known that dry weight per unit fresh weight is often reduced in rust-free tissues on infected plants, due, for example, to reduced carbohydrate contents. In groundsel the percentage dry weight content of the roots may be almost halved in heavily rusted plants (e.g. 6·2% at the onset of flower-production as opposed to 11·8% in controls; N. D. Paul, unpublished). If such changes occurred in the experiment in which nutrient relations were studied, the observed increases in nitrogen and phosphorus concentrations per unit dry weight could have occurred even if concentrations per unit fresh weight were substantially reduced by infection. Changes in dry weight content per unit fresh weight may also explain reports of elevated nutrient concentrations per unit dry weight in the uninfected tissues of plants infected with powdery mildew (see above) and have also been invoked to explain increased phosphorus concentration in plants infected with mycorrhizas (Stribley, Tinker & Rayner, 1980).

In biotroph-infected tissues dry weight per unit fresh weight increases and hence increased nitrogen and phosphorus concentrations expressed on a dry weight basis will underestimate changes in concentrations in the aqueous phase. Accumulations in infected tissues reflect concentrations both in host tissues and in the parasite itself and may result from increased import and/or decreased export. These mechanisms and the distribution of nutrients between host and parasite are discussed by Walters (Chapter 7). While such changes are central to increased nutrient concentrations in infected tissues they obviously cannot account for the increased nutrient concentrations of the whole plant which often follows infection (see above): this requires increases in the uptake of nutrients by the roots of rusted or mildewed plants.

Nutrient uptake

In most studies of the effects of plant pathogens, nutrient uptake by the host has been derived, by growth analysis, from differences in nutrient contents at successive harvests. In diseased plants, one limitation of this technique is the unknown loss of nutrients as spores, which is not easily quantified over long periods. Ahmad *et al.*, (1982) found that urediospores of *Puccinia hordei* were rich in phosphorus and nitrogen and while losses to spores may be of little importance in short term studies (e.g. Murray & Ayres, 1986), investigations extending over a number of weeks (e.g. Paul & Ayres, 1988a & b; Roberts & Walters, 1988b) are inevitably liable to underestimate actual uptake. Perhaps the ideal method of studying the nutrient relations of infected plants is the flowing nutrient solution technique (e.g. Clement *et al.*, 1974) which enables long-term uptake to be measured by extraction from the medium, avoiding errors due to loss of spores, as well as by growth analysis. Unfortunately, to date this method has not been applied to plants with foliar infections and in the following discussion of uptake the possibility of underestimation should be borne in mind.

Studies of the effects of powdery mildews on nitrogen uptake have largely been confined to one system, *Erysiphe graminis* on barley. While ammonium uptake was unaffected by mildew infection (Walters & Ayres, 1980), the rate of nitrate uptake by the roots of mildewed barley was inhibited before the onset of sporulation (Walters & Ayres, 1980; Murray & Ayres, 1986) and stimulated after sporulation (Murray & Ayres, 1986). The rates of phosphate uptake by the roots of mildewed plants increased in wheat (Comhaire, 1963) as well as barley (Walters & Ayres, 1981a, 1982) although in the latter Fric (1978) reported decreased phosphorus uptake in heavily infected seedlings. Rust infection in groundsel and leek increased the rates of uptake of both phosphorus and nitrogen when expressed on the basis of root dry weight (Paul & Ayres, 1988a; Roberts & Walters, 1988b). Although no stimulation of phosphate uptake was observed in barley infected with *Puccinia hordei* (Ahmad *et al.*, 1982), the rate of uptake of nitrogen was enhanced.

The range of responses in the uptake of different ions to infection in the various pathosystems studied has discouraged any unified explanation of the underlying mechanisms, especially since the effects may differ at different times within the same system. However, an overall view shows that any decrease in phosphorus or nitrogen (nitrate or ammonium) uptake by plants infected with foliar biotrophs is the exception rather than the rule (Table 8.1), while the uptake of cations such as potassium seems to be invariably increased. A plethora of mechanisms have been proposed for specific ions in particular systems (see Walters, 1985) but, as has often been noted, increased rates of uptake present a dilemma in that they are concurrent with reduced assimilate supply to the roots of infected plants (see Ayres, 1984), a relationship not easily reconciled with the energy requirements of active ion uptake. This anomaly might be explained through a coordinated response of ion uptake to alterations in plant growth regulators (PGRs) in the roots of infected plants (e.g. Ahmad *et al.*, 1982; Walters, 1985). However, our knowledge of the effects of endogenous PGRs on nutrient uptake in healthy plants is rather patchy and superimposed on this is an almost complete absence of reliable data of pathogen-induced changes in these compounds. Thus, while the involvement of PGRs cannot be excluded, any evidence for such a role is, for the present at least, tenuous. The effects of reduced assimilate might be buffered in the roots of infected plants by changes in partitioning between 'growth' and 'maintenance' respiration (Penning de Vries, 1975). Growth respiration has been shown to be more sensitive to reduced assimilate supply than maintenance in healthy plants and shoot infections are known to reduce not only root dry weight accumulation but also cell division in the root tip (Martin & Hendrix, 1974; Walters & Ayres, 1981b). Furthermore, respiration in the roots appears to be largely via the energetically inefficient 'alternative-pathway' (Lambers, 1980) and it is possible that this might be disengaged in infected plants, as it is under conditions of environmental stress, so improving the efficiency with which available assimilate is utilised.

Table 8.1. Summary of the effects of rusts and powdery mildews on the nitrogen and phosphorus relations of the host.

		Host and parasite		
		barley/mildew	barley/rust	leek/rust
Content per plant	nitrogen	increased[1]	decreased[5]	increased[7]
	phosphorus	increased[2]	decreased[5]	increased[7]
Concentration in infected tissues	nitrogen	unchanged[3]	decreased[5]	increased[7,8]
	phosphorus	increased[2]	increased[5,6]	increased[7]
Concentration in uninfected tissues	nitrogen	increased[3]	increased[5]	increased[6]
	phosphorus	increased[2]	decreased[5]	increased[7]
Rate of uptake	nitrogen	incr[1],decr[2,3]	increased[5]	increased[7]
	phosphorus	incr[2],decr[4]	unchanged[5,6]	increased[7]

References: 1, Walters & Ayres, 1980; 2, Murray & Ayres, 1986; 3, Walters & Ayres, 1981a; 4, Fric, 1978; 5, Ahmad *et al.*, 1982; 6, Ahmad *et al.*, 1984; 7, Roberts & Walters, 1987; 8, Roberts & Walters, 1988.

The mechanisms through which changes in nutrient uptake occur in pathogen infected plants may best be considered in the light of recent ideas concerning uptake in intact, healthy plants. That uptake from solutions of low nutrient concentrations can be described in terms equivalent to the Michaelis-Menten equation derived for enzyme kinetics has been known for over 30 years (see Clarkson, 1974). It is also well established that the kinetics of uptake at higher concentrations are more complex but may be described as the sum of two or more terms, each showing Michaelis-Menten type kinetics (Clarkson, 1974). However, despite the widespread acceptance of models based on multiphasic active uptake, there is also a range of evidence which reveals the existence of a linear component to uptake and Borstlap (1981) has shown that many examples of apparently multiphasic uptake are better described as the sum of one (or two) Michaelis-Menten type term(s) and a linear component. Since the saturable uptake mechanism reaches its maximum rate at relatively low external ion concentrations, the linear term assumes increasing importance as concentrations increase but may account for about 30% of uptake even at low concentrations (e.g. Kochian & Lucas, 1982). In considering mechanisms underlying the observed increases in ion uptake in infected plants it is worth noting that all existing measurements were made on plants grown at high nutrient concentrations, in nutrient solutions, sand culture or, in the case of Roberts & Walters (1988b), a peat potting compost. In all cases the ion concentrations were much higher than that at which the saturable component reaches its maximum and, especially in view of the growth media used, passive uptake would be expected to represent a major component of the total. The two uptake mechanisms show very different properties. From the point of view of the potential effects of pathogens on uptake, perhaps the most important difference is that the mechanism which exhibits Michaelis-Menten kinetics is active, being critically dependent on the energy status of the roots whereas the linear, non-saturable component of uptake is largely, but not entirely independent of energy supply (Kochian & Lucas, 1982). Active uptake may be regulated by ion concentrations in the external medium or in the root tissue (see Drew & Saker, 1984). However, in intact plants where both active and passive uptake is occurring, regulatory mechanisms appear more complex and depend on the nutrient status of the whole plant not just the roots. Drew & Saker (1984) have suggested that the limiting step is the control of ion release from the symplast to the xylem, which may be influenced not only by water flux to the xylem, and hence by transpirational flux (see Pitman, 1982), but also by nutrient concentrations in the phloem. Concentrations in the phloem might in turn be determined by retranslocation from the shoot and, with transpiration, provides a means by which root and shoot activity might be integrated.

An insight into the seemingly contradictory changes in ion uptake and assimilate supply in infected plants may come from studies with healthy plants, particularly those investigating the effects of low root temperatures on nutrient uptake (e.g. MacDuff *et al.*, 1986; Clarkson, Hopper & Jones, 1986; White, Clarkson & Earnshaw, 1987), which have highlighted the capacity of the plant to adapt to inhibition of root metabolism and the potential importance of passive uptake in this response. Active uptake shows a temperature dependence similar to that of many metabolic systems and in this sense low root temperature is analogous to the reduced assimilate supply to the roots of infected plants, Furthermore, intact plants grown with the root temperature lower than that of the shoots show progressive increases in the shoot: root ratio (S:R), as do plants infected with foliar pathogens (Ayres, 1984). The magnitude of the interaction between the effects of temperature on uptake and on S:R have been elegantly demonstrated by Clarkson *et al.* (1986). During an initial period after its imposition, low root temperature suppressed the rates of uptake of both nitrogen and phosphorus (Fig. 8.1) as expected from metabolic considerations. However, the duration of this phase was short (less than 9 days) and was proscribed by the low-temperature induced increase in S:R. Once adjustments of S:R had oc-

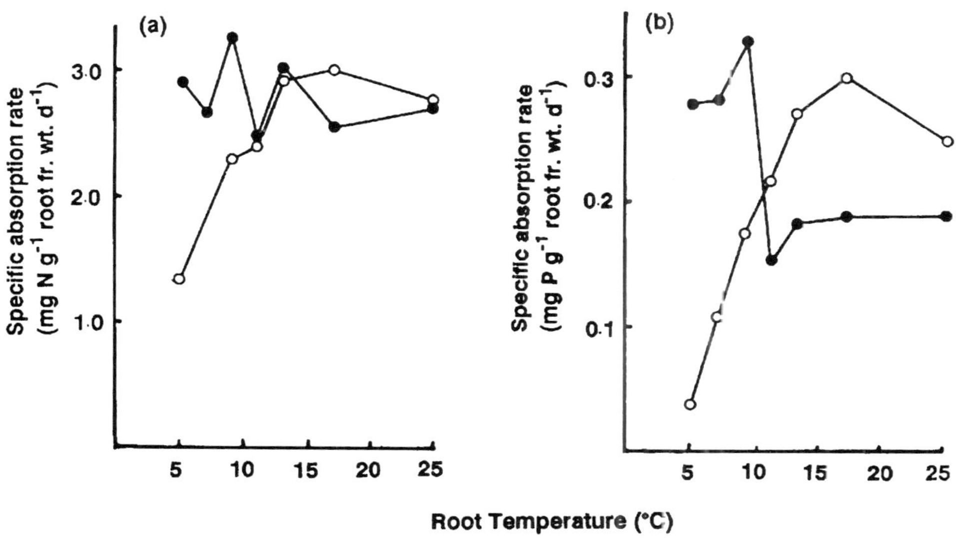

Fig. 8.1. Changes in the rates of uptake of (a) nitrogen and (b) phosphorus in barley grown at a range of root temperatures. In the initial period following the imposition of low root temperatures (open symbols) rates of uptake decline with temperature. However, after longer exposure to low root temperature (closed symbols) the ratio of shoot:root increased and the rates of uptake became insensitive to temperature. Modified from Clarkson *et al.* (1986).

curred, root temperatures down to 5°C caused no reduction in either nitrogen or phosphorus uptake (Fig. 8.1) and the linear, passive component of uptake assumed primary importance. As Clarkson *et al.* (1986) state 'the adjustment of root size ... had a profound effect on the rate of ion absorption per unit root weight'. The increase in S:R results in a requirement for each unit of root to supply relatively more resources to the shoot, i.e. 'shoot demand' on the roots is increased (see Drew & Saker, 1984; Clarkson *et al.*, 1986; White *et al.*, 1987).

The relationship between S:R and uptake per unit root dry weight (SAR_W) was investigated in rust-infected groundsel (Paul & Ayres, 1988b) where a close positive correlation was found to exist for both phosphorus and nitrogen (Fig. 8.2). Such results, together with those from healthy plants, suggest that the frequently observed increases in SAR_Ws in infected plants may result from increased shoot demand, and hence that it is the passive component of uptake which is of paramount importance. Thus, increased nutrient uptake per unit root would be the expected outcome whenever foliar pathogens preferentially inhibit root growth. On this basis, it is those cases where uptake is inhibited by shoot infections that raise the most intriguing questions. However, variations in shoot demand may be based on more complex mechanisms than changes in S:R, as is evident, for example, in the deviation of the S:R—SAR_W relationship from linear in the case of nitrogen (Fig. 8.2a) and increased shoot demand as the basis of increased uptake subsumes a number of component mechanisms in addition to S:R. Two such mechanisms are modified recycling of nutrients from shoot to root and altered transpirational fluxes. The role of decreased shoot-root recycling in increased phosphorus and nitrogen uptake in mildewed barley has been acknowledged before (Walters & Ayres, 1981a; Murray & Ayres, 1986). Increased shoot demand for water, especially in rusts where transpiration is stimulated after sporulation, may be especially effective in increasing ion uptake if the hydraulic conductivity of the root system is increased (White *et al.*, 1987), as is indeed the case in mildewed barley (Walters & Ayres, 1982) and rusted beans (Tissera & Ayres, 1988).

Studies of healthy plants emphasise the potential importance of root morphology as well as physiology in responses to increased

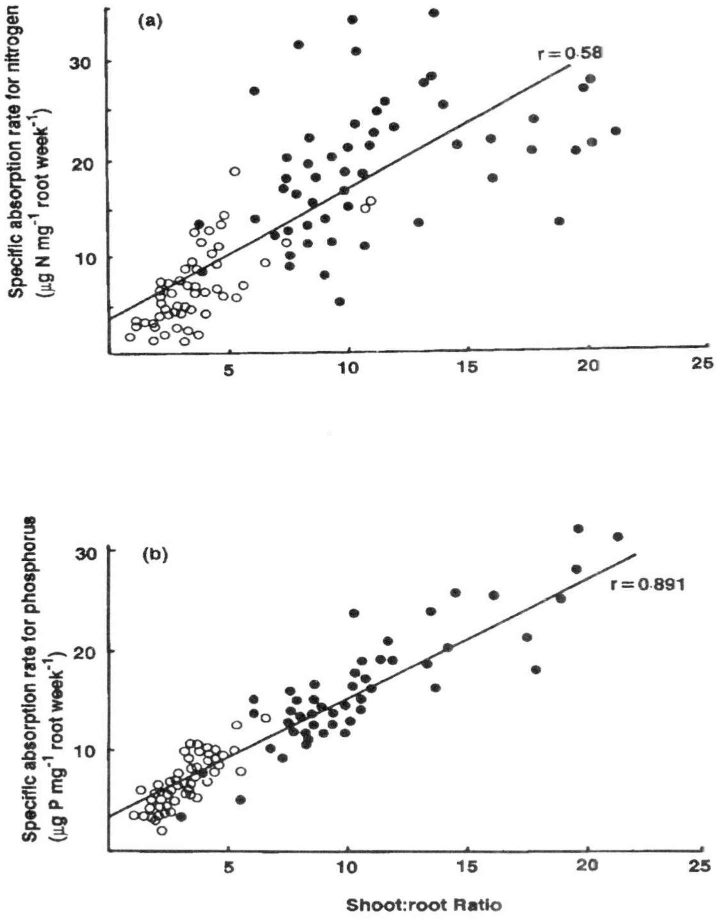

Fig. 8.2. The relationship between shoot:root ratio (S:R) and the rate of uptake of (a) nitrogen and (b) phosphorus on the basis of root dry weight in groundsel infected with *Puccinia lagenophorae* (closed symbols) or uninoculated (open symbols). The wide range of S:R were obtained by growing plants at different dilutions of a nutrient solution. The relationships between S:R and rates of uptake were not changed by infection. Thus, while the rates of uptake of both nitrogen and phosphorus were increased in infected plants, this increase was no greater than expected from their higher shoot:root ratio. Redrawn from Paul & Ayres (1988b).

shoot demand (MacDuff *et al.*, 1986). In rusted wheat (Martin & Hendrix, 1974), bean (Tissera & Ayres, 1988) and groundsel (Fig. 8.3 and see Paul & Ayres, 1986) the length of root per unit dry weight (specific root length, SRL) has been found to increase. In the last system, increased SRL was the key mechanism underlying increases in SAR_W, since if the rates of uptake of nitrogen and phosphorus were expressed on the basis of length no rust-induced increases were apparent (Paul & Ayres, 1988b). To date no study has examined the influence of powdery mildews on SRL, although reductions in root radius have been reported (Walters, 1981). While changes in root morphology are clearly important for plants in solution or sand culture, their significance is likely to be even greater for plants growing in soil in the field (see below).

Clearly, changes in shoot demand as an explanation for the effects of pathogens on nutrient uptake give scope for the complex patterns of change which have been observed, since it encompasses an integrated and dynamic response in both the physiology and morphology of the plant to the effects of infection. It remains to be seen whether increased shoot demand is of general importance in plants infected with foliar pathogens, especially as most of the hosts studied to date share the same nutritional strategy, that of species adapted to nutrient rich habitats (see Chapin, 1980). Leek (Roberts & Walters, 1988b) is an exception as it appears to share much in common with slow-growing wild species adapted to conditions of comparative nutrient poverty (Chapin, 1980). One outstanding feature of leek when infected by rust is the stability of root growth (Roberts & Walters, 1988b). This may reflect the alternative strategy in which S:R varies little in response to environmental stimuli (see Chapin, 1980). However, that changes in nutrient uptake in leek following rust infection were broadly similar to those of the other species studied may indicate that the pattern of responses to infection already reported may be common to a wide range of wild, as well as crop species.

Nutrient relations of plants infected with necrotrophic pathogens

Investigations of changes in nitrogen, phosphorus and sulphur caused by necrotrophs have been confined to pathogens infecting the roots or vascular tissues. For example, root infection by

Phytophthora cinnamomi was found to decrease concentrations of sulphur in the shoots of *Isopogon ceratophylla* (Weste *et al.*, 1980) and *Eucalyptus* spp. (Cahill *et al.*, 1986) and of phosphorus and sulphur in the leaves of Avocado (Whiley *et al.*, 1987). Severe take-all disease of cereals, which is caused by root infection by *Gaeumannomyces graminis*, also causes reductions in shoot phosphate (e.g. Fitt & Hornby, 1978). Such reductions in shoot concentrations may seem the inevitable consequence of root decay. However, in many cases visual assessment of root symptoms are unreliable guides to changes in root function as the cortical tissues of older roots may be lost without impairment of

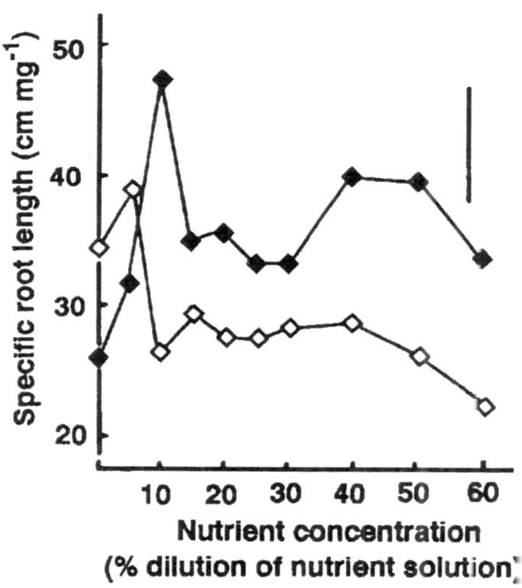

Fig. 8.3. The effect of *Puccinia lagenophorae* on the specific root length (SRL, root length per unit root dry weight) of groundsel grown over a wide range of dilutions of a standard nutrient solution. The SRL of rusted plants (closed symbols) was substantially higher than that of controls (open symbols) at all but extremely low nutrient concentrations. Redrawn from Paul & Ayres (1986a). Points are means of five plants and the bar represents the LSD at $P = 0{\cdot}05$.

root function (Clarkson & Sanderson, 1971). In wheat a number of pathogens which caused extensive discoloration of the root cortex had no effect on shoot phosphate concentrations (Fitt & Hornby, 1978). Similarly, *G. graminis* causes no reduction in phosphate uptake, despite extensive discoloration of the cortex, until phloem tissues are colonised (Clarkson *et al.*, 1975; Fitt & Hornby, 1978). Once the phloem is invaded by *G. graminis*, regions of the root distal to the site of infection rapidly show decreased phosphorus uptake as a result of the acute assimilate starvation of these tissues (Fitt & Hornby, 1978). The cessation of phloem transport beyond sites of infection by *G. graminis* (or by *Aureobasidium bolleyi*) is very different from the chronic assimilate deprivation found in roots of plants with biotrophic infections of the shoot, and many of the compensatory mechanisms discussed above are unlikely to occur. However, some compensation does take place in regions proximal to *G. graminis* where phosphorus uptake may increase slightly (Clarkson *et al.*, 1975), perhaps because shoot demand on these still functioning tissues is increased. When take-all is mild, compensatory increases in uptake by uninfected roots may buffer the plant from deleterious changes occurring in infected roots (Hornby & Goring, 1972). Compensatory changes in pathogen-free tissues may not occur in all root infections, for example, *Cochliobolus sativus* and *Fusarium culmorum*, caused decreased concentrations of phosphorus in wheat shoots in the absence of phloem colonization (Fitt & Hornby, 1978), possibly because uptake was impaired by phytotoxins.

Nutrient uptake and the growth and development of root systems

In nature it is often limitations on nutrient flux in the soil, rather than in the root, which is the paramount factor limiting uptake, especially of the more immobile ions, such as phosphate (Nye, 1977; Robinson, 1986). Thus, in soil those physiological characteristics of the root discussed above, which stimulated nutrient uptake by rusted or powdery mildew-infected plants in solution or sand culture may be less important than morphological characteristics, both of individual roots and of the complete root system, which allow the plant to explore the maximum volume of soil most efficiently (Fitter, 1987).

Of the morphological features of individual roots, the diameter is perhaps the most important, influencing the efficiency of uptake through effects on both gross exploitation of the soil and its cost, in terms of dry weight investment (Fitter, 1987). Nutrient inflow per unit root length is greater in roots with greater diameters but conversely the volume of soil exploited by a given volume (or weight) of root is greater for finer roots (Fitter, 1987). Thus, especially for immobile ions, the return for a given investment of dry weight is greatest when root radius is low (when SRL is high). The reductions in root diameter, or increase in SRL, found in plants infected with rusts or mildews have already been described (Fig. 8.3). Thus, infected plants may be able to compensate for the limited assimilate available for root growth by enhancing the efficiency with which dry weight is exploited for ion uptake. The effective root diameter for uptake, especially of immobile ions, may be increased by the production of root hairs or by mycorrhizal associations which increase the zone of soil exploited with comparatively small investments in dry weight (Clarkson, 1985). The length of root hairs has been found to increase in plants subjected to low root temperatures (MacDuff *et al.*, 1986) and might also be expected to form part of a complex of adaptions in root morphology in response to foliar biotrophs. However, the effects of foliar pathogens on root hairs remain wholly uninvestigated.

The absence of any information concerning interactions between foliar pathogens and mycorrhizas is another major lacuna in our understanding of nutrient relations of infected plants, especially in view of the increasing awareness of the key role of mycorrhizas in natural systems (see Chapters 9, 10 & 11). Mycorrhizas are known to be sensitive to reduced photosynthesis, for example due to low light intensities, indeed mycorrhizal infection may reduce plant growth when irradiance is low and soil phosphorus content high (e.g. Son & Smith, 1988). Foliar pathogens, like low irradiance, will reduce the supply of photosynthate to the roots and hence it might be expected that shoot infection would reduce or reverse the normal stimulation of plant growth resulting from mycorrhizal infection. Indeed, the effects of foliar parasites may be greater than those of reducing light intensity, since infection sites and mycorrhizas will represent competing sinks for the limited supply of photoassimilate.

While it is the morphology of individual roots which has the greatest influence on immobile ions, it is the architecture of whole root systems which is paramount for mobile nutrients such as nitrate (Fitter, 1987). Root system architecture is complex and as yet poorly understood, despite major advances in its study in recent years (Fitter, 1987). Architecture is determined by metric components, such as the frequency and length of branches and branch angle, and non-metric aspects or topology (Fitter, 1987). Analysis of root branching patterns in infected plants are limited to two concerning powdery mildew of barley and one with *Puccinia recondita tritici* on wheat and, in all three, only metric components of root architecture were considered. In the former, Visarova & Minarcic (1974) found that the seminal roots were shorter in infected plants than in controls and Walters & Ayres (1981b) found that the lengths of root axes and primary laterals were reduced in both the nodal and seminal roots of mildewed seedlings but that the length of secondary laterals was substantially increased. While the total production of root branches was found to be reduced in both studies, it was evident from the results of Walters & Ayres (1981b) that the number of branches per centimetre was greater in infected than control plants. This more compact and relatively more branched root system may compromise the ability of mildewed plants to absorb efficiently mobile nutrients, such as nitrate, in the field. By contrast, stripe rust reduced branching in wheat roots both in seedlings and mature plants, especially in proximal parts of the root system (Martin & Hendrix, 1974). Thus, it would seem that in rusted wheat the limited resources available for root growth are directed towards maximising the volume of soil explored. The contrasting changes in the root branching patterns of plants infected with rusts and powdery mildews is a paradox but may reflect a greater demand of rusted plants for mobile soil resources, water as well as mobile nutrients. Clarification of such changes awaits the application of newly available techniques of analysis to a wider range of host-parasite combinations.

Finally, it must be noted that superimposed on the nutrient foraging strategy of the whole root system is the capacity of roots to proliferate locally in areas of nutrient enrichment (e.g. Robinson & Rorison, 1983). This capacity may be critical in soil, where areas of localised enrichment often occur against a low background nutrient supply, and may represent an alternative strategy to decreased S:R (Crick & Grime, 1987). Such mechanisms remain poorly understood even in healthy plants but preliminary studies have shown that, when grown with the root system divided between separate containers of nutrient-rich and nutrient-poor media, rusted groundsel showed preferential root growth in the former so that root dry weight was 3·7 times that in the latter (N. D. Paul, unpublished). Control groundsel showed less differential root growth (2·6:1) but a lower S:R (3·4 as opposed to 4·2). Thus, rusted groundsel was able to compensate for an increased S:R by flexible partitioning within the root system. The capacity of infected plants to restrict reduced root growth to areas of resource depletion has been shown for water in pea (*Pisum sativum*) infected with powdery mildew (Ayres, 1981) and rusted bean (Tissera, 1986). Thus, increases in S:R in infected plants may be a poor guide of the functional capacity of the root system in natural soils.

Nitrogen, Phosphorus and Sulphur and populations of host and pathogen

It is evident, from numerous reports in the literature, that the severity of crop infection varies substantially in response to applications of fertilizers, especially nitrogen. However, as stated in the review by Zentmyer & Bald (1977), 'the reactions to nutrition are complex and cannot be described by simple rules or principles'. Thus, while the application of nitrogen fertilizers is generally considered to increase the severity of many pathogens in a wide range of crops, the form of nitrogen applied is also critical (Huber & Watson, 1974). In cereal crops, the severity of rusts and powdery mildews is generally increased by the application of nitrate, the preferred nitrogen source of the host, but may be decreased by ammonium. Conversely, of the examples cited by Huber & Watson (1974) there are many where the application of the preferred nitrogen source of the host decreases the severity of diseases caused by necrotrophic fungi. This contrast may reflect differing pathogen responses to host vigour, strong growth often favouring infection and growth of biotrophs (see below) but increasing the capacity of the host to replace tissues destroyed by necrotrophs.

Whether fertilizers alter disease severity by a direct effect of host nutrition on the growth and reproduction of the pathogen or through changes in crop growth and canopy structure, which may affect spore dispersal and the microclimate for infection, has been the subject of debate for at least three decades (e.g. Last, 1962). The relative importance of the two mechanisms appears to vary between pathogens. In the case of *Rhynchosporium secalis* applications of nitrogen in the glasshouse, where changes in canopy characteristics were insignificant, inhibited lesion production (Jenkyn & Griffiths, 1976). Applications of nitrogen fertilizers result in increased leafblotch in the field, where indirect effects of nitrogen on canopy structure thus appear to be paramount (Jenkyn, 1977). By contrast, Bainbridge (1974) showed in the glasshouse that, while conidiospore germination and appressorial formation in *Erysiphe graminis* f. sp. *hordei* were unaffected by nitrogenous fertilizers, the percentage of appressoria which gave rise to infections, the size of colonies and the number of conidia produced per unit area of the pustule were all significantly increased, and the generation interval reduced. Similarly, Tomesh & Structmeyer (1979) found that nitrogen deficiency in cucumber inhibited the area of colonies of *E. cichoracearum*. Thus, stimulation of powdery mildews by nitrogenous fertilizers in the field is probably due to the direct effects of nitrogen on the pathogen. Sulphur stress had no effect on the colony area of *E. cichoracearum* on cucumber but conidiophore production per colony was increased: severe phosphorus deficiency decreased both these measures of pathogen growth (Tomesh & Structmeyer, 1979). The very low level of phosphorus supplied by Tomesh & Structmeyer (1979) may explain the contrast between this report and that from in the field where the application of phosphate fertilizers reduced the severity of mildew infection in barley (Last, 1962). In addition to modifying the growth and spread of the pathogen, nutrient supply may also alter the effects of a given amount of disease on the growth and physiology of the host. *Puccinia lagenophorae* caused relatively smaller reductions in leaf expansion, dry weight growth, relative growth rate and net assimilation rate in groundsel subjected to severe nutrient deficiency than in plants freely supplied with nutrients, even though there were no differences in the severity of infection (Paul & Ayres, 1986b).

Whatever the underlying mechanisms, interactions between diseases and fertilizers, especially nitrogen, have major consequences for crop growth and yields. Powdery mildews in particular have minimal effects on cereal growth, total yield and its components when no nitrogen is applied (Jenkyn, 1977; Darwinkel, 1980). One important consequence of this response is that in the absence of disease control increases in crop yields following fertilizer applications are greatly reduced (Fig. 8.4a). The same principle can be restated to the effect that fungicide application is less necessary when little or no nitrogen is applied. Similarly, Shaner & Finney (1977) observed that slow-mildewing resistance in wheat reduced the severity of infection only when nitrogen was applied at high rates. It seems that the vulnerability of agricultural crops to damaging disease, and the concomitant requirement for efficient control, may be partly the result of the high inputs of fertilizers typical of modern agriculture.

The abundant supply of nutrients available in such crops is obviously remote from that of most natural plant communities, which are often subject to some degree of nutrient limitation (Fitter, 1987). Information on pathogen-nutrient interactions, comparable to that for cereals is completely absent for populations of non-crop species, indeed, our understanding of the effects of pathogens on the population biology of wild plants is extremely limited (Burdon, 1987). However, some idea of the potential interaction between nutrient supply and pathogens comes from the Park Grassland Experiment at Rothamsted, in which areas of grassland have been subjected to defined levels of nitrogen application for almost 100 years. When *Anthoxanthum odoratum* populations adapted to these different nitrogen regimes were grown under uniform conditions in a garden plot (Snaydon & Davies, 1971), there was a close, negative correlation between nitrogen supply in the native habitat and the severity of infection by powdery mildew (Fig. 8.4b). Snaydon & Davies (1971) concluded that ecotypes adapted to abundant nitrogen were more resistant to mildew because they had been subject to more intense selection over many generations by the greater severity of infection under high nitrogen conditions. This result supports the conclusion that the variation in the severity of disease observed in crops will also be apparent in natural systems. If this is indeed the case then the

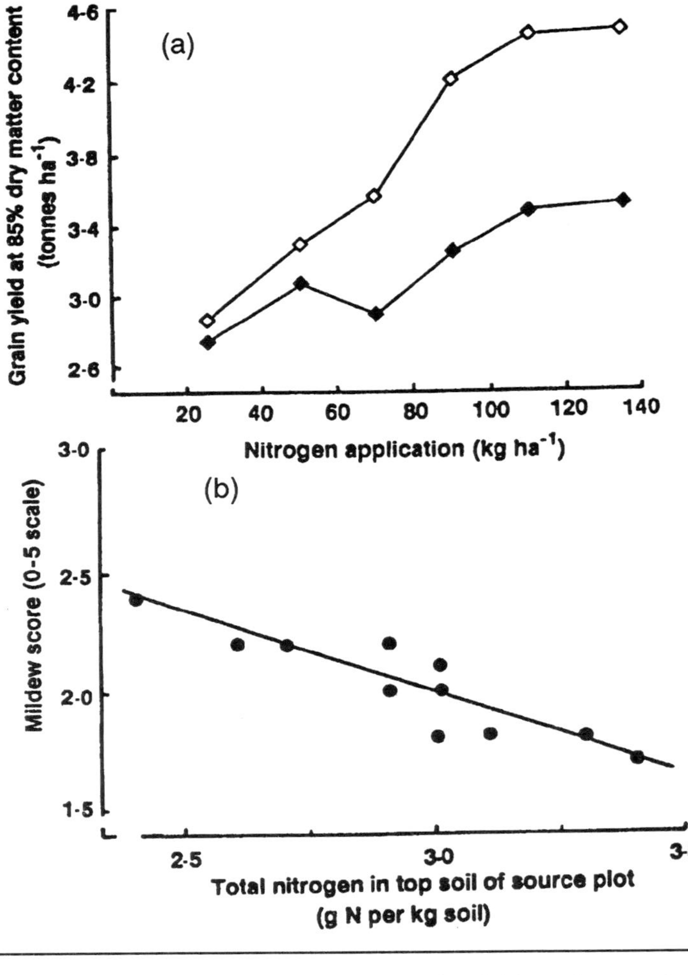

Fig. 8.4. (a) The response of barley grain yield to nitrogen fertilizer in the presence (open symbols) and absence (closed symbols) of powdery mildew control. Without adequate control of mildew, the response of yield to nitrogen application is small. Redrawn from Jenkyn (1977). (b) The severity of powdery mildew infection on forms of *Anthoxanthum odoratum* adapted to differing levels of soil nitrogen. When grown under standard garden conditions, forms adapted to high levels of soil nitrogen were less severely infected than those adapted to a limited nitrogen supply. The apparent greater resistance of forms from nitrogen-rich sites may be attributable to a more intense selection for resistance due to the greater prevalence of mildew. Redrawn from Snaydon & Davies (1971).

lower nutrient supply and tissue concentrations (Chapin, 1980) of wild plants might be expected to limit the prevalence and effect of phytopathogens in natural vegetation, especially in nutrient poor habitats, with major implications not only for the dynamics but also the genetics of populations of wild species.

In natural vegetation in particular, the effects of nutrient supply on competition between the host and uninfected neighbours may be of great importance. For plants in mixtures far more than for spaced individuals any reduction in the size of the root system of infected plants will be profoundly damaging and the relative stability of root length may be critical in maintaining the capacity of the infected host to compete for nutrients. However, since ion uptake seems less sensitive to foliar pathogens than shoot growth and activity, infected plants might be expected to compete comparatively better when nutrients are limiting.

Conclusions

As already noted, our knowledge of the effects of pathogens, especially biotrophs, on host nutrient relations is founded almost solely on studies of plants grown in artificial media in controlled environments. Such methods are not inappropriate for elucidating nutrient accumulation or cycling within the host and the increased nutrient content of infected tissues found under such conditions have also been confirmed in a few field studies. However, our dependence on results obtained under such artificial conditions will inevitably bias our understanding of nutrient uptake. The rates of uptake of nitrogen or phosphorus per unit root dry weight are consistently increased in plants infected with rusts or powdery mildews when grown in solution or sand culture. I have proposed here that such increases are driven by the greater 'shoot demand' (often increased S:R) of infected plants and that it is principally the non-saturable, 'passive' component of ion uptake which is enhanced. The mechanism and regulation of non-saturable uptake is not fully characterised in healthy plants but factors such as transpiration and recycling from the shoot are important and changes in these processes in infected plants are consistent with increased passive uptake. Other, potentially important factors such as counter-ion transport and the ionic balance of the xylem await investigation in infected plants. Regardless of the underlying mechanisms, there remains some doubt as to whether such increases in ion uptake detected under laboratory conditions will also occur under natural conditions in the field. There is an almost complete absence of experimental data concerning the effects of pathogens on many aspects of plant physiology and morphology which determine nutrient uptake in the field. These include aspects of root system growth and architecture, already noted, the effects of infection on nutrient leaching from leaves, on recycling from senescent leaves, and on specific physiological mechanisms which increase nutrient availability at the root surface in soil, all of which may be important to nutrient budgets in the field (see Chapin, 1980). In addition, the increasing evidence of the key role of mycorrhizas, rhizobia and free-living nitrogen-fixing bacteria in the nutrient relations of natural systems, suggests that interactions between mutualistic and plant-parasitic microorganisms may form yet another unknown but important aspect of pathogen-nutrient interactions in wild species.

In view of these many lacunae speculation concerning the overall role of nutrient-pathogen interactions is premature. Nevertheless, results from crops show that interactions not only occur in the field but that they may have major consequences for growth and yield. Furthermore, that pathogen-nutrient interactions in nature may prove to be diverse and far reaching is already evident, for example from nutrient-mediated pathogen-herbivore interactions (Peining, 1974; White, 1984). Clearly, studies of plant nutrition form part of the much wider challenge presented to plant pathology and ecology by our ignorance of the place of pathogens in natural vegetation.

Acknowledgements I am grateful to NERC for financial support and to D. R. Walters, A. J. S. Murray and especially P. G. Ayres for their comments on the manuscript and for numerous valuable discussions.

References

Ahmad, I., Farrar, J. F. Owera, S. A. P. & Whitbread, R. (1982). The distribution of five major nutrients in barley plants infected with brown rust. *Physiological Plant Pathology*, **21**, 335-346.

Ahmad, I., Farrar, J. F. & Whitbread, R. (1984). Fluxes of phosphorus in leaves of barley infected with brown rust. *New Phytologist*, **98**, 361-375.

Ayres, P. G. (1981). Root growth and solute accumulation in pea in response to water stress and powdery mildew. *Physiological Plant Pathology*, **19**, 169-180.

Ayres, P. G. (1984). Effects of infection on root growth and function; consequences for plant nutrient and water relations. In *Plant Diseases: Infection Damage and Loss*, eds. R. K. S. Wood & G. J. Jellis, pp. 105-117. Blackwell, Oxford.

Ayres, P. G. & Paul, N. D. (1986). Foliar pathogens alter the water relations of their hosts with consequences for both host and pathogen. In *Water, Fungi and Plants*, eds, P. G. Ayres & L. Boddy, pp. 267-285. Cambridge University Press, Cambridge.

Bainbridge, A. (1974). Effect of nitrogen nutrition of the host on barley powdery mildew. *Plant Pathology*, **23**, 160-161.

Borstlap, A. C. (1981). Invalidity of the multiphasic concept of ion absorption in plants. *Plant, Cell and Environment*, **4**, 189-195.

Burdon, J. J. (1987). *Diseases and Plant Population Biology*. Cambridge University Press, Cambridge.

Cahill, D., Wookey, C., Weste, G. & Rouse, J. (1986). Changes in the mineral content of *Eucalyptus marginata* and *E. calophylla* grown under controlled conditions and inoculated with *Phytophthora cinnamomi*. *Journal of Phytopathology*, **116**, 18-29.

Chapin, F. S. (1980). The mineral nutrition of wild plants. *Annual Review of Ecology and Systematics*, **11**, 233-260.

Clarkson, D. T. (1974). *Ion Transport and Cell Structure in Plants*. McGraw-Hill, London.

Clarkson, D. T. (1985). Factors affecting mineral nutrient acquisition by plants. *Annual Review of Plant Physiology*, **36**, 77-115.

Clarkson, D. T. & Sanderson, J. S. (1971). Relationship between the anatomy of cereal roots and the absorption of nutrients and water. *Agricultural Research Council Letcombe Laboratory Report, 1970*, 16-25.

Clarkson, D. T., Drew, M. C., Ferguson, I. B. & Sanderson, I. (1975). The effect of take-all fungus *Gaeumannomyces graminis* on the transport of ions by wheat plants. *Physiological Plant Pathology*, **6**, 75-84.

Clarkson, D. T., Hopper, M. J. & Jones, L.H.P. (1986). The effect of root temperature on the uptake of nitrogen and the relative size of the root system in *Lolium perenne*. I. Solutions containing both NH_4^+ and NO_3^-. *Plant, Cell and Environment*, **9**, 535-545.

Clement, C. R., Hopper, M. J., Canaway, R. J. & Jones, L. H. P. (1974). A system for measuring the uptake of ions by plants from flowing solutions of controlled composition. *Journal of Experimental Botany*, **25**, 81-99.

Comhaire, F. (1963). Study of the absorption and repartition of phosphorus in *Triticum vulgare* infected by *Erysiphe graminis*. *Leujeunia N.S.*, **15**, 1-87.

Crick, J. C. & Grime, J. P. (1987). Morphological plasticity of and mineral nutrient capture in two herbaceous species of contrasted ecology. *New Phytologist*, **107**, 403-414.

Darwinkel, A. (1980). Grain production of winter wheat in relation to nitrogen and diseases. II. Relationship between nitrogen dressings and mildew infection. *Journal of Agronomy and Crop Science*, **109**, 309-317.

Drew, M. C. & Saker, L. R. (1984). Uptake and long-distance transport of phosphate, potassium and chloride in relation to internal ion concentrations in barley: evidence for non-allosteric regulation. *Planta*, **160**, 500-507.

Fitt, B. D. L. & Hornby, D. (1978). Effects of root infecting fungi on wheat transport processes and growth. *Physiological Plant Pathology*, **13**, 335-346.

Fitter, A. H. (1987). An architectural approach to the comparative ecology of plant root systems. *New Phytologist*, **106** (Suppl), 61-77.

Fric, F. (1978). Absorption and translocation of phosphate in barley infected with powdery mildew. *Phytopathologische Zeitschrift*, **91**, 23-32.

Hornby, D. & Goring, C. A. I. (1972). Effects of ammonium and nitrate nutrition on take-all disease of wheat in pots. *Annals of Botany*, **70**, 225-231.

Huber, D. M. & Watson, R. D. (1974). Nitrogen form and plant disease. *Annual Review of Phytopathology*, **12**, 139-165.

Jenkyn, J. F. (1977). Nitrogen and leaf diseases of spring barley. In *Fertiliser Use and Plant Health, Proceedings of the 12th Colloquium of the International Potash Institute*, ed. K. Mengel, pp. 119-128. International Potash Institute: Berne, Switzerland.

Jenkyn, J. F. & Griffiths, E. (1976) Some effects of nutrition on *Rhynchosporium secalis*. *Transactions of the British Mycological Society*, **66**, 329-332.

Kochian, L. V. & Lucas, W. J. (1982). Potassium transport in corn roots. I. Resolution of kinetics into a saturable and linear component. *Plant Physiology*, **70**, 1723-1731.

Lambers, H. (1980). The physiological significance of cyanide-resistant respiration in higher plants. *Plant, Cell and Environment*, **3**, 293-302.

Last, F. T. (1962). Effects of nutrition on the incidence of barley powdery mildew. *Plant Pathology*, **11**, 133-135.

MacDuff, J. H., Wild, A., Hopper, M. J. & Dhanoa, M. S. (1986). Effects of temperature on parameters of root growth relevant to nutrient uptake: measurements on oilseed rape and barley grown in flowing nutrient solution. *Plant and Soil*, **94**, 321-332.

Martin, P. J. & Hendrix, J. W. (1974). Anatomical and physiological responses of Baart wheat roots affected by stripe rust. *Washington Agricultural Experimental Station, Technical Bulletin*, No. 77.

Murray, A. J. S. & Ayres, P. G. (1986). Uptake and translocation of nitrogen by mildewed barley seedlings. *New Phytologist*, **104**, 355-365.

Nye, P. H. (1977). The rate limiting step in plant nutrient absorption. *Soil Science*, **123**, 292-297.

Paul, N. D. & Ayres, P. G. (1986a). The effects of nutrient deficiency and rust infection on the relationship between root dry weight and length in groundsel (*Senecio vulgaris* L.). *Annals of Botany*, **57**, 353-360.

Paul, N. D. & Ayres, P. G. (1986b). The effects of rust (*P. lagenophorae*) on the growth of groundsel (*Senecio vulgaris* L.) cultivated under a range of nutrient concentrations. *Annals of Botany*, **58**, 321-331.

Paul, N. D. & Ayres, P. G. (1988a). Nutrient relations of groundsel (*Senecio vulgaris* L.) infected by rust (*Puccinia lagenophorae* Cooke) at a range of nutrient concentrations. I. Concentrations, contents and distribution of N, P and K. *Annals of Botany*, **61**, 489-498.

Paul, N. D. & Ayres, P. G. (1988b). Nutrient relations of groundsel (*Senecio vulgaris* L.) infected by rust (*Puccinia lagenophorae* Cooke) at a range of nutrient concentrations. II. Uptake of N, P and K and shoot-root interactions. *Annals of Botany*, **61**, 499-506.

Penning de Vries, F. W. T. (1975). The cost of maintenance processes in plant cells. *Annals of Botany*, **39**, 77-92.

Peining, L. G. (1972). Effects of leaf rust on nitrate in rye. *Canadian Journal of Plant Science*, **52**, 842-843.

Pitman, M. G. (1982). Transport across plant roots. *Quarterly Review of Biophysics*, **15**, 481-554.

Raggi, V. (1974). Free and protein amino acids in rusted bean leaves. *Phytopathologische Zeitschrift*, **81**, 289-300.

Raggi, V. (1976). Amino acids in mycelium of *Sphaerotheca pannosa* var *persicae* and in infected and surrounding tissues of peach leaves. *Phytopathologia Mediterranea*, **15**, 110-114.

Roberts, A. M. & Walters, D. R. (1988a). Nitrogen assimilation and metabolism in rusted leek leaves. *Physiological and Molecular Plant Pathology*, **32**, 229-235.

Roberts, A. M. & Walters, D. R. (1988b). Root:shoot interrelationships in leeks infected by the rust, *Puccinia allii* Rud.: growth and nutrient relations. *New Phytologist*, in press.

Robinson, D. (1986). Limits to nutrient inflow in roots and root systems. *Physiologia Plantarum*, **68**, 551-559.

Robinson, D. & Rorison, I. H. (1983). A comparison of the responses of *Lolium perenne* L., *Holcus lanatus* L. and *Deschampsia flexuosa* (L.) Trin. to a localised supply of nitrogen. *New Phytologist*, **94**, 263-273.

Shaner, G. & Finney, R. E. (1977). The effect of nitrogen fertilisation on the expression of slow-mildewing resistance in Knox wheat. *Phytopathology*, **67**, 1051-1056.

Snaydon, R. W. & Davies, M. S. (1971). Rapid population differentiation in a mosaic environment. II. Morphological variation in *Anthoxanthum odoratum*. *Evolution*, **26**, 390-405.

Son, C. L. & Smith, S. E. (1988). Mycorrhizal growth responses: interactions between photon irradiance and phosphorus nutrition. *New Phytologist*, **108**, 305-314.

Stribley, D. P., Tinker, P. B. & Rayner, J. H. (1980). Relation of internal phosphorus concentration and plant weight in plants infected with vesicular-arbuscular mycorrhizas. *New Phytologist*, **86**, 261-266.

Tissera, P. (1986). Physiological interactions between rust (*Uromyces viciae-fabae*) and water-stress in faba bean (*Vicia faba*). Ph.D. Thesis, University of Lancaster.

Tissera, P. & Ayres, P. G. (1988). Hydraulic conductivity and anatomy of roots of *Vicia faba* L. infected by *Uromyces viceae-fabae* (Pers.) Shroet. *Physiological and Molecular Plant Pathology*, **32**, 199-207.

Tomesh, R. J. & Structmeyer, B. E. (1979). Effects of nutrient stress on elemental concentrations and powdery mildew growth in cucumber. *Journal of the American Society for Horticultural Science*, **104**, 70-74.

Visarova, G. & Minarcic, P. (1974). The influence of powdery mildew upon the cytokinins and root morphology of barley roots. *Phytopathologische Zeitschrift*, **81**, 49-55.

Walters, D. R. (1981). Root function in mildewed barley. Ph.D. Thesis, University of Lancaster.

Walters, D. R. (1985). Shoot:root interrelationships: the effects of obligately biotrophic fungal pathogens. *Biological Reviews*, **60**, 47-79.

Walters, D. R. & Ayres, P. G. (1980). Effects of powdery mildew disease on the uptake and metabolism of nitrogen by roots of infected barley. *Physiological Plant Pathology*, **17**, 369-379.

Walters, D. R. & Ayres, P. G. (1981a). Phosphate uptake and transport by roots of powdery mildew infected barley. *Physiological Plant Pathology*, **18**, 195-205.

Walters, D. R. & Ayres, P. G. (1981b). Growth and branching patterns of roots of powdery mildew infected barley. *Annals of Botany*, **47**, 159-163.

Walters, D. R. & Ayres, P. G. (1982). Water movement through root systems excised from healthy and mildewed barley: relationship to phosphate transport. *Physiological Plant Pathology*, **20**, 275-284.

Weste, G., Chaudri, M. A., Haukka, M. & Vithanage, K. (1980). The influence of root infection by *Phytophthora cinnamomi* on mineral content of certain native species. *Phytopathologische Zeitschrift*, **99**, 205-214.

Whiley, A. W., Pegg, K. G., Saranah, I. B. & Lagdon, P. W. (1987). Influence of *Phytophthora* root rot on mineral nutrient concentrations in avocado leaves. *Australian Journal of Experimental Agriculture*, **27**, 173-177.

White, P. J., Clarkson, D. T. & Earnshaw, M. J. (1987). Acclimation of potassium influx in rye (*Secale cereale*) to low root temperatures. *Planta*, **171**, 377-385.

White, T. C. R. (1984). The abundance of invertebrate herbivores in relation to the availability of nitrogen in stressed food plants. *Oecologia*, **63**, 90-103.

Yarwood, C. E. & Jacobson, L. (1954) Accumulation of chemicals in diseased areas of leaves. *Phytopathology*, **45**, 43-48.

Zentmyer, G. A. & Bald, J. G. (1977). Management of the environment. In *Plant Disease. I. How Disease is Managed*, eds. I. G. Horsfall & E. B. Cowling, pp. 121-144. Academic Press, New York.

Chapter 9

The nitrogen nutrition of mycorrhizal fungi and their host plants

D. J. Read, J. R. Leake & A. R. Langdale

Department of Plant Sciences, University of Sheffield, Sheffield S10 2TN, UK

Introduction

A number of reviewers have dealt with mycorrhizas and nitrogen nutrition in recent years (Bowen & Smith, 1981; Alexander, 1983; France & Reid, 1983; Harley & Smith, 1983; Martin *et al.*, 1987). That being the case, justification for a further examination of the subject can be made only if recent advances in some part either of the topic itself, or of a related field, have been of sufficient interest to necessitate its re-appraisal. Some justification for such a re-appraisal now exists on both counts. Within the topic there is increasing evidence of the importance of ericoid and ectomycorrhizal fungi in providing access to organic sources of nitrogen. Since the host plants themselves have little or no access to these resources their fungal associates are seen potentially to play an even more crucial role in the mutualism than has hitherto been accepted (Read, 1987). In the related field of plant physiological ecology it has become widely acknowledged that availability of nitrogen is the key factor determining the photosynthetic capabilities and hence productivities of natural plant communities (Field & Mooney, 1986). This is because the maximum photosynthetic rates (P_{max}) of C_3 plants, under field conditions with saturating light intensities and at normal CO_2 levels, are determined by the concentration of nitrogen in the leaf. Summarising data from studies of a wide range of natural plant communities Field & Mooney (1986) have shown that practically all the data points relating P_{max} as a function of nitrogen cluster around a single line of regression. This suggests a fundamental relationship that is relatively insensitive to differences among species or growth conditions and demonstrates that over a wide diversity of C3 plants photosynthetic capacity is strongly regulated by leaf nitrogen content. Physiologically, the dependance of photosynthetic rate upon nitrogen content arises directly from the use of the element in the carbon fixing enzymes which contain a large proportion of total leaf nitrogen. It is this interdependence of nitrogen and P_{max} which, on a world scale, makes nitrogen the element which most limits plant growth.

In all cases, except those in which prokaryotic organisms fix dinitrogen gas, the element becomes available to plants through the activities of heterotrophic populations which recycle organic matter. In view of the importance of nitrogen for autotrophs it can be predicted that any environment in which the turnover of nitrogen is restricted will generate powerful pressures leading to the selection of associations between autotrophs and those heterotrophs which are sufficiently specialised to maintain access to nitrogen supplies. In an earlier review (Read , 1985) a relationship was recognised between increasing latitude or altitude, accumulation of organic matter and the occurrence of distinct mycorrhizal types. It was suggested that these types have been selected primarily on the basis of their ability to provide the autotroph with access to the particular type of resource which is most likely to be limiting the growth in each part of the environmental gradient. In the warmer parts of the gradient mineralisation of organic matter takes place rapidly and the relatively high pH characteristic of soils in such systems enables nitrification to proceed with the release of mobile nitrate ions. Under these circumstances the relative immobility of the phosphate ion leads to the development of phosphorus depletion zones around roots, and all the evidence suggests that the vesicular-arbuscular (VA) type of infection found most frequently in this type of environment is primarily involved with provision of increased access to PO_4 ions (see Chapters 10 & 11). While there are reports of enhancement of plant nitrogen content (Possingham & Groot Obbink, 1971), of nitrogen inflow rates (Smith *et al.*, 1986) in VA infected plants as well as of nitrogen transfer through VA mycorrhizal mycelium to the plant (Ames *et al.*, 1983) such functions are justifiably considered to be of secondary importance in this mycorrhizal type and are not considered further here.

In cooler climates, in contrast, where organic matter is turned over more slowly, nitrogen mobilization becomes progressively more restricted. Thus, while low phosphorus availability may still be a factor limiting plant growth, access to nitrogen becomes increasingly important as greater proportions of the element are sequestered in organic combinations. In quantitative terms the plant requirement for nitrogen is approximately ten times greater than that for phosphorus so that any restriction of access to the former very rapidly begins to have adverse effects upon growth and physiology. Evidence will be presented here which confirms that many of those mycorrhizal fungi which are characteristic of habitats in which organic matter accumulates have at least some of the attributes which would enable them to gain access to the nitrogenous components of such materials. Prior to a consideration of these aspects, however, our knowledge of the uptake and assimilation of mineral nitrogen is briefly surveyed.

Utilization and assimilation of inorganic nitrogen

Utilization of inorganic nitrogen by mycorrhizal fungi in pure culture

As indicated above, soils in which the most extensive development of ericoid and ectomycorrhizal systems occur have high organic content, and, partly as a result of the accumulation of this organic material, a predominance of carboxyl groups gives rise to an acid pH. Under these circumstances nitrification is inhibited and ammonium is the main form of mineral nitrogen. Most ectomycorrhizal fungi readily use ammonium as a nitrogen source. In the less extreme parts of the environmental gradient, however, nitrification does take place and it is perhaps not surprising, therefore, that the ericoid endophyte and many ectomycorrhizal fungi can utilise nitrate as well as ammonium.

In a comparative analysis of four ecto-fungi France & Reid (1984) obtained higher dry weight yield of *Pisolithus tinctorius* on nitrate than on ammonium, the reverse situation in *Cenococcum geophilum* and *Thelephora terrestris*, and an equal facility for use of both ions in *Suillus granulatus*. This pattern reflects to some extent the distribution of the fungi in the natural environment, *P. tinctorius* being predominantly a fungus of the arid part of the environmental gradient where nitrification becomes an important process, while *C. geophilum* and *T. terrestris* occur characteristically on organic soils and *S. granulatus* occurs right across the gradient. In some cases, for example *Lactarius rufus*, it has been shown that the fungus completely failed to use nitrate (Bigg & Alexander, 1981). There is, of course, considerable genetic variability amongst individual strains of ectomycorrhizal fungi. Thus, Laiho (1970) demonstrated clearly that whereas some strains of *Paxillus involutus* produced much higher yields on ammonium than on nitrate the reverse was true with other strains. This pattern was largely confirmed for a range of fungi by Lundeberg (1970). Such observations highlight the genetic plasticity of ectomycorrhizal fungi and make clear the need for careful description of the races and origins of organisms being used in assays of nutrient utilization. Indeed, the plasticity in terms of NH_4 and NO_3 utilization appears to be so great as to make generalization about a single species of doubtful validity.

All strains of the ericoid endophyte *Hymenoscyphus ericae* that have so far been examined are able to utilise both NH_4 and NO_3 (Pearson & Read, 1975; Bajwa & Read, 1986) and while there is an initial lag in growth on nitrate, which probably represents a period during which nitrate reductase is induced, final yields on this nitrogen source can be as high as those on ammonium.

Utilization of mineral nitrogen by the mycorrhizal root

In some of the first experimental studies of nitrogen transfer in mycorrhizal systems, Melin & Nilsson (1952) fed ^{15}N-labelled ammonium to mycorrhizal mycelium and detected its transfer to infected pine seedlings. Carrodus (1966) demonstrated that excised beech mycorrhizas absorbed ammonium more rapidly than nitrate. A largely similar situation has since been shown to prevail in a number of important coniferous species (France & Reid, 1983; Rygiewitz *et al.*, 1984 a & b). The latter workers observed that mycorrhizal infection enhanced the uptake of nitrate in only one of the species examined.

Whereas in ericoid mycorrhizas ammonium is a better source of nitrogen than nitrate, infection can significantly improve the levels of nitrate utilization relative to that seen in uninfected plants (Bajwa & Read, 1986).

Assimilation of absorbed inorganic nitrogen

In his studies of beech mycorrhizas Carrodus (1967) showed that absorbed ammonium was rapidly assimilated into glutamine. While glutamate was the major product of nitrate assimilation, glutamine and aspartate were also formed. The principle route of ammonium assimilation in beech ecto-mycorrhizas has since been shown by Martin *et al.* (1986) to be that involving glutamine synthetase and glutamate synthase. This study, using $^{15}NH_4$, confirmed that glutamine was the major product of ammonium assimilation and suggested that alanine and aspartate were labelled secondarily as a result of amino transferase activity. The evidence suggests that most of these assimilation processes take place in the fungal component of the mycorrhiza, and much interest surrounds the nature of the compounds subsequently transferred to the host plant.

Studies of Reid & Lewis and of France (see France & Reid, 1983) have indicated that glutamine is the major transfer product following assimilation of ammonium by excised beech mycorrhizas. Finlay *et al.* (1988) have recently studied the assimilation of ammonium in intact mycorrhizal systems of pine infected with *Rhizopogon roseolus*, *Suillus bovinus*, *Pisolithus tinctorius* and *Paxillus involutus* in natural unsterile peat. Ammonium labelled with ^{15}N was supplied to mycelial fans of the fungi which had grown across a barrier into a root-free portion of a transparent chamber, and incorporation of label into amino-compounds of the mycelia, mycorrhizal roots and shoots of pine was determined after 72 hours. By this time large proportions of the glutamate/glutamine, aspartate/asparagine and alanine in the mycelia of *R. roseolus* and *S. bovinus* were labelled with ^{15}N (Fig 9.1). A similar situation was found in *Pisolithus tinctorius*, but *Paxillus involutus* contained no asparagine/aspartate. Lower proportions of labelled serine, threonine, tyrosine, lysine, ornithine and arginine were also found in the mycelia. As would be expected, the amount of enrichment declined throughout the transport pathway though the percentage contribution of labelled amino-acid remained relatively high in the

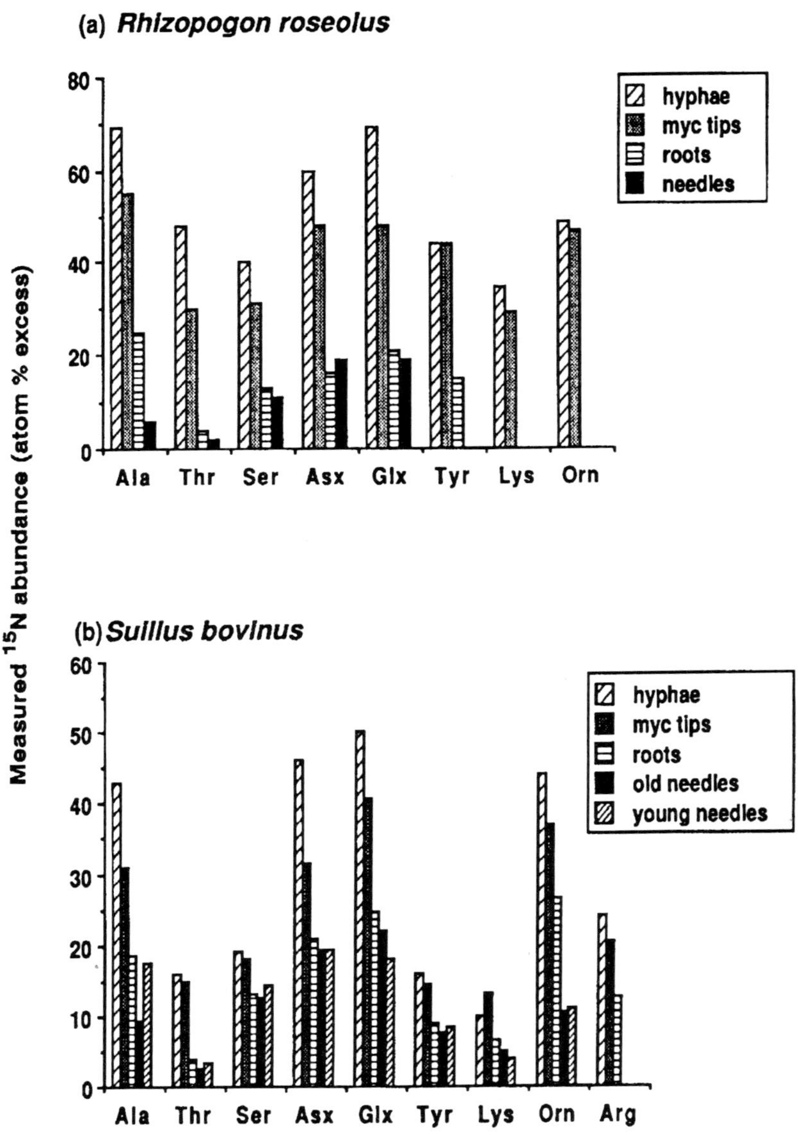

Fig. 9.1. Distribution of ^{15}N amongst amino-acids in different regions of mycorrhizal systems 72 hours after feeding $^{15}NH_4Cl$ to mycelium of two ectomycorrhizal fungi growing from seedlings of *Pinus* in a root chamber; myc = mycorrhizal. From Finlay *et al.*, 1988.

mycorrhizal root tips (30-60%), and main root axes (20-40%). Values of between 5 & 30% were detected in the needles.

The free amino acid pools of the fungal mycelia were similar in composition to those reported by Booker (1980), Genetet *et al.* (1984), and Martin *et al.* (1986), as were those of the shoots when compared with data from coniferous needles (Mackenzie & Holme, 1984). Studies such as this of intact systems have the great advantage that normal source-sink relations are maintained so enabling realistic estimates of nutrient fluxes to be obtained. Further studies of this type with sequential harvests being taken at regular short intervals from the time of ^{15}N application should provide definition both of the primary products of ammonium assimilation and of the earliest transferase activities.

Utilization of organic nitrogen

Ericoid systems

Growing at the extreme end of the environmental gradient which leads to the accumulation of recalcitrant mor humus, plants with ericoid mycorrhizas dominate ecosystems in which, by definition, mineralisation processes are slow. There is increasing evidence that their success in these conditions is dependent upon the activities of their mycorrhizal fungi. While the early studies of the function of ericoid mycorrhizas revealed that infection could enhance uptake of ammonium when the mineral ion was the only nitrogen source present (Stribley & Read, 1976), it was also shown (Stribley & Read, 1974) that when ^{15}N labelled ammonium was added to organic heathland soil the amount of label in mycorrhizal plants was less than that in their non-mycorrhizal counterparts even though the mycorrhizal plants contained more nitrogen. This study suggested that infected plants had access to an alternative source of nitrogen, which in the absence of nitrate, was assumed to be organic. In an initial attempt to determine whether infection did, in fact, provide plants with access to such sources, the pattern of uptake of amino-acids by mycorrhizal and non-mycorrhizal plants was examined (Stribley & Read, 1980). This work demonstrated conclusively that whereas uninfected plants could make little use of amino compounds, their infected counterparts assimilated most of the compounds as readily as they did ammonium. Detailed analyses of the nitrogen fractions of heathland soil

(Abuarghub & Read 1988a & b) have since revealed that 'free' amino-acids constitute a significant proportion of the available nitrogen pool, a particularly large concentration occurring in autumn (Fig. 9.2). Mycorrhizal infection will thus provide access

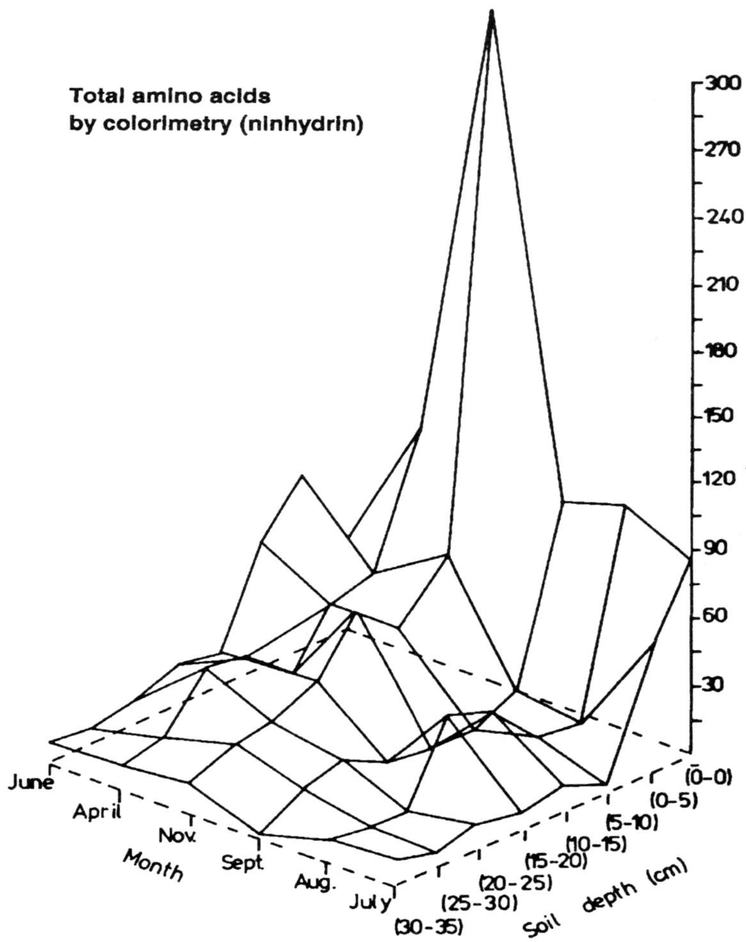

Fig. 9.2. Three dimensional presentation of the 'free' amino acids extracted by dilute ammonium acetate at different depths and times of year under a pure *Calluna* sward from an upland peaty-gley. From Abuarghub & Read, 1988.

to a major component of soil nitrogen which is otherwise unavailable to the plant.

Amino-acids may themselves, however, only be breakdown products of polymeric nitrogen sources so it is necessary to consider the question of the ability of the mycorrhizal fungus to provide access to more complex organic nitrogen compounds. It has now been established that the ericoid mycorrhizal endophyte will utilise not only amino-acids but also peptides (Bajwa & Read, 1985) and proteins (Bajwa, Abuarghub & Read, 1985) as nitrogen sources, and that the assimilated nitrogen is rapidly transferred to infected plants. Production of an extra-cellular proteinase by *Hymenoscyphus ericae* was reported by Read & Bajwa (1985) and Spinner & Hasewandter (1985). Subsequent studies have confirmed and elaborated upon these observations. A more detailed characterization of the factors influencing its activity has since been provided by Leake & Read (1987, 1988). The enzyme has a decidedly acid pH optimum (Fig. 9.3) and a K_m of 0·1 mg protein ml^{-1} which is comparable with those of some acid and alkaline microbial proteinases reported previously, for example 0·15 - 0·54 mg ml^{-1} (Nannipieri *et al.*, 1982) and 0·2 - 0·5 mg ml^{-1} (Chaloupka *et al.*, 1975) but lower than the value of 2·0 - 2·5 mg ml^{-1} obtained for an acid proteinase of the protozoan parasite *Plasmodium* (Levy & Chou, 1974).

Sequential analyses of aliquots of culture filtrates taken while *H. ericae* was growing on media containing protein as a sole nitrogen source have enabled the progress of degradation of the macromolecule to be followed. Growth of the endophyte over the first days lead to rapid decline of the substrate and concomitant increase of amino-acid content of the medium (Fig. 9.4). After 24 days most of the protein had been utilised and at this point the size of the amino-acid pool began to decline. No ammonium was detected in the solution. In view of the proven ability of the endophyte to utilise amino-acids as nitrogen sources, the cleavage of protein into these units is, in terms of energy requirement, the most economical metabolic pathway of nitrogen acquisition. The possible ecological and physiological significance of these observations is discussed later.

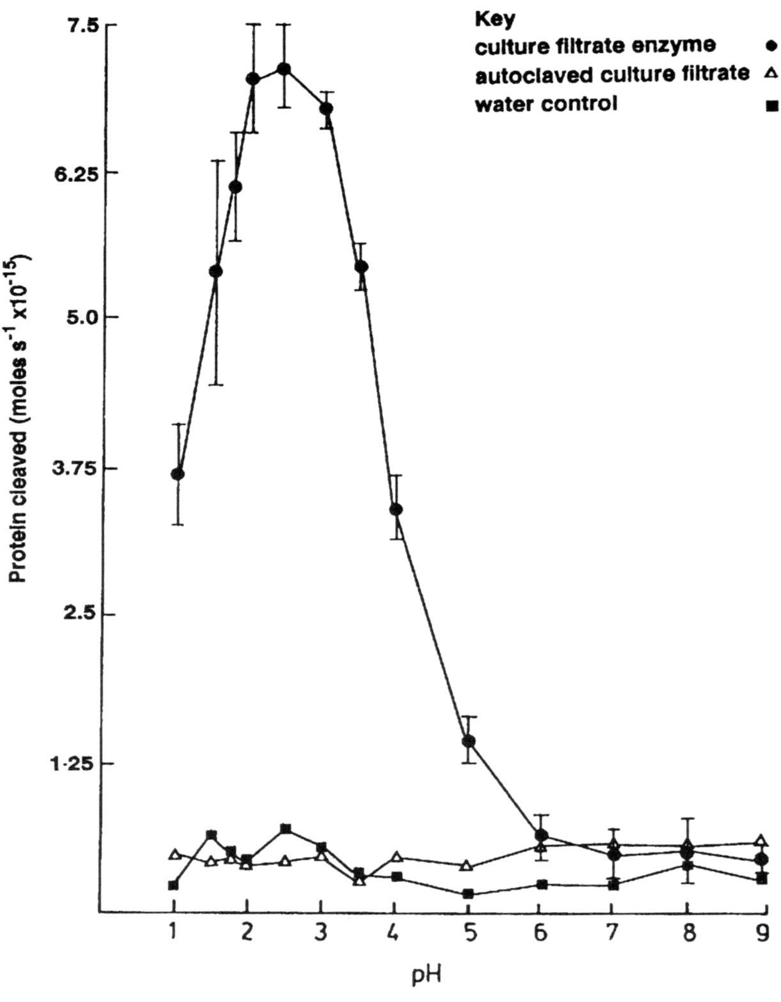

Fig. 9.3. Cleavage of protein by free proteinase enzyme of *Hymenoscyphus ericae* assayed over a pH range with FITC labelled bovine serum albumen (BSA) as substrate. The fungus was first grown on a liquid medium containing unlabelled protein as sole nitrogen source and sterile filtrate was obtained from these cultures at 21 days for enzyme assay.

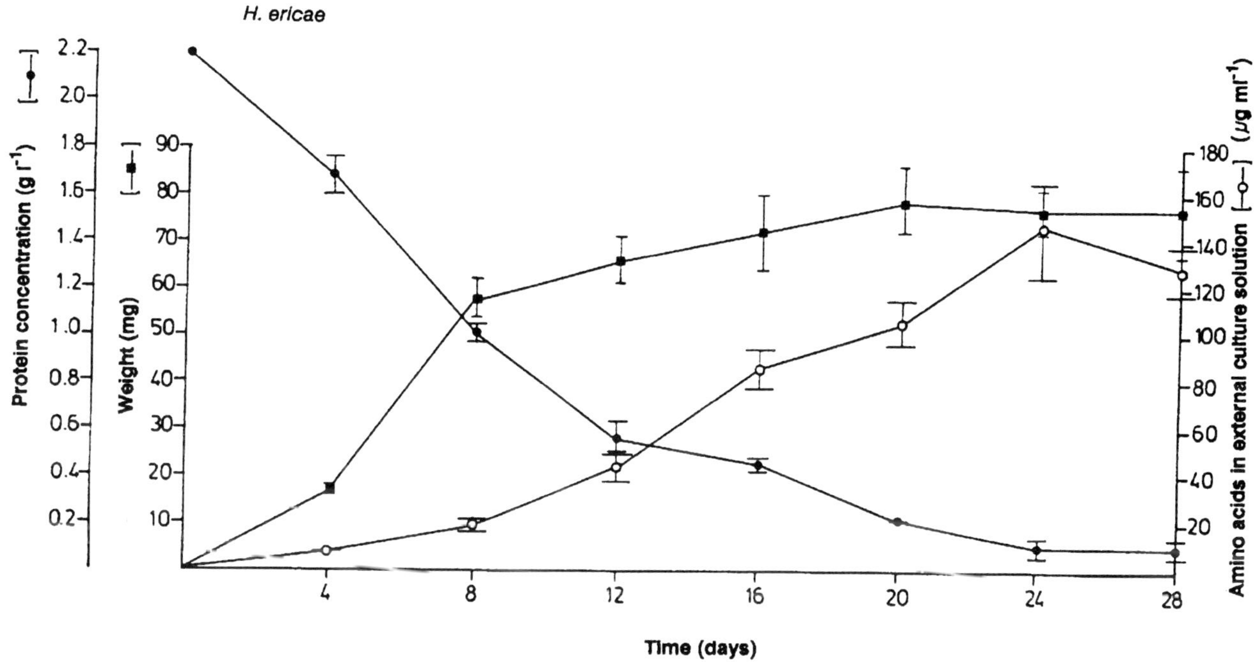

Fig. 9.4. Time course analysis of yield increase, utilization of protein, and release of amino acids into the culture medium when *Hymenoscyphus ericae* was grown over a 28 day period with pure protein as the sole nitrogen source.

Ectomycorrhizal systems

Early studies of Melin (1925) showed that some ectomycorrhizal fungi could utilise amino acids, protein hydrolysate and even, though to a lesser extent, the protein legumin. Later, Melin & Nilsson (1953) demonstrated that ^{15}N labelled glutamate was absorbed by the mycelium of *S. granulatus* and that the nitrogen was transferred to the shoots of pine seedlings infected by the fungi in aseptic culture. The absorption of glutamate, aspartate and their amides by excised beech mycorrhiza was reported by Carrodus (1966) and, using ^{14}C labelled glycine, Sangwanit & Bledsoe (1985) showed that excised mycorrhizas of *Pseudotsuga menziesii* infected with *Hebeloma crustuliniforme* rapidly absorbed the acid which, within 60 minutes, was converted into several other amino compounds.

We have fed ^{14}C-labelled alanine to intact mycorrhizal mycelia growing in association with pine seedlings in unsterile forest soil (Fig. 9.5a) and, by means of autoradiography, followed the distribution of label through the mycelium (Fig. 9.5b). Within 48 hours of commencement of feeding the isotope, label was detected throughout the mycelial network, there being accumulation in the mycorrhizal roots of the interconnected seedlings. Qualitative and quantitative aspects of the transfer of amino-nitrogen and of amino acid carbon in mycorrhizal systems are currently being investigated in more detail.

Despite the early observations of Frank (1894), that mycorrhizal roots were intimately associated with organic material, utilization of such complex sources of nitrogen by ectomycorrhizal fungi has been little studied until recently. The work of Lundeberg (1970) which suggested that ecto fungi were not capable of using humic nitrogen or degrading protein has widely been interpreted to mean that no significant access to polymeric form of nitrogen can be expected in soil. However, it has been shown in numerous studies that ectomycorrhizal roots in forest ecosystems are more intimately associated with the litter than the humified horizon (Harley, 1940; Werlich & Lyr, 1957; Mikola & Laiho, 1962; Meyer 1973; Harvey *et al.*, 1976; Deans, 1979; Alexander & Fairley 1983). On the basis of such observations it has been suggested (Abuzinadah *et al.*, 1986; Read, 1987) that it is ecologically more meaningful to consider the ability of ectomycorrhizal fungi to use the less recalcitrant forms of nitrogen likely to be present in such material. Using the pure protein bovine serum albumin as a model compound, Abuzinadah & Read (1986) showed that a wide range of protein degrading capabilities occurred in ecto fungi. Some, which were termed 'protein fungi' produced higher yield on this substrate than on ammonium, while others termed 'non protein fungi' failed to degrade the macromolecule. When protein fungi such as *Suillus bovinus*, *Amanita muscaria* and *Hebeloma crustuliniforme* were grown in mycorrhizal association with pine, spruce or birch it was shown that nitrogen derived from protein was rapidly transferred to the infected plants. Uninfected plants, in contrast, had

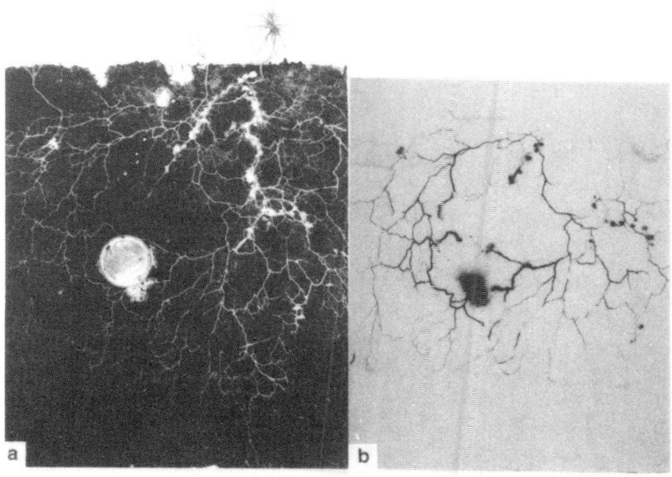

Fig. 9.5. (a) Mycelial network of *Suillus bovinus* interconnecting seedlings of *Pinus sylvestris*, showing feeding dish containing ^{14}C-labelled alanine into which are placed the cut ends of individual mycelial strands. (b) Autoradiograph of 9.5a taken 48 hours after commencement of ^{14}C feeding showing distribution of isotope throughout the network and its accumulation in mycorrhizal roots of the interconnected plants.

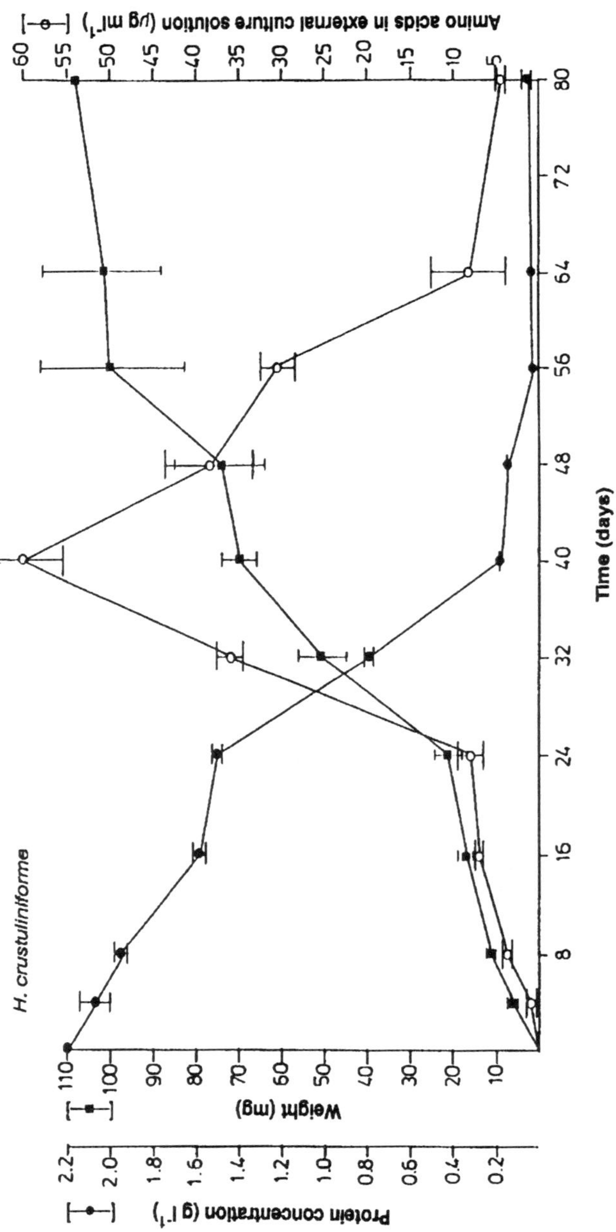

Fig. 9.6. Time course analysis of yield increase, utilization of protein, and release of amino acid into the culture medium when the ectomycorrhizal fungus *Hebeloma crustuliniforme* was grown over an 80 day period with pure protein as the sole nitrogen source.

no access to this nitrogen. It has since been confirmed that, as in the case of the ericoid endophyte, ectomycorrhizal 'protein fungi' produce a free acid proteinase when grown in solution with protein as a sole nitrogen source and that the major products of protein degradation are amino acids (Fig. 9.6), many of which are known from a parallel study of Abuzinadah & Read (1988) to be readily assimilable by the fungi which are releasing them. Again, as in the case of *Hymenoscyphus ericae*, ammonium is not a product of the main phase of protein degradation. In contrast to the situation reported above, Botton *et al.* (1986) could detect only alkaline proteinase activity in *Cenococcum geophilum*. It may be that the strain used was adapted to alkaline soil but details of its origin were not provided.

Discussion

Current evidence indicates that abilities to utilise complex organic sources of nitrogen are far better developed in ericoid and ectomycorrhizal systems than has hitherto been recognised. Such observations justify a re-evaluation both of the function of the mycorrhizal symbiosis and of the overall structure of the nitrogen cycle in heathland and forest ecosystems.

The conventional view of mycorrhizal function sees the fungus as an extension of the root system which, by exploring a volume of soil beyond the depletion zone around roots, increases access to mineral nutrient ions. Simple observations of mycorrhizal structure reveal, however, that in many cases such increases of absorptive capability may not be obtained. The root system of many plants with ericoid mycorrhizas, for example, is itself extremely finely divided into dense wefts of 'hair roots' the net effect being a situation in which all rhizosphere depletion 'shells' overlap. Most of the biomass of the mycorrhizal fungus is located within or close to the root (Read 1983, 1985). In such a situation it is inconceivable that the infection functions as conventionally envisaged. In ectomycorrhizas also, Harley (1978) pointed out that many ensheathed roots are entirely smooth so apparently having few, if any, mycelial connections with soil. It is difficult to see how such a structure can increase absorptive efficiency of a root. If, on the other hand, the fungi were producing diffusible enzymes which alone could provide the plant with access to a growth limiting resource such as nitrogen as it accumulated around the root, this would explain the heavy investment of host carbon in a mycorrhizal structure that did not fulfil a scavenging role. The evidence reported here shows that ericoid and some ectomycorrhizal fungi have the ability to produce such diffusible enzymes and that they readily assimilate the breakdown products of protein degradation.

So far the work has been largely confined to pure culture and there is a need to extend it to soil situations. While many proteins, for example those being released from senescing microbial cells, may be in relatively pure form, some will be rendered inaccessible as a result of chemical binding to polyphenolic compounds, and others will be protected by physical barriers formed by cell walls and lignified structures. A suite of enzymes is likely to be involved in releasing proteinaceous substrates from both chemically bound and physically isolated conditions. Among them polyphenol oxidases, cellulases, peroxidases and lignases are likely to be important. In a preliminary study of mycorrhizal mycelial mats in the decomposition horizon of Douglas fir forests Griffiths *et al.* (1987) have shown a significantly enhanced level of peroxidase and lignase as well as of proteinase activity associated with the fungal component as compared to surrounding soil material. There is obviously an urgent need to examine the field situation in more detail both from the point of view of determining the level of substrate availability and of fungal activity. The revelation of markedly seasonal pattern of amino acid release in heathland soil suggests that there may be temporal as well as spatial heterogeneity in the production and degradation of protein-rich substrate. Whatever the precise pattern of resource capture it is likely that the presence, under the litter layer, of a mat of mycorrhizal roots which have proteolytic capability, will have a profound impact upon nitrogen cycling processes.

In place of the conventional notion that the mycorrhizal fungi are dependent upon release of ammonium by a saprotrophic population, we now contemplate that there will be competition between the two groups for the organic nitrogenous resources. It was suggested by Abuzinadah *et al.* (1986) that the mycorrhizal fungi would be at an advantage in such a competitive situation because they alone were attached to a continuous supply of free carbon with which to synthesise the necessary proteinase enzyme.

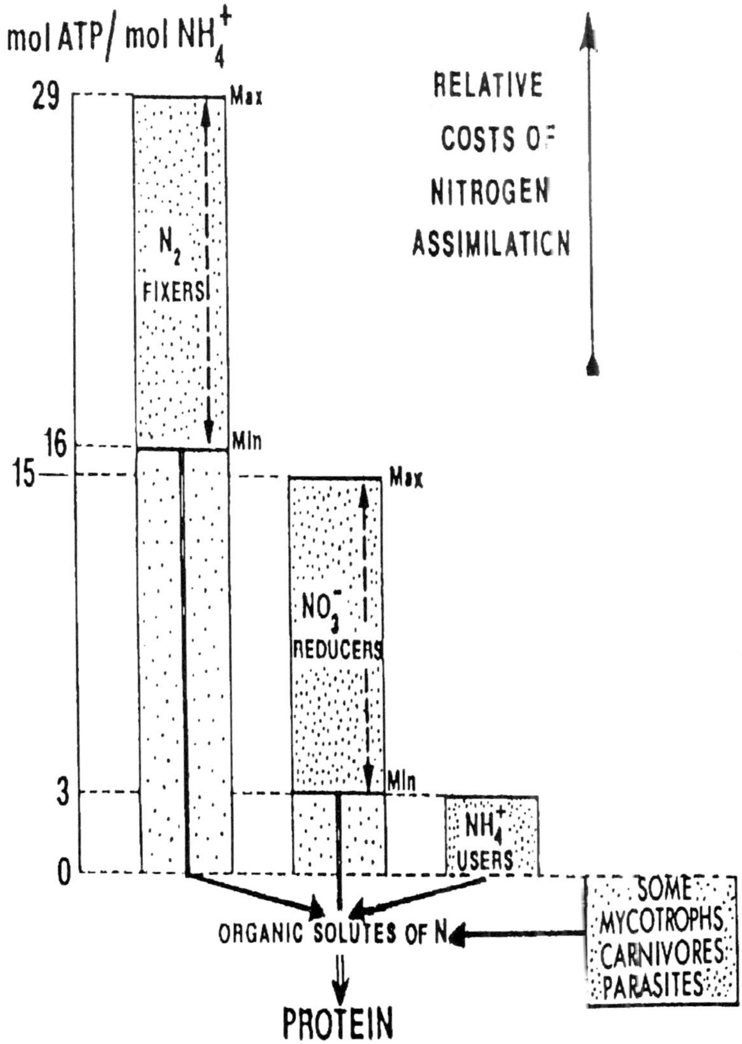

Fig. 9.7. Theoretically based costs of assimilation of different forms of nitrogen by different trophic classes of plants. Ericoid and ectomycorrhizal plants which assimilate nitrogen largely in the ammonium or organic form are seen to suffer the lowest energy costs. Modified from Pate, 1986.

The carbon balance is likely to be of critical importance in determining the outcome of competitive interactions in soils. Dighton *et al.* (1987) investigated the relative rates of breakdown of organic substrates, including hide powder protein, when mycorrhizal seedlings were grown with and without the presence of a saprotrophic fungus in irradiated peat. While their results suggest that the saprotroph reduced the degradative abilities of the mycorrhizal fungus, they must be treated with caution because the interactions were studied in microcosms to which a nutrient solution containing free carbon and mineral salts was added.

In energetic terms the assimilation of simple organic moieties is a highly efficient process. Pate (1986) has calculated the energy costs of assimilation of different types of nitrogen (Fig. 9.7). Assimilation of amino acids is the least energy demanding since they can be absorbed directly and do not require reduction. Ammonium assimilation is, in turn, more energy efficient than is that of nitrate or dinitrogen gas, both of which require several reduction steps. Energetic considerations alone, therefore, suggest that there may be selection for use of amino compounds in situations where organic materials predominate. To be set against this advantage are the costs, as yet unquantified, of synthesis of proteolytic enzymes.

The likely pathways of nitrogen transfer, based upon observations made to date, are shown in Fig. 9.8. The organic nitrogen pathway would be predicted to predominate under circumstances in which mineralisation processes were slow or non existent, whereas pathways of the more traditional kind involving ammonium assimilation would be expected to become more important as turnover of organic matter becomes more rapid. As rates of mineralisation and nutrient release become faster the role of mycorrhizas in nitrogen assimilation probably becomes less critical. In this case, if mycorrhizas provide access to another growth limiting resource, for example phosphorus, infection may be retained. Alternatively, as in most fertile situations, where it is likely to provide the plant with little advantage, infection may be lost altogether.

Emphasis in mycorrhizal research has for a long time been concentrated on the question of the uptake and transfer of the phos-

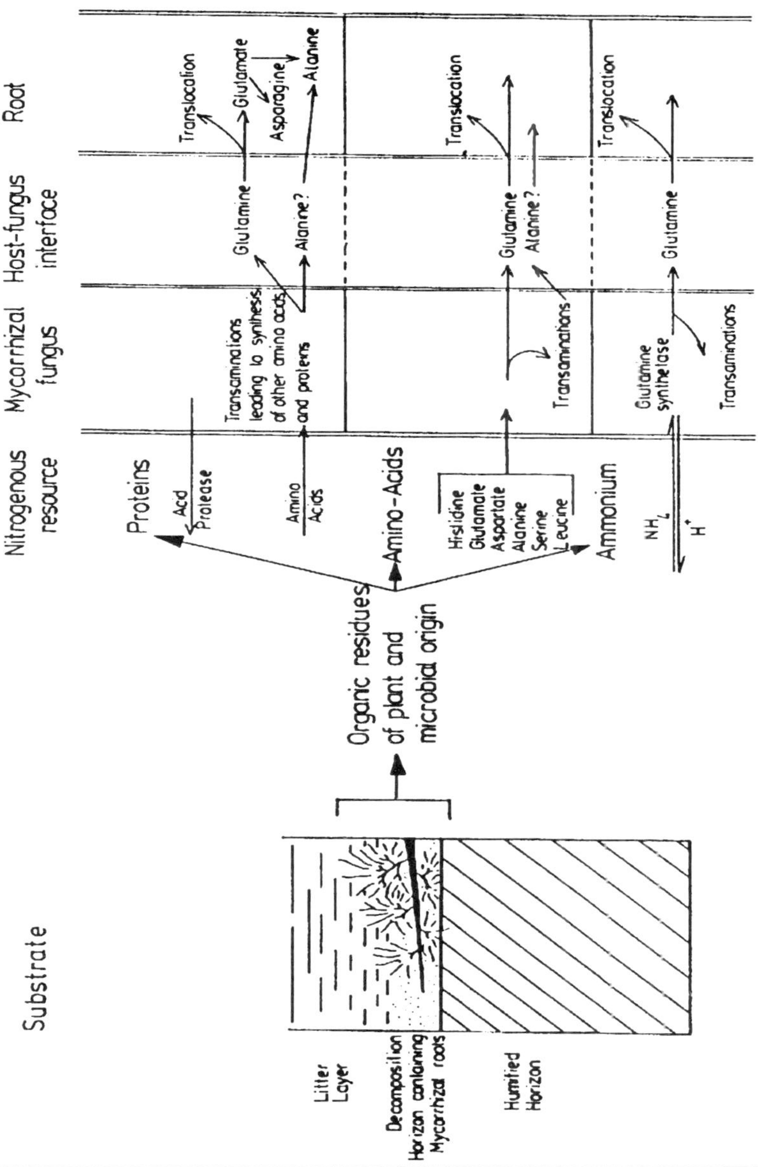

Fig 9.8. The distribution of mycorrhizal roots in relation to the potential nitrogenous substrates in soil and the likely pathways of nitrogen assimilation in ericoid and ectomycorrhizal associations.

phate ion and mycorrhizas have been considered to function simply as extensions of the root system which increase the efficiency of foraging for mineral nutrients. While this function is well documented and, indeed, may constitute the major role in VA mycorrhizas (see Chapters 10 & 11) there is now sufficient evidence to support the view, originally expressed by Frank (1894), that some mycorrhizas have a more fundamental function since their activities are central to the process of mobilization of nutrients from organic polymers. Ericoid and some ectomycorrhizal associations are thus seen to play a crucial role in the nutrient cycling process since through their activities the single most important element, nitrogen, together probably with phosphorus, can be released from otherwise inaccessible complexes and transported directly to the plants. If we are to understand the nutrient dynamics of forest and heathland ecosystems a fuller knowledge of these key nutrient mobilising processes will be essential.

References

Abuarghub, S. M. & Read, D. J. (1988a). The biology of mycorrhiza in the Ericaceae XI. The distribution of nitrogen in soil of a typical upland *Callunetum* with special reference to the 'free' amino acids. *New Phytologist*, **108**, 425-431.

Abuarghub, S. M. & Read, D. J. (1988b). The biology of mycorrhiza in the Ericaceae XII. Quantitative analysis of individual 'free' amino acids in relation to time and depth in the soil profile. *New Phytologist*, **108**, 433-441.

Abuzinadah, R. A. & Read, D. J. (1986a). The role of proteins in the nitrogen nutrition of ectomycorrhizal plants. I. Utilization of peptides and proteins by ectomycorrhizal fungi. *New Phytologist*, **103**, 481-493.

Abuzinadah, R. A. & Read, D. J. (1986b). The role of proteins in the nitrogen nutrition of ectomycorrhizal plants. III. Protein utilization by *Betula*, *Picea* and *Pinus* in mycorrhizal association with *Hebeloma crustuliniforme*. *New Phytologist*, **103**, 507-514.

Abuzinadah, R. A. & Read, D. J. (1988). Amino acids as nitrogen sources for ectomycorrhizal fungi. *Transactions of the British Mycological Society*, **91**, 473-479.

Abuzinadah, R. A., Finlay, R. D. & Read, D. J. (1986). The role of proteins in the nitrogen nutrition of ectomycorrhizal plants. II. Utilization of protein by mycorrhizal plants of *Pinus contorta*. *New Phytologist*, **103**, 495-506.

Alexander, I. J. (1983). The significance of ectomycorrhizas in the nitrogen cycle. In *Nitrogen as an Ecological Factor*, eds. J. A. Lee, S. McNeill & I. H. Rorison, pp. 69-93. Blackwell Scientific Publications, Oxford.

Alexander, I. J. & Fairley, R. I. (1983). Effects of N fertilization on populations of fine roots and mycorrhizas in spruce humus. *Plant and Soil*, **71**, 49-53.

Ames, R. M., Reid, C. P. P., Porter, L. K. & Cambardella, C. (1983). Hyphal uptake and transport of nitrogen from two ^{15}N labelled sources by *Glomus mosseae*, a vesicular-arbuscular mycorrhizal fungus. *New Phytologist*, **95**, 381-396.

Bajwa, R. & Read, D. J. (1985). The biology of mycorrhiza in the Ericaceae. IX. Peptides as nitrogen sources for the ericoid endophyte and for mycorrhizal and non-mycorrhizal plants. *New Phytologist*, **101**, 459-467.

Bajwa, R. & Read, D. J. (1986). Utilization of mineral and amino N sources by the ericoid mycorrhizal endophyte *Hymenoscyphus ericae* and by mycorrhizal and non-mycorrhizal seedlings of *Vaccinium*. *Transactions of the British Mycological Society*, **87**, 269-277.

Bajwa, R., Abuarghub, S. & Read, D. J. (1985). The biology of mycorrhiza in the Ericaceae. X. The utilization of proteins and the production of proteolytic enzymes by the mycorrhizal endophyte and by mycorrhizal plants. *New Phytologist*, **101**, 469-486.

Bigg, W. L. & Alexander, I. J. (1981). A culture unit for the study of nutrient uptake by intact mycorrhizal plants under aseptic conditions. *Soil Biology & Biochemistry*, **13**, 77-78.

Booker, C. E. (1980). Free and bound amino acids in the ectomycorrhizal fungus *Pisolithus tinctorius*. *Mycologia*, **72**, 868-881.

Botton, B., El Badaoui, K. & Martin, F. (1985). Induction of extracellular proteases in the ascomycete *Cenococcum geophilum*. In *Proceedings of the 1st European Symposium on Mycorrhizas*, eds. V. Gianinazzi-Pearson & S. Gianinazzi, pp. 403-406. INRA, Paris.

Bowen, G. D. & Smith, S. E. (1981). The effects of mycorrhizas on nitrogen uptake by plants. In *Terrestrial Nitrogen Cycles*, eds. F. E. Clark & T. Rosswall, pp. 237-247. Volume 33 of the Ecological Bulletin, Stockholm.

Carrodus, B. B. (1966). Absorption of nitrogen by mycorrhizal roots of beech. I. Factors affecting the assimilation of nitrogen. *New Phytologist*, **65**, 358-371.

Carrodus, B. B. (1967). Absorption of nitrogen by mycorrhizal roots of beech. II. Ammonium and nitrate as sources of nitrogen. *New Phytologist*, **66**, 1-4.

Chaloupka, J., Obdrzalek, V., Kreckova, P., Nesmeyanova, M. A. & Zalabak, V. (1975). Protease activity in cells of *Bacillus megaterium* during derepression. *Folia Microbiologia*, **20**, 277-288.

Deans, J. D. (1979). Fluctuations of the soil environment and fine root growth in a young sitka spruce plantation. *Plant and Soil*, **52**, 195-205.

Dighton, J., Thomas, E. D. & Latter, P. M. (1987). Interactions between tree roots, mycorrhizas, a saprotrophic fungus and the decomposition of organic substrates in a microcosm. *Biology & Fertility of Soils*, **4**, 145-150.

Field, C. B. & Mooney, H. A. (1986). The photosynthesis-nitrogen relationship in wild plants. In *On the Economy of Plant Form and Function*, ed. T. J. Givnish, pp. 25-55. Cambridge University Press, Cambridge.

Finlay, R. D., Ek, H., Odham, G. & Söderström B. (1988). Mycelial uptake, translocation and assimilation of nitrogen from ^{15}N-labelled ammonium by *Pinus sylvestris* plants infected with four different ectomycorrhizal fungi. *New Phytologist*, **110**, 59-66.

France, R. C. & Reid, C. P. P. (1983). Interactions of nitrogen and carbon in the physiology of ectomycorrhizae. *Canadian Journal of Botany*, **61**, 964-984.

France, R. C. & Reid, C. P. P. (1984). Pure culture growth of ectomycorrhizal fungi on inorganic nitrogen sources. *Microbial Ecology*, **10**, 187-195.

Frank, A. B. (1894). Die Bedeutung der Mykorrhizapilze für die gemeine Kiefer. *Forstwissenschaftliches Centraalblatt*, **16**, 1852-1890.

Genetet. I., Martin. F. & Stewart, G. R. (1984). Nitrogen assimilation in mycorrhizas. Ammonia assimilation in the N-starved ectomycorrhizal fungus *Cenococcum graniforme*. *Plant Physiology*, **76**, 395-399.

Griffiths. R. P., Caldwell, B. A., Cromack, K., Castellano, M. A., & Morita, R. Y. (1987). A study of chemical and microbial variables ia forest soils colonised with *Hysterangum setchelii* rhizomorphs. In *Proceedings of the 7th North American Conference on Mycorrhizae*, eds. D. M. Sylvia, L. L. Hung & J. H. Graham, p. 196. Institute of Food and Agricultural Sciences, University of Florida, Gainesville, Florida, USA.

Harley, J. L. (1940). A study of the root system of the beech in woodland soils with especial reference to mycorrhizal infection. *Journal of Ecology*, **28**, 107-117.

Harley, J. L. (1978). Ectomycorrhizas as nutrient absorbing organs. *Proceedings of the Royal Society of London, Series B*, **203**, 1-21.

Harley, J. L. & Smith, S. E. (1983). *Mycorrhizal Symbiosis*. Academic Press, London.

Harvey, A. E., Larsen, M. J. & Jurgensen, M. F. (1976). Distribution of ectomycorrhizae in a mature Douglas fir/larch forest system in Western Montana. *Forest Science*, **22**, 393-398.

Laiho, O. (1970). *Paxillus involutus* as a mycorrhizal symbiont of forest trees. *Acta Forestalia Fennica*, **106**, 1-65.

Leake, J. R. & Read, D. J. (1987). Studies on free acid protease of the ericoid endomycorrhizal fungus. In *Proceedings of the 7th North American Conference on Mycorrhizae*, eds. D. M. Sylvia, L. L. Hung & J. H. Graham, pp. 333-334. Institute of Food and Agricultural Sciences, University of Florida, Gainesville, Florida, USA.

Leake, J. R. & Read, D. J. (1988). The biology of mycorrhiza in the Ericaceae. XIII. Some characteristics of the extracellular protease activity of the ericoid endophyte *Hymenoscyphus ericae*. *New Phytologist*, in press.

Levy, M. R. & Chou, S. C. (1974). Some properties and susceptibility to inhibitors of partially purified acid proteases from *Plasmodium berghei* and from ghosts of mouse red cells. *Biochimica et Biophysica Acta*, **334**, 423-430.

Lundeberg, G. (1970). Utilization of various nitrogen sources, in particular bound soil nitrogen, by mycorrhizal fungi. *Studia Forestalia Sueccica*, **79**, 1-95.

MacKenzie, S. L. & Holme, K. R. (1984). Analysis of conifer leaf free amino acids by gas liquid chromatography. *Journal of Chromatography*, **299**, 387-396.

Martin, F. (1985). ^{15}N-NMR studies of nitrogen assimilation and amino acid biosynthesis in the ectomycorrhizal fungus *Cenococcum graniforme*. *FEBS Letters*, **182**, 350-354.

Martin, F., Stewart, G. R., Genetet, I. & Le Tacon, F. (1986). Assimilation of $^{15}NH_4{}^+$ by beech (*Fagus sylvatica* L.) ectomycorrhizas. *New Phytologist*, **102**, 85-94.

Martin, F., Ramstedt, M. & Söderhall, K. (1987). Carbon and nitrogen metabolism in ectomycorrhizal fungi and ectomycorrhizas. *Biochemie*, **69**, 569-581.

Melin, E. (1925). *Untersuchungen über die Bedeutung der Baummykorrhize*. G. Fischer, Jena.

Melin. E. & Nilsson, H. (1952). Transport of labelled nitrogen from an ammonium source to pine seedlings through mycorrhizal mycelium. *Svensk Botanisk Tidskrift*, **46**, 281-285.

Melin, E. & Nilsson, H. (1953). Transfer of labelled nitrogen from glutamic acid to pine seedlings through the mycelium of *Boletus variegatus* (Sw.) Fr. *Nature*, **171**, 134.

Meyer, F. H. (1973). Distribution of ectomycorrhizae in native and man-made forests. In *Ectomycorrhizae*, eds. G. C. Marks & T. T. Kozlowski, pp. 79-105. Academic Press, New York.

Mikola, P. & Laiho, O. (1962). Mycorrhizal relations in the raw humus layer of northern spruce forests. *Communicationes Instituti Forestalia Fenniae*, **55**, 1-13.

Nannipieri, P., Ceccanti, B., Cervelli, S. & Conti, C. (1982). Hydrolases extracted from soil, kinetic parameters of several enzymes catalysing the same reaction. *Soil Biology & Biochemistry*, **14**, 429-432.

Pate, J. (1986). Economy of symbiotic nitrogen function. In *On the Economy of Plant Form and Function*, ed. T. J. Givnish, pp. 299-325. Cambridge University Press, Cambridge.

Pearson, V. & Read, D. J. (1975). The physiology of the mycorrhizal endophyte of *Calluna vulgaris*. *Transactions of the British Mycological Society*, **64**, 1-7.

Possingham, J. V. & Groot Obbink, J. (1971). Endotrophic mycorrhiza and the nutrition of grape vines. *Vitis*, **10**, 120-130.

Read, D. J. (1983). The biology of mycorrhiza in the Ericales. *Canadian Journal of Botany*, **61**, 985-1004.

Read, D.J. (1985). The structure and function of the vegetative mycelium of mycorrhizal roots. In *The Ecology and Physiology of the Fungal Mycelium*, eds. D. H. Jennings & A. D. M. Rayner, pp. 215-240. Cambridge University Press. Cambridge.

Read, D. J. (1987). In support of Frank's organic nitrogen theory. *Angewandt Botanik*, **61**. 25-37.

Read, D. J. & Bajwa, R. (1985). Some nutritional aspects of the biology of ericaceous mycorrhizas. *Proceedings of the Royal Society of Edinburgh*, **85B**, 317-332.

Rygiewitz, P. T., Bledsoe, G. S. & Zasoski, R. J. (1984a). Effects of ectomycorrhizae and solution pH on ^{15}N ammonium uptake by coniferous seedlings. *Canadian Journal of Forest Research*, **14**, 885-892.

Rygiewitz, P. T., Bledsoe, G. S. & Zasoski, R. J. (1984b). Effects of ectomycorrhizae and solution pH on ^{15}N nitrate uptake by coniferous seedlings. *Canadian Journal of Forest Research*, **14**, 893-899.

Sangwanit, V. & Bledsoe, C. (1985). Organic nitrogen uptake by axenically-grown mycorrhizal coniferous roots. In *Proceedings of the 6th North American Conference on Mycorrhizas*, ed. R. Molina, p. 346. Forest Research Laboratory, Oregon State University, Corvallis, Oregon, USA.

Smith, F. A., Smith, S. E., St John, B. J., & Nicholas, D. J. D. (1986). Inflow of N & P into roots of mycorrhizal and non-mycorrhizal onions. In *Proceedings of the 1st European Symposium on Mycorrhizas*, eds. V. Gianinazzi-Pearson & S. Gianinazzi, pp. 371-375. INRA, Paris.

Spinner, S. & Haselwandter, K. (1985). Proteins as nitrogen sources for *Hymenoscyphus (= Pezizella) ericae*. In *Proceedings of the 6th North American Conference on Mycorrhizae*, ed. R. Molina, p. 422. Forest Research Laboratory, Oregon State University, Corvallis, Oregon, USA.

Stribley, D. P. & Read, D. J. (1974). The biology of mycorrhiza in the Ericaceae. IV. The effect of mycorrhizal infection on uptake of ^{15}N from labelled soil by *Vaccinium macrocarpon* Ait. *New Phytologist*, **73**, 1149-1155.

Stribley, D. P. & Read, D. J. (1976). The biology of mycorrhiza in the Ericaceae VI. The effects of mycorrhizal infection and the concentration of ammonium on growth of cranberry (*Vaccinium macrocarpon* Ait.) in sand culture. *New Phytologist*, **77**, 63-72.

Stribley, D. P. & Read, D. J. (1980). The biology of mycorrhiza in the Ericaceae. VIII. The relationship between mycorrhizal infection and the capacity to utilize simple and complex organic nitrogen sources. *New Phytologist*, **86**, 365-371.

Werlich, I. & Lyr, H. (1957). Über die Mykorrhizaausbildung von Kiefer (*Pinus sylvestris* L.) und Buche (*Fagus sylvatica* L.) auf verschiedenen Standorten. *Archiv Forstwiss*, **61**, 1-23.

Chapter 10

Acquisition of phosphorus by VA mycorrhizal fungi and the growth responses of their host plants

B. A. D. Hetrick

Department of Plant Pathology, Kansas State University, Throckmorton Hall, Manhatten, Kansas 665016, USA

Introduction

While phosphorus is an extremely important mineral for plant nutrition, it is only sparingly available in most soils. The majority of phosphorus is present in soil in insoluble inorganic forms. At low pH, iron and aluminium phosphates are common while apatite, hydroxyapatite or possibly tricalcium phosphates occur at higher pH (Barber, 1984). Phosphorus can also be present in organic forms, the availability of which is controlled by mineralization rates in soil. The phosphorus content of soils averages approximately 0·05% by weight, but that of the soil solution is much lower. For example, Barber (1984) compared a wide range of soils and found the average phosphorus content of soil solution to be 0·05 mg P l^{-1}.

As plants remove phosphorus from the soil solution, a depletion zone of 1-2 mm may develop around roots (Bhat & Nye, 1974; Lewis & Quirk, 1967). Following development of this depletion zone, availability of phosphorus to the plant is not so much determined by the concentration of phosphorus in the soil, but rather by the rate of movement of phosphorus through the soil to replenish the soil solution surrounding the root. Phosphorus availability to plants is often considered to be diffusion limited because phosphorus is virtually immobile in soil.

Nye & Tinker (1977) suggested that diffusion of phosphorus in soil is related to volumetric water content, tortuosity (a measure of how difficult a pathway nutrients must take to reach a plant root), the diffusion coefficient of orthophosphate in water, and the buffering capacity of the soil. The buffering capacity of soil is a function of the intensity of phosphorus (or phosphorus in soil solution) and the total quantity of phosphorus in a soil. In soils fertilized with a readily available form of phosphorus, the level of phosphorus in the soil solution increases and phosphorus uptake by plants may be most limited by the rate at which the plant can absorb the element. In this situation, mycorrhizal fungi may offer little benefit to plant phosphorus uptake.

In contrast, soils amended with relatively insoluble fertilizer forms, such as rock phosphate, show little change in solution phosphorus but the overall quantity of phosphorus increases. In these soils and in native, nonfertilized soils which are naturally low in available phosphorus, phosphorus uptake by the plant is strongly diffusion-limited and plant acquisition of the element depends on replenishment of the depletion zone around the roots. It is in these cases that mycorrhizal fungi may make an important contribution to phosphorus acquisition by plants.

Since the pioneering research of Mosse (1957) it has been demonstrated repeatedly, and it is widely accepted, that mycorrhizal plants contain considerably higher concentrations of phosphorus than do nonmycorrhizal plants. While a number of mechanisms have been proposed to explain this phenomenon (Tinker, 1975), the main mechanism is considered to be absorption of phosphorus from soil by hyphae of mycorrhizal fungi and translocation of this nutrient through hyphae into roots colonized by the mycorrhizal fungus. In this way the soil outside the root's depletion zone is explored for nutrient and the ability of mycorrhizal plants to obtain phosphorus is extended far beyond the capability of a nonmycorrhizal plant (Rhodes & Gerdemann, 1975). It has been estimated that the hyphae external to the root may virtually double the absorbing surface of plants (Gianinazzi-Pearson & Gianinazzi, 1983). These external hyphae absorbed and transported phosphorus over 7 cm distance to a root while nonmycorrhizal plant roots could not absorb phosphorus even 1 cm away from the root (Rhodes & Gerdemann, 1975). In addition to this physical absorbing surface provided by mycorrhizal fungi, Cress, Throneberry & Lindsey (1979) have suggested that external hyphae have a greater affinity (lower 'K_m') for phosphorus than do nonmycorrhizal roots. This mechanism was questioned by Karunaratne, Baker & Barker (1986) who observed no such increased affinity of mycorrhizas for phosphorus. They concluded that mycorrhizal infection simply increases the number of uptake sites per unit area of root.

The greater phosphorus uptake of mycorrhizal plants can lead to considerable increases in plant growth (Mosse, 1973; Menge, 1983; Hayman, 1983). For example, biomass increases of 306% for corn, 156% for subterranean clover, 80-100% for white clover, 29-34% for ryegrass, 48-3155% for onions, 250% for strawberries, and 220% for wheat have been observed (Menge, 1983). These levels of growth response are generally maxima achieved by assessing growth response under ideal conditions. Thus, these levels of yield increase may represent the potential rather than the actual benefit from the mycorrhizal symbiosis which can be realized under field conditions. Indeed, responses to mycorrhizal symbiosis in the field are notoriously difficult to predict (Hayman, 1983), perhaps because the bulk of research to date has been conducted with sterilized soils and fails to consider the significant impact of the soil microflora. Also, most experiments have studied single plants in single pots and fail to consider the influence of inter- or intra-specific plant competition.

This chapter examines phosphorus acquisition by mycorrhizal fungi and the resulting growth enhancement for host plants as an interaction of phosphorus availability in soil, the dependence of the plant on the symbiosis for phosphorus acquisition and the ability of the fungus to absorb and translocate phosphate. The modifying effect of the soil microflora and interplant competition on this interaction will also be discussed.

Phosphorus availability in soil

Microbial solubilization of inorganic forms of phosphorus

While phosphorus availability in soil is intimately related to the parent material of which a soil is composed, mineralization and solubilization are microbially related processes which can directly control phosphorus availability (Tiessen, Stewart & Cole, 1984). The relatively recent observations that the rhizosphere microflora of mycorrhizal plants differs from that of nonmycorrhizal plants (Meyer & Linderman, 1986; Ames, Reid & Ingham, 1984) invites speculation about the impact of this difference in microbial composition on phosphorus availability in the rhizosphere of mycorrhizal roots.

It is clear that rock phosphate-solubilizing bacteria can themselves increase plant yield (Bajpai & Rao, 1971; Kabesh, Saber & Young, 1975), and the synergistic interaction between these bacteria and mycorrhizal fungi may further improve plant growth (Azcon, Barea & Hayman, 1976; Raj, Bagyaraj & Manjunath, 1981). However, contradictory results have also been reported by Lee & Bagyaraj (1986) who observed that phosphorus-solubilizing bacteria did not increase shoot dry weight or phosphorus uptake, whether these bacteria were applied alone or in combination with mycorrhizal fungi. Also, Lee & Bagyaraj (1986) observed no benefit from sulphur-oxidizing *Thiobacillus* bacteria which are known to solubilize phosphorus during sulphur oxidation. Based on these results it is impossible to conclude whether or not phosphorus-solubilizing bacteria play any significant role in availability of phosphorus to mycorrhizal fungi. Since most studies to date have examined the reintroduction of phosphorus-solubilizing microorganisms into sterilized soil, little information is available on the role of these microbes in mycorrhizal plant growth under natural conditions. It is also unclear whether phosphorus-solubilizing microorganisms are more common in the rhizosphere of mycorrhizal plants than nonmycorrhizal plants.

The ability of mycorrhizal plants themselves to obtain phosphorus from relatively unavailable (inorganic) phosphates has been sharply contested for many years. It was generally believed that mycorrhizal and nonmycorrhizal plants use the same phosphate sources (Sanders & Tinker, 1971; Mosse, Hayman & Arnold, 1973; Gianinazzi-Pearson *et al.*, 1981; Brechet & LeTacon, 1984). This conclusion, reached by comparing the specific activity of mycorrhizal and nonmycorrhizal plants following ^{32}P addition to soil, assumes that added ^{32}P remains in the labile pool and does not exchange with unavailable forms of phosphorus. However, Bolan *et al.* (1984) demonstrated that precipitation or exchange of ^{32}P

with phosphorus sources unavailable to plant growth does occur. Hence, lack of difference in specific activity of mycorrhizal and nonmycorrhizal plants does not preclude use of unavailable phosphorus forms by mycorrhizal plants. Subsequently, Bolan, Robson & Barrow (1987) amended soil with three inorganic phosphorus sources of varied solubility. Mycorrhizal plants benefitted more than nonmycorrhizal plants from these fertilizers but the degree of difference was greatest with the most insoluble phosphorus fertilizer. In other words, the less soluble the phosphorus source, the greater the benefit from mycorrhizal fungus inoculation. This is not surprising since phosphorus availability to plants is still diffusion-limited following fertilization with insoluble forms of phosphorus. It is in precisely these situations that the benefits of mycorrhizal infections are maximised.

The data of Bolan *et al.* (1987) and others (Daft & Nicholson, 1966; Hall, 1975; Powell & Daniel, 1978), suggest that mycorrhizal plants may be able to use relatively unavailable phosphorus sources. However, since mycorrhizal inoculation in most experiments is accomplished with inoculum consisting of rhizosphere sievings, nonaxenic spores, or soil containing spores, infected root fragments and hyphae, it is unclear whether the benefit to mycorrhizal plants arising from application of insoluble phosphorus is a result of solubilization of the phosphorus by the fungi themselves or by their accompanying microflora.

Microbial mineralization of organic forms of phosphorus

Mineralization of phosphorus from organic matter is another means by which the element becomes available for plant growth. It has been suggested that if mineralization rates are high, mycorrhizal colonization could be shed by the plant in response to higher levels of available phosphorus (Baylis, 1967; Daft & Nicholson, 1969; Crush, 1973). Nutrient release by mineralization could certainly explain why microorganisms other than mycorrhizal fungi improve plant growth when added to steamed soil (Hetrick *et al.*, 1988a). Furthermore, if the need for mycorrhizas decreases as a consequence of mineralization, this could partially explain why mycorrhizal growth response and root colonization are reduced in nonsterile soil (Hetrick, *et al.*, 1988a). However, since plant growth in nonsterile soil is still reduced as compared to that in steamed soil, it seems most realistic to assume that both mineralization and microbial competition for available nutrients may occur more or less simultaneously.

Tallgrass prairie soils are highly productive but contain relatively little plant-available phosphorus as predicted by soil tests (Halm, Stewart & Halstead, 1972). Instead, organic phosphorus is abundant and mineralization appears to be rate-limiting (Cole *et al.*, 1978; Blair, Till & Smith, 1977). Plants native to tallgrass prairie are highly dependent on mycorrhizal symbiosis (Hetrick, Kitt & Wilson, 1988c), suggesting that mycorrhizal fungi, or their associated microflora, play a significant role in mineralization of organic matter. That the fungi themselves mineralize organic matter is suggested by the superior growth of mycorrhizal plants in steamed soil from which the soil microflora has been virtually eliminated (Hetrick *et al.*, 1988a). Mycorrhizal fungi are also found in association with organic matter in soils (Dowding, 1959; Mosse, 1959; Nicolson, 1959; Koske, Sutton & Shepard, 1975; Nicolson & Johnston, 1979). St. John, Coleman & Reid (1983) observed that, while mycorrhizal fungal hyphae did not appear to grow toward organic matter deposits, hyphal length was greater around organic matter, probably because of increased hyphal branching in these sites. This suggests that hyphae adopted an absorptive growth pattern when confronted with a nutrient source.

Mineralization of organic matter by mycorrhizal fungi is further suggested by Warner & Mosse (1980), Hepper & Warner (1983), and Warner (1984), who observed that saprotrophic growth by mycorrhizal fungi was related to soil organic matter. More recently, it was demonstrated that acid phosphatase activity is higher in the rhizosphere of mycorrhizal plants than nonmycorrhizal plants (Dodd *et al.*, 1987). It is impossible to conclude from these data whether it is mycorrhizal fungi themselves or the associated microflora which actually effect mineralization of nutrient. It is entirely possible that mineralization is accomplished by the soil microflora and that mycorrhizal hyphae merely take advantage of the released nutrient by positioning themselves and proliferating near organic matter. It is also possible that mycorrhizal plants, by altering their rhizosphere microflora (Barea, Azcon & Hayman, 1975; Azcon *et al.*, 1976; Hayman, 1978; Bagyaraj & Menge, 1978;

Krishna, Balakrishna & Bagyaraj, 1982; Ames *et al.*, 1984; Meyer & Linderman, 1986; Ames, Mihara & Bethlenfalvay, 1987) actually increase mineralization rates in soil. That mycorrhizal fungi exert some direct control on their rhizosphere is implied by the results of Morandi, Bailey & Gianinazzi-Pearson (1984), who found mycorrhizal plants to contain higher levels of isoflavonoids, glyceollin I, coumestrol, and diadzein than nonmycorrhizal plants. Although Morandi *et al.* (1984) suggest that these levels of phenolic compounds explain the greater resistance of mycorrhizal plants to plant pathogens, these compounds could also be expected to impact on nonpathogenic microorganisms if they leak from roots or are released from killed plant cells. If mycorrhizas stimulate mineralization by differentially increasing the population of soil microorganisms involved in mineralization, this would be an extremely important role for these fungi in nutrient cycling. However, the impact of mycorrhizal symbiosis on mineralization has not been studied and is a subject of speculation.

Acquisition of phosphorus and plant response

Role of soil fertility

Phosphorus acquisition by mycorrhizal fungi and subsequent benefit to plant growth, or growth response, is related to the degree to which plants rely on the symbiosis for nutrient absorption. Gerdemann (1975) used the term mycorrhizal dependence (MD) to describe this concept, defining it as 'the degree to which a plant is dependent on the mycorrhizal condition to produce its maximum growth or yield at a given level of soil fertility.' The reliance of a plant on the symbiosis was defined quantitatively by Menge *et al.* (1982) as percentage growth response:

$$\mathrm{MD} = \frac{\text{dry wt of mycorrhizal plant}}{\text{dry wt of nonmycorrhizal plant}} \times 100\%$$

Subsequently, Plenchette, Fortin & Furlan (1983) proposed that mycorrhizal dependence should more accurately be expressed as a percentage value termed relative field mycorrhizal dependency (RFMD):

$$\mathrm{RFMD} = \frac{\text{(dry wt mycorrhizal plant)-(dry wt nonmycorrhizal plant)}}{\text{dry wt mycorrhizal plant}} \times 100\%$$

The latter assessment of dependency recognizes that a plant cannot be more than 100% dependent on the symbiosis.

Differences in the mycorrhizal dependence of plants probably relate largely to innate genetically determined characteristics such as root morphology, growth rate or carbohydrate allocation. However, soil fertility can determine what portion of the potential benefit from mycorrhizal symbiosis is realized. Recognising this, Plenchette *et al.* (1983) proposed that the RFMD value should carry a superscript to designate the soil phosphorus level at which it was derived. However, available soil phosphorus levels are not the sole determinants of mycorrhizal dependence. For example, Menge *et al.* (1982) demonstrated that although dependence of Troyer citrange (citrus rootstock) varied in 26 soils tested, each was positively correlated with pH and inversely correlated with extractable soil P, Zn, Mn, Cu, percentage organic matter, and cation exchange capacity. Using regression analysis, soil P, Mn, and Zn were shown to have the greatest predictive value for mycorrhizal dependence. In contrast, in a similar study using big bluestem (*Andropogon gerardii* Vitman) in 15 native prairie and agricultural soils, mycorrhizal dependence was not strongly correlated with any individual soil parameter, but a combination of parameters (K, Ca, Cu, and Fe) allowed soils resulting in high or low dependence to be differentiated (Kitt, Hetrick & Wilson, 1988). Thus, soil fertility is an important determinant of mycorrhizal phosphorus uptake and of a plant's dependence on mycorrhizas.

Role of root morphology

A relationship between root morphology and mycorrhizal dependence was first elucidated by Baylis (1970). He suggested that root hairs and mycorrhizas were alternative means for obtaining phosphorus with obligatory mycotrophy occurring in plants with poorly developed roothairs. Subsequently, Baylis (1974) observed that there is a 'sharply defined level of available phosphorus, related to root morphology, above which growth without mycorrhizas proceeds steadily and below which the species is mycotrophic.' Since this hypothesis was originally proposed, the relationship between root morphology, available phosphorus level, and mycorrhizal dependence has been confirmed for a wide range of plant species (St. John, 1980; Chilvers & Daft, 1981; Saif, 1987).

In support of this theory, it was suggested that the root hair density of certain 'nonhost' plants (plants which normally do not form mycorrhizas) is actually a physical deterrent to root colonization (Ocampo, Martin & Hayman, 1980) and that root hairs and mycorrhizas may, in fact, be mutually exclusive (Chilvers & Daft, 1981). In contrast, Pope *et al.* (1983) suggested that root length and fineness of the root system might better predict mycorrhizal dependence than root hair density. However, in tropical forage species, root production, fibrosity, and geometry were more closely related to dependence than was root hair length (Saif, 1987). Throughout these studies, the shorter and coarser (less finely branched) the root system, the greater was the mycorrhizal dependence of the host plants. In five species of orange, root fineness (length per unit weight of dry root) was also an important predictor of mycorrhizal dependence, perhaps because a fine root system contained greater surface area for absorption of water (Graham & Syvertsen, 1985). Since ability to absorb water is related to ability to take up phosphorus, they suggested an inverse relationship between hydraulic conductivity and mycorrhizal dependence.

These studies demonstrate a significant relationship between mycorrhizal dependence and root morphology. While the growth-altering impact of ectomycorrhizal fungi on host root morphology is well documented (Harley & Smith, 1983), similar changes in root morphology are not generally attributed to vesicular-arbuscular mycorrhizal fungi (Gerdemann, 1971). Only recently have any differences in the geometry of mycorrhizal and nonmycorrhizal plant roots been observed. Fitter (1985a), using graph theory to model root topology, found that both mycorrhizal and nonmycorrhizal plants grew initially in an ordered pattern, presumably facilitating efficient exploration of soil. However, nonmycorrhizal plants began to branch and change to a more random growth pattern sooner than mycorrhizal plants. Presumably, the latter growth pattern facilitates more effective absorption of nutrients from soil. Mycorrhizal plants stayed in the less-branched, exploratory growth pattern longer, perhaps because mycorrhizal hyphae performed absorption for the plants and branching of roots was less necessary. Whether such changes in growth pattern reflected nutritional differences between mycorrhizal and nonmycorrhizal plants was unclear.

Since VAM fungi produce hormones (Allen, Moore & Christensen, 1980) or may stimulate hormone production by the host plant (Edress, Davis & Burger, 1984), mycorrhizal fungi may alter rooting strategy in ways not equalled by phosphorus fertilization, although mycorrhizal change in root architecture might still be a response to host nutrition. In support of this hypothesis, Hetrick *et al.* (1988b) have demonstrated that the mycorrhizal condition significantly reduces root branching in big bluestem grass. By maintaining a more elongate, exploratory root growth pattern, mycorrhizal plants exploit a greater soil volume for nutrients and have access to larger nutrient pools. They also limit energy loss due to unnecessary and redundant root maintenance if mycorrhizal hyphae perform absorption functions for the plant. Since plants with coarser root systems are generally more dependent on mycorrhizal symbiosis, Hetrick *et al.* (1988b) have further suggested that, by reducing root branching in host plants, mycorrhizal fungi can actually increase mycorrhizal dependence. The changes in root branching were not explained by differences in phosphorus availability but seemed to be directly controlled by the fungi. In the presence of the soil microflora, however, root branching was further decreased, suggesting that mycorrhizal plants can explore an even greater soil volume for nutrients when microbial competition further limits nutrient availability. Thus, while phosphorus availability alone could not explain the observed differences in root topology of mycorrhizal plants, Hetrick *et al.* (1988b) suggest that the mycorrhizal fungi control root topology in response to soil conditions.

Role of host plant physiology

While root morphology is an important determinant of mycorrhizal dependence, host plant physiology also contributes. Variation in plant growth rate affects mycorrhizal dependence, with slow growing plants having generally lower dependence (Hall, 1975; Graham & Syvertson, 1985). Root exudation also affects mycorrhizal dependence. When mycotrophic and nonmycotrophic plant species were compared, plants which formed mycorrhizal associations had higher rates of root exudation, but no specific qualitative differences in the exudate of these plants were detected (Schwab, Leonard & Menge, 1984). This relationship between root exudation and mycorrhizal symbiosis was also demonstrated by Schwab, Johnson & Menge (1982) who found that a normally nonmycorrhizal plant species (*Chenopodium quinona*) became weakly mycorrhizal following application of sublethal dosages of simazine, a herbicide which increased root exudation. Apparently, some minimum rate of exudation is necessary to initiate and maintain mycorrhizal symbiosis. The greater exudation rate of mycorrhizal plants may result from selective pressure to supply substrate for the fungus, while, alternatively, the lower exudation rate of nonmycorrhizal or weakly mycorrhizal plants could result from selection pressure to conserve photosynthate (Schwab *et al.*, 1984). Certainly, loss of photosynthate to a fungal symbiont when that symbiont is not necessary to plant growth and survival would be counterproductive.

A relationship between root exudation and mycorrhizal dependence was also examined by Azcon & Ocampo (1981), who observed that wheat cultivars which displayed higher levels of mycorrhizal dependence also had relatively high rates of root exudation. While the degree to which plants exude nutrients is, in part, genetically determined, it is obvious from comparisons of mycotrophic and nonmycotrophic plant species that root exudation may reflect variation in the ability of a plant's root system to absorb nutrients adequately without mycorrhizal symbiosis. However, subtle, more temporal differences in root exudation also occur and have been related to the phosphorus content of host plant tissues (Ratnayake, Leonard & Menge, 1978; Graham, Leonard & Menge, 1981). Presumably, phosphorus mediates the membrane permeability of roots and therefore stimulates or inhibits the mycorrhizal symbiosis. In this way the host plant can control loss of photosynthate and decrease mycorrhizal dependence when nutrient is not limited and the symbiosis is unnecessary, i.e. when adequate amounts of nutrients can be supplied by the root system itself. From this perspective, root morphology and root exudation may be interrelated and in concert define a plant's mycorrhizal dependence and growth response.

This interrelationship of root exudation and root morphology led Hetrick *et al.* (1988b) to propose a model in which plant and fungus possess interrelated controls on the symbiosis. According to this model, in response to the nutritional status of the plant, root exudation rate changes and thereby determines whether or not the fungus will colonize and function in the plant roots (Ratnayake *et al.*, 1978; Graham *et al.*, 1981). Once the symbiosis is established, in response to fertility and rhizosphere conditions encountered in the soil, the mycorrhizal fungus alters the host plant root branching pattern to control the volume of soil that will be explored for nutrient (Hetrick *et al.*, 1988b).

Ability of mycorrhizal fungi to acquire phosphorus

Impact of the soil microflora

Acquisition of phosphorus by mycorrhizal fungi is directly related to the development and survival of mycorrhizal fungal hyphae in soil. External hyphae of mycorrhizal fungi are subject to a wide range of synergistic and antagonistic interactions with the soil microflora which can contribute to or reduce mycorrhizal phosphorus acquisition. The potential impact of these interactions is considerable and led Daniels & Menge (1980) to suggest that non-pathogenic soil microorganisms could be viewed as plant pathogens if they limit mycorrhizal nutrient uptake and growth response.

Certainly, hyperparasites of mycorrhizal fungi may limit the potential contribution of mycorrhizas to plant growth by reducing inoculum levels in soil (Daniels & Menge, 1980). Similarly, collembola and microarthropods which graze on mycorrhizal fungal spores and hyphae probably reduce mycorrhizal phosphorus uptake. The impact of these grazers is potentially considerable, since they could destroy hyphal bridges between plants and limit hyphal growth in soil. Since hyphal extension in soil is critical to nutrient uptake by the fungi, Fitter (1985b) has suggested that, under field conditions, mycorrhizal symbiosis may be of more limited benefit to plant growth than is implied by pot experiments.

Fitter's (1985b) contention is supported by the fact that growth response to mycorrhizas and root colonization by the fungi in non-sterile soil frequently does not equal that achieved in sterilized soil to which only mycorrhizal fungi are returned (Mosse, 1977; Mosse, Hayman & Ide, 1969; Habte & Aziz, 1985; Hetrick, Kitt & Wilson,

1986). This suppression of mycorrhizal growth response and root colonization in nonsterile soil is attributable to the actions of the soil microorganisms rather than to any chemical or physical differences because addition of 1 or 10% nonsterile soil to pasteurized soil results in suppression similar to that occurring in 100% nonsterile soil (Hetrick *et al.*, 1986). While this evidence for microbial involvement in the suppression phenomenon is indirect, more direct evidence was found when bacteria and fungi, isolated from nonsterile soil by dilution plating, were returned to sterilized soil (Hetrick *et al.*, 1988a). By this method, bacteria which directly suppress mycorrhizal phosphorus uptake and subsequent growth response were detected. Although Hetrick *et al.* (1988a) observed no deleterious effects of soil fungi on growth response, at least some fungi probably also limit growth response and root colonization, since application of the fungicide metalaxyl to soil increases mycorrhizal root colonization (Groth & Martinson, 1983).

While hyperparasites or VAM-grazers could explain some of these observations, another mechanism has also been proposed. The soil microflora may compete with mycorrhizal hyphae for available nutrient. This seems likely, since addition of phosphorus to infertile soils can eliminate the plant growth depression observed in the presence of the soil microflora (Hetrick *et al.*, 1988b). Also, higher phosphorus levels are tolerated, without eliminating mycorrhizal growth response, when the soil microflora is present (Hetrick *et al.*, 1988b). Wilson, Hetrick & Kitt (1988) observed that mycorrhizal fungal spore germination was lower in nonsterile phosphorus-deficient soil than in steamed soil. Spore germination in nonsterile soil was stimulated by addition of phosphorus while a similar amendment to steamed soil inhibited germination. Whether nutrient is directly necessary for spore germination or indirectly stimulates spore germination by encouraging growth of the soil microflora adjacent to the spore is unknown. However, reduced spore germination as a result of microbial competition for nutrient could explain the lower root colonization levels and mycorrhizal growth response observed in the presence of the soil microflora.

The relationship of soil nutrient levels to suppression of plant growth in nonsterile soil was further studied by Kitt *et al.* (1988)

who observed plant growth in 15 soils and determined that the magnitude of suppression could be described by regression equations using soil Na, Zn, Mn, organic matter, Mg, and NH_4. High levels of suppression occurred when levels of Na, Mg, and organic matter were low and Zn, Mn, and NH_4 were abundant. Thus, while competition for nutrients may limit growth response and root colonization, nutrients other than phosphorus may also be important in determining the degree of suppression which occurs in a particular soil.

Impact of interplant competition

As previously mentioned, mycorrhizal fungal acquisition of phosphorus has been studied almost exclusively in greenhouse pot experiments which ignore intra- and inter-specific competitive interactions of plants. Plant density and spatial pattern effects, potential influences of the above and below-ground microclimate of a natural plant stand, and the impact of these interactions on growth response have also been ignored. As plant roots and mycorrhizal fungi compete for soil nutrients, it seems likely that phosphorus availability will be modified in ways not predicted by chemical soil tests of nutrient availability. Above-ground competition between plants could also influence below-ground interactions and growth response (Harper, 1977).

Most studies involving mycorrhizal host plant effects on other host plants have been conducted with the facultative mycotroph, ryegrass (*Lolium perenne*). For example, Fitter (1977) observed that ryegrass dry weight and root length decreased when grown in competition with Yorkshire fog (*Holcus lanatus*). Root colonization by the mycorrhizal fungus increased in ryegrass, however, in response to that competition. Presumably, *H. lanatus* was better able than *L. perenne* to take advantage of the symbiosis, and mycorrhizal colonization was ultimately deleterious to *L. perenne*. Thus, even in low phosphate soil, mycorrhizal symbiosis may be deleterious if the plant is a poor competitor for resources (Fitter, 1977).

Similar results were obtained when clover (*Trifolium repens*) and ryegrass were cropped together. Hall (1978) observed that, while clover grown alone benefitted from mycorrhizal symbiosis only at lower phosphorus levels, clover grown with ryegrass benefitted from the symbiosis at higher phosphorus levels. However, clover dry weight and phosphorus content were still reduced in the presence of ryegrass. Mycorrhizal symbiosis improved the ability of clover to compete with ryegrass, and this competitive advantage was attributed to its superior ability to compete for phosphorus. Subsequently, Buwalda (1980), also using clover and ryegrass, observed that the competitive advantage of clover was reduced at high nitrogen and phosphorus levels, confirming that clover's competitive advantage in low fertility soils is related to its superior nutrient uptake ability. Mycorrhizal symbiosis, which at low fertility functions more efficiently in clover than in ryegrass, is apparently a major factor influencing competition between these plants.

The potential role of mycorrhizal symbiosis in influencing outcomes of interspecific competition has also been confirmed with two tallgrass prairie plant species, big bluestem (*Andropogon gerardii*) (obligately mycorrhizal) and Junegrass (*Koeleria pyranidata*) (facultatively mycorrhizal) (B. A. D. Hetrick *et al.*, unpublished). In greenhouse studies, dry weight of mycorrhizal big bluestem was not influenced by the presence of Junegrass but dry weight of nonmycorrhizal big bluestem was significantly reduced when grown with Junegrass, despite phosphorus fertilization. In contrast, dry weight of mycorrhizal Junegrass was significantly lower in the presence of mycorrhizal big bluestem. In the absence of the symbiosis, Junegrass was not affected by the presence of big bluestem. Hetrick, Kitt & Wilson (1988c) have demonstrated that cool-season (C_3) grasses like Junegrass are significantly less dependent on mycorrhizal symbiosis than are warm-season (C_4) grasses like big bluestem. They have also confirmed a relationship between root morphology and mycorrhizal dependence for these plants. The detrimental effect of the symbiosis on the competitive ability of facultative mycotrophs like Junegrass may have resulted in selection for temporal niche divergence (shifting growth phenology to earlier in the season), thus reducing competition and allowing the stable coexistence of these two groups.

Janos (1980) has suggested that obligate mycotrophs dominate in plant communities established in nutrient-deficient soils because the absorption of nutrients via hyphae affords the greatest exploration of the soil for nutrient. Janos (1980) further suggests that, in these nutrient deficient soils, obligate mycotrophs would have a competitive advantage over nonmycotrophs or facultative mycotrophs because development of abundant root hairs or fine root systems, adaptations for nutrient acquisition without the symbiosis, are a costly expenditure for the plant, particularly if they are ineffective. Root hairs are less effective than hyphae in absorbing phosphorus from nutrient-poor soils because phosphorus depletion zones around root hairs overlap and subsequently, root hairs may compete with each other for nutrients (Janos, 1980).

Conclusions

Mycorrhizal symbiosis may be most important to plant growth when phosphorus availability in soil is diffusion-limited. In these situations absorption of nutrients by mycorrhizal fungal hyphae and translocation of these nutrient through hyphae to plants can significantly improve plant growth. Phosphorus acquisition by mycorrhizal fungi and the resulting growth response in plants is envisioned as an interaction of phosphorus availability in soil, the plant's reliance on the symbiosis and the ability of the fungus to absorb and translocate phosphorus.

Phosphorus availability in soil is microbially mediated. A growing body of circumstantial evidence suggests that mycorrhizal fungi or their accompanying microflora may improve soil phosphorus availability by solubilizing inorganic forms of phosphorus or by mineralization of organic phosphorus.

The dependency of plants on mycorrhizal fungi for phosphorus acquisition is influenced by soil fertility, root morphology and host plant physiology, Apparently, the plant controls the presence or absence of the symbiosis in response to internal phosphorus concentrations. The fungus then controls plant root branching patterns and, in effect, the volume of soil explored for nutrient.

The ability of mycorrhizal fungi to acquire phosphorus is influenced by the soil microflora. Hyperparasites and grazers may directly limit mycorrhizal fungus inoculum level and hyphal development in soil, but reduced phosphorus uptake by mycorrhizal fungi may also result from microbial competition for phosphorus. Competition between plants of differing mycorrhizal dependencies also affects mycorrhizal phosphorus acquisition. Mycorrhizal symbiosis may be detrimental to plants with lower mycorrhizal dependency, although the mechanism controlling these competitive interactions needs further research.

References

Allen, M. F., Moore, T. S., Jr. & Christensen, M. (1980). Phytohormone changes in *Bouteloua gracilis* infected by vesicular-arbuscular mycorrhizae. I. Cytokinin increases in the host plant. *Canadian Journal of Botany*, **58**, 371-374.

Ames, R. N., Mihara, K. L. & Bethlenfalvay, G. J. (1987). The establishment of microorganisms in vesicular-arbuscular mycorrhizal and control treatments. *Biology and Fertility of Soils*, **3**, 217-223.

Ames, R. N., Reid, C. P. P. & Ingham, E. R. (1984). Rhizosphere bacterial population responses to root colonization by a vesicular-arbuscular mycorrhizal fungus. *New Phytologist*, **96**, 555-563.

Azcon, R., Barea, J. M. & Hayman, D. C. (1976). Utilization of rock phosphate in alkaline soils by plants inoculated with mycorrhizal fungi and phosphate-solubilizing bacteria. *Soil Biology and Biochemistry*, **8**, 135-138.

Azcon, R. & Ocampo, J. A. (1981). Factors affecting the vesicular-arbuscular infection and mycorrhizal dependency of thirteen wheat cultivars. *New Phytologist*, **87**, 677-685.

Bagyaraj, D. C. & Menge, J. A. (1978). Interaction between a VA mycorrhizal and *Azotobacter* and their effects on rhizosphere microflora and plant growth. *New Phytologist*, **80**, 567-573.

Bajpai, P. D. & Rao, W. V. B. S. (1971). Phosphate solubilising bacteria. Part III. Soil inoculation with phosphorus solubilising bacteria. *Soil Science and Plant Nutrition*, **17**, 46-53.

Barber, S. A. (1984). *Soil nutrient bioavailability: A mechanistic approach*. John Wiley & Sons, New York.

Barea, J. M., Azcon, R. & Hayman, D. S. (1975). Possible synergistic interaction between *Endogone* and phosphate-solubilizing bacteria in low phosphate soils. In *Endomycorrhizas*, eds. F. E. Sanders, B. Mosse & P. B. Tinker, pp. 409-417. Academic Press, London.

Baylis, G. T. S. (1967). Experiments on the ecological significance of phycomycetous mycorrhizas. *New Phytologist*, **66**, 231-243.

Baylis, G. T. S. (1970). Root hairs and phycomycetous mycorrhizas in phosphorus deficient soil. *Plant and Soil*, **33**, 713-716.

Baylis, G. T. S. (1974). The evolutionary significance of phycomycetous mycorrhizas. In *Mechanisms of Regulation of Plant Growth*, bulletin 12, pp. 191-193. The Royal Society of New Zealand.

Bhat, K. K. S. & Nye, P. H. (1974). Diffusion of phosphate to plant roots in soil. II. Uptake along the roots at different times and effect of different levels of phosphorus. *Plant Soil*, **41**, 365-382.

Blair, G. J., Till, A. R. & Smith, R. C. G. (1977). The phosphorus cycle — What are the sensitive areas? In *Reviews in Rural Science III*, Proceedings of a Symposium held at the University of New England, ed. G. J. Blair, pp. 9-19. The University of New England, Armidale, NSW, Australia.

Bolan, N. S., Robson, A. D., Barrow, N. J. & Alymore, L. A. G. (1984). Specific activity of phosphorus in mycorrhizal and non-mycorrhizal plants in relation to the availability of phosphorus to plants. *Soil Biology and Biochemistry*, **16**, 299-304.

Bolan, N. S., Robson, A. D. & Barrow, N. J. (1987). Effects of vesicular-arbuscular mycorrhiza on the availability of iron phosphates to plants. *Plant and Soil*, **99**, 401-410.

Brechet, C. & LeTacon, F. (1984). Responses of endomycorrhizal and non-mycorrhizal plants of *Acer pseudoplatanus* to different levels of soluble and non-soluble phosphorus. *European Journal of Forestry Pathology*, **14**, 68-77.

Buwalda, J. G. (1980), Growth of a clover-ryegrass association with VA mycorrhizas. *New Zealand Journal of Agricultural Research*, **23**, 379-383.

Chilvers, M. T. & Daft, M. J. (1981). Mycorrhizas of the Lilii florae. II. Mycorrhiza formation and incidence of root hairs in field grown *Narcissus* L., *Tulipa* L. and *Crocus* L. cultivars. *New Phytologist*, **89**, 247-261.

Cole, C. V., Elliott, E. T., Hunt, H. W. & Coleman, D. C. (1978). Trophic interactions in soils as they affect energy and nutrient dynamics. V. Phosphorus transformations. *Microbial Ecology*, **4**, 381-387.

Cress, W. A., Throneberry, G. O. & Lindsey, D. L. (1979). Kinetics of phosphorus absorption by mycorrhizal and nonmycorrhizal tomato roots. *Plant Physiology*, **64**, 484-487.

Crush, J. R. (1973). Significance of endomycorrhizas in tussock grassland in Otago, New Zealand. *New Zealand Journal of Botany*, **11**, 645-660.

Daft, M. J. & Nicholson, T. H. (1966). Effect of *Endogone* mycorrhiza on plant growth. *New Phytologist*, **65**, 343-350.

Daft, M. J. & Nicholson, T. H. (1969). Effect of *Endogone* mycorrhiza on plant growth. II. Influence of soluble phosphate on endophyte and host in maize. *New Phytologist*, **68**, 945-952.

Daniels, B. A. & Menge, J. A. (1980). Hyperparasitization of vesicular-arbuscular mycorrhizal fungi. *Phytopathology*, **70**, 584-588.

Dodd, J. C., Burton, C. C., Burns, R. G. & Jefferies, P. (1987). Phosphatase activity associated with the roots and the rhizosphere of plants infected with vesicular-arbuscular mycorrhizal fungi. *New Phytologist*, **107**, 163-172.

Dowding, E. S. (1959). Ecology of *Endogone*. *Transactions of the British Mycological Society*, **42**, 449-457.

Edris, M. H., Davis, R. M. & Burger, D. W. (1984). Influence of cytokinin production in sour orange. *Journal of the American Society of Horticultural Science*, **109**, 587-590.

Fitter, A. H. (1977). Influence of mycorrhizal infection on competition for phosphorus and potassium by two grasses. *New Phytologist*, **79**, 119-125.

Fitter, A. H. (1985a). Functional significance of root morphology and root system architecture. In *Ecological Interactions in Soil*, eds. A. H. Fitter, D. Atkinson, D. J. Read & M. B. Usher, pp. 87-106. Blackwell Scientific Publications, Oxford.

Fitter, A. H. (1985b). Functioning of vesicular-arbuscular mycorrhizas under field conditions. *New Phytologist*, **99**, 257-265.

Gerdemann, J. W. (1971). Fungi that form the vesicular-arbuscular type of endomycorrhizae. In *Mycorrhizae*, ed. E. Hacskaylo, pp. 9-18. Publication number 1189 of the USDA Forest Service, Washington D. C., USA.

Gerdmann, J. W. (1975). Vesicular-arbuscular mycorrhizae. In *The Organization and Structure of Roots*, eds. J. G. Torrey & D. T. Clarkson, pp. 575. Academic Press, London, New York, San Francisco.

Gianinazzi-Pearson, V., Fardeau, J., Asimi, S. & Gianinazzi, S. (1981). Source of additional phosphorus absorbed from soil by vesicular-arbuscular mycorrhizal soybeans. *Physiologie Végétale*, **19**, 33-43

Gianinazzi-Pearson, V. & Gianinazzi, S. (1983). The physiology of vesicular-arbuscular mycorrhizal roots. *Plant Soil*, **71**, 197-209.

Graham, J. H., Leonard, R. T. & Menge, J. A. (1981). Membrane-mediated decrease in root exudation responsible for phosphorus inhibition of vesicular-arbuscular mycorrhiza formation. *Plant Physiology*, **68**, 548-552.

Graham, J. H. & Syvertsen, J. P. (1985). Host determinants of mycorrhizal dependency of citrus rootstock seedlings. *New Phytologist*, **101**, 667-676

Groth, D. E. & Martinson, C. A. (1983). Increased endomycorrhizal infection of maize and soybeans after soil treatment and metalaxyl. *Plant Disease*, **67**, 1377-1378.

Habte, M. & Aziz, T. (1985). Response of *Sesbania grandiflora* to inoculation of soil with vesicular-arbuscular mycorrhizal fungi. *Applied and Environmental Microbiology*, **50**, 701-703.

Hall, I. R. (1975). Endomycorrhizas of *Metrosideros umbellata* and *Weinmannia racemosa*. *New Zealand Journal of Botany*, **13**, 463-472.

Hall, I. R. (1978). Effects of endomycorrhizas on the competitive ability of white clover. *New Zealand Journal of Agricultural Research*, **21**, 509-515.

Halm, B. J., Stewart, J. W. B. & Halstead, R. H. (1972). The phosphorus cycle in a native grassland ecosystem. In *Proceedings of a Symposium on Isotopes and Radiation in Soil-Plant Relationships including Forestry*, pp. 571-589. International Atomic Energy Agency, Vienna.

Harley, J. L. & Smith, S. E. (1983). *Mycorrhizal Symbiosis*. Academic Press, London.

Harper, J. L. (1977). *Population Biology of Plants*. Academic Press, New York.

Hayman, D. S. (1978). Endomycorrhizae. In *Interactions between Nonpathogenic Soil Microorganisms and Plants*, eds. Y. R. Dommergues & S. V. Krupa, pp. 401-442. Elsevier, Amsterdam, Oxford & New York.

Hayman, D. S. (1983). The physiology of vesicular-arbuscular endomycorrhizal symbiosis. *Canadian Journal of Botany*, **61**, 944-963.

Hepper, C. & Warner, A. (1983). Role of organic matter in growth of a vesicular-arbuscular mycorrhizal fungus in soil. *Transactions of the British Mycological Society*, **81**, 155-156.

Hetrick, B. A. D., Kitt, D. G. & Wilson, G. T. (1986). The influence of phosphorus fertilization, drought, fungal species and soil microorganisms on mycorrhizal growth response in tallgrass prairie plants. *Canadian Journal of Botany*, **64**, 1199-1203.

Hetrick, B. A. D., Wilson, G. W. T., Kitt, D. G. & Schwab, A. P. (1988a). Effects of soil microorganisms on mycorrhizal contribution to growth of big bluestem grass in nonsterile soil. *Soil Biology and Biochemistry*, **20**, 501-507.

Hetrick, B. A. D., Leslie, J. F., Wilson, G. W. T. & Kitt, D. G. (1988b). Physical and topological assessment of effects of a vesicular-arbuscular mycorrhizal fungus on root architecture of big bluestem. *New Phytologist*, **110**, 85-96.

Hetrick, B. A. D., Kitt, D. G. & Wilson, G. W. T. (1988c). Mycorrhizal dependence and growth habit of warm-season and cool-season tallgrass prairie plants. *Canadian Journal of Botany*, **66**, 1376-1380.

Janos, D. P. (1980). Mycorrhizae influence tropical succession. *Biotropica*, **12**, 56-64.

Kabesh, M. O., Saber, M. S. & Young, M. (1975). Effect of phosphate dissolving bacteria and phosphate fertilization on the P-uptake by pea plants cultivated in a calcareous soil. *Zeitschrift für Pflanzenernahrung und Bodekunde*, **6**, 506-511.

Karunaratne, R. S., Baker, J. H. & Barker, A. V. (1986). Phosphorus uptake by mycorrhizal and nonmycorrhizal roots of soybean. *Journal of Plant Nutrition*, **9**, 1303-1313.

Kitt, D. G., Hetrick, B. A. D. & Wilson, G. W. T. (1988). Relationship of soil fertility to suppression of the growth response of mycorrhizal big bluestem in nonsterile soil. *New Phytologist*, **109**, 473-481.

Koske, R. E., Sutton, J. C. & Sheppard, B. R. (1975). Ecology of *Endogone* in Lake Huron sand dunes. *Canadian Journal of Botany*, **53**, 87-93.

Krishna, K. R., Balakrishna, A. N. & Bagyaraj, D. J. (1982). Interaction between a vesicular-arbuscular mycorrhizal fungus and *Streptomyces cinnamomeous* and their effects on finger millet. *New Phytologist*, **92**, 401-405.

Lee, A. & Bagyaraj, D. J. (1986), Effect of soil inoculation with vesicular-arbuscular mycorrhizal fungi and either phosphate rock dissolving bacteria or thiobacilli on dry matter production and uptake of phosphorus by tomato plants. *New Zealand Journal of Agricultural Research*, **29**, 525-531.

Lewis, D. G. & Quirk, J. P. (1967). Phosphate diffusion in soil and uptake by plants. III. P32 movement and uptake by plants as indicated by P32 autoradiography. *Plant Soil*, **26**, 445-453.

Menge, J. A. (1983). Utilization of vesicular-arbuscular mycorrhizal fungi in agriculture. *Canadian Journal of Botany*, **61**, 1015-1024.

Menge, J. A., Jarrell, W. M., Labanauskas, C. K., Ojala, J. C., Haszar, C., Johnson, E. L. V. & Sibert, D. (1982). Predicting mycorrhizal dependency of Troyer citrange on *Glomus fasciculatus* in California citrus soils and nursery mixes. *Soil Science Society of America Journal*, **46**, 762-768.

Meyer, J. R. & Linderman, R. G. (1986). Selective influence on populations of rhizosphere or rhizoplane bacteria and actinomycetes by mycorrhizas formed by *Glomus fasciculatus*. *Soil Biology and Biochemistry*, **18**, 191-196.

Morandi, D., Bailey, J. A. & Gianinazzi-Pearson, V. (1984). Isoflavonoid accumulation in soybean roots infected with vesicular-arbuscular mycorrhizal fungi. *Physiological Plant Pathology*, **24**, 357-364.

Mosse, B. (1957). Growth and chemical composition of mycorrhizal and nonmycorrhizal apples. *Nature*, **179**, 922-924.

Mosse, B. (1959). Observations on the extramatrical mycelium of a vesicular-arbuscular endophyte. *Transactions of the British Mycological Society*, **42**, 439-448.

Mosse, B. (1973). Advances in the study of vesicular-arbuscular mycorrhiza. *Annual Review of Phytopathology*, **11**, 171-195.

Mosse, B. (1977). Plant growth responses to vesicular-arbuscular mycorrhiza. X. Responses of *Stylosanthes* and maize to inoculation in unsterile soils. *New Phytologist*, **78**, 277-288.

Mosse, B., Hayman, D. S. & Arnold, D. (1973). Plant growth responses to vesicular-arbuscular mycorrhiza. V. Phosphate uptake from 32P labelled soil solution by three plant species. *New Phytologist*, **72**, 809-815.

Mosse, B., Hayman, D. S. & Ide, G. J. (1969). Growth responses of plants in unsterilized soil to inoculation with vesicular-arbuscular mycorrhiza. *Nature*, **224**, 1031-1032.

Nicolson, T. H. (1959). Mycorrhiza in the Gramineae. I. Vesicular-arbuscular endophytes, with special reference to the external phase. *Transactions of the British Mycological Society*, **42**, 421-438.

Nicolson, T. H. & Johnston, C. (1979). Mycorrhiza in the Gramineae. III. *Glomus fasciculatus* as the endophyte of pioneer grasses in a maritime sand dune. *Transactions of the British Mycological Society*, **72**, 261-268.

Nye, P. H. & Tinker, P. B. (1977). *Solute Movement in the Soil-Root System*. Blackwell Scientific Publishers, Oxford.

Ocampo, J. A., Martin, J. & Hayman, D. S. (1980). Influence of plant interactions on vesicular-arbuscular mycorrhizal infections. I. Host and non-host plants grown together. *New Phytologist*, **84**, 27-35.

Plenchette, C., Fortin, J. A. & Furlan, V. (1983). Growth responses of several plant species to mycorrhizae in a soil of moderate phosphorus fertility. I. Mycorrhizal dependency under field conditions. *Plant and Soil*, **70**, 199-209.

Pope, P. E., Chaney, W. R., Rhodes, J. D. & Woodhead, S. H. (1983). The mycorrhizal dependency of four hardwood tree species. *Canadian Journal of Botany*, **61**, 412-417.

Powell, C. L. & Daniel, J. (1978). Mycorrhizal fungi stimulate uptake of soluble and insoluble phosphate fertilizer from a phosphate-deficient soil. *New Phytologist*, **80**, 351-358.

Raj, J., Bagyaraj, D. J. & Manjurath, A. (1981). Influence of soil inoculation with vesicular-arbuscular mycorrhiza and a phosphate-dissolving bacterium on plant growth and ^{32}P uptake. *Soil Biology and Biochemistry*, **13**, 105-108.

Ratnayake, M., Leonard, R. T. & Menge, J. A. (1978). Root exudation in relation to supply of phosphorus and its possible relevance to mycorrhizal formation. *New Phytologist*, **81**, 543-552.

Rhodes, L. H. & Gerdemann, J. W. (1975). Phosphate uptake zones of mycorrhizal and nonmycorrhizal onions. *New Phytologist*, **75**, 555-561.

Saif, S. R. (1987). Growth responses of tropical forage plant species to vesicular-arbuscular mycorrhizae. *Plant and Soil*, **97**, 25-35.

Sanders, F. E. & Tinker, P. B. (1971). Mechanism of absorption of phosphate from soil by *Endogone* mycorrhizas. *Nature*, **233**, 278-279.

Schwab, S. M., Johnson, E. L, V. & Menge, J. A. (1982). Influence of simazine on formation of vesicular-arbuscular mycorrhizae in *Chenopodium quinona* Willd. *Plant and Soil*, **64**, 283-287.

Schwab, S. M., Leonard, R. T. & Menge, J. A. (1984). Quantitative and qualitative comparison of root exudates of mycorrhizal and nonmycorrhizal plant species. *Canadian Journal of Botany*, **62**, 1227-1231.

St. John, T. V. (1980). Root size, root hairs and mycorrhizal infection: a reexamination of Baylis's hypothesis with tropical trees. *New Phytologist*, **84**, 483-487.

St. John, T. V., Coleman, D. C. & Reid, C. P. P. (1983). Association of vesicular-arbuscular mycorrhizal hyphae with soil organic particles. *Ecology*, **64**, 957-959.

Tiessen, H., Stewart, J. W. B. & Cole, C. V. (1984). Pathways of phosphorus transformations in soils of different pedogenesis. *Soil Science Society of America Journal*, **48**, 853-858.

Tinker, P. B. H. (1975). Effects of vesicular-arbuscular mycorrhizas on higher plants. *Symposium of the Society for Experimental Biology*, **29**, 325-349.

Warner, A. (1984). Colonization of organic matter by vesicular-arbuscular mycorrhizal fungi. *Transactions of the British Mycological Society*, **82**, 252-354.

Warner, A. & Mosse, B. (1980). Independent spread of vesicular-arbuscular mycorrhizal fungi in soil. *Transactions of the British Mycological Society*, **74**, 407-446.

Wilson, G. W. T., Hetrick, B. A. D. & Kitt, D. G. (1988). Suppression of VA mycorrhizal fungus spore germination by nonsterile soil. *Canadian Journal of Botany*, in press.

Chapter 11

Phosphorus metabolism in mycorrhizas

V. Gianinazzi-Pearson & S. Gianinazzi

Laboratoire de Phytoparasitologie, Station d'Amélioration des Plantes, INRA, BV 1540, 21034 DIJON CEDEX, France

Introduction

Because of the fundamental role of phosphorus in biological systems (Westheimer, 1987), organisms have developed specialised mechanisms which enable them to obtain this element from their environment. For many land plants mycorrhizal associations are one way of guaranteeing adequate phosphorus absorption from reserves in the soil.

The central role of the fungal symbiont in the provision of greater efficiency of phosphate absorption has been shown both *in vitro* and *in vivo* (see Harley & Smith, 1983, for references; Hale & Sanders, 1982; Jakobsen, 1986). Modifications of phosphate uptake properties of root systems arising from mycorrhizal infection will, consequently, depend largely upon (a) development of extramatrical hyphae in soil and their absorption of phosphorus (see also Hetrick, Chapter 10), (b) translocation of phosphorus through hyphae, and (c) mycelial development within root tissues and transfer of phosphorus from fungus to root cells (Smith & Gianinazzi-Pearson, 1988).

Phosphorus sources exploited by mycorrhizal systems

There is now no doubt that mycorrhizas readily absorb soluble phosphate from the labile pool of phosphorus in soils (see Harley & Smith, 1983, for references; Thomas *et al.*, 1982; Gianinazzi-Pearson, 1986), but their role in the removal or breakdown of insoluble, condensed or complexed forms of phosphate has still not been clearly elucidated.

Mycorrhizal plants sometimes respond more than nonmycorrhizal plants to application of insoluble or poorly soluble phosphate fertilizers (Pairunan, Robson & Abbott, 1980; Cabala-Rosand & Wild, 1982; Harley & Smith, 1983)(Table 11.1), and we have recently been investigating the mechanisms involved in the growth enhancement of vesicular-arbuscular (VA) infected plants. In two tropical acid soils amended with a poorly soluble rock phosphate and labelled with ^{32}P in the soil solution, VA-mycorrhizal palms grew better and used more fertilizer than those which were nonmycorrhizal (Table 11.1). However, similar values for specific activity and percentage phosphorus derived from fertilizer were observed in all the plants, indicating that they had adsorbed from the same pool of phosphorus in the soil. It is highly probable, therefore, that the response of the mycorrhizal plants to the fertilizer was due to their more efficient absorption of those phosphate ions drawn into solution following natural dissolution of the rock phosphate by the low soil pH. This is confirmed by the fact that application of soluble triple superphosphate fertilizer gave identical results to those obtained with rock phosphate (Table 11.1). In experiments using ^{32}P-labelled tricalcium phosphate, Azcon-Aguilar *et al.* (1986) showed that in a neutral soil, where this fertilizer remains insoluble, VA mycorrhiza formation did not, in fact, improve plant utilization of this form of phosphate. Ecto- and ericoid mycorrhizal fungi can cause acidification of culture media (see Harley & Smith, 1983; Bajwa & Read, 1985; St John *et al.*, 1985), but whether this is of significance for phosphate dissolution in soils by corresponding mycorrhizal root systems has not been determined.

Mycorrhizal infection may, however, increase the ability of root systems to break down other condensed or complexed forms of phosphate since it is known that the activity of root surface phytases and acid phosphatases can be enhanced by mycorrhiza formation (Williamson & Alexander, 1975; Antibus *et al.*, 1981; Gianinazzi-Pearson & Gianinazzi, 1981, 1986; Doumas *et al.*, 1986; Dodd *et al.*, 1987). Ectomycorrhizas can also affect the activity of surface phosphatases which break down inorganic pyrophosphate and tripolyphosphate (Doumas *et al.*, 1986). Although stimulation of enzyme activity can vary with the fungal species in the mycorrhizal complex (Antibus *et al.*, 1981; Gianinazzi-Pear-

Table 11.1. Utilization of two phosphate fertilizers (66 ppm P) by oil palms in a ^{32}P-labelled sandy clay acid soil* from Soubré (Ivory Coast).

Plant treatment	Fertiliser	Dry weight (g shoot^{-1})	Fertiliser utilisation coefficient (%)	% P derived from fertiliser	Specific activity ($^{32}P/^{31}P$)
Non-Mycorrhizal	none	0·6	-	-	-
	rock phosphate	1·5	4·3	65	1·7
	triple superphosphate	1·6	5·0	75	1·9
Mycorrhizal	none	2·4	-	-	-
	rock phosphate	3·3	16·8	76	1·9
	triple superphosphate	3·1	13·8	74	2·0

* Soil water pH = 5·2; ppm P (Olsen) = 3·9. Unpublished data of Blal, Morel, Gianinazzi-Pearson, Fardeau & Gianinazzi.

son & Gianinazzi, 1986; Dodd *et al.*, 1987), it is not possible to say whether the effects are host-mediated and/or due to fungal enzymes. In the case of ecto- and ericoid mycorrhizas, the fungal symbionts are known to produce and excrete acid phosphatases and phytases in pure culture (Pearson & Read, 1975; Ho & Zak, 1979; Calleja *et al.*, 1980; Dighton, 1983; Bousquet, Mousain & Salsac, 1986; Doumas *et al.*, 1986; Mousain & Salsac, 1986; Straker & Mitchell, 1986), and electron microscope studies have revealed that hyphae of the mycorrhizal fungi growing on the surface of host roots have intense acid phosphatase activity associated with their walls (Dexheimer *et al*,, 1986; Gianinazzi-Pearson, Bonfante-Fasolo & Dexheimer, 1986).

Straker & Mitchell (1986) purified and characterised acid phosphatases of an ericoid mycorrhizal fungus from *Erica hispidula*. They found that a large proportion of the enzyme activity was associated with a high molecular weight wall-bound phosphatase which could use a wide range of phosphate esters as substrates, including, as with ectomycorrhizas, pyrophosphate, This phosphatase has now been purified further and used to raise polyclonal antibodies (Straker et al., 1989). Ultrastructural localization of acid phosphatase using cytochemical techniques and indirect immunogold labelling with the anti-phosphatase antiserum confirmed that the enzyme is associated with the hyphal wall of the ericoid fungus in pure culture (Fig. 11.1 a & b). Using the same antiserum it has been possible to detect the enzyme on hyphae associated with the host root (Fig. 11.1d). Preliminary light and electron microscope studies (V. Gianinazzi-Pearson, D. Mousain, J. Dexheimer, J. C. Cleyet-Marel, J. Lei & S. Gianinazzi, unpublished data) using polyclonal antibodies against phosphatase of *Pisolithus tinctorius* (Bousquet, 1988) indicate that, as for ericoid mycorrhizas, acid phosphatase associated with the fungal hyphae in ectomycorrhizas is also the same as that produced by the fungus in pure culture.

It is highly probable, therefore, that ericoid and ectomycorrhizal fungi can play a role in the mobilization of phosphorus from certain condensed or complexed forms of organic and inorganic phosphate in the soil. Such a role for VA fungi has yet to be demonstrated and more research in this area is clearly necessary as phosphate mobilising enzymes could be of considerable ecological importance in natural soils (Woolhouse, 1969; MacLachlan, 1976).

Fig. 11.1. (*Facing page*) Electron (a-c) and light (d & e) micrographs of cytochemical localisation of acid phosphatase in an ericoid mycorrhizal fungus from *Erica hispidula*.

(a) β-glycerophosphatase activity (pH 5·0) and (b) immunogold labelling (anti-phosphatase antiserum 1/10000) of wall phosphatase (arrows) in hyphae growing in pure culture;

(d) vizualization of antigenic sites of the enzyme (arrows) using the silver enhancement technique (Vanderbosch, 1986), on hyphae associated with a host root;

(c) and (e) controls prepared without anti-phosphatase antiserum with no labelling (open arrows) of fungus.

Unpublished data of V. Gianinazzi-Pearson, C. J. Straker & S. Gianinazzi.

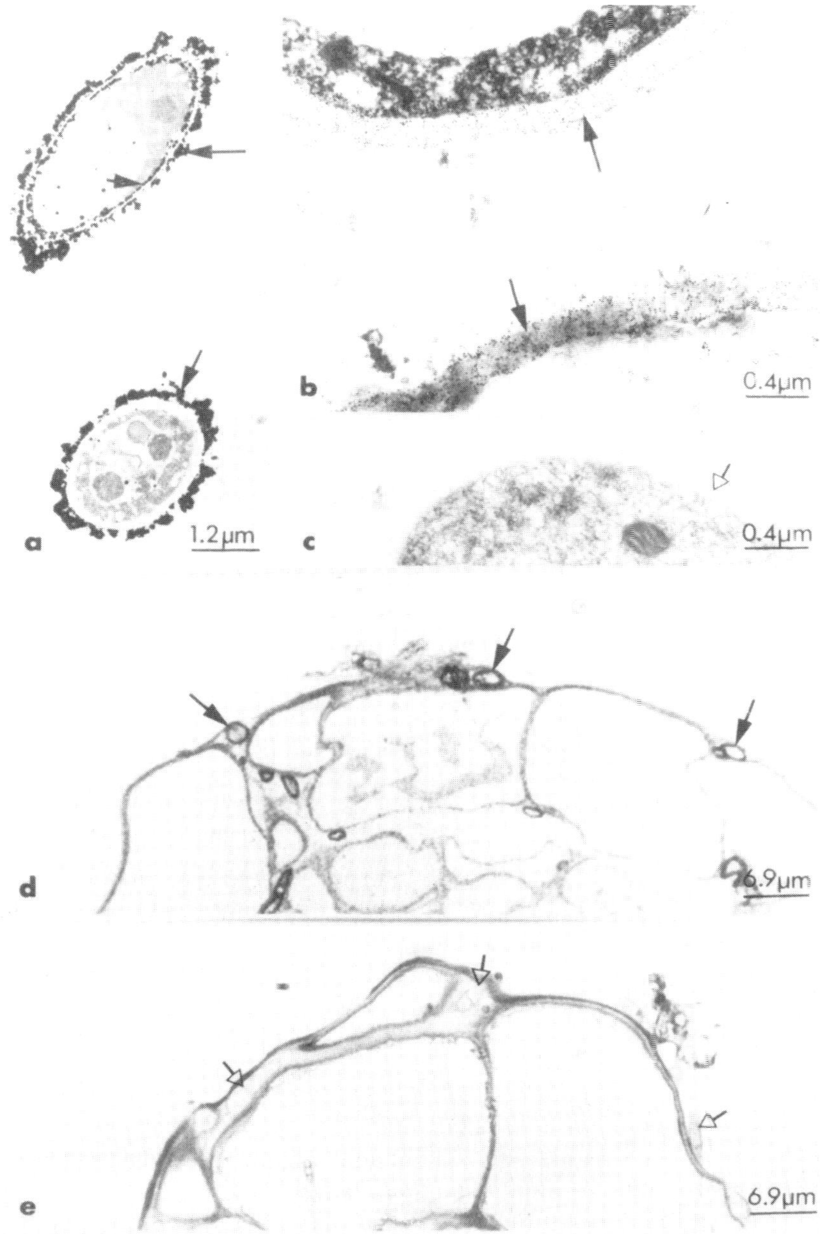
a
1.2μm
b
0.4μm
c
0.4μm
d
6.9μm
e
6.9μm

Phosphorus accumulation and transport by mycorrhizal fungi

Whatever the source of phosphate used and whatever the mycorrhizal association, the extramatrical hyphae of mycorrhizal fungi absorb, accumulate and translocate phosphate ions to the host root (see Harley & Smith, 1983). Absorption is an active, metabolically dependent process (Gianinazzi-Pearson & Gianinazzi, 1986) and hyphae can accumulate large amounts of phosphorus which would otherwise interfere with cell metabolism, by converting it into condensed, osmotically inactive polyphosphate, of which at least a part is compartmentalised in the fungal vacuoles (see Harley & Smith, 1983; Martin *et al.*, 1985; Straker & Mitchell, 1985; Ashford *et al.*, 1986; Jennings, Chapter 1). The ability of ectomycorrhizal fungi to mobilise rapidly phosphorus from their polyphosphate store under conditions of phosphate shortage (Harley & Brierley, 1954; Martin *et al.*, 1985) makes the sheath tissue a potentially important storage organ for ectomycorrhizal host plants.

Polyphosphate has also been suggested to be involved in the mechanism of phosphorus translocation by mycorrhizal fungi. Translocation along hyphae of VA and ectomycorrhizal fungi occurs at high rates and is metabolically dependent (Skinner & Bowen, 1974; Pearson & Tinker, 1975; Cooper & Tinker, 1978; Finlay & Read, 1986). For the moment, experimental evidence suggests that it is mainly driven by rapid cytoplasmic streaming which mixes and carries phosphate down a concentration gradient in the cytoplasm (Tinker, 1975; Harley & Smith, 1983). This gradient will depend on phosphorus uptake rates by external hyphae, phosphate removal rates from hyphae in the roots and the efficiency of stirring by cytoplasmic currents along the pathway of translocation. As the polyphosphate-containing vacuoles are carried and/or mixed by the latter, the continuous loading and unloading of their contents would maintain the concentration of phosphate in the cytoplasm and so ensure its rapid displacement in the cytoplasmic currents.

Polyphosphate turnover and rapid rates of phosphorus translocation are not, however, distinguishing features of mycorrhizal fungi and they can be found in saprotrophic and parasitic fungal groups (Beever & Burns, 1980). What does appear to be specific to mycorrhizal fungi is the persistent functional compatibility they establish with host tissues (Gianinazzi-Pearson, 1984), and their continuous transfer of phosphorus to the phytosymbiont.

Phosphorus transfer from myco- to phytosymbiont

Large areas of surface contact between fungus and host cells are features of all mycorrhizas but processes occurring at the fungus-host interface are poorly understood. It is not known what triggers phosphate loss from hyphae in the presence of host root cells. The value for phosphorus flux of 780 nmol m^{-2} min^{-1} calculated, by Cox & Tinker (1976), to occur across the fungus-plant interface in arbuscules of VA mycorrhizas is similar to those for phosphorus influx into fungal hyphae and giant algal cells (Beever & Burns, 1980). Cells do not normally have high rates of phosphorus efflux and so phosphorus loss to the host in mycorrhizas must result from modifications in fungal efflux, possibly brought about through host-induced increase in hyphal membrane permeability. However, no mechanisms for such effects are known. Efflux of phosphorus from fungal hyphae can be greatly increased by treatments, like high external sugar concentrations, which affect plasmalemma integrity (Beever & Burns, 1980), and it has been previously suggested that phosphate transfer from fungus to host in mycorrhizas may somehow be coupled to carbohydrate uptake from the host pool in which they bathe (Harley & Smith, 1983). Phosphate release would lead to reduced cytoplasmic concentrations of phosphorus in hyphae and consequently mobilization of polyphosphate stores. Vacuolar polyphosphate granules disappear from the terminal hyphae developing in host cells in VA mycorrhizas (Cox *et al.*, 1975; Strullu *et al.*, 1981) (Fig. 11.2) and polyphosphatases are characteristically localised within the internal mycelium (Capaccio & Callow, 1982).

Structural modifications occurring in host and fungus in the region of cellular contact between the mycorrhizal symbionts lead to the formation of a simplified interface which is highly favourable for the exchange of metabolites between them (Scannerini & Bonfante-Fasolo, 1983; Gianinazzi-Pearson, 1984; Duddridge & Read, 1984). Furthermore, when physiologically active cells of the symbionts come into contact with each other, ATPase activity is

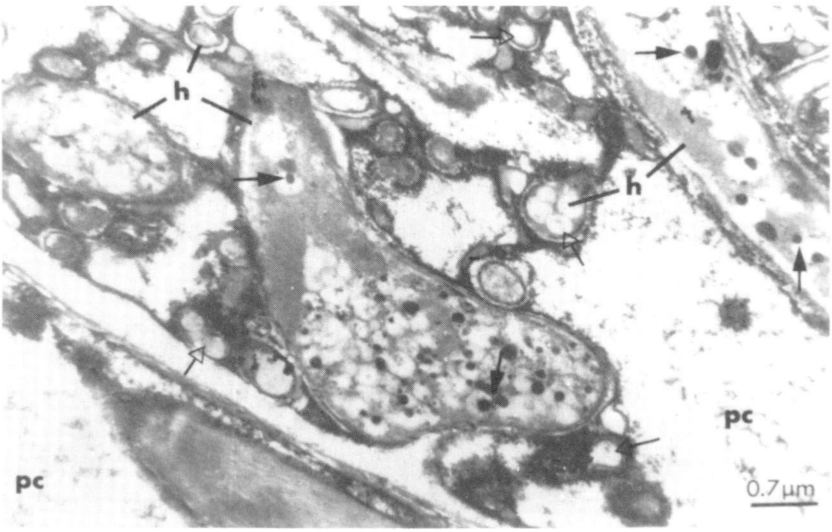

Fig. 11.2. Electron micrograph of arbuscular hyphae (h) of *Glomus fasciculatum* in a parenchymal cell (pc) of the root cortex of *Rubus idaeus*; electron dense polyphosphate granules are present (arrows) in vacuoles of larger hyphae and absent (open arrows) from fine arbuscule branches (illustration courtesy of D. Morandi).

associated with the plasmalemma of both (Marx *et al.*, 1982; Gianinazzi *et al.*, 1983; Lei & Dexheimer, 1988), indicating the existence of an energy generating system for transmembrane transport of metabolites by either or both partners (Smith & Smith, 1986). In the case of VA mycorrhizas, localization of host plasmalemma-bound ATPase activity around the fungus only occurs in the parenchyma cells of the root cortex and in the presence of living branches of the arbuscules (Gianinazzi-Pearson & Gianinazzi, 1988; Gianinazzi-Pearson *et al.* 1988). It is not known whether this specific localization involves activation of the enzyme, or results from induced changes in genome expression of the host cell with *de novo* synthesis of ATPase; these possibilities are presently being investigated using immunocytochemical techniques. The overall result of VA mycorrhiza formation is, consequently, that the parenchyma cortical cells acquire a new function of phosphate absorption, generally restricted to the outer root cells, and which is directly linked to the activity of the fungus in channelling phosphate to these cells.

It is the persistence of this cellular and functional compatibility between host and fungal cells as mycorrhizal infections develop which is a particular distinguishing feature of this type of mutualism. Mutualistic phases have been observed in the very early stages of pathogen infections (Lucas, 1977; O'Connell, 1987) but these are transient, short lived features which rapidly degenerate into pathogenic, fungus dominated infections or resistance.

Effects of external phosphate on mycorrhiza development and phosphorus metabolism

Mycorrhizas are most effective in phosphate nutrition at low to moderate levels of phosphorus in the growth medium and, in general, mycorrhizal effects decrease with increasing levels of available phosphate (Harley & Smith, 1983; Gianinazzi-Pearson, 1986; Hetrick, Chapter 10). Although this is mainly explained by the better phosphorus nutrition of nonmycorrhizal plants, increasing phosphate can also affect development and physiology of the mycorrhizal infection.

Abbott, Robson & De Boer (1984) have shown that increases in soluble phosphate reduce the development of extramatrical hyphae in VA mycorrhizal systems well before significantly affecting percentage infection levels within roots. Arbuscule development in tissues is also reduced before overall infection levels drop (Trouvelot, Kough & Gianinazzi-Pearson, 1986) so that the essential phases for absorption and transfer of phosphorus are the first to be affected by increasing soil phosphate concentrations. It has been suggested that such phosphate effects may be due to membrane-mediated changes in root exudation or in metabolite concentrations, which would reduce the availability of metabolites to the fungus (Graham, Leonard & Menge, 1982; Schwab, Menge & Leonard, 1983), but this remains to be confirmed. In recent experiments by Son & Smith (1988) on effects of phosphate and light supplies on phosphate metabolism in mycorrhizal onions (*Allium cepa*), it was shown that in plants growing at high light intensities phosphate fertilizer reduced overall infection levels but phosphorus inflow rates remained considerably higher into mycorrhizal than non mycorrhizal plants. However, phosphorus inflow rates decreased in mycorrhizal plants with similar infection development at the highest phosphorus level when irradiance was reduced, whilst there was no effect on inflow into non mycorrhizal plants. Reductions of inflow were, therefore, probably due to reduced activity of the fungal pathway of phosphate transfer to the plant, rather than to reduced infection or to direct effects on host phosphorus uptake. This reduced fungal activity at low levels of light supply and high phosphate concentrations could be due to reduced growth of external mycelium and/or arbuscule formation (which were not measured), or due to modifications in the nutrient transport and/or transfer processes in the mycorrhizal system.

Very little is known about the effect that high phosphate concentrations may have on the phosphate physiology of mycorrhizal fungi. In VA fungi, phosphate can affect rates of cytoplasmic streaming in hyphae developing from germinating spores (Pons & Gianinazzi-Pearson, 1984) and alkaline phosphatase activity in VA fungal mycelium developing within roots (Gianinazzi, Gianinazzi-Pearson & Dexheimer, 1979; Gianinazzi-Pearson & Gianinazzi, 1986) is considerably reduced by applications of phosphate fertilizer to soil (Gianinazzi-Pearson & Gianinazzi, 1983).

Conclusions

The distinguishing feature of mycorrhizal fungi is not so much their ability to accumulate and move large quantities of phosphate along their hyphae since certain pathogenic and saprotrophic fungi also do this, but rather their ability to sustain functional connections between soil and plant tissues which is coupled with the establishment of a durable phase of compatibility with host cells, enabling the transfer of phosphate to take place. The uptake and translocation of phosphate by the fungi, together with transport processes at the host-fungus interface, must be rate-limiting steps which determine the supply of phosphorus to mycorrhizal plants. Much more effort is needed to further our understanding of the molecular basis of compatibility at the host-fungus interface in mycorrhizas, and of the mechanisms whereby fungal and root cells interact to control nutrient uptake and transfer between the symbionts.

References

Abbott, L. K., Robson, A. D. & De Boer, G. (1984). The effect of phosphorus on the formation of hyphae in soil by the vesicular-arbuscular mycorrhizal fungus *Glomus fasciculatum*. *New Phytologist*, **97**, 437-446.

Antibus, R. K., Croxdale, J. G., Miller, O. K. & Linkins, A. E. (1981). Ectomycorrhizal fungi of Salix rotundifolia. III. Resynthesized mycorrhizal complexes and their surface phosphatase activities. *Canadian Journal of Botany*, **59**, 2458-2465.

Ashford, A. E., Peterson, R. L., Dwarte, D. & Chilvers, G. A. (1986). Polyphosphate granules in eucalypt mycorrhizas: determination by energy dispersive X-ray microanalysis. *Canadian Journal of Botany*, **64**, 677-687.

Azcon-Aguilar, C., Gianinazzi-Pearson, V., Pardeau, J. C. & Gianinazzi, S. (1986). Effect of vesicular-arbuscular mycorrhizal fungi and phosphate solubilising bacteria on growth and nutrition of soybean in a neutral-calcareous soil amended with ^{32}P-^{45}Ca tricalcium phosphate. *Plant and Soil*, **96**, 3-15.

Bajwa, R. & Read, D. J. (1985). The biology of mycorrhiza in the Ericaceae. IX. Peptides as nitrogen sources for the ericoid endophyte and for mycorrhizal and non-mycorrhizal plants. *New Phytologist*, **101**, 459-467.

Beever, R. E. & Burns, D. J. W. (1980). Phosphorus uptake storage and utilisation by fungi. *Advances in Botanical Research*, **8**, 128-219.

Bousquet, N. (1988). Étude immunoenzymatique de phosphatases acides de champignons ectomycorhiziens. Doctorat Thesis, Université des Sciences et Techniques de Languedoc, Montpellier, France.

Bousquet, N., Mousain, D. & Salsac, L. (1986). Influence de l'orthophosphate sur les activités phosphatases de *Suillus granulatus* en culture *in vitro*. *Physiologie Végétale*, **24**, 153-162.

Cabala-Rosand, P. & Wild, A. (1982). Direct use of low grade phosphate rock from Brazil as fertilizer. II. Effects of mycorrhizal inoculation and nitrogen source. *Plant and Soil*, **65**, 363-373.

Calleja, M., Mousain, D., Lecouvreur, B. & d'Auzac, J. (1980). Influence de la carence phosphatée sur les activités phosphatases acides de trois champignons mycorhiziens: *Hebeloma edurum* Metrod., *Suillus granulatus* (L. ex Fr.) O. Kuntze et *Pisolithus tinctorius* (Pers.) Cooker et Couch. *Physiologie Végétale*, **18**, 489-504.

Capaccio, L. C. M. & Callow, J. A. (1982). The enzymes of polyphosphate metabolism in vesicular-arbuscular mycorrhizas. *New Phytologist*, **91**, 81-91.

Cooper, K. M. & Tinker, P. B. (1978). Translocation and transfer of nutrients in vesicular-arbuscular mycorrhizas. II. Uptake and translocation of phosphorus, zinc and sulphur. *New Phytologist*, **81**, 43-52.

Cox, G. & Tinker, P. B. (1976). Translocation and transfer of nutrients in vesicular-arbuscular mycorrhizas. I. The arbuscule and phosphorus transfer: a quantitative ultrastructural study. *New Phytologist*, **77**, 371-378.

Cox, G., Sanders, F. E., Tinker, P. B. & Wild, J. A. (1975). Ultrastructural evidence relating to host-endophyte transfer in a vesicular-arbuscular mycorrhiza. In *Endomycorrhizas*, eds. F. E. Sanders, B. Mosse & P. B. Tinker, pp. 297-312. Academic Press: London & New York.

Dexheimer, J., Aubert-Dufresne, M. P., Gérard, J., Letacon, F. & Mousain, D. (1986). Étude de la localisation ultrastructurale des activités phosphatasiques acides dans deux types d'ectomycorhizes: *Pinus nigra nigricans/Hebeloma crustu-*

liniforme et *Pinus pinaster/Pisolithus tinctorius*. *Bulletin de la Société Botanique Française*, **133**, 343-352.

Dighton, J. (1983). Phosphatase production by mycorrhizal fungi. *Plant and Soil*, **71**, 455-462.

Dodd, J. C., Burton, C. C., Burns, R. G. & Jeffries, P. (1987). Phosphatase activity associated with the roots and the rhizosphere of plants infected with vesicular-arbuscular mycorrhizal fungi. *New Phytologist*, **107**, 163-172.

Doumas, P., Berjaud, C., Calléja, M., Coupé, M., Espiau, C. & d'Auzac, J. (1986). Phosphatases extracellulaires et nutrition phosphatases chez les champignons ectomycorhiziens et les plantes-hôtes. *Physiologie Végétale*, **24**, 173-184.

Duddridge, J. A. & Read, D. J. (1984). The development and ultrastructure of ectomycorrhizas. II. Ectomycorrhizal development on pine *in vitro*. *New Phytologist*, **96**, 575-582.

Finlay, R. D. & Read, D. J. (1986). The structure and function of the vegetative mycelium of ectomycorrhizal plants. II. The uptake and distribution of phosphorus by mycelial strands of interconnecting plants. *New Phytologist*, **103**, 157-165.

Gianinazzi, S., Gianinazzi-Pearson, V. & Dexheimer, J. (1979). Enzymatic studies on the metabolism of vesicular-arbuscular mycorrhiza. III. Ultrastructural localisation of acid and alkaline phosphatase in onion roots infected by *Glomus mosseae* (Nicol. Gerd.). *New Phytologist*, **82**, 127-132.

Gianinazzi, S., Dexheimer, J., Gianinazzi-Pearson, V. & Marx, C. (1983). Role of the host-arbuscule interface in the symbiotic nature of VA mycorrhizal associations: ultracytological studies of processes involved in phosphate and carbohydrate exchange. *Plant and Soil*, **71**, 211-215.

Gianinazzi-Pearson, V. (1984). Host-fungus specificity, recognition and compatibility in mycorrhizae. In *Genes Involved in Microbe Plant Interactions. Advances in Plant Gene Research, Basic Knowledge and Application*, eds. E. S. Dennis, B. Hohn, Th. Hohn, P. King, I. Schell & D. P. S. Verma, pp. 225-253. Springer-Verlag: Vienna & New York.

Gianinazzi-Pearson, V. (1986). Mycorrhizae: a potential for better use of phosphate fertilizer. *Fertilizers and Agriculture*, **92**, 3-12.

Gianinazzi-Pearson, V. & Gianinazzi, S. (1981). Role of endomycorrhizal fungi in phosphorus cycling in the ecosystem. In *The Fungal Community; its Organization and Role in the Ecosystem*, eds. D. T. Wicklow & G. C. Carrol, pp. 637-652, Marcel Dekker Inc., New York.

Gianinazzi-Pearson, V. & Gianinazzi, S. (1983). The physiology of vesicular-arbuscular mycorrhizal roots. *Plant and Soil*, **71**, 197-209.

Gianinazzi-Pearson, V. & Gianinazzi, S. (1986). The physiology of improved phosphate nutrition in mycorrhizal plants. In Physiological and Genetical Aspects of Mycorrhizae, eds. V. Gianinazzi-Pearson & S. Gianinazzi, pp. 101-109. INRA Publications, Paris.

Gianinazzi-Pearson, V. & Gianinazzi, S. (1988). Morphological integration and functional compatibility between symbionts in vesicular-arbuscular endomycorrhizal associations. In *Cell to Cell Signals in Plant, Animal and Microbial Symbiosis*, NATO ASI series vol. 17, eds. S. Scannerini, D. C. Smith, P. Bonfante-Fasolo & V. Gianinazzi-Pearson, pp. 73-84. Springer-Verlag, Berlin.

Gianinazzi-Pearson, V., Bonfante-Fasolo, P. & Dexheimer, J. (1986). Ultrastructural studies of surface interactions during adhesion and infection by ericoid endomycorrhizal fungi. In *Recognition in Microbe-Plant Symbiotic and Pathogenic Interactions*, NATO ASI series vol. 4, ed. B. Lugtenberg, pp. 273-282. Springer-Verlag, Berlin.

Gianinazzi-Pearson, V., Gianinazzi, S., Dexheimer, J., Morandi, D., Trouvelot, A. & Dumas, E. (1988). Recherche sur les mécanismes intervenant dans les interactions symbiotiques plant-champignons endomycorhizogènes VA. *Cryptogamie Mycologie*, **9**, 201-209.

Graham, J. H., Leonard, R. T. & Menge, J. A. (1982). Interactions of light intensity and soil temperature with phosphorus inhibition of vesicular-arbuscular mycorrhiza formation. *New Phytologist*, **91**, 683-690.

Hale, K. A. & Sanders, F. E. (1982). Effects of benomyl on vesicular-arbuscular mycorrhizal infection of red clover (*Trifolium pratense* L.) and consequences for phosphorus inflow. *Journal of Plant Nutrition*, **5**, 1355-1367.

Harley, J. L. & Brierley, J. K. (1954). The uptake of phosphate by excised mycorrhizal roots of the beech. VI. Active transport of phosphate from the fungal sheath into the host tissue. *New Phytologist*, **53**, 240-252.

Harley, J. L. & Smith, S. E. (1983). *Mycorrhizal Symbiosis*. Academic Press: London & New York.

Ho, I. & Zak, B. (1979). Acid phosphatase activity of six ectomycorrhizal fungi. *Canadian Journal of Botany*, **57**, 1203-1205.

Jakobsen, I. (1986). Vesicular-arbuscular mycorrhiza in field-grown crops. III. Mycorrhizal infection and rates of phosphorus inflow in pea plants. *New Phytologist*, **104**, 573-581.

Lei, J. & Dexheimer, J. (1988). Ultrastructural localisation of ATPase activity in the *Pinus silvestris/Laccaria laccata* ectomycorrhizal association. *New Phytologist*, **108**, 329-334.

Lucas, R. L. (1977). The movement of nutrients through fungal mycelium. *Transactions of the British Mycological Society*, **69**, 1-9.

MacLachlan, K. D. (1976). Comparative responses in plants to a wide range of phosphorus situations. *Australian Journal of Agriculture Research*, **27**, 323-341.

Martin, F., Marchal, J. P., Timinska, A. & Canet, D. (1985), The metabolism and physical state of polyphosphate in ectomycorrhizal fungi. A ^{31}P nuclear magnetic resonance study. *New Phytologist*, **101**, 275-290.

Marx, C., Dexheimer, J., Gianinazzi-Pearson, V. & Gianinazzi, S. (1982). Enzymatic studies on the metabolism of vesicular-arbuscular mycorrhiza. IV. Ultra-cytoenzymological evidence (ATPase) for active transfer processes in the host-arbuscule interface. *New Phytologist*, **90**, 37-43.

Mousain, D. & Salsac, L. (1986). Utilisation du phytate et activités phosphatases acides chez *Pisolithus tinctorius*, basidiomycète mycorhizien. *Physiologie Végétale*, **24**, 193-200.

O'Connell, R. J. (1987). Absence of a specialised interface between intracellular hyphae of *Colletotrichum lindemuthianum* and cells of *Phaseolus vulgaris*. *New Phytologist*, **107**, 725-734.

Pairunan, A. K., Robson, A. D. & Abbott, L. K. (1980). The effectiveness of vesicular-arbuscular mycorrhizas in increasing growth and phosphorus uptake of subterranean clover from phosphorus sources of different solubilities. *New Phytologist*, **84**, 327-338.

Pearson, V. & Read, D. J. (1975). The physiology of the mycorrhizal endophyte of *Calluna vulgaris*. *Transactions of the British Mycological Society*, **64**, 1-7.

Pearson, V. & Tinker, P. B. (1975). Measurement of phosphorus fluxes in the external hyphae of endomycorrhizas. In *Endomycorrhizas*, eds. F. E. Sanders, B. Mosse & P. B. Tinker, pp. 277-287. Academic Press: London.

Pons, F. & Gianinazzi-Pearson, V. (1984). Influence du phosphore, du potassium, de l'azote et du pH sur le comportement *in vitro* de champignons endomycorhizogenes à vésicules et arbuscules. *Cryptogamie Mycologie*, **5**, 87-100.

Scannerini, S. & Bonfante-Fasolo, P. (1983). Comparative ultrastructural analysis of mycorrhizal associations. *Canadian Journal of Botany*, **61**, 917-943.

Schwab, S. M., Menge, J. A. & Leonard, R. T. (1983). Quantitative and qualitative effects of phosphorus on extracts and exudates of sudangrass in relation to vesicular-arbuscular mycorrhiza formation. *Plant Physiology*, **73**, 761-765.

Skinner, M. F. & Bowen, G. D. (1974). The uptake and translocation of phosphate by mycelial strands of pine mycorrhizae. *Soil Biology and Biochemistry*, **6**, 53-56.

Smith, F. A. & Smith, S. E. (1986). Movement across membranes: physiology and biochemistry. In *Physiological and Genetical Aspects of Mycorrhizae*, eds. V. Gianinazzi-Pearson & S. Gianinazzi, pp. 75-84. INRA Publications, Paris.

Smith, S. E. & Gianinazzi-Pearson, V. (1988). Physiological interactions between symbionts in vesicular-arbuscular mycorrhizal plants. *Annual Review of Plant Physiology and Plant Molecular Biology*, **39**, 221-244.

Son, C. L. & Smith, S. E. (1988). Mycorrhizal growth responses: interactions between photon irradiance and phosphorus nutrition. *New Phytologist*, **108**, 305-314.

St. John, B. J., Smith, S. E., Nicholas, D. J. D. & Smith, F. A. (1985). Enzymes of ammonium assimilation in the mycorrhizal fungus *Pezizella ericae* (Read). *New Phytologist*, **100**, 579-584.

Straker, C. J. & Mitchell, D. T. (1985). The characterization and estimation of polyphosphates in endomycorrhizas of the Ericaceae. *New Phytologist*, **99**, 431-440.

Straker, C. J. & Mitchell, D. T. (1986). The activity and characterization of acid phosphatases in endomycorrhizal fungi of the Ericaceae. *New Phytologist*, **104**, 243-256.

Straker, C. J., Gianinazzi-Pearson, V., Gianinazzi, S., Cleyet-Marel, J. C. & Bousquet, N. (1989). Electrophoretic and immunological studies on acid phosphatase from a mycorrhizal fungus of *Erica hispidula* L. *New Phytologist*, **111**, in press.

Strullu, D. G., Gourret, J. P., Garrec, J. P. & Fourcy, A. (1981). Ultrastructure and electron-probe microanalysis of the metachromatic vacuolar granules occurring in *Taxus* mycorrhizas. *New Phytologist*, **87**, 537-545.

Thomas, G. W., Clarke, C. A., Mosse, B. & Jackson, R. M. (1982). Source of phosphate taken up from two soils by mycorrhizal (*Thelephora terrestris*) and non-mycorrhizal *Picea sitchensis* seedlings. *Soil Biology and Biochemistry*, **14**, 73-75.

Tinker, P. B. (1975). Effects of vesicular-arbuscular mycorrhizas on higher plants. *Symposium of the Society for Experimental Biology*, **29**, 325-329.

Trouvelot, A., Kough, J. & Gianinazzi-Pearson, V. (1986). Mesure du taux de mycorhization VA d'un système radiculaire. Recherche de méthodes d'estimation ayant une signification fonctionnelle. In *Physiological and Genetical Aspects of Mycorrhizae*, eds. V. Gianinazzi-Pearson & S. Gianinazzi, pp. 217-221. INRA Publications, Paris.

Vanderbosch, K. A. (1986). Light and electron microscopic visualization of uricase by immunogold labelling of sections of resin-embedded soybean nodules. *Journal of Microscopy*, **143**, 187-197.

Westheimer, F. H. (1987). Why nature chose phosphates. *Science*, **235**, 1173-1178.

Williamson, B. W. & Alexander, I. J. (1975). Acid phosphatase localised in the sheath of beech mycorrhiza. *Soil Biology and Biochemistry*, **7**, 195-198.

Woolhouse, H. W. (1969). Differences in the properties of the acid phosphatases of plant roots and their significance in the evolution of edaphic ecotypes. In *Ecological Aspects of the Mineral Nutrition of Plants*, ed. I. H. Rorison, pp. 357-380. Blackwell, Oxford.

Chapter 12

Nitrogen relations of mat-forming lichens

P. D. Crittenden

Department of Botany, University of Nottingham, Nottingham NG7 2RD, UK

Introduction

Of the three principal types of mutualistic symbiosis found in terrestrial environments – lichens, mycorrhizas and legumes/actinorhizas, the lichens are the least well understood. Interactions between the symbionts in mycorrhizas and legumes are known to involve bilateral exchanges of nutrients: photosynthate moving from photobiont to microsymbiont and mineral nutrients from microsymbiont to photobiont. These nutrient fluxes have been studied intensively, mainly in attempts to determine costs and benefits to the vascular plant hosts. In contrast, our view of the interplay between symbionts in lichens has been coloured almost entirely by the results of studies on carbon nutrition (cf. Smith & Douglas, 1987). With the exception of nitrogen metabolism in cyanobacterial lichens, about which there is now a growing body of information, knowledge of nitrogen, phosphorus, potassium and sulphur relationships in lichens is comparatively rudimentary.

This situation has arisen partly because lichenology has not benefited from an economic impetus comparable to that driving research on the legume and mycorrhizal symbioses, but it also reflects difficulties posed by lichens as experimental systems. For example, experiments in which photobionts are grown either in the presence or absence of a heterotrophic symbiont have been of paramount importance in elucidating interactions in higher plant symbioses. In the case of lichens this approach is blocked because, among other things, lichens do not readily grow under laboratory conditions. Moreover, frequent reports of 'negative' results, such as absence of clear seasonal changes in thallus nutrient concentrations (Lewis Smith, 1978; Puckett, 1985) and the apparent insensitivity of growth and photosynthetic rates to inorganic nutrient additions (Armstrong, 1977; Carstairs & Oechel, 1978; Bailey & Larson, 1982; Matthes-Sears, Nash & Larson, 1987), may have done much to create an impression that lichen mineral nutrient relations is an unrewarding field of enquiry. These problems make the subject all the more challenging.

There are several crucial questions on which quantitative information is needed: (i) at what rates and from what sources do mineral elements arrive at lichen thalli; (ii) how efficiently do lichens capture nutrients from these supplies; (iii) does uptake satisfy potential growth requirement; (iv) by which pathways in the thallus do mineral nutrients reach the photobiont? These questions are discussed in this chapter with particular reference to nitrogen and to the results of recent work on the ecology of mat-forming lichens in subarctic woodlands. Ecological and physiological parallels between lichens and mycorrhizas are briefly discussed.

Aspects of the ecology of subarctic lichen woodland

Lichen woodlands occur extensively on well-drained terrain throughout the northern boreal and forest-tundra zones. They consist of open stands of trees with lichen-dominated ground cover and often develop as seral communities on sites of forest fires (Ahti, 1977; Kershaw, 1977; Auclair, 1983; Foster, 1985) but may also arise in the absence of fire to form mature and stable vegetation (Auclair, 1983; Rowe, 1984; Payette *et al.*,1985). Lichen woodland soils are acid, oligotrophic podzols or brunisols (Kershaw, Rouse & Bunting, 1975; Ahti, 1977; Moore, 1980; Ugolini *et al.*, 1981; Sepponen, 1985; Rowe & Acton, 1985). The most important mat-forming lichens in the ground cover are species of *Cladonia*, *Cetraria*, *Stereocaulon* and *Alectoria* but three species, *Cladonia stellaris*, *C. mitis* and *S. paschale*, are especially abundant and each frequently forms continuous carpets. The abundance of *S. paschale* is of particular interest because it is a tripartite lichen containing the cyanobacterium *Stigonema* in cephalodia and achieves appreciable rates of nitrogen fixation (Kallio & Kallio, 1975; Huss-Danell, 1977, 1978; Crittenden & Kershaw, 1978, 1979).

Fig. 12.1. Birch-lichen woodland near Ai'bmejav'ri, Finnmarksvidda, Norway (69° 05′N 24° 50′E): open stands of *Betula pubescens* ssp. *tortuosa* with interjacent lichen mats composed almost exclusively of *Cladonia stellaris*. The birch trees, each with an understorey of dwarf shrubs (principally *Empetrum hermaphroditum*, *Vaccinium vitis-idaea* and *B. nana*), appear as green islands in aerial view (b). *Betula nana* is seen in the foreground of (a).

In Europe, lichen woodland is best represented in the pine (*Pinus sylvestris*) and subarctic mountain birch (*Betula pubescens* ssp. *tortuosa*) forests of northernmost Fennoscandia (Hämet-Ahti, 1963; Oksanen & Ahti, 1982). In Finnmarksvidda, in particular, there are large tracts of open birch-*Cladonia stellaris* woodlands (Fig. 12.1). Beneath the canopy of each tree, where litter fall has resulted in comparatively deep accumulation of fibrous organic matter (cf. F horizon) (Fig. 12.2a), communities of dwarf shrubs occur comprised in the main of *Empetrum hermaphroditum*, *Vaccinium vitis-idaea* and *Betula nana* (Fig. 12.1). The reasons for the limited invasion of the inter-tree areas by these shrubs are not fully understood but probably include low soil nutrient status, periodic summer drought and phytotoxicity of lichen products (cf. Rowe, 1984; Bakaeva & Galanin, 1985). The suggestion that soil nutrient availability may restrict dwarf shrub distribution in these habitats is supported by the results of a bioassay in which growth of mycorrhizal *V. vitis-idaea* in a controlled environment was found to be poor in surface organic soil collected beneath lichen mats (cf. H horizon) and the periphery of birch canopies compared to that in organic topsoil taken closer to tree bases (Fig. 12.2b). There is also evidence that lichen secondary products may inhibit seed germination and plant growth (Lawrey, 1984, 1986; Kershaw, 1985) but most work on this topic has involved the use of ecologically improbable concentrations of metabolites and hence the ecological relevance of many of the reported effects remains to be critically assessed. Although it is indeed likely that lichen metabolites are washed into soil during rainfall (Malicki, 1965; García-Junceda & Xavier Filho, 1986; P. D. Crittenden & R. P. Beckett, unpublished data) they decompose comparatively rapidly and do not accumulate substantially (Dawson, Hrutfiord & Ugolini, 1984; Vainstein & Ravinskaya, 1984). At first sight the view of Brown & Mikola (1974) and Fisher (1979) that lichen ground cover can inhibit the formation of mycorrhizas does not appear to be borne out here since a cursory examination revealed both birch trees and dwarf shrubs to have well developed mycorrhizal associations even in the case of roots that extend out under lichen mats.

Read (1983, 1984) has suggested that vesicular arbuscular, ecto- and ericoid mycorrhizas are each associated with a specific set of

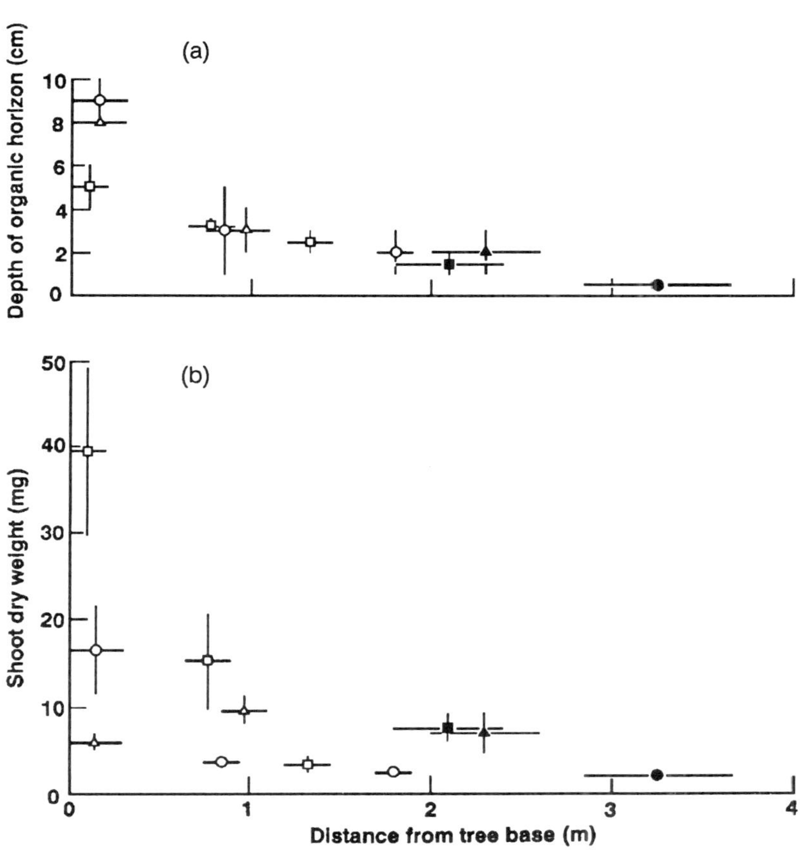

Fig. 12.2. (a) Relationship between distance from tree base and approximate depth of the organic soil horizon. Data for three birch trees each at different sites in Finnmarksvidda, Norway: *Cladonia stellaris*-rich woodland at Ai'bmejav'ri (circles, 69° 05'N 24° 50'E) and Av'zi (triangles, 69° 01'N 23° 15'E); *Stereocaulon paschale*-rich woodland, Kautokeino (squares, 69° 01'N 23° 10'E). Horizontal bars indicate areas in which measurements were made and samples of surface soil organic matter collected (see below) beneath both birch canopies (open symbols) and lichen mats (closed symbols). Vertical bars give ranges of values. (b) Shoot dry weight of *Vaccinium vitis-idaea* grown for 15 weeks in a controlled environment growth room in organic soils identified in (a). Surface sterilized seed of *V. vitis-idaea* was germinated on distilled water agar. When the cotyledons were expanded fully (45 days after sowing) seedlings were each transfered to 60 ml of soil contained in nylon mesh (150 μm aperture) bags embedded in washed silica sand in 150 ml capacity free-draining pots; the sand and soil surfaces were covered in opaque plastic beads to inhibit growth of algae and bryophytes. Sufficient deionized water was added to the sand at 1 or 2 day intervals to maintain the soil/sand system at field capacity with minimum leaching losses. Growth-room conditions: 330-365 μmol m^{-2} s^{-1} photosynthetically active radiation at soil level; 16/8h, 16/10°C light/dark cycle; c. 78% relative humidity. Mean values (n = 20) are plotted together with 95% confidence intervals. All plants were found to be heavily infected with mycorrhizal endophyte at harvest.

soil circumstances typical of a broad latitudinal and altitudinal zone. He considers ericoid mycorrhizas to be characteristic of marginal ecological situations at higher latitudes and altitudes where the capacity of the endophytes to exploit organic forms of nitrogen and phosphorus in nutrient poor heathland soils underlies their success (Read, Leake & Langdale, Chapter 9). This scheme might perhaps be extended further to include large areas of drier terrain in the subarctic and arctic where soil conditions appear to be even more extreme than those to which ericoid mycorrhizas are well adapted. In these habitats the potential for nutrient uptake from soil is so small and the vigour of vascular plants so reduced that yet another distinct type of mutualistic symbiosis between fungi and photoautotrophs becomes predominant. This involves the association of microalgae (and/or cyanobacteria) with highly differentiated aerial fungal mycelia to form lichen thalli. Since the fungal partner invests the cells of the photobiont in a sheath-like manner, there are certain structural similarities between lichen thalli and ectomycorrhizal roots. However, unlike mycorrhizas these mat-forming lichens probably derive relatively few nutrients from soil, the strategy developed instead is one of exploiting nutrient deposition from the atmosphere.

Scavenging nitrogen from the atmosphere

The widely held belief that lichens derive nutrients from the atmosphere is founded on two main lines of evidence. First, there are often good correlations between element concentrations in lichen thalli and those measured or expected in atmospheric deposits. Thus, when exposed to air pollution lichens may accumulate, among other things, sulphur (Gilbert, 1969; Pyatt, 1973; Tomassini *et al.*, 1976; Pakarinen, 1981b; Zakshek, Puckett & Percy, 1986), heavy metals (e.g. Nieboer *et al.*, 1972; Tomassini *et al.*, 1976; Pilegaard, 1978, 1979; Addison & Puckett, 1980; Garty, Galun & Hochberg, 1986; Zakshek *et al.*, 1986) and radionuclides (e.g. Tuominen & Jaakkola, 1973; Hutchison-Benson, Svoboda & Taylor, 1984; Sheard, 1986; Ellis & Smith, 1987), and at coastal sites they may accumulate chloride (Brown & Di Meo, 1972; Larson, Matthes-Sears & Nash, 1985; Matthes-Sears *et al.*, 1987). Secondly, lichens rapidly and actively absorb key ions, such as ammonium, nitrate and phosphate from solutions similar in concentration to rainwater (Smith, 1962; Farrar, 1976; Lang, Reiners & Heier, 1976; Fletcher, 1976; Hällbom & Bergman, 1983; Reiners & Olson, 1984; Nieboer *et al.*, 1984). On the basis of kinetic studies on phosphorus uptake in *Hypogymnia physodes*, Farrar (1976) predicted that input in rainfall would be sufficient to satisfy the growth requirement for phosphorus in this species. There is also unequivocal evidence of accumulation of minerals from substrates, notable examples being crustaceous lichens growing on calcareous rocks amassing calcium (Syers, Birnie & Mitchell, 1967) and those on metal-rich substrata amassing elements such as iron, copper, cadmium and zinc (Brodo, 1973; Tuominen & Jaakkola, 1973; Purvis, 1984). However, since the substrata on which lichens grow are characteristically severely deficient in available nitrogen and phosphorus it is unlikely that lichens can derive significant quantities of these elements from this source. It should be noted that correlations between the chemical composition of lichens and that of their substrata can also result from atmospheric processes viz. deposition of aeolian particles derived from local soils or bedrock which then become entrapped in the thallus (Nieboer, Richardson & Tomassini, 1978; Nieboer & Richardson, 1981; Puckett, 1985).

Data concerning accumulation of atmospheric deposits usually reveal little about rates of deposition and retention and, while there is much literature relating to elements such as cadmium, zinc and lead, there is a dearth of information on the essential macronutrients. Data on ion uptake from solutions are usually obtained using procedures the ecological relevance of which is questionable; immersing or spraying lichens with nutrient solutions may not adequately simulate rainfall and nutrient delivery under field conditions. For example, abrupt rewetting of air-dry lichen thalli can result in an efflux of metabolites into the bathing solution (Farrar & Smith, 1976; Lang *et al.*, 1976; Millbank, 1978; Buck & Brown, 1979), a process for which there is little evidence in nature (Crittenden, 1983).

Recently, we have attempted to determine the efficiency with which mats of *Stereocaulon paschale* and *Cladonia stellaris* capture nitrogen from incident rainfall in subarctic birch woodlands (P. D. Crittenden & R. P. Beckett, unpublished data). The experimental procedures have been detailed elsewhere (Crittenden,

Table 12.1. Net uptake from rainfall of inorganic nitrogen and potassium by reconstructed lichen mats.

	% Retention					
	NO_3^-		NH_4^+		K^+	
	range	mean	range	mean	range	mean
Stereocaulon paschale	86 to 100	95	40 to 99	86	-37 to +90	42
Cladonia stellaris	62 to 99	83	50 to 97	77	-978 to +65	-154

Data are based on analyses of bulked (i.e. unreplicated) collections from 10 replicate funnels of each type for 19 rainfall events between 27 May and 17 August 1984 (152 mm total rainfall). Total input (per 10 funnels): 460 μmol NO_3^-, 621 μmol NH_4^+, and 385 μmol K^+. Rainfall in event 19 was insufficient to exceed interception loss in both species (i.e. there was no lichen throughfall) resulting in 100% net uptake; this information is not included in the ranges and mean values. Net losses of potassium were greatest during the early rainfall events: mean net uptake for events 7 to 19 was 48% for *S. paschale* and -46% for *C. stellaris*. Potassium was determined by atomic absorption spectrophotometry; for other analytical methods see Table 12.3. Mean dry weights of lichen per funnel at the end of the experiment were 39·9±0·9 and 14·8±0·2 g for *S. paschale* and *C. stellaris* respectively.

1983). In brief, lichen thalli were arranged vertically on stainless steel grilles (3×3 mm mesh) so as to form single-layered lichen canopies of each species and at densities equivalent to those of undisturbed lichen mats. Only the well-pigmented upper 20-35 mm of podetia (*C. stellaris*) and pseudopodetia (*S. paschale*) were used in these reconstructions. Lichen-covered grilles were superimposed on funnels supported in trenches in the ground, with the mouth of the funnels (253 mm i.d.) and the lichen mats at ground level. During periods of rain throughfall from the lichen canopies, and unmodified rainfall from funnels covered by steel grilles only, were collected quantitatively and analyzed for ammonium, nitrate and potassium. The results for 19 consecutive rainfall events (Table 12.1) reveal that, on average, 80% or more of inorganic nitrogen delivered in rainfall was retained by these species, even during thunderstorms with comparatively high rainfall intensities. Results obtained from sequential sampling during individual rainfall events (Fig. 12.3) were in good accord with those obtained from bulked water samples and suggest that nitrogen escape in

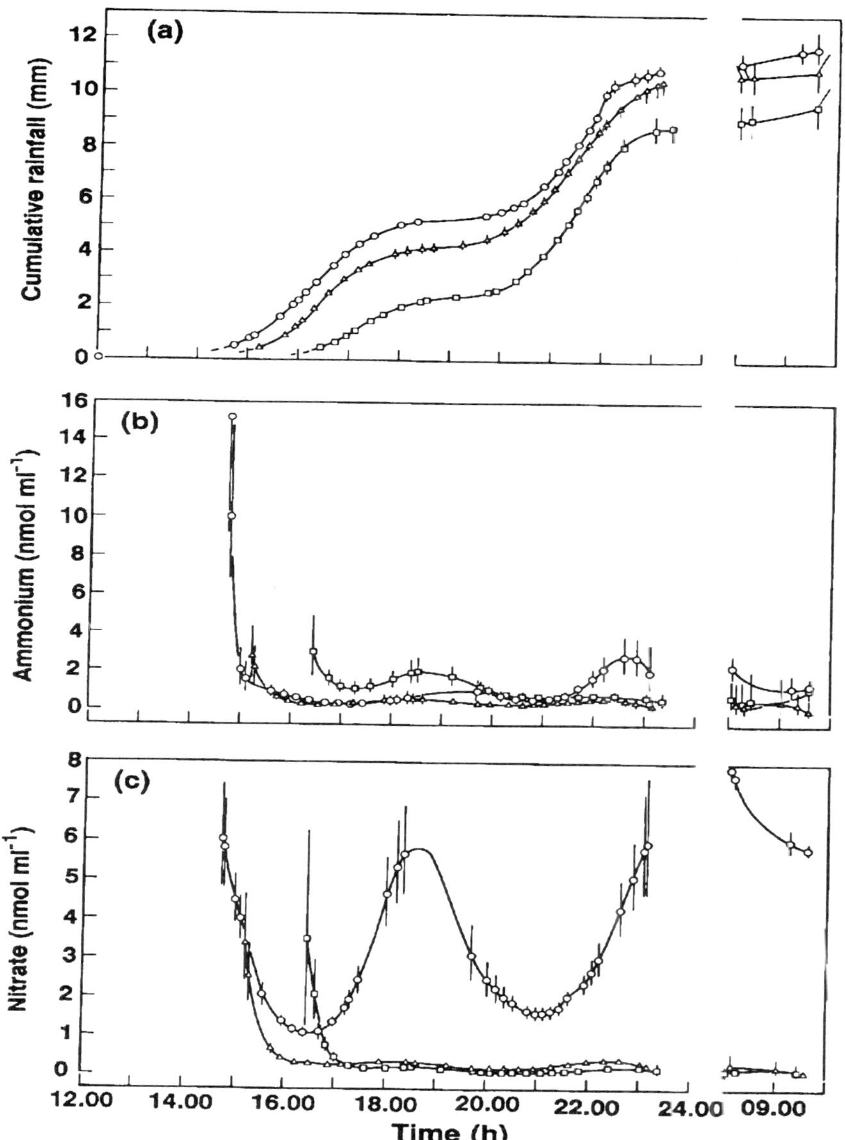

Fig. 12.3. Progress curves in rainfall event 14 (17-18 July, 1984) of cumulative rainfall (a) and the concentration of ammonium (b) and nitrate (c). Rainfall (circles), *Cladonia* throughfall (triangles) and *Stereocaulon* throughfall (squares) were collected sequentially and quantitatively in 20-30 ml aliquots from 5 replicate funnels of each type. Smooth lines were fitted using the computer program of Parsons & Hunt (1981) and fitted values (n = 5) are plotted together with 95% confidence limits (all values have been back-transformed from logarithms). Air temperature at ground level ranged between 9-14°C during the collection period. For analytical methods see Table 12.3.

throughfall was to a large extent due to low uptake efficiency during the early deliveries of rainfall.

Stereocaulon paschale and *C. stellaris* are clearly efficient at scavenging nitrogen from wet deposits. Occupying the inter-tree areas the lichen mats intercept rainfall as opposed to throughfall, although during wind-driven rain the quantities of rainfall received may be reduced in the lee of trees. Nitrogen inputs from dry deposition and mist (occult precipitation) were not quantified.

Nitrogen supply and demand

In the study reported above lichens suspended over rain gauges increased in biomass during the course of the experiment. Tagged thalli of *S. paschale* and *C. stellaris*, placed within the reconstructed lichen mats in the early part of the investigation, had increased in dry eight by 24·3 ± 1·5% (mean ± SEM; n = 10) and 26·4 ± 1·1% (n = 7) respectively by the end of the experiment (the respective mean relative growth rates were 0·0036 ± 0·0002 and 0·0039 ± 0·0001 d^{-1}). The observed growth and the absence of a major potassium efflux (Table 12.1) suggest that neither species was damaged by the experimental treatment. However, the concentration of thallus nitrogen was significantly lower ($P < 0{\cdot}05$) in experimental material at the termination of the study than in undisturbed field material collected at the start and end of the work and in the following spring (Table 12.2). Net loss of soluble organic nitrogen during rainfall (Table 12.3) was too small to explain the differences and it was assumed that thallus nitrogen had been diluted by growth in the absence of an adequate nitrogen supply.

Table 12.3 gives observed and expected increments in nitrogen capital in the lichen mats during the experimental period derived from (i) the final total dry weight of lichen per funnel (Table 12.1), (ii) the growth increments, and (iii) thallus nitrogen concentration at the start and end (Table 12.2). The expected increment is that needed to maintain the starting thallus nitrogen concentration and is a measure of the growth demand for this element. The observed nitrogen gain in *S. paschale*, over and above that attributable to atmospheric deposits, provides a gravimetric estimate of nitrogen fixation. This value (669 mg N m^{-2}) compares favourably with rates of nitrogen fixation in this species derived from acetylene reduction assays by Crittenden & Kershaw (1979). Despite the large er-

Table 12.2. Total nitrogen conentration (% dry weight) in field and experimental material of *Stereocaulon paschale* and *Cladonia stellaris*.

	Samples from undisturbed lichen mats			Material from steel grilles at end of experiment
	May 1984	August 1984	June 1985	
S. paschale	1·08 ± 0·10	1·15 ± 0·04	1·11 ± 0·09	0·96 ± 0·06
C. stellaris	n.d.	0·40 ± 0·04	0·42 ± 0·05	0·35 ± 0·03

The top 25 mm of pseudopodetia and podetia were analyzed for total Kjeldahl nitrogen by the method of Bremner & Breitenbeck (1983). In the case of experimental material thalli of this length were selected; in the case of field material thalli were pruned to this length. Mean values (n=10) appear with 95% confidence intervals. n.d. = not determined.

Table 12.3. Increments, inputs and losses of nitrogen in reconstructed lichen mats during an 82 day growth period (mg N m^{-2} pure lichen cover).

	Stereocaulon paschale	*Cladonia stellaris*
Increment in total N capital[1]		
expected	1717 ± 380	252 ± 53
observed	758 ± 245	95 ± 51
Inorganic N in rainfall[2]		
deposited	31	31
retained	27	25
Net loss of soluble organic N[3]	18·7	11·4
Unaccountable N (error and/or dry deposition)	(81)	81[†]
Estimated N_2-fixation	669[††]	-

[1]Derived from mean values of 7 to 10 measurements (derived mean ± SEM).
[2, 3]Water samples were filtered through prewashed cellulose acetate membrane filters (0·22 μm pore size) then ammonium was determined using a modification of the indophenol blue method given in Allen *et al.* (1974), nitrate by HPLC with uv detection, soluble organic N by a modification of Johnson's (1941) method as described by Crittenden (1983).
[†]Observed N increment - net N uptake from rainfall; it is assumed that the same value applies in the case of *S. paschale*.
[††]Observed N increment - (net N uptake from rainfall + unaccountable N).

rors associated with the estimated nitrogen demands an inescapable fact emerges: nitrogen scavenged from rainfall accounts for a surprisingly small fraction of the apparent nitrogen requirement in these lichens (c. 2·6% in the case of *S. paschale*, over and above that provided by nitrogen fixation, and c. 10% in *C. stellaris*). Accordingly, the possibility that other sources of nitrogen are available to undisturbed lichen mats *in situ* must be considered.

In transplanting lichen to the metal grilles two potentially important sources of nutrients had been removed: the soil and senescent basal parts of lichen thalli. On first consideration it seems unlikely that significant quantities of nitrogen could diffuse or be translocated upwards from the soil. These lichens do not make intimate contacts with the underlying soil and, after all, the most obvious explanation for the poor growth of mycorrhizal *Vaccinium vitis-idaea* in organic surface soils gathered from beneath lichen mats is that they are deficient in available forms of both inorganic and organic nitrogen. However, in the selection of 'healthy' (well-pigmented) apical parts of *C. stellaris* podetia and *S. paschale* pseudopodetia for lichen mat reconstruction roughly equivalent weights of discoloured basal thallus material were discarded. Nutrients in these older regions may represent an internal resource that can be recycled to support new growth. In Fig. 12.4 the vertical distribution of thallus nitrogen in experimental and field material of both species is compared. Nitrogen concentration in *S. paschale* progressively decreases with increasing distance from the thallus tip. In all depth intervals nitrogen concentration is lower in experimental material grown on steel grilles compared to previously undisturbed field material but the difference is least at the apex (although still significant at $P<0{\cdot}05$) where new growth is assumed to have occurred. These results are consistent with the proposition that nitrogen is remobilized in older regions of the thallus and moves to the growing apices. Data for *C. stellaris* is incomplete but shows some consistency with that for *S. paschale*. Thus the pronounced vertical gradients in nitrogen, phosphorus and potassium reported by Pakarinen (1981a) in *C. arbuscula* growing on bogs may owe more to tight internal nutrient recycling than to decomposition involving the soil microflora.

There is no reason to believe that a reduced supply of nitrogen (and perhaps other nutrients) retarded thallus growth in the rec-

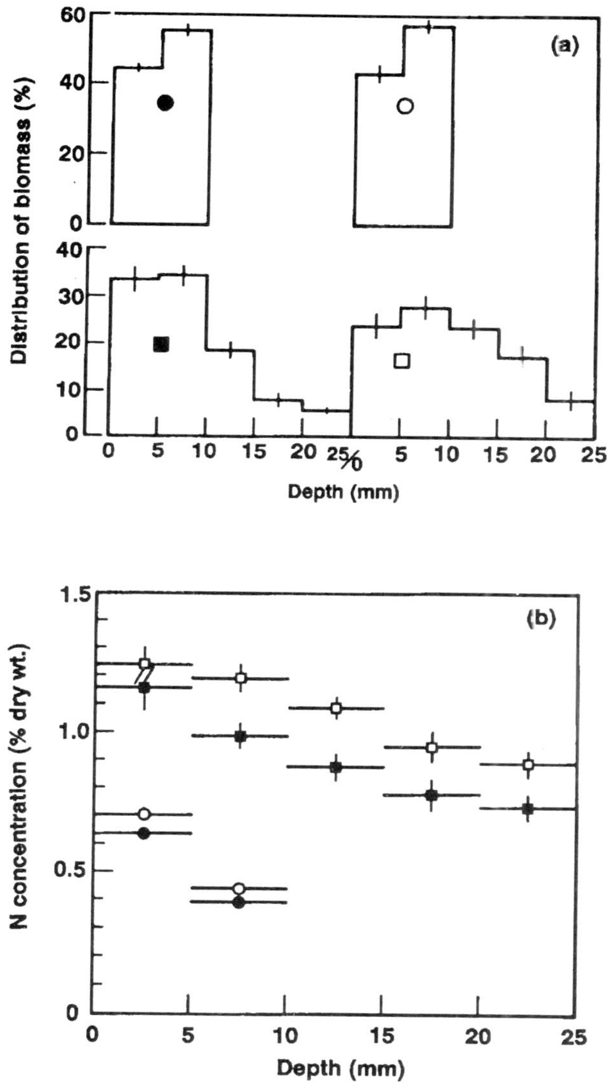

Fig. 12.4. Vertical distribution of biomass (a) and vertical variation in total nitrogen concentration (b) in *Stereocaulon paschale* (squares) and *Cladonia stellaris* (circles) taken from steel grilles at the end of the experiment (closed symbols) and from undisturbed lichen mats (open symbols). In the case of experimental material of *S. paschale*, pseudopodetia 25 mm in length were selected for analysis; in the case of field material pseudopodetia were pruned to this length. Because of the limited availability of material, the top 10 mm only of *C. stellaris* podetia were cut from samples not standardized for length. Total Kjeldahl nitrogen was determined by the method of Bremner & Breitenbeck (1983). Mean values (n = 10) are plotted together with 95% confidence limits.

onstructed lichen mats. The relative growth rates for *C. stellaris* and *S. paschale* reported above agree well with values given by Kärenlampi (1971) for the same species at the same site. Armstrong (1979, 1984) also found that the radial extension rate of peripheral lobes of the foliose lichen *Parmelia conspersa* was unaffected by removal of the older central areas of thalli. However, it is clear from the above results that growth rate (and it may follow, photosynthetic rate) is not a good indicator of the adequacy of mineral nutrient supply as may have been anticipated previously (cf. Carstairs & Oechel, 1978; Bailey & Larson, 1982; Armstrong, 1984). At the same time it would be unwise to assume that large scale internal recycling of nutrients is a universal phenomenon in lichens. In species such as *C. stellaris* and *S. paschale* that grow vertically upwards and characteristically form dense mats such a mechanism will allow much faster growth rates than could otherwise be sustained in a chronically oligotrophic environment. This could have several advantages. It would help to maintain a large mass of structurally intact basal litter giving depth to the mats and casting deep shade, a process by which these species are able to exert dominance *sensu* Grime (1979). It may also promote rapid lateral spread by regrowth of damaged or fragmented thalli and by branching or budding (cf. Yarranton, 1975). Rapid remobilization of resources and production of dead thallus material would be less of an advantage to foliose or crustose lichens that grow horizontally. In slow growing species nutrient requirement for new growth may perhaps be met entirely by new input as was suggested by Farrar (1976) in the case of *Hypogymnia physodes*.

Mutualism or controlled parasitism?

In a classic series of experiments Smith and his collaborators elucidated the fundamental processes involved in lichen carbon nutrition. They demonstrated that during periods of photosynthetic activity a large fraction of the total carbon assimilated by the photobiont moves in the form of single specific simple carbohydrates to the fungus where it accumulates in fungal products (Smith, Muscatine & Lewis, 1969; Smith, 1975, 1980). In so doing, they verified the hypothesis, first advanced by de Bary and Schwendener in the nineteenth century, that lichen fungi are sustained by assimilate provided by their algal symbionts (Quispel,

1943). However, as Smith has repeatedly pointed out (Smith, 1975, 1976, 1980; Smith & Douglas, 1987) there is only very limited evidence to support the long standing supposition that there is bilateral nutrient exchange, the fungus supplying the photobiont with mineral elements and/or growth factors. In view of this, Ahmadjian & Jacobs (1981, 1983) and Ahmadjian (1982) proposed that the symbiotic relationship in lichens is one of controlled parasitism rather than mutualism.

The data presented above for *C. stellaris* and *S. paschale* have been interpreted in terms of internal recycling and translocation of nitrogen from older parts of the thallus through mycobiont pathways to the apices. The question of whether nitrogen might pass from fungus to photobiont in these regions where growth and multiplication of algal cells are most active (Honegger, 1987) now deserves attention. An argument in favour of this proposal can be developed from a consideration of temporal relationships between nutrient deposition and growth in lichen mats (Fig. 12.5). With the onset of rainfall, mats are rewetted and respiratory and assimilatory processes recover rapidly (Ried, 1960; Smith & Molesworth, 1973; Farrar & Smith, 1976; Crittenden & Kershaw, 1979; Brown, MacFarlane & Kershaw, 1983; Groulx & Lechowicz, 1987; Coxson, 1988; Fritz-Sheridan & Coxson, 1988). Thallus water contents sufficient for growth may obtain for substantial periods after rain ceases. Indeed, pseudopodetia of *S. paschale* can measurably increase in dry weight during individual 'wet events' comparable to that depicted in Fig. 12.5 (K. A. Kershaw & P. D. Crittenden, unpublished). Nutrients are deposited in rainfall but the major inputs of most ions are often contained in the first deliveries of rain (Seymour & Stout, 1983; Crittenden, 1983; Pellett, Bustin & Harriss, 1984; P. D. Crittenden & R. P. Beckett, unpublished). Consequently, growth may occur during comparatively long periods when there is little nutrient input and when intercellular thallus water is depleted of key mineral ions. Under these conditions, rapid growth and multiplication of algae in thallus apices may be dependent on nitrogen and other nutrients provided by the fungus.

The photobiont in *C. stellaris* and *S. paschale* is probably a species of *Trebouxia* (Ahmadjian, 1980). In the symbiotic state *Trebouxia* reproduces by autosporulation and this occurs predomi-

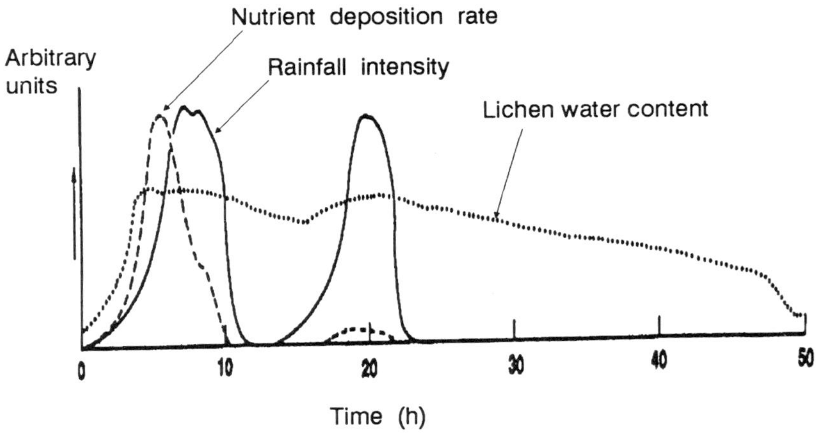

Fig. 12.5. Temporal variation in rainfall intensity, nutrient deposition rate and lichen water content during and following a hypothetical rainfall event. Time scale is indicated as an example. This scheme is based on the author's experience in northern boreal forests; data sets in partial support of it are given by Crittenden & Kershaw (1979) and Coxson (1987).

nantly in the growing apices of lichen thalli (Honegger, 1987). Young autospores are contacted by mycobiont hyphae that penetrate the old mother cell wall (Scott & Larson, 1984; Honegger, 1985). It is tempting to speculate that these contacts at an early growth stage of individual algal cells might establish symplastic pathways essential for mineral nutrient exchange.

Indirect evidence of a requirement for nitrogen in growing photobiont populations in lichen thalli has come from studies on nitrogen assimilating enzymes in cyanobacterial lichens of the genus *Peltigera*. Rowell, Rai & Stewart (1985) have shown that in *P. canina* the ammonium assimilating enzymes glutamate dehydrogenase (GDH) and glutamine synthetase (GS) have somewhat complementary distributions. GDH occurs throughout the thallus, located mainly in fungal hyphae lying close to the *Nostoc* photobiont, but activity is lower at the growing margins. GS, which is exclusive to the photobiont, is located chiefly in the young marginal regions although at levels of activity very much lower than those known in free-living *Nostoc*. Nitrogenase activity (mg thallus protein^{-1}) occurs in all parts of the thallus but declines progressively with increasing distance from the margin (note that on a unit thallus area basis nitrogenase activity increases with increasing distance from the growing margin (Millbank, 1972)). These results suggest that the capacity to assimilate photobiont generated ammonium exists in the fungus in all parts of the thallus but that there may be significant retention of fixed nitrogen by *Nostoc* in growing regions. Lallemant & Savoye (1985) noted similar differences between the distributions of GS and GDH in the mature thallus and actively growing lobules of *P. praetextata*. Studies on the tripartite lichen *P. aphthosa* have revealed that nitrogen can pass from fungus to alga: ^{15}N supplied as dinitrogen is fixed by *Nostoc* in surface cephalodia, rapidly assimilated by the cephalodial fungus (Rowell *et al.*, 1985), transferred to the main thallus and appears slowly in the *Coccomyxa* photobiont (Millbank & Kershaw, 1969; Kershaw & Millbank, 1970; Rai, Rowell & Stewart, 1981; Kershaw, 1985). The question of whether movement of label might be channelled towards new growth was not addressed.

The possible pathways by which mineral nutrients may reach the photobiont in a lichen thallus are summarized in Fig. 12.6 adapted from Harley & Smith's (1983) scheme for ectomycorrhizas. The model tentatively suggested here is similar to that proposed for sheathing mycorrhizas; acquisition of mineral nutrients by the photobiont is mainly from living fungal cells (B), diffusive movement through the external fungal tissue (A3) being of minor importance. If this is not the situation then it is to be assumed that the photobiont absorbs inorganic nutrients from surrounding intercellular water in competition with the fungus (cf. Smith 1975, 1976, 1980; Smith & Douglas 1987). Since the growing apices of *C. stellaris* and *S. paschale* form the uppermost surface of lichen mats they are well placed to intercept a large proportion of the incoming nutrients. Farrar & Smith (1976) and Smith (*op. cit.*) believe that leakage of metabolites from lichen symbionts during the rewetting phase at the onset of rainfall, and subsequent reabsorption, may provide an opportunity for exchange of nutrients in fa-

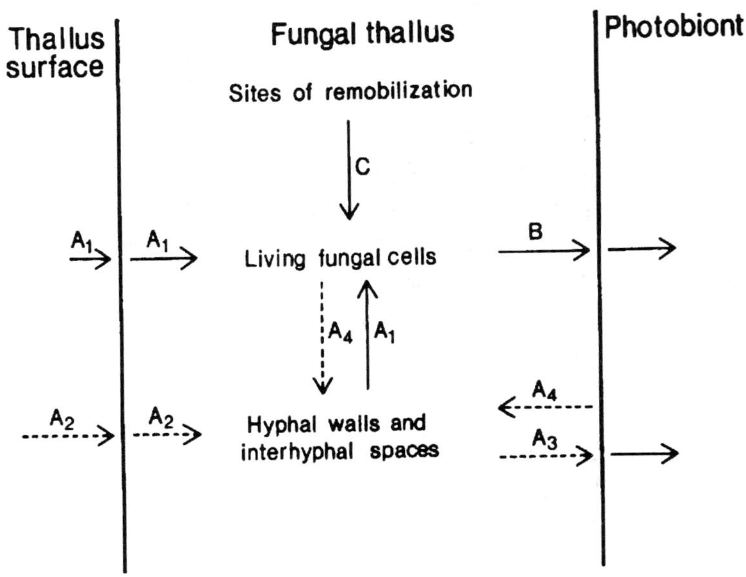

Fig. 12.6. Diagram showing potential pathways for mineral nutrient movement through fungal tissue in a lichen thallus (cf. sheath of ectomycorrhizas). Broken arrows show movement by diffusion or mass flow; solid arrows show metabolically dependent movement. Mineral nutrients move by diffusion or mass flow in surface water films. They are absorbed by metabolically dependent reactions A_1 into hyphae at the thallus surface and may either remain incorporated in the fungal biomass or be released to the photobiont by metabolic reaction B. Nutrients in older parts of the thallus are mobilized by reaction C and move to regions of growth (internal recycling). Nutrients may also pass non-metabolically from surface water films through the hyphal cell walls and interhyphal spaces of the cortex (*Stereocaulon*) or plectenchyma (*Cladonia stellaris* and other species of the subgenus *Cladina*; Ahti, 1984)(A_2) from where they may be accumulated into the living fungal cells (A_1) or into photobiont cells (A_3). During rewetting of dry lichen thalli at the onset of rainfall nutrients from both symbionts may be released passively into the free space (A_4). Adapted from Harley & Smith's (1983) scheme for ectomycorrhizas; cf. also Honegger's (1985) scheme for water transport in lichen thalli.

vour of the photobiont; again, such a process may be most evident in the thallus apices. Richardson & Nieboer (1980) argue that the growing regions in a lichen may develop a particularly favourable electrochemical gradient for ion uptake during the rewetting phase resulting in ion migration and accumulation in the apices. Resolving which of these pathways (Fig. 12.6) operate will necessitate a combination of astute observation and ingenious experimental approach.

Finally, recent evidence for pathway B has been provided by quite a different line of enquiry. Ahmadjian & Jacobs (1987) used axenic cultures of symbionts isolated from species of *Cladonia* to resynthesize lichens on mica strips. They maintained the cultures on mica in closed vessels at about 95% relative humidity for more than a year with no addition of nutrients. Not only did individual squamules survive this long period but new squamules formed. The authors concluded that nutrients for the new growth of both fungus and alga must have been derived largely from old mycelium that made up the original inoculum.

Concluding remarks

There are features common to the mineral nutrient relationships of ectomycorrhizas (as described by Harley & Smith, 1983) and to those proposed above for the lichens *C. stellaris* and *S. paschale*. These are (i) exploitation of periodic nutrient flushes, (ii) nutrient storage in fungal tissues, and (iii) remobilization of stored reserves to meet the growth requirements of the photobiont. In the case of lichens, inorganic nutrients are absorbed during rainfall which results in a flush of mineral ions in the form of wet deposits, and dry deposits brought into solution; nutrients in older parts of thalli are recycled and move to the growing apices (to this extent the thallus functions as a nutrient store) where there is nutrient exchange from fungus to alga promoting autosporulation. Ectomycorrhizas are believed to trap efficiently mineral nutrients delivered in seasonal flushes associated with periods of leaf senescence, fall and decomposition; nutrients in excess of the immediate demands of the host are stored in fungal sheaths and can subsequently be remobilized and passed to the photobiont during periods of nutrient shortage. Interestingly, polyphosphate storage granules characteristic of phosphorus accumulation in fungal sheaths of ectomycorrhizas have also been observed in large numbers in lichen fungi (Chilvers, Ling-Lee & Ashford, 1978).

Using ectomycorrhizas and coelenterates as examples, Lewis (1973) ascribed the relative success of mutualistic symbioses in nutrient poor environments to the possession therein of mechanisms for tight nutrient cycling; 'a tightness which is assisted by antibiotic production and the partial elimination of competitors'. Preliminary evidence reported here indicates that mat-forming lichens conform closely to this general model.

Acknowledgements. I am particularly indebted to the staff of the Kevo Subarctic Research Station (University of Turku) for the generous provision of facilities during the course of the fieldwork and to Dr R. P. Beckett (Department of Botany, University of Natal) for permitting me to present some of our previously unpublished data . Mamoru Matsuki kindly helped with collecting soils which were brought into the UK under MAFF Licence No. PHF 77/108 (86). The work was supported by a grant from the Erna and Victor Hasselblad Foundation.

References

Addison, P. A. & Puckett, K. J. (1980). Deposition of atmospheric pollutants as measured by lichen element content in the Athabasca oil sands area. *Canadian Journal of Botany*, **58**, 2323-2334.

Ahmadjian, V. (1980). Separation and artificial synthesis of lichens. In *Cellular Interactions in Symbiosis and Parasitism*, eds. C. B. Cook, P. W. Pappas & E. D. Rudolph, pp. 3-29. Ohio State University Press, Columbus, Ohio, USA.

Ahmadjian, V. (1982). Algal/fungal symbioses. In *Progress in Phycological Research*, vol 1, eds. F. E. Round & D. J. Chapman, pp. 179-233. Elsevier Biomedical Press, Amsterdam.

Ahmadjian, V. & Jacobs, J. B. (1981). Relationship between fungus and alga in the lichen *Cladonia cristatella* Tuck. *Nature*, **289**, 169-172.

Ahmadjian, V. & Jacobs, J. B. (1983). Algal-fungal relationships in lichens: recognition, synthesis, and development. In *Algal Symbiosis*, ed. L. J. Goff, pp. 147-172. Cambridge University Press, Cambridge, UK.

Ahmadjian, V. & Jacobs, J. B. (1987). Studies on the development of synthetic lichens. In *Progress and Problems in Lichenology in the Eighties*, ed. E. Peveling, pp. 47-58. J. Cramer, Berlin.

Ahti, T. (1977). Lichens of the boreal coniferous zone. In *Lichen Ecology*, ed. M. R. D. Seaward, pp. 145-181. Academic Press, London.

Ahti, T. (1984). The status of *Cladina* as a genus segregated from *Cladonia*. *Nova Hedwigia* Beiheft 79, 25-61.

Allen, S. E., Grimshaw, H. M., Parkinson, J. A. & Quarmby, C. (1974). *Chemical Analysis of Ecological Materials*. Blackwell Scientific Publications, Oxford, UK.

Armstrong, R. A. (1977). The response of lichen growth to additions of distilled water, rainwater and water from a rock surface. *New Phytologist*, **79**, 373-376.

Armstrong, R. A. (1979). Growth and regeneration of lichen thalli with the central portions artificially removed. *Environmental and Experimental Botany*, **19**, 175-178.

Armstrong, R. A. (1984). Growth of experimentally reconstructed thalli of the lichen *Parmelia conspersa*. *New Phytologist*, **98**, 497-502.

Auclair, A. N. D. (1983). The role of fire in lichen-dominated tundra and forest-tundra. In *The Role of Fire in Northern Circumpolar Ecosystems*, eds. R. W. Wein & D. A. MacLean, pp. 235-256. J. Wiley & Sons, Chichester.

Bailey, C. & Larson, D. W. (1982). Water quality and pH effects on *Umbilicaria mammulata* (Ach.) Tuck. *Bryologist*, **85**, 431-437.

Bakaeva, M. V. & Galanin, A. V. (1985). Ecological role of the lichen cover in the white-moss pine forests of the central Vychegda. *Soviet Journal of Ecology*, **16**, 80-84.

Bremner, J. M. & Breitenbeck, G. A. (1983). A simple method for determination of ammonium in semimicro-Kjeldahl analysis of soils and plant materials using a block digester. *Communications in Soil Science and Plant Analysis*, **14**, 905-913.

Brodo, I. M. (1973). Substrate Ecology. In *The Lichens*, eds. V. Ahmadjian & M. E. Hale, pp. 401-441. Academic Press, London.

Brown, D., MacFarlane, J. D. & Kershaw, K. A. (1983). Physiological-environmental interactions in lichens. XVI. A re-examination of resaturation respiration phenomena. *New Phytologist*, **93**, 237-246.

Brown, D. H. & Di Meo, J. A. (1972). Influence of local maritime conditions on the distribution of two epiphytic lichens. *Lichenologist*, **5**, 305-310.

Brown, R. T. & Mikola, P. (1974). The influence of fruticose soil lichens upon the mycorrhizae and seedling growth of forest trees. *Acta Forestalia Fennica*, **141**, 1-23.

Buck, G. W. & Brown, D. H. (1979). The effect of desiccation on cation location in lichens. *Annals of Botany*, **44**, 265-277.

Carstairs, A. G. & Oechel, W. C. (1978). Effects of several microclimatic factors and nutrients on net carbon dioxide exchange in *Cladonia alpestris* (L.) Rabh. in the Subarctic. *Arctic and Alpine Research*, **10**, 81-94.

Chilvers, G. A., Ling-Lee, M. & Ashford, A. E. (1978). Polyphosphate granules in the fungi of two lichens. *New Phytologist*, **81**, 571-574.

Crittenden, P. D. (1983). The role of lichens in the nitrogen economy of subarctic woodlands: nitrogen loss from the nitrogen-fixing lichen *Stereocaulon paschale* during rainfall. In *Nitrogen as an Ecological Factor*, eds. J. A. Lee, S. McNeill & I. H. Rorison, pp. 43-68. Blackwell Scientific Publications, Oxford, UK.

Crittenden, P. D. & Kershaw, K. A. (1978). Discovering the role of lichens in the nitrogen cycle in boreal-arctic ecosystems. *Bryologist*, **81**, 258-267.

Crittenden, P. D. & Kershaw, K. A. (1979). Studies on lichen-dominated systems. XXII. The environmental control of nitrogenase activity in *Stereocaulon paschale* in spruce-lichen woodland. *Canadian Journal of Botany*, **57**, 236-254.

Coxson, D. S. (1987). Photoinhibition of net photosynthesis in *Stereocaulon virgatum* and *S. tomentosum*, a tropical-temperate comparison. *Canadian Journal of Botany*, **65**, 1707-1715.

Coxson, D. S. (1988). Recovery of net photosynthesis and dark respiration on rehydration of the lichen, *Cladonia mitis*, and the influence of prior exposure to sulphur dioxide while desiccated. *New Phytologist*, **108**, 483-487.

Dawson, H. J., Hrutfiord, B. F. & Ugolini, F. C. (1984). Mobility of lichen compounds from *Cladonia mitis* in arctic soils. *Soil Science*, **138**, 40-45.

Ellis, K. M. & Smith, J. N. (1987). Dynamic model for radionuclide uptake in lichen. *Journal of Environmental Radioactivity*, **5**, 185-208.

Farrar, J. F. (1976). The uptake and metabolism of phosphate by the lichen *Hypogymnia physodes*. *New Phytologist*, **77**, 127-134.

Farrar, J. F. & Smith, D. C. (1976). Ecological physiology of the lichen *Hypogymnia physodes*. III. The importance of the rewetting phase. *New Phytologist*, **77**, 115-125.

Fisher, R. F. (1979). Possible allelopathic effects of reindeer-moss (*Cladonia*) on jack pine and white spruce. *Forest Science*, **25**, 256-260.

Fletcher, A. (1976). Nutritional aspects of marine and maritime lichen ecology. In *Lichenology: Progress and Problems*, eds. D. H. Brown, D. L. Hawksworth & R. H. Bailey, pp. 359-384. Academic Press, London

Foster, D. R. (1985). Vegetation development following fire in *Picea mariana* (black spruce) – *Pleurozium* forests of south-eastern Labrador, Canada. *Journal of Ecology*, **73**, 517-534.

Fritz-Sheridan, R. P. & Coxson, D. S. (1988). Nitrogen fixation on a tropical volcano, La Soufrière (Guadeloupe): the interaction of temperature, moisture, and light with net photosynthesis and nitrogenase activity in *Stereocaulon virgatum* and response to periods of insolation shock. *Lichenologist*, **20**, 63-81.

García-Junceda, E. & Xavier Filho, L. (1986). Solubilization of lichen phenolics from *Cladonia sprucei* by simulated rainfall. *Lichen Physiology and Biochemistry*, **1**, 61-69.

Garty, J., Galun, M. & Hochberg, Y. (1986). The accumulation of metals in *Caloplaca aurantia* growing on concrete roof tiles. *Lichenologist*, **18**, 257-263.

Gilbert, O. L. (1969). The effect of SO_2 on lichens and bryophytes around Newcastle upon Tyne. In *Air Pollution, Proceedings of the First European Congress on the Influence of Air Pollution on Plants and Animals, Wageningen 1968*, pp. 223-235. Centre for Agricultural Publishing and Documentation, Wageningen, The Netherlands.

Grime, J. P. (1979). *Plant Strategies and Vegetation Processes*. John Wiley & Sons, Chichester.

Groulx, M. & Lechowicz, M. J. (1987). Net photosynthetic recovery in subarctic lichens with contrasting water relations. *Oecologia*, **71**, 360-368.

Hällbom, L. & Bergman, B. (1983). Effects of inorganic nitrogen on C_2H_2 reduction and CO_2 exchange in the *Peltigera praetextata-Nostoc* and *Peltigera aphthosa-Coccomyxa-Nostoc* symbioses. *Planta*, **157**, 441-445.

Hämet-Ahti, L. (1963). Zonation of the mountain birch forests in northernmost Fennoscandia. *Annales Botanici Societatis Zoologicæ Botanicæ Fennicæ 'Vanamo'*, **34** (No. 4), 1-127.

Harley, J. L. & Smith, S. E. (1983). *Mycorrhizal Symbiosis*. Academic Press, London.

Honegger, R. (1985). Fine structure of different types of symbiotic relationships in lichens. In *Lichen Physiology and Cell Biology*, ed. D. H. Brown, pp. 287-302. Plenum Press, New York.

Honegger, R. (1987). Questions about pattern formation in the algal layer of lichens with stratified (heteromerous) thalli. In *Progress and Problems in Lichenology in the Eighties*, ed. E. Peveling, pp. 59-71. J. Cramer, Berlin.

Huss-Danell, K. (1977). Nitrogen fixation by *Stereocaulon paschale* under field conditions. *Canadian Journal of Botany*, **55**, 585-592.

Huss-Danell, K. (1978). Seasonal variation in the capacity for nitrogenase activity in the lichen *Stereocaulon paschale*. *New Phytologist*, **81**, 89-98.

Hutchison-Benson, E., Svoboda, J. & Taylor, H. W. (1984). The latitudinal inventory of ^{137}Cs in vegetation and topsoil in northern Canada, 1980. *Canadian Journal of Botany*, **63**, 784-791.

Johnson, M. J. (1941). Isolation and properties of a pure yeast polypeptidase. *Journal of Biological Chemistry*, **137**, 575-586.

Kallio, S. & Kallio, P. (1975). Nitrogen fixation in lichens at Kevo, north-Finland. In *Fennoscandian Tundra Ecosystems, Part 1. Plants and Micro-organisms (Ecological Studies, 16)*, ed. F. E. Wielgolaski, pp. 292-304. Springer-Verlag, Berlin.

Kärenlampi, L. (1971). Studies on the relative growth rate of some fruticose lichens. *Reports from the Kevo Subarctic Research Station*, **7**, 33-39.

Kershaw, K. A. (1977). Studies on lichen-dominated systems. XX. An examination of some aspects of the northern boreal lichen woodlands in Canada. *Canadian Journal of Botany*, **55**, 393-410.

Kershaw, K. A. (1985). *Physiological Ecology of Lichens*. Cambridge University Press, Cambridge, UK.

Kershaw, K. A. & Millbank, J. W. (1970). Nitrogen metabolism in lichens. II. The partition of cephalodial-fixed nitrogen between the mycobiont and phycobionts of *Peltigera aphthosa*. *New Phytologist*, **69**, 75-79.

Kershaw, K. A., Rouse, W. R. &. Bunting, B. T. (1975). *The Impact of Fire on Forest and Tundra Ecosystems*, INA Publication No. QS-8038-000-EE-A1. Ministry of Indian and Northern Affairs, Ottawa, Canada.

Lallemant, R. & Savoye, D. (1985). Lectins and morphogenesis: facts and outlooks. In *Lichen Physiology and Cell Biology*, ed. D. H. Brown, pp.335-350. Plenum Press, New York.

Lang, G. E., Reiners, W. A. & Heier, R. K. (1976). Potential alteration of precipitation chemistry by epiphytic lichens. *Oecologia*, **25**, 229-241.

Larson, D. W., Matthes-Sears, U. & Nash IIIrd, T. H. (1985). The ecology of *Ramalina menziesii*. I. Geographical variation in form. *Canadian Journal of Botany*, **63**, 2062-2068.

Lawrey, J. D. (1984). *Biology of Lichenized Fungi*. Praeger Publishers, New York.

Lawrey, J.D. (1986). Biological role of lichen substances. *Bryologist*, **89**, 111-122.

Lewis, D. H. (1973). The relevance of symbiosis to taxonomy and ecology, with particular reference to mutualistic symbioses and the exploitation of marginal habitats. In *Taxonomy and Ecology*, ed. V. H. Heywood, pp. 151-172. Academic Press: London.

Lewis Smith, R. I. (1978). Summer and winter concentrations of sodium, potassium and calcium in some maritime Antarctic cryptogams. *Journal of Ecology*, **66**, 891-909.

Malicki, J. (1965). The effect of lichen acids on the soil microorganisms. Part 1. The washing down of the acids into the soil. *Annales Universitatis Mariae Curie - Skl odowska, Sectio C*, **20**, 239-248.

Matthes-Sears, U., Nash IIIrd, T. H. & Larson, D. W. (1987). Salt loading does not control CO_2 exchange in *Ramalina menziesii* Tayl. *New Phytologist*, **106**, 59-69.

Millbank, J. W. (1972). Nitrogen metabolism in lichens. IV. The nitrogenase activity of the *Nostoc* phycobiont in *Peltigera canina*. *New Phytologist*, **71**, 1-10.

Millbank, J. W. (1978). The contribution of nitrogen fixing lichens to the nitrogen status of their environment. In *Environmental Role of Nitrogen-fixing Blue-green Algae and Asymbiotic Bacteria (Ecological Bulletins, 26)*, ed. U. Granhall, pp. 260-265. Swedish Natural Science Research Council, Stockholm.

Millbank, J. W. & Kershaw, K. A. (1969). Nitrogen metabolism in lichens. 1. Nitrogen fixation in the cephalodia of *Peltigera aphthosa*. *New Phytologist*, **68**, 721-729.

Moore, T. R. (1980). The nutrient status of subarctic woodland soils. *Arctic and Alpine Research*, **12**, 147-160.

Nieboer, E., Ahmed, H.M., Puckett, K. J. & Richardson, D. H. S. (1972). Heavy metal content of lichens in relation to distance from a nickel smelter in Sudbury, Ontario. *Lichenologist*, **5**, 292-304.

Nieboer, E., Padovan, D., Lavoie, P. & Richardson, D. H. S. (1984). Anion accumulation by lichens. II. Competition and toxicity studies involving arsenate, phosphate, sulphate and sulphite. *New Phytologist*, **96**, 83-93.

Nieboer, E. & Richardson, D. H. S. (1981). Lichens as monitors of atmospheric deposition. In *Atmospheric Pollutants in Natural Waters*, ed. S. J. Eisenreich, pp. 339-388. Ann Arbor Science Publishers Inc., Ann Arbor, Michigan, USA.

Nieboer, E., Richardson, D. H. S. & Tomassini, F. D. (1978). Mineral uptake and release by lichens: an overview. *Bryologist*, **81**, 226-246.

Oksanen, J. & Ahti, T. (1982). Lichen-rich pine forest vegetation in Finland. *Annales Botanici Fennici*, **19**, 275-301.

Pakarinen, P. (1981a). Nutrient and trace metal content and retention in reindeer lichen carpets of Finnish ombrotrophic bogs. *Annales Botanici Fennici*, **18**, 265-274.

Pakarinen, P. (1981b). Regional variation of sulphur concentrations in *Sphagnum* mosses and *Cladonia* lichens in Finnish bogs. *Annales Botanici Fennici*, **18**, 275-279.

Parsons, I.T. & Hunt, R. (1981). Plant growth analysis: a program for the fitting of lengthy series of data by the method of *B*-splines. *Annals of Botany*, **48**, 341-352.

Payette, S., Filion, L., Gauthier, L. & Boutin, Y. (1985). Secular climate change in old-growth tree-line vegetation of northern Quebec. *Nature*, **315**, 135-138.

Pellet, G. L., Bustin, R. & Harriss, R. C. (1984). Sequential sampling and variability of acid precipitation in Hampton, Virginia. *Water, Air, and Soil Pollution*, **21**, 33-49.

Pilegaard, K. (1978). Airborne metals and SO_2 monitored by epiphytic lichens in an industrial area. *Environmental Pollution*, **17**, 81-92.

Pilegaard, K. (1979). Heavy metals in bulk precipitation and transplanted *Hypogymnia physodes* and *Dicranoweisia cirrata* in the vicinity of a Danish steelworks. *Water, Air, and Soil Pollution*, **11**, 77-91.

Puckett, K.J. (1985). Temporal variation in lichen element levels. In *Lichen Physiology and Cell Biology*, ed. D. H. Brown, pp. 211-225. Plenum Press, New York.

Purvis, O. W. (1984). The occurrence of copper oxalate in lichens growing on copper sulphide-bearing rocks in Scandinavia. *Lichenologist*, **16**, 197-204.

Pyatt, F. B. (1973). Plant sulphur content as an air pollution gauge in the vicinity of a steelworks. *Environmental Pollution*, **5**, 103-115.

Quispel, A. (1943). The mutual relations between algae and fungi in lichens. *Recueil des travaux botaniques néerlandais*, **40**, 413-541.

Rai, A. N., Rowell, P. & Stewart, W. D. P. (1981). $^{15}N_2$ incorporation and metabolism in the lichen *Peltigera aphthosa* Willd. *Planta*, **152**, 544-552.

Read, D. J. (1983). The biology of mycorrhiza in the Ericales. *Canadian Journal of Botany*, **61**, 985-1004.

Read, D. J. (1984). The structure and function of the vegetative mycelium of mycorrhizal roots. In *The Ecology and Physiology of the Fungal Mycelium*, eds. D. H. Jennings & A. D. M. Rayner, pp. 215-240. Cambridge University Press, Cambridge, UK.

Reiners, W. A. & Olson, R. K. (1984). Effects of canopy components on throughfall chemistry: an experimental analysis. *Oecologia*, **63**, 320-330.

Richardson, D. H. S. & Nieboer, E. (1980). Surface binding and accumulation of metals in lichens. In *Cellular Interactions in Symbiosis and Parasitism*, eds. C. B. Cook, P. W. Pappas & E. D. Rudolph, pp. 75-94. Ohio State University Press, Columbus, Ohio, USA.

Ried, A. (1960). Stoffwechsel und Verbreitungsgrenzen von Flechten. II. Wasser- und Assimilationshaushalt, Entquellungs- und Submersionsresistenz von Krustenflechten benachbarter Standorte. *Flora, Jena*, **149**, 345-385.

Rowe, J. S. (1984). Lichen woodland in northern Canada. In *Northern Ecology and Resource Management*, eds. R. Olson, R. Hastings & F. Geddes, pp 225-237. University of Alberta Press, Edmonton, Alberta, Canada.

Rowe, J. S. & Acton, D. F. (1985). Taproots of jack pine (*Pinus banksiana* Lamb.) and soil tongues in Saskatchewan. *Canadian Journal of Forest Research*, **15**, 646-650.

Rowell, P., Rai, A. N. & Stewart, W. D. P. (1985). Studies on the nitrogen metabolism of the lichens *Peltigera aphthosa* and *Peltigera canina*. In *Lichen Physiology and Cell Biology*, ed. D. H. Brown, pp. 145-160. Plenum Press, New York.

Scott, M. G. & Larson, D. W. (1984). Comparative morphology and fine structure of a group of *Umbilicaria* lichens. *Canadian Journal of Botany*, **62**, 1947-1964.

Sepponen, P. (1985). The ecological classification of sorted forest soils of varying genesis in northern Finland. *Communicationes Instituti Forestalis Fenniae*, **129**, 1-77.

Seymour, M. D. & Stout, T. (1983). Observations on the chemical composition of rain using short sampling times during a single event. *Atmospheric Environment*, **17**, 1483-1487.

Sheard, J. W. (1986). Distribution of uranium series radionuclides in upland vegetation of northern Saskatchewan. I. Plant and soil concentrations. *Canadian Journal of Botany*, **64**, 2446-2452.

Smith, D., Muscatine, L. & Lewis, D. (1969). Carbohydrate movement from autotrophs to heterotrophs in parasitic and mutualistic symbiosis. *Biological Reviews*, **44**, 17-90.

Smith, D. C. (1962). The biology of lichen thalli. *Biological Reviews*, **37**, 537-570.

Smith, D. C. (1975). Symbiosis and the biology of lichenised fungi. In *Symbiosis*, Symposia of the Society for Experimental Biology, No. 29, pp. 373-405. Cambridge University Press, Cambridge, UK.

Smith, D. C. (1976). A comparison between the lichen symbiosis and other symbioses. In *Lichenology: Progress and Problems*, eds. D. H. Brown, D. L. Hawksworth & R. H. Bailey, pp. 497-513. Academic Press, London.

Smith, D. C. (1980). Mechanisms of nutrient movement between the lichen symbionts. In *Cellular Interactions in Symbiosis and Parasitism*, eds. C. B. Cook, P. W. Pappas & E. D. Rudolph, pp. 197-227. Ohio State University Press, Columbus, Ohio, USA.

Smith, D. C. & Douglas, A. E. (1987). *The Biology of Symbiosis*. Edward Arnold, London.

Smith, D. C. & Molesworth, S. (1973). Lichen Physiology. XIII. Effects of rewetting dry lichens. *New Phytologist*, **72**, 525-533.

Syers, J. K., Birnie, A. C. & Mitchell, B. D. (1967). The calcium oxalate content of some lichens growing on limestone. *Lichenologist*, **3**, 409-414.

Tomassini, F. D., Puckett, K. J., Nieboer, E., Richardson, D. H. S. & Grace, B. (1976). Determination of copper, iron, nickel, and sulphur by X-ray fluorescence in lichens from the Mackenzie Valley, Northwest Territories, and the Sudbury District, Ontario. *Canadian Journal of Botany*, **54**, 1591-1603.

Tuominen, Y. & Jaakkola, T.(1973). Absorption and accumulation of mineral elements and radioactive nuclides. In *The Lichens*, eds. V. Ahmadjian & M. E. Hale, pp. 185-223. Academic Press, London.

Ugolini, F. C., Reanier, R. E., Rau, G. H. & Hedges, J. I. (1981). Pedological, isotopic, and geochemical investigations of the soils at the boreal forest and alpine tundra transition in northern Alaska. *Soil Science*, **131**, 359-374.

Vainstein, E. A. & Ravinskaya, A. P. (1984). Biological degradation of lichen acids in the soil. *Botanicheskii zhurnal*, **69**, 1347-1351.

Yarranton, G. A. (1975). Population growth in *Cladonia stellaris* (Opiz.) Pouz. and Vezda. *New Phytologist*, **75**, 99-110.

Zakshek, E. M., Puckett, K. J. & Percy, K. E. (1986). Lichen sulphur and lead levels in relation to deposition patterns in Eastern Canada. *Water, Air, and Soil Pollution*, **30**, 161-169.

Chapter 13

Role of fungi in nitrogen, phosphorus and sulphur cycling in temperate forest ecosystems

J. Dighton & Lynne Boddy

Institute of Terrestrial Ecology, Merlewood Research Station, Grange-over-Sands, Cumbria LA11 6JU, UK & School of Pure and Applied Biology, University of Wales, College of Cardiff, Cardiff CF2 1XH, UK

Introduction

The continuing flow of energy through any ecosystem depends upon maintenance of nutrient supplies to the primary producers and since there is only a limited input of nutrients from outside, a balanced cycle within the ecosystem is essential. Three main types of cycle (Fig. 13.1) provide nutrients to the primary producer. The ***external geological cycle*** involves inputs to the ecosystem from the atmosphere (from natural sources and from pollutants), by weathering of soil parent material and from the addition of fertilizers, whereas losses result from leaching, burning and harvesting. The ***biological cycle*** involves exchange between soil, plants, herbivores and carnivores and the decomposer subsystem. ***Internal cycling*** occurs within these subsystems and within individual organisms, particularly trees. Nutrient cycling within forests has received detailed treatment by a number of authors, including Ovington (1962), Duvigneaud & Denaeyar de-Smet (1970), Borman & Likens (1979), Heal (1979), Miller (1979), Whittaker *et al.* (1979), Swank & Waid (1980), Clark & Rosswall (1981), Cole & Rapp (1981), Heal, Swift & Anderson (1981), yet the specific role of fungi has received much less attention (but see Lindberg, 1981).

The crucial role of fungi in the cycling of nutrients lies in the fact that fluxes of nitrogen, phosphorus and sulphur within the decomposer subsystem and other subsystems are mediated largely by

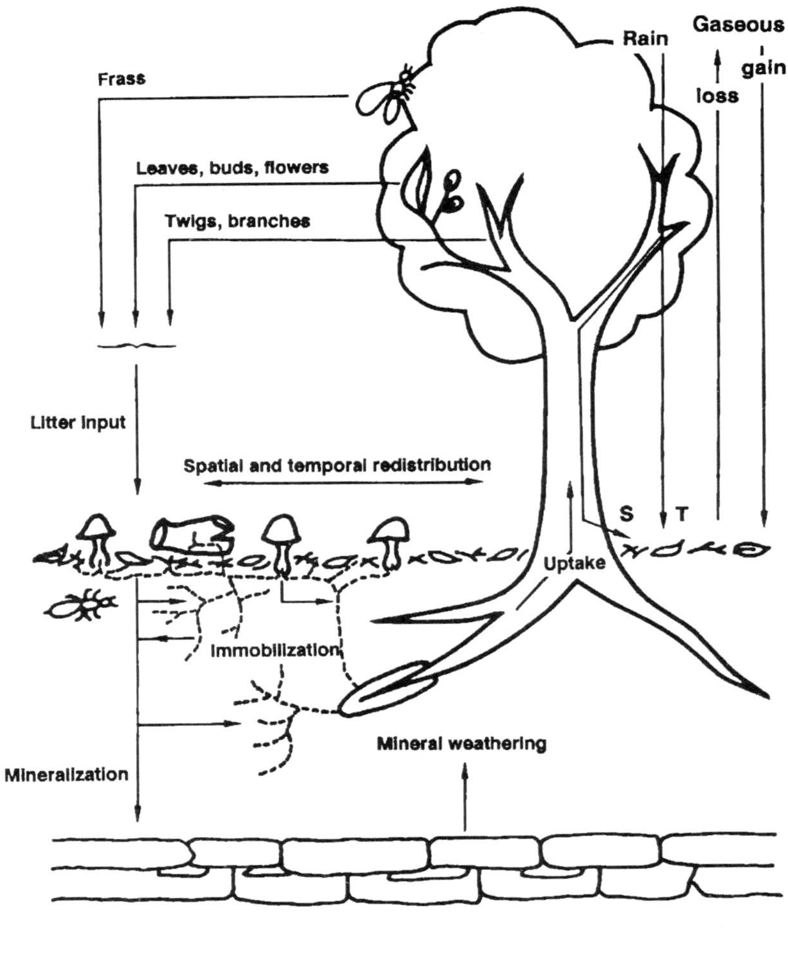

Fig. 13.1. Diagrammatic representation of nutrient inputs, immobilization and mineralization in a forest ecosystem. (S = stem flow; T = throughfall).

them. Fungi, as a result of saprotrophic nutrition, are the major decomposers of organic substrata in terrestrial ecosystems and consequently are responsible for the return of nutrients immobilized in dead plant, animal and microbial tissues to the soil pool. Fungi may also sequester nutrients which have 'leaked' from plants in, for example, the phylloplane and the rhizosphere, and obtain them directly by parasitic interactions. They also directly influence the nutrient relations of autotrophs by the mutualistic association of mycorrhizal fungi and plant roots, and of certain fungi and algae which form lichens. Nutrient fluxes between parasitic fungi and plants are explored by Walters (Chapter 7) and Paul (Chapter 8) and of lichens by Crittenden (Chapter 12), and will not be considered here. The role of mycorrhizas in nitrogen and phosphorous nutrition of plants is treated by Read, Leake & Langdale (Chapter 9), Hetrick (Chapter 10) and Gianinazzi-Pearson & Gianinazzi (Chapter 11), and will only be considered here in terms of their role in saprotrophic decomposition and overall ecosystem nutrient cycling. Emphasis in this chapter will be placed on the role of saprotrophic fungi in nitrogen and phosphorus cycling within temperate forests. Relatively little is known of the involvement of fungi in the sulphur cycle of forest ecosystems but some consideration is given to the possible impact of sulphur depositions arising from atmospheric pollution (see also Wainwright, Chapter 4).

Forest ecosystem development: changes associated with succession

Forests are static neither in space nor time; under suitable climatic conditions, bare ground will be successively covered by herbaceous vegetation and then by woody forms, ultimately forming climax forest communities. The latter are often of uneven age and may be patchy because of localized loss and re-establishment of the canopy. Thus, when considering the role of organisms in nitrogen and phosphorus cycling, it is essential to be aware of the discontinuous and dynamic nature of the ecosystem.

As the seral succession of plant species proceeds, important changes in many aspects of the structure and functioning of the ecosystem occur. Net primary production (NPP) progressively increases and then declines as the climax community becomes established. Associated with the initial increase in NPP is an increase in the total biomass of the system, with an increasing proportion of NPP going into the production of perennial above- and below-ground woody tissues. Associated with the shifts in plant species composition and production of perennial tissues are changes in the quality and quantity of nutrient resources. Thus the standing crop and litter input of herbaceous ground flora decreases, whilst the relative contribution of perennial components increases (Fig. 13.2). This provides increasing and often diversifying opportunities for the decomposer community resulting in a seral succession of the fungal community which will accompany that of the plant community.

Clearly then, dynamics of nitrogen, phosphorus and sulphur cycling processes in an ecosystem will differ according to the successional stage of the system, and changes in the supply of and demand for these elements by the forest (Miller *et al.*, 1979; Miller, 1981; Dighton, 1987; J. Dighton & A. F. Harrison, unpublished) may have significant effects on the physiology of the fungal components of the decomposer community.

Role of saprotrophic fungi in forest nutrient cycling

Input, standing crop and turnover of plant litter and hence nitrogen, phosphorus and sulphur: general considerations

Input and standing crop in temperate forests range, respectively, between 3,700-5,500 kg ha^{-1} yr^{-1} and 11,500-44,500 kg ha^{-1} which represents a large investment of nutrient capital (Table 13.1; Vogt, Grier & Vogt, 1986). Whilst input is largely determined by productivity of green plants, standing crop represents the ratio between input and turnover time, the latter depending on the decomposer organisms, resource quality and microclimate. Above-ground litter comprises various fractions which, in temperate woodlands, are approximately 55% leaves, 10% fruits, buds, flowers, etc., 20% twigs, 10% branches, 5% insect frass and miscellaneous (e.g. Bray & Gorham, 1964; Brown, 1974; Boddy & Swift, 1983). The amount of wood entering the decomposer system has, however, usually been underestimated and a more realistic figure is 40% (Swift, 1977). The input of below-ground

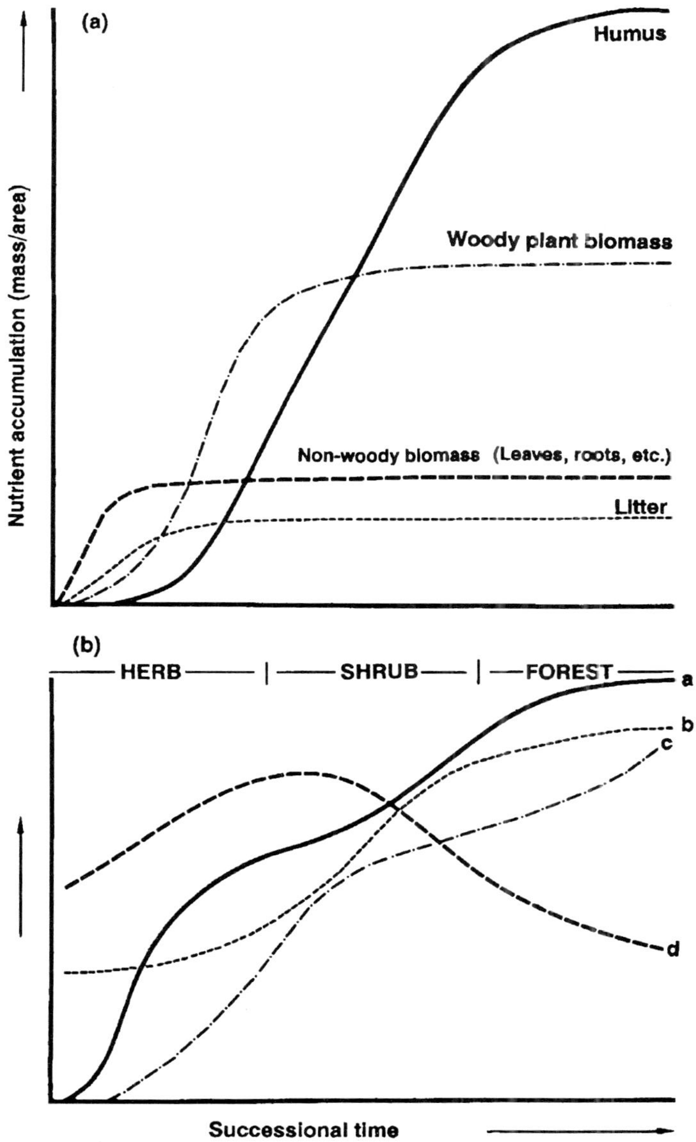

Fig. 13.2. (a) Generalized pattern of changes in nutrient pools during successional time in a forest ecosystem. (b) Hypothetical changes in resource quality and decomposition rates during primary succession in a forest ecosystem. (a = litter-fall mass, b = litter fall C:N, c = decomposition of forest floor organic matter (excluding litter), d = litter decomposition). (After Heal & Dighton, 1986).

Table 13.1. Above ground standing crop and input of nutrients and biomass in woody and non-woody litter in temperate woodlands.

	Forest floor ($kg\ ha^{-1}$)			Mean residence time on forest floor (yr)			Non-woody litter-fall ($kg\ ha^{-1}\ yr^{-1}$)			Woody litter-fall ($kg\ ha^{-1}\ yr^{-1}$)		
	biomass	N	P	biomass	N	P	biomass	N	P	biomass	N	P
Warm temperate												
broadleaf deciduous	11,480	163	11·9	2·7	4·8	4·3	4,236	36	3·8	891	2·6	0·8
broadleaf evergreen	19,148	60	3·8	3·1	1·0	2·2	6,484	55	3·7	-	-	-
needleleaf evergreen	20,026	362	25·1	4·6	13·8	11·0	4,432	28	2·7	1,107	2·5	0·2
Cold temperate												
broadleaf deciduous	32,207	624	50·2	10·2	19·1	11·1	3,854	43	4·6	1,046	3·7	0·2
broadleaf evergreen	13,900	200	10·5	3·9	-	-	3,590	-	-	-	-	-
needleleaf evergreen	44,574	504	44·9	17·9	32·8	22·1	3,144	26	3·2	602	1·1	0·1

After Vogt, Grier & Vogt (1986).

biomass is difficult to determine and has frequently been omitted from calculations, but this input can be considerable and must not be ignored (Vogt *et al.*, 1986). Vogt *et al.* (1983), for example, have shown that fine root biomass input to the decomposer community is 200 to 300% greater than leaf litter fall in *Abies amabilis* forests. Similarly, Fogel & Hunt (1983) estimated that for Douglas fir (*Pseudotsuga menziesii*), the return of organic matter to soil by fine roots and mycorrhizas ranged from 78-84% of total tree return, which accounted for over 80% of the total tree return of nitrogen and phosphorus (Table 13.2). Return of nitrogen and phosphorus from mycorrhizas was also estimated to be 4-5 times greater than from other root components (Fogel & Hunt, 1983).

The greatest nitrogen and phosphorus input to the decomposer subsystem is from the non-woody components (Table 13.1) which may sometimes contribute over 35 times that of the woody components. Return of nitrogen and phosphorus immobilized in non-woody litter is also rapid compared to woody components: complete decomposition of most deciduous litter takes between 9

Table 13.2. Standing crops and flux of biomass and nutrients into the decomposer system of a Douglas fir stand.

Component	Standing crop ($kg\ ha^{-1}$)	Throughput ($kg\ ha^{-1}\ yr^{-1}$)	Turnover time (yr)	Release and return of nutrients ($kg\ ha^{-1}$) [%] N	P
Tree					
foliage	14,732	2,410	6	15·3 [5]	4 [7]
wood	243,396	407	-	3 [1]	2 [3·5]
roots	64,303	≈10,000	1·6	103 [34]	37 [65]
Forest floor	19,034	3,032	6·3	38 [12·5]	2 [3·5]
Fungi					
mycorrhizas	12,794	8,262	1·5		
sporocarps	65	65	1·0	145·5 [47·5]	12 [21]
hyphae	7,035	6,991	1·1		
Stand Total	361,359	31,167		305	57

After Fogel & Hunt (1979, 1983).

months and 3 years, depending on species, with coniferous needle litter taking slightly longer, about 7 years, whereas complete decomposition of twigs, branches and small logs often takes 3 to 20 years and whole trees can take as long as 400 years (Rayner & Boddy, 1988). However, woody material, by virtue of this relatively slow decay rate, forms a large standing crop of biomass. For example, Harmon *et al.* (1986) estimated that the standing crop of large woody components on the floor of deciduous forests was 11 to 38 $\times$ 10^3 kg ha^{-1}, and in coniferous forests was 10 to 511 $\times$ 10^3 kg ha^{-1} which represents a large reserve of nitrogen and phosphorus. As such, it has been suggested that wood may act as a buffer when systems are perturbed (Ausmus, 1977; Swift, 1977).

Contribution of fungi to decomposition processes and hence to turnover of nitrogen and phosphorus

Decomposition of organic substrata results from a variety of interacting biotic and abiotic factors which can be considered as four distinct processes: non-enzymatic chemical reactions, leaching and/or volatilization, comminution and catabolism (Swift, Heal & Anderson, 1979; Boddy, 1986). Non-enzymatic chemical reactions, such as photosensitized oxidation by sunlight, weakening or direct cleavage of chain bonds by ultraviolet light are probably only significant in exposed situations and may, therefore, be unimportant in woodlands.

Leaching of inorganic ions and, to a lesser extent, soluble organic materials, occurs when water passes through and/or over dead organic substrata and living organisms. Leachates may be captured further down the canopy or soil profile or lost from the ecosystem. Some authors have not found leaching to result in significant weight loss of litter, although others have indicated that losses of 10 to 30% may occur (see Swift *et al.*, 1979; Boddy, 1986). Its significance relative to other components of the decomposition process probably depends on climate (particularly rainfall and temperature), soluble components of different substrata, size and extent of decay, and comminution. With regard to nutrients *per se*, losses can be quite large (Berg & Staaf, 1981). For example, in a mixed deciduous woodland in Belgium, 2 and 20% respectively of the total input of nitrogen and phosphorus at litter fall was attributable to leachate (Duvigneaud & Denaeyer de Smet, 1970). Gosz, Likens & Bormann (1973) found that up to 25% of total nitrogen was leached from *Fagus grandifolia* leaf litter in the field. Volatilization results in loss of oils, waxes, and resins, which will be of significance only in plant litters containing large quantities of these materials, e.g. creosote bush, eucalypts, etc. (Whitford *et al.*, 1981).

Comminution is the reduction in particle size of organic substrata which can be brought about by physical phenomena, such as wetting/drying and freezing/thawing cycles and wind abrasion, but in most ecosystems, including forests and woodlands, invertebrates are the main agents (Swift *et al.*, 1979). Amongst other effects, particle size influences susceptibility to leaching and alters the pattern of microbial colonization with small particles tending to select for unicellular as opposed to mycelial forms (Hanlon & Anderson, 1980; Swift & Boddy, 1984).

Catabolism is the component of decomposition processes which is entirely biologically mediated. Invertebrates catabolize both simple and complex organic molecules, either as a result of their own metabolism or in association with microorganisms, but *quan-*

titatively, their contribution is small, accounting for between about 1 to 15% of total heterotrophic metabolism (Edwards, Reichle & Crossley, 1970), although other invertebrate activities are directly or indirectly much more significant to nutrient cycling. Most of the heterotrophic metabolism in forests, therefore, results from microbial activity. For example, Reichle (1977) estimated that over 85% of the living heterotrophic biomass in a temperate *Liriodendron* forest was microbial. Frankland, (1982) estimated that approximately two-thirds of the total microbial biomass of the litter layer and soil of a temperate deciduous woodland was saprotrophic fungi (i.e. not mycorrhizal nor pathogenic) and amounted to approximately 75 kg ha^{-1}. This figure is, however, not necessarily related to activity, since turnover of bacteria is greater than for fungi. Furthermore, the ability to utilize different components of the litter varies between and within fungal and bacterial groups. With regard to the above- and below-ground woody litter, decomposition is brought about almost entirely by fungi, in particular by basidiomycetes and xylariacious ascomycetes, since they have the capacity to break down the lignocellulose complex, but microfungi are also important at early and late stages of decomposition (Rayner & Boddy, 1988).

Immobilization and mineralization of nitrogen and phosphorus by saprotrophic fungi

It is evident then that the heterotrophic microflora of the forest floor are ultimately responsible for releasing large quantities of nitrogen and phosphorus which would otherwise be retained in plant tissues. 'Ultimately' is, however, the operative word, since before these nutrient elements become mineralized (i.e. made available to the soil pool as soluble inorganic forms) they become immobilized in the fungal tissue of the saprotrophic community. The amounts of nutrients immobilized are dependent upon the group of fungi effecting decomposition of a specific substratum, the resource quality of that substratum and the season. Frankland (1982) estimated that some 30-40% of phosphorus and 18% of nitrogen from *Fraxinus* and *Betula* leaf litter was immobilized by *Mycena galopus* hyphae during incubation for six months. She also demonstrated that immobilization into wood decay fungi can be even greater. Stark (1972) showed that hyphae had 193 to 272%

greater nitrogen content and 104 to 223% greater phosphorus content than needle litter in a temperate *Pinus* forest, which indicated immobilization of these elements. Similarly, Bääth & Söderström (1979) showed that 10%, or 3·8 g m^{-2}, of Norway spruce (*Picea abies*) forest floor nitrogen and 20% of the phosphorus was immobilized into the fungal component.

Nitrogen, phosphorus and sulphur are released from fungal tissues on their death, following lysis or autolysis and during their life by excretion, secretion of extracellular enzymes, leaching and grazing by animals. The lifespan of an organism is, therefore, a major determinant of the duration of the immobilization phase. The lifespan of bacteria and ruderal fungi utilizing ephemeral organic compounds may be only a few days, hence they immobilize nutrients for relatively short periods of time. However, these nutrients may be rapidly sequestered and immobilized by other fungi, particularly if the latter caused the death and lysis of the former. At the other extreme, there are many fungi, almost exclusively basidiomycetes and xylariaceous ascomycetes, which utilize recalcitrant woody substrata and whose individual mycelial systems can persist for several years. Obvious examples include the perennial fruit bodies of certain of the Aphyllophorales, rhizomorphs of certain *Armillaria* spp. and mycelial cords of, for example, *Phanerochaete velutina*, *Phallus impudicus*, and *Tricholomopsis platyphylla* (see Thompson & Rayner, 1982; Dowson, Rayner & Boddy, 1988).

Not only are nitrogen and phosphorus immobilized in fungal structures, but they are also probably relocated within the ecosystem over considerable distances via fungi. Thus, systems of mycelial cords can ramify through leaf litter forming connections between woody and other resources, covering areas of over 2 m^2 (Thompson & Rayner, 1982; Dowson *et al.*, 1988), and individuals of the same genotype of *Armillaria bulbosa* and of *Phanerochaete velutina* have been found decaying dead trees over 50 m apart (Thompson & Boddy, 1983). Large fluxes and rapid rates of translocation (sometimes exceeding 25 cm h^{-1}) of carbon and mineral nutrients have been demonstrated from substrata through cord systems in the laboratory, in the comprehensive studies on the dry rot fungus *Serpula lacrimans* (which is not found in nature)(Watkinson, 1971; Brownlee & Jennings, 1982a & b), and in *Armillaria* sp. (Granlund, Jennings & Thompson, 1985), *Phallus impudicus*, *Mutinus caninus* and *Phanerochaete velutina* (Cairney, 1986; J. M. Wells, L. Boddy & J Dighton, unpublished). Rigorous field studies have, however, not been performed.

The implications are that nitrogen and phosphorus are translocated out of organic substrata by cord systems and into substrata which are being colonized. Indeed, such movement of nutrients is likely to occur in all fungi that are capable of mycelial extension out of the substrata in which they are growing. That nitrogen is imported into organic substrata is borne out by the fact that *absolute* concentrations (i.e. amount of nutrient expressed as a percentage of the original quantity or of the volume of the substratum, rather than as a percentage of weight of organic substratum) within leaf and wood litter often increase at early stages of decomposition, particularly in substrata with initially low nitrogen concentrations (Figs. 13.3 & 13.4; e.g. Aber & Melillo, 1980; Berg & Staff, 1981; Melillo, Aber & Muratore, 1982; Swift & Boddy, 1984). However, rigorous field studies are required to determine whether translocation of phosphorus and sulphur to new resources is significant.

Spore dispersal and subsequent colonization is another means of relocating nutrients. For example, spores of *Fomes applanatus* contain 3·05% nitrogen (by dry weight), and it has been estimated by Merrill & Cowling (1966) that all of the nitrogen in 33·7 kg of *Betula alleghaniensis* is required to supply the annual production of the spores of a typical sporophore. The nitrogen in 5·75 kg of wood would be required to produce the sporophore.

The resource as a determinant of rate of mineralization of nitrogen and phosphorus

It was implied above that the lifespan of fungi may be closely related to the degradability of the resource. Hence release of nutrients to the soil pool may largely be governed by the resource itself. The carbon : nitrogen and lignin : nitrogen ratios are known to be determinants of the resistance of substrata to decomposition, and the ultimate release of nutrients has also been explained in terms of these ratios (e.g. Swift *et al.*, 1979; Melillo *et al.*, 1982; Paustian & Schnurer, 1987). At early stages of decomposition the relative concentration (i.e. expressed as % dry weight) of nitrogen and phosphorus tends to increase as the process proceeds, or expressed another way the carbon : nutrient ratio decreases (Figs. 13.3 & 13.4). This occurs irrespective of whether nutrients are imported, because carbon is lost, as CO_2 in respiration, but nitrogen and phosphorus are incorporated into microbial biomass. It has been suggested (e.g. Swift *et al.*, 1979) that net mineralization does

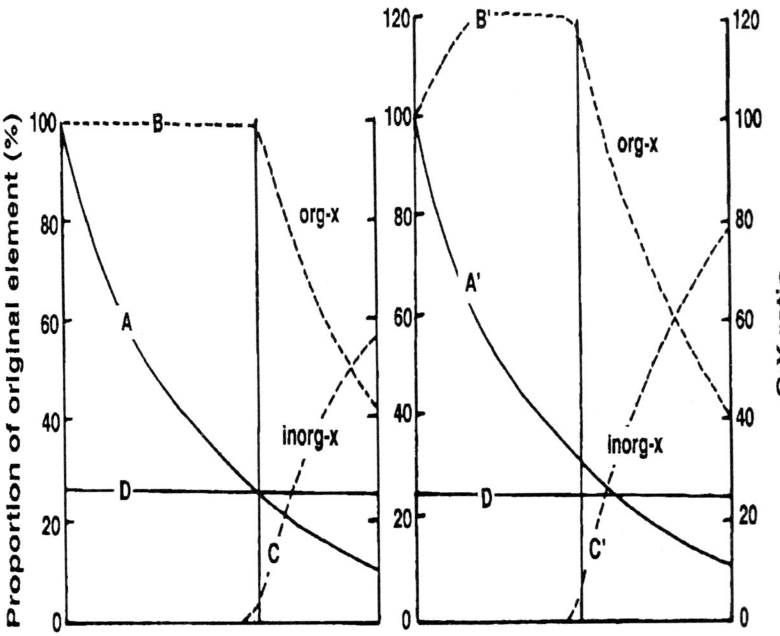

Fig. 13.3. A simple time-course model to illustrate the importance of C:nutrient ratios as indices of the equilibrium between immobilization and mineralization of an element X. Curve A = C:X ratio of an organic resource having an initial ratio of 800:1. Curve B = proportion of X in organic form (immobilised), (Curve B' = case where there is import of X from sources external to the resources being decomposed). Curve C = proportion of X in inorganic form (mineralized), (Curve C' = mineralization in response to alteration of C:X ratio by import of X as in case B'). D = C:X ratio of decomposer organism. After this point net mineralization occurs and curve A flattens off although carbon loss may continue on the same gradient. (From Swift, Heal & Anderson, 1979).

not occur until the carbon : nutrient ratio of the substratum reaches a threshold level which is approximately equivalent to the carbon : nutrient element ratio of the fungus (which is around 15 : 1 for phosphorus and 6 : 1 for nitrogen)(Fig. 13.3; Swift, 1977).

The initial carbon : nitrogen and carbon : phosphorus ratios vary considerably between different litter components. The carbon : nitrogen ratio of undecayed wood typically lies between 350 : 1 and 500 : 1, but can be as high as 1250 : 1, and the carbon : phosphorus ratio is >3500 : 1. In leaf litter these ratios range respec-

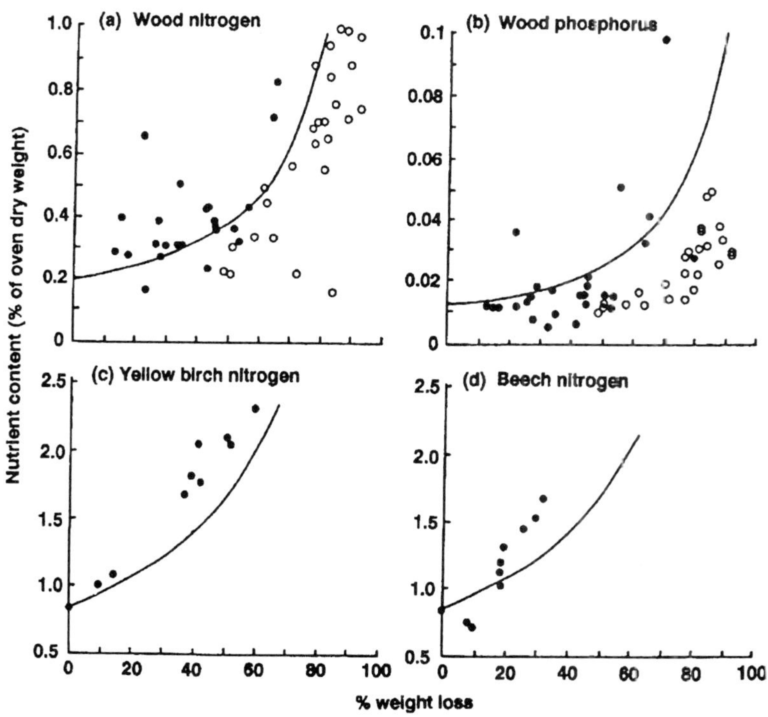

Fig. 13.4. The relationship between mineral nutrient concentration (as % of oven dry weight) and weight loss (% of original dry weight) of wood uninvaded (closed circles) and invaded (open circles) by invertebrates, and of leaf litter (panels c & d). The line on each graph indicates the predicted change in elemental concentration as decomposition proceeds, assuming no net gain or loss. Points above the line therefore represent increase in absolute concentration. [(a) and (b) from Swift & Boddy, 1984; data recalculated from Swift, 1977. Panels (c) and (d) redrawn from Aber & Melillo, 1980; original study by Gosz, Likens & Bormann, 1973].

tively between 25 : 1 to 100 : 1 and 450 : 1 to 1850 : 1 (Cowling, 1970; Swift *et al.*, 1979). Where the ratio is high, as in wood, the decomposer system becomes conservative and with increasing carbon : nutrient ratio there is an increasing trend for the fungi to immobilize nitrogen and phosphorus into their own biomass rather than to effect mineralization. Although the validity of the carbon : nitrogen ratio as a criterion for nitrogen supply is questionable (see Park, 1976; Dowding, 1981; Rayner & Boddy, 1988), it is likely that strong selective pressures act on wood decay fungi for 'economic' use of nitrogen (and phosphorus). As fungi may translocate nutrients it is likely that resources are recycled from 'redundant' hyphae to more active regions. Circumstantial evidence for this possibility is provided by the fact that several wood decay species can utilize their own dead mycelium, and that of other species, as sole source of nitrogen (Levi, Merrill & Cowling, 1968; Jennings, Chapter 1). Cord-forming fungi probably depend largely on 'exploitive' mycelium within the resource to supply the 'exploratory' systems with nitrogen and phosphorus, although it has been suggested that their requirements may be partly met by scavenging nutrients from the soil (Clipson, Cairney & Jennings, 1986). A consequence of efficient internal recycling, at least with cord-forming fungi, is that nitrogen and phosphorus will not necessarily be released to the soil pool when the carbon : nutrient ratio of the wood and the mycelium become similar, but rather will be conserved and translocated elsewhere.

Nitrogen and phosphorus availability may, however, be one factor governing the production of exit structures (i.e. outgrowth of mycelium, cords, rhizomorphs, sporocarps). Although conservation of nutrients is well developed in these fungi this is the part of the system which is 'leaky'. As mentioned earlier, large quantities of nitrogen are removed from wood by production of spores, and since few of these will establish new colonies, much of the nitrogen will be mobilised. Cord systems are also likely to 'leak'. Although no leakage into the soil was detected from major cords of *P. velutina* (L. Boddy & E. M. Owens, unpublished), considerable regression (i.e. dying back) of mycelium has been shown to occur during foraging outgrowth from resources (Dowson, *et al.*, 1986). In particular, mature cords develop following hyphal aggregation within mycelial fronts, with lysis of all hyphae and finely branched cords between the regions of greatest aggregation. Also, when exploratory systems encounter new resources there is often regression of non-connective cords and mycelium which emanate from the 'old' resource. Whilst some of the nutrients may be translocated back into the food base, the remainder will enter the soil pool. Quantitative studies are now necessary before the nutrient dynamics of such systems can be understood.

While carbon : nutrient ratio may partly regulate production of exit structures, at least some species, including cord-forming fungi, produce exit structures from relatively undecayed wood, providing the microclimate is suitable (e.g. Dowson *et al.*, 1986, 1988; Dowson, Boddy & Rayner, 1989). Since some lysis of these reproductive or vegetative structures occurs relatively rapidly, then some mineralization may occur when the carbon : nutrient ratio is still high.

Importance of interactions between soil fauna and fungi in mineralization

The soil fauna is a major determinant of the balance between mineralization and immobilization in organic substrata and microbial biomass, but rather than playing a direct role through catabolism or excretion it influences mineralization by interactions with the microflora (e.g. Anderson & Ineson, 1984). Thus grazing may result directly in release of nutrients as well as altering the relative contribution of different fungal species to the decomposition process (Parkinson, Visser & Whittaker, 1979; Newell, 1984a & b; Swift & Boddy, 1984). At present, however, the functional consequences of alterations in species composition are unclear.

The direct effects of grazing on mineralization and on stimulation/inhibition of fungal respiration and biomass production appear to be variable, probably at least partly because grazing interactions are subject to predator/prey regulation. For example, Visser, Whittaker & Parkinson (1981) found no evidence of enhanced nitrogen mineralization in systems grazed by collembola, and Seastedt & Crossley (1980) suggested that greater immobilization could result from the stimulation of fungal growth by microarthropod grazing. On the other hand there are several examples of increased mobilization of nutrients resulting from animal activity. In a microcosm experiment, Ineson, Leonard & Anderson (1982) found that fungal grazing by the collembolan *Folsomia candida*, on leaf litter, reduced fungal biomass and increased the mineralization of ammonium and nitrate nitrogen (Fig. 13.5). Similarly, consumption of litter (which contains fungi) by larger invertebrates also enhances mineralization. For example, Anderson, Ineson & Huish (1983) investigated the amount of leachable ammonium from microcosms containing oak

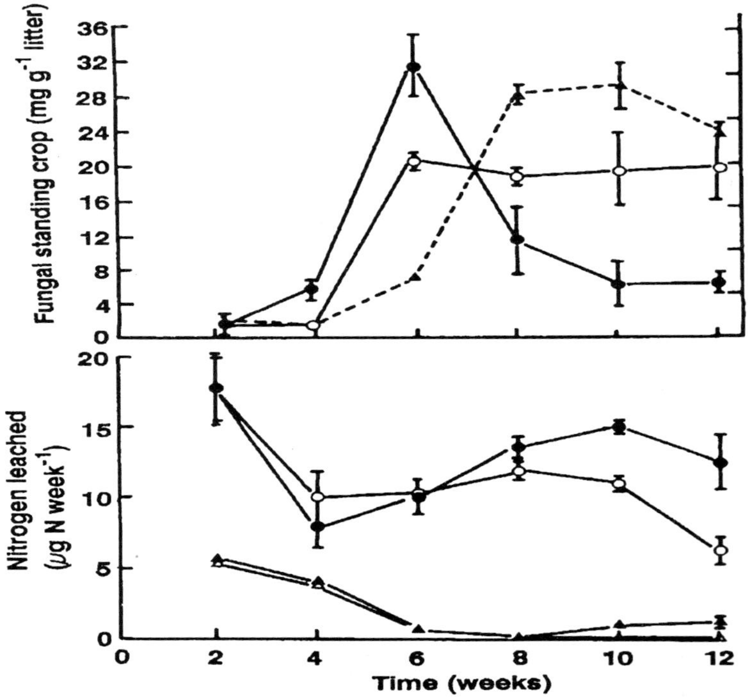

Fig. 13.5. Effect of collembolan feeding activities on fungal standing crop and nitrogen mobilization. (a) fungal standing crop on leaf litter with (closed circles) and without (open circles) collembola, and changes in collembolan populations (triangles); (b) mobilization of nitrogen as ammonium (circles) and nitrate (triangles) in presence (closed symbols) and absence (open symbols) of collembola. Values shown are means ± one SE; n = 4. From Ineson, Leonard & Anderson, 1982.

leaf litter colonised by one of six species of fungi, with and without the millipede *Glomeris marginata*. In all cases mineralization of nitrogen was enhanced in the presence of the millipede, but was strongly correlated with the different abilities of the fungi to effect mineralization. Likewise, enhanced mineralization of sulphur during grazing by the isopod *Oniscus asellus* has been demonstrated by Morgan & Mitchell (1987). Release of nitrogen from wood is strongly correlated with invasion by invertebrates, although appreciable phosphorus appears to occur before animal invasion (Fig. 13.4; Swift, 1977; Swift & Boddy, 1984).

Influence of mycorrhizas on nitrogen and phosphorus dynamics in the soil/plant system

Since much of the assimilable nitrogen and phosphorus in forest ecosystems is immobilized in living and dead plant biomass, the availability of nutrients for further plant growth depends upon recycling within the plant and mineralization by saprotrophic and pathogenic fungi. However, as implied earlier, nitrogen and phosphorus may be immobilized in fungal tissue for considerable periods before becoming available to plants. At early successional stages, nutrients may be relatively more available compared to mature woodlands, especially on poor soils, where the growth of trees is highly dependent on the release of nutrients from organic matter (Miller, 1979; Dighton, 1987; J. Dighton & A. F. Harrison, unpublished). In mature woodland tree roots tend to form associations with ectomycorrhizal fungi (Read, 1986). These not only take up inorganic nitrogen and phosphorus compounds, but can also utilize simple organic molecules (see Read, Leake & Langdale, Chapter 9). More importantly, some mycorrhizal species may be able to act as decomposers and thus short circuit the mineralization/nutrient uptake cycle (Went & Stark, 1968).

There is recent evidence that some ectomycorrhizal fungi have the ability to decompose organic forms of phosphorus and nitrogen by producing extracellular enzymes such as polyphenol oxidases (Giltrap, 1982), phosphatases (Alexander & Hardy, 1981; Ho & Zak, 1979; Dighton, 1983), cellulases (Linkins & Antibus, 1981) and proteinases (Abuzinadah & Read, 1986a & b; Abuzinadah, Finlay & Read, 1986; Dighton, Thomas & Latter, 1987). The ability of ectomycorrhizal fungi to produce degradative enzymes is probably related to the availability of nutrients in the soil. In general, coniferous forests planted on nutrient poor soils increasingly immobilise nutrients as the crop grows. There is a peak of immobilization around canopy closure where nitrogen and phosphorus are not released from organic complexes in the soil at a sufficient rate to supply the tree's demand (Miller, 1979; J. Dighton & A. F. Harrison, unpublished). With an increase in the organic : inorganic ratio of nutrients, the induction of decomposer activity in the mycorrhizal fungal flora would be of importance to the tree crop in the direct procurement of nutrients from organic sources.

It has been suggested by Dighton & Mason (1985) that change in the organic : inorganic ratio of soil nutrients may be one of the forces influencing mycorrhizal fungal succession in that the late stage fungi with '*K*-' strategies have the enzymatic capabilities that the '*r*-' fungi do not. Supporting evidence comes from the data of Abuzinadah & Read (1986a) of growth of eight mycorrhizal fungi on protein, where it was the late stage fungi *Suillus bovinus*, *Rhizopogon roseolus* and *Amanita muscaria* which grew better than the early stage fungi *Laccaria laccata* and *Lactarius rufus*. Similarly, in a gnotobiotic experiment, the decomposition of organic substrates (hide powder and cellulose) was greater in the presence of *Suillus luteus* than *Hebeloma crustuliniforme* (Dighton *et al.*, 1987).

Counter to the opinion that mycorrhizal fungi may be involved in the direct release of nutrients from organic sources, Janos (1983) suggested that it is the juxtaposition of mycorrhizal and saprotrophic hyphae which enables mycorrhizas to remove mineralized nitrogen and phosphorus before they can enter the measurable soil nutrient pool. It is important, therefore, to consider the role of mycorrhizas in the decomposition of organic substrata and acquisition of nutrients from these and recalcitrant inorganic forms in relation to the total soil microbial population. These interactions were explored by Gadgil & Gadgil (1971, 1975) who demonstrated that the presence of mycorrhizal roots suppressed decomposition of needle litter in a *Pinus radiata* forest. However, Berg & Lindberg (1980) repeated these experiments but were unable to detect suppression of litter decomposition in the presence of mycorrhizal roots in a *Pinus sylvestris* forest.

Other interactions between mycorrhizas, saprotrophic organisms and soil fauna are also important. For example, where ectomycorrhizal fungi develop together with bacteria in controlled experiments, they synergistically enhance phosphate uptake by the host pines from recalcitrant inorganic and organic forms (Chakly & Berthelin, 1982; Laheurte & Berthelin, 1986). In an interaction study of the ability of mycorrhizal fungi to decompose organic resources in the presence of a saprotrophic fungus, Dighton *et al.* (1987) showed that the saprotroph inhibited the decomposition potential of the mycorrhizal fungus. Recently, Gransaull & Brown (1987) have shown the synergistic effect of adding both mycorrhizas and *Frankia*: these significantly increased the growth of *Pinus* and *Casuarina* species compared with plants with either mycorrhizas or nitrogen fixing actinomycetes alone. However, much work on such interactions is necessary to evaluate their importance in nutrient mineralization and nutrient sequestering by plants.

Effects of exogenous inputs of pollutants on nutrient cycling

Effects of exogenous nutrient inputs on the forest ecosystem fall into three main types: damage to plants; interactions with fungi; and direct nutrient effects. Pollutant gases may cause physical damage to the canopy foliage and premature leaf fall. This causes a temporal shift in the pulse of nutrient input into the soil system and can also affect the resource quality of the litter. These may in turn influence the rate of decomposition and microbial community structure and consequently nutrient cycling. For example, P. A. Wookey (personal communication) has shown that increasing the atmospheric SO_2 concentration in the range of 10 to 50 ppb significantly reduced microbial respiration in both mixed deciduous and pine leaf litter and altered the rate of mineralization of major nutrients. Sulphur dioxide has also been shown to affect differentially the growth of fungi, e.g. at concentrations of 15 to 50 ppb *Trichoderma* spp. increased, whereas both *Cladosporium cladosporioides* and *Coniothyrium olivaceum* decreased. In addition, atmospheric SO_2 can be used by fungi as a source of sulphur, with different species immobilizing sulphur at different rates (Craker & Manning, 1974; Wainwright, Chapter 4). However, these processes are only just beginning to receive attention. Atmospheric pollutants also affect plants, which act as nutrient sinks, by influencing the species composition and physiology of mycorrhizas (Ulrich, Mayer & Khanna, 1979; Hütterman, 1982; Becker, 1982; Stroo & Alexander, 1985; Reich *et al.*, 1985; Jansen, Dighton & Bresser, 1988). For example, 'acid rain' caused soil chemical changes (Dighton & Skeffington, 1987) and acidified stemflow (Kumpfer & Heyser, 1986), which was correlated with a decreased frequency of corralloid mycorrhizas. Although the nutritional consequences of this imbalance have not been studied in detail, this community shift may represent a movement from '*K*-' to '*r*-' selected fungi and consequently a potential loss in the ability of the mycorrhizal system to sequester nutrients from organic resources in the soil. These effects may be due to mobilization of aluminium and Thompson & Medve (1984) demonstrated that aluminium concentrations of 146 μM could significantly reduce linear growth of agar cultures of the mycorrhizal fungi *Cenococcum*, *Pisolithus*, and *Thelephora*. Also, Oelbe-Farivar (1985) showed that aluminium toxicity reduced protein synthesis in mycorrhizal fungi.

Similarly, there is indirect evidence from the Netherlands that atmospheric nitrogen pollution can significantly affect the succession of mycorrhizal fungi under coniferous trees. By comparing recent fungal fruit body surveys with historical fungal foray records, Arnolds (1985, 1988) has demonstrated a trend of reduction in frequency of occurrence of late stage mycorrhizal fungal fruit bodies, which he attributes to an increase in nitrogen deposition. Also, numbers of *Russula* spp. fruit bodies in beech woods have been found to decline following the application of NPK fertilizer (Hall, 1978). Although care must be taken when interpreting such data (increasing nitrogen content of growth medium *in vitro* typically inhibits fruiting, but not production of mycelial biomass), the effects on plant nitrogen uptake of any such decline in mycorrhizas may be balanced by exogenous inputs of nitrogen which ameliorate conditions on poor soils. If, however, other major nutrients, such as phosphorus, do not increase concomitantly, imbalances could occur, resulting in a reduced yield of the tree crop.

Fungal conversion of inorganic N, P and S compounds

In addition to their saprotrophic abilities, some fungi are able to convert inorganic nitrogen, phosphorus and sulphur compounds from one form to another. Nitrogen exists in a number of inter-convertible inorganic forms:

$$N_2 \leftrightarrow N_2O \leftrightarrow (NO) \leftrightarrow NO_2^- \leftrightarrow NO_3^- \leftrightarrow NH_4^+$$

and there is evidence that certain fungi can perform nitrification (i.e. convert NH_4^+ to NO_3^-) e.g. *Penicillium citrinum*, *P. nigricans*, *Cladosporium herbarum*, *Verticillium lecanii* and several members of the genus *Aspergillus*, of which *A. flavus* has received most attention (e.g. Balasubramanya & Patel, 1980; Killham, 1986; Lang & Jagnow, 1986). Several different pathways of heterotrophic nitrification have been postulated and although it is uncertain which operates, it seems unlikely that energy is gained from the process (Killham, 1986). Thus, the benefits of nitrification are unclear. However, it has been suggested that nitrification may provide competitive advantages as certain of the products are known to be toxins, mutagens and microbial growth factors. Whilst it has been shown that the relevant fungi have the ability to nitrify following inoculation into autoclaved soil (Doxtader & Rovira, 1968), there does not appear to be any direct evidence that fungi play a significant role in nitrification in natural soils (Wainwright, 1988). Bacteria are assumed to be the major agents of nitrification (Keunen & Robertson, 1988), however, in acid coniferous forest soils the microbial biomass is often predominately fungal and fungi are believed to play a role in these situations (Killham, 1986).

Some fungi are also capable of nitrate reduction (i.e. convert NO_3^- to NH_4^+) (see also Jennings, Chapter 1; Tomsett, Chapter 2), e.g. *Aspergillus nidulans* and *Neurospora crassa* possess dissimilatory nitrate reduction pathways. Also, a number of species in the genera *Acremonium*, *Aspergillus* and *Fusarium* can perform various steps in the denitrification process (i.e. convert NO_3^- to other nitrogen oxides and dinitrogen) (Bleakley & Tiedje, 1982). These fungi cannot necessarily effect all steps of the process, for example, *Fusarium oxysporum* and *F. solani* can reduce nitrite to nitrous oxide at low oxygen tensions but are unable to reduce nitrate (Bollag & Tung, 1972). As with nitrification, fungi are only likely to make a significant contribution to denitrification in acid forest soils (Bleakley & Tiedje, 1982).

It has in the past been suggested that fungi can fix atmospheric dinitrogen, but Millbank (1969) reappraised the evidence and came to the conclusion that this was not so. More recently, work of Ginterova & Gallon (1979) indicated that the wood-rotting fungus *Pleurotus ostreatus* could reduce dinitrogen to ammonia. However, their results are open to several criticisms and the ability to fix dinitrogen remains unproven (Wainwright, 1988). Even if some species can fix dinitrogen it is likely to be quantitatively insignificant, but they may influence fixation in other ways. For example, Jenson & Holm (1975) have demonstrated that a number of soil fungi stimulate bacterial dinitrogen fixation.

Oxidation of inorganic sulphur compounds is analogous to nitrification of ammonia and nitrate, and a number of fungi are capable of oxidizing elemental sulphur and reduced sulphur compounds. The significance of these transformations to the organisms involved and to nutrient cycling remain unclear, but possibilities are considered by Wainwright (Chapter 4).

With regard to inorganic phosphorus, some fungi are able to release phosphate from insoluble inorganic phosphates, for example, species of *Aspergillus*, *Fusarium*, *Penicillium* and *Sclerotium* produce organic acids and/or chelating components which act as solubilising agents (Beever & Burns, 1980; Wainwright, 1981). Some mycorrhizal fungi are also known to have this ability (Gianinazzi-Pearson & Gianinazzi, Chapter 11). There is some evidence that some fungi, e.g. *Aspergillus niger*, *Alternaria* sp., *Chaetomium* sp. and *Penicillium notatum*, but not others, e.g. *Pythium debaryanum* and *Rhizoctonia solani*, can utilize phosphite as a source of phosphorus. The quantitative significance is again unclear but probably small, particularly as phosphites and hypophosphites are found only in small quantities in soils (Beever & Burns, 1980). Condensed phosphates are widespread in soils arising from physicochemical processes, living organisms and some fertilizers and detergents (Beever & Burns, 1980). Many fungi, e.g. *Aspergillus niger*, *Coprinus lagopus* and *Phytophthora parasitica*, are capable of utilizing pyrophosphate as sole sources of phosphorus.

Whether fungi can utilize condensed phosphates containing three or more phosphorus units is unclear (Beever & Burns, 1980).

Conclusions

Fungi are the major agents of organic matter decomposition in temperate forests and as such play a key role in releasing inorganic nutrients through mineralization. They do, however, immobilize nutrients, often for long periods, in their own biomass and the balance between immobilization and release is often very much dependent on interactions with other organisms, e.g. bacteria, fauna and plant roots. Fungi also participate in inorganic conversions and are affected by exogenous inputs. However, there is little quantitative information on the role of fungi in the dynamics of nutrient cycling. It is essential to the management of forest soil as resource for future crops that this situation is remedied.

Acknowledgment. We would like to thank John Wells for comments on a draft of this chapter.

References

Aber, J. D. & Melillo, J. M. (1980). Litter decomposition: measuring relative contributions of organic matter and nitrogen to forest soils. *Canadian Journal of Botany*, **58**, 416-421.

Abuzinadah, R. A. & Read, D. J. (1986a). The role of proteins in the nitrogen nutrition of ectomycorrhizal plants. I. Utilization of peptides and proteins by ectomycorrhizal fungi. *New Phytologist*, **103**, 481-493.

Abuzinadah, R. A. & Read, D. J. (1986b). The role of proteins in the nitrogen nutrition of ectomycorrhizal plants. III. Protein utilization by *Betula*, *Picea* and *Pinus* in mycorrhizal association with *Hebeloma crustuliniforme*. *New Phytologist*, **103**, 507-514.

Abuzinadah, R. A., Finlay, R. D. & Read, D. J. (1986). The role of proteins in the nitrogen nutrition of ectomycorrhizal plants. II. Utilization of protein by mycorrhizal plants *of Pinus contorta*. *New Phytologist*, **103**, 495-506.

Alexander, I. J. & Hardy, K. (1981). Surface phosphatase activity of Sitka spruce mycorrhizas from a serpentine soil. *Soil Biology and Biochemistry*, **13**, 301-303.

Anderson, J.M. & Ineson, P. (1984). Interactions between microorganisms and soil invertebrates in nutrient flux pathways of forest ecosystems. In *Invertebrate Microbial Interactions*, eds. J. M. Anderson, A. D. M. Rayner & D. W. H. Walton, pp. 59-88. Cambridge University Press, Cambridge, UK.

Anderson, J. M., Ineson, P. & Huish, S. A. (1983). Nitrogen and cation mobilization by soil fauna feeding on leaf litter and soil organic matter from deciduous woodlands. *Soil Biology and Biochemistry*, **15**, 463-467.

Arnolds, E. J. M. (ed.) (1985). Veranderingen in de paddestoelenflora. *Wetenschappelijke Mededeling van de Koninklijke Nederlandse Natuarhistorische Vereniging, 167,* Hoogwoud.

Arnolds, E. J. M. (1988). The changing macromycete flora in The Netherlands. *Transactions of the British Mycological Society*, **90**, 391-406.

Ausmus, B. (1977). Regulation of wood decomposition rates by arthropod and annelid populations. In *Soil Organisms as Components of Ecosystems*, Ecological Bulletin 25, eds. U. Lohm & T. Persson, pp. 180-192. Swedish Natural Science Research Council, Stockholm.

Bääth, E. & Söderström, B. (1979). Fungal biomass and fungal immobilization of plants nutrients in Swedish coniferous forest soils. *Revue d'Ecologie et de Biologie du Sol*, **16**, 477-489.

Balasubramanya, R. H. & Patel, R. B. (1980). Heterotrophic nitrification by microorganisms capable of degrading carboxin and oxycarboxin. *Indian Journal of Microbiology*, **20**, 294-297.

Becker, von A. (1982). Aussaatersuch mit Bucheckern im Gewächshaus. In *LOLF - Mitteilungen, Landesanstalt für Ökologie, Landschaftsentwicklung und Forstplanung Nordrhein - Westfalen*, pp. 37-42.

Beever, R. E. & Burns, D. J. W. (1980). Phosphorus uptake, storage and utilization by fungi. *Advances in Botanical Research*, **8**, 128-220.

Berg, B. & Lindberg, T. (1980). Is litter decomposition retarded in the presence of mycorrhizal roots in forest soil? *Swedish Coniferous Project Internal Report, 95,* Uppsala.

Berg, B. & Staff, H. (1981). Leaching, accumulation and release of nitrogen in decomposing forest litter. In *Terrestrial Nitrogen Cycles*, Ecological Bulletin 33, eds. F. E. Clark & T. Rosswall, pp. 163-178. Swedish Natural Science Research Council, Stockholm.

Bleakley, B. H. & Tiedje, J. M. (1982). Nitrous oxide production by organisms other than nitrifiers or denitrifiers. *Applied and Environmental Microbiology*, **44**, 1342-1348.

Boddy, L. (1986). Water and decomposition processes in terrestrial ecosystems. In *Water, Fungi and Plants*, eds. P. G. Ayres & L. Boddy, pp. 375-398, Cambridge University Press, Cambridge, UK.

Boddy, L. & Swift, M. J. (1983). Wood decomposition in an abandoned beech and oak coppiced woodland in S.E. England. I. Patterns of wood-litter fall. *Oikos*, **6**, 320-332.

Bollag, J. M. & Tung, G. (1972). Nitrous oxide release by soil fungi. *Soil Biology and Biochemistry*, **4**, 271-276.

Bormann, F. H. & Likens, G. E. (1979). *Pattern and Process in a Forested Ecosystem.* Springer, New York.

Bray, J. R. & Gorham, E. (1964). Litter production in forests of the world. *Advances in Ecological Research*, **2**, 101-157.

Brown, A. H. F. (1974). Nutrient cycles in oakwood ecosystems in N. W. England. In *The British Oak*, eds. M. G. Morris & F. H. Perring, pp. 141-161, BSBI, Classey, Faringdon.

Brownlee, C. & Jennings, D. H. (1982a). Long distance translocation in *Serpula lacrimans*: velocity estimates and the continuous monitoring of induced perturbation. *Transactions of the British Mycological Society*, **79**, 143-148.

Brownlee, C. & Jennings, D. H. (1982b). The pathway of translocation in *Serpula lacrimans. Transactions of the British Mycological Society*, **79**, 401-407.

Cairney, J. W. G. (1986). Basidiomycete linear mycelial structures as nutrient absorbing and translocating organs in soil. Ph.D. thesis, University of Liverpool.

Chakly, M. & Berthelin, J. (1982). Rôle d'une ectomycorhize *Pisolithus tinctorius - Pinus caribea* et d'une bactérie rhizosphérique sur la mobilisation du phosphore de phosphates minéraux et organiques insolubles. In *Les Mycorhizes: Biologie et Utilisation*, eds. S. Gianinazzi, V. Gianinazzi-Pearson & A. Trouvelot, pp. 215-220. INRA, Dijon.

Clark, F. E. & Rosswall, T. (eds.) (1981). *Terrestrial Nitrogen Cycles*, Ecological Bulletin 33. Swedish Natural Science Research Council, Stockholm.

Clipson, N. J. W., Cairney, J. W. G. & Jennings, D. H. (1986). The physiology of basidiomycete linear organs. I. Phosphate uptake by cords and mycelium in the laboratory and the field. *New Phytologist*, **104**, 444-458.

Cole, D. W. & Rapp, M. (1981). Elemental cycling in forest ecosystems. In *Dynamic Properties of Forest Ecosystems*, ed. D. E. Reichle, pp. 341-409. Cambridge University Press, Cambridge, UK.

Cowling, E. B. (1970). Nitrogen in forest trees and its role in wood deterioration. *Acta Universitatius Upsalliensis*, **164**.

Craker, L. E. & Manning, W. J. (1974). SO_2 uptake by soil fungi. *Environmental Pollution*, **6**, 309-311.

Dighton, J. (1983). Phosphatase production by mycorrhizal fungi. *Plant and Soil*, **71**, 455-462.

Dighton, J. (1987). Ecology and management of mycorrhizas in the U.K. In *Mycorrhizae in the Next Decade: Practical Applications and Research Priorities*, Proceedings of the 7th North American Conference on Mycorrhizae, eds. D. M. Sylvia, L. L. Hung & J. H. Graham, pp. 75-77. Institute of Food & Agricultural Science, University of Florida, Gainesville, Florida, USA.

Dighton, J. & Mason, P. A. (1985). Mycorrhizal dynamics during forest tree development. In *Developmental Biology of Higher Fungi*, eds. D. Moore, L. A. Casselton, D. A. Wood, & J. C. Frankland, pp. 117-139. Cambridge University Press, Cambridge, UK.

Dighton, J. & Skeffington, R. A. (1987). Effects of artificial acid precipitation on the mycorrhizas of Scots pine seedlings. *New Phytologist*, **107**, 191-202.

Dighton, J., Thomas, E. D. & Latter, P. M. (1987). Interactions between tree roots, mycorrhizas, a saprotrophic fungus and the decomposition of organic substrates in a microcosm. *Biology and Fertility of Soils*, **4**, 145-150.

Dowding, P. (1981). Nutrient uptake and allocation, during substrate exploitation by fungi. In *The Fungal Community, Its Organization and Role in the Ecosystem*, eds. D. T. Wicklow & G. C. Carroll, pp. 621-635. Marcel Dekker, New York & Basel.

Dowson, C. G., Boddy, L. & Rayner, A. D. M. (1989). Development and extension of mycelial cords in soil at different temperatures and moisture contents. *Mycological Research*, in press.

Dowson, C. G., Rayner, A. D. M. & Boddy, L. (1986). Outgrowth patterns of mycelial cord-forming basidiomycetes from and between woody resource units in soil. *Journal of General Microbiology*, **121**, 203-211.

Dowson, C. G., Rayner, A. D. M. & Boddy, L. (1988). Outgrowth patterns of mycelial cord-forming basidiomycetes into woodland soils. II. Resource capture and persistence. *New Phytologist*, **109**, 343-349.

Doxtader, K. G. & Rovira, A. D. (1986). Nitrification by *Aspergillus flavus* in sterilized soil. *Australian Journal of Soil Research*, **6**, 141-147.

Duvigneaud, P. & Denaeyer-de-Smet, S. (1970). Biological cycling of minerals in temperate deciduous forests. In *Analysis of Temperate Forest Ecosystems*, ed. D. E. Reichle, pp. 199-225. Springer-Verlag, New York.

Edwards, C. A., Reichle, D. E. & Crossley, D. A. (1970). The role of soil invertebrates in turnover of organic matter and nutrients. In *Ecological Studies. Analysis and Synthesis*, Vol. 1., ed. D. E. Reichle, pp. 147-172. Springer-Verlag, Berlin.

Fogel, R. & Hunt, G. (1979). Fungal and arboreal biomass in a western Oregon Douglas fir ecosystem: distribution patterns and turnover. *Canadian Journal of Forest Research*, **9**, 245-256.

Fogel, R. & Hunt, G. (1983). Contribution of mycorrhizae and soil fungi to nutrient cycling in a Douglas fir ecosystem. *Canadian Journal of Forest Research*, **13**, 219-232.

Frankland, J. C. (1982). Biomass and nutrient cycling by decomposer basidiomycetes, In *Decomposer Basidiomycetes: their Biology and Ecology*, eds. J. C. Frankland, J. N. Hedger & M. J. Swift, pp. 241-261. Cambridge University Press, Cambridge, UK.

Gadgil, R. L. & Gadgil, P. D. (1971). Mycorrhiza and litter decomposition. *Nature*, **233**, 133.

Gadgil, R. L. & Gadgil, P. D. (1975). Suppression of litter decomposition by mycorrhizal roots of *Pinus radiata*. *New Zealand Journal of Forest Science*, **5**, 33-41.

Giltrap, N. J. (1982). Production of polyphenol oxidases by ectomycorrhizal fungi with special reference to *Lactarius* spp. *Transactions of the British Mycological Society*, **78**, 75-81.

Ginterova, A. & Gallon, J. (1979). *Pleurotus ostreatus*: a nitrogen fixing fungus? *Biochemical Society Transactions*, **7**, 1293-1295.

Gosz, J. R., Likens, G. E. & Bormann, F. H. (1973). Nutrient release from decomposing leaf and branch litter in the Hubbard Brook Forest, New Hampshire. *Ecological Monographs*, **43**, 173-191.

Granlund, H. I., Jennings, D. H. & Thompson, W. (1985). Translocation of solutes along rhizomorphs of *Armillaria mellea*. *Transactions of the British Mycological Society*, **84**, 111-119.

Gransaull, R. E. & Brown, R. T. (1987). Relationships among mycorrhizae, nitrogen fixing actinomycetes and host plants. In *Mycorrhizae in the Next Decade, Practical Application and Research Priorities*, Proceedings of the 7th North American Conference on Mycorrhizae, eds. D. M. Sylvia, L. L. Hung & J. H. Graham, p. 21. Institute of Food & Agricultural Science, University of Florida, Gainesville, Florida, USA.

Hall, D. (1978). Experiments with NPK fertilizers in relation to the growth of toadstools in beechwoods. 1. Preliminary study. *Vasculum*, **63**, 25-41.

Hanlon, R. D. G. & Anderson, J. M. (1980). Influence of macroarthropod feeding activities on microflora in decomposing oak leaves. *Soil Biology and Biochemistry*, **12**, 255-261.

Harmon, M. E., Franklin, J. F., Swanson, F. J., Sollins, P., Gregory, S. V., Lattin, J. D., Anderson, N. H., Cline, S. P., Aumen, N. G., Seddell, J. R., Lienkaemper, G. W., Cromack Jr, K. & Cummins, K. W. (1986). Ecology of course woody debris in temperate ecosystems. *Advances in Ecological Research*, **15**, 133-302.

Heal, O. W. (1979). Decomposition and nutrient release in even-aged plantations, In *The Ecology of Even-Aged Plantations*, eds. E. D. Ford, D. C. Malcolm & J. Atterson, pp. 257-291. Institute of Terrestrial Ecology, Cambridge, UK.

Heal, O. W. & Dighton, J. (1986). Nutrient cycling and decomposition in natural terrestrial ecosystems. In *Microfloral and Faunal Interactions in Natural and Agro-ecosystems*, eds. M. J. Mitchell & J. P. Nakas, pp. 14-73. Martinus Nijhoff/Dr. W. Junk, Dordrecht.

Heal, O. W., Swift, M. J. & Anderson, J. M. (1981). Nitrogen cycling in United Kingdom forests: the relevance of basic ecological research. *Philosophical Transactions of the Royal Society of London, B*, **276**, 427-444.

Ho, I. & Zak, B. (1979). Acid phosphatase activity of six ectomycorrhizal fungi. *Canadian Journal of Botany*, **37**, 1203-1205.

Hütterman, A. (1982). Frühdiagnose von Immisionsschaden in Würzelbereich von Waldbäumen. In *LOLF - Mitteilungen, Landesanstalt für Ökologie, Landschaftsentwicklung und Forstplanung Nordrhein - Westfalen*, pp. 26-31.

Ineson, P., Leonard, M. A. & Anderson, J. M. (1982). Effect of collembolan grazing on nitrogen and cation leaching from decomposing leaf litter. *Soil Biology and Biochemistry*, **14**, 601-605.

Janos, D. P. (1983). Tropical mycorrhizas, nutrient cycles and plant growth. In *Tropical Rain Forest: Ecology and Management*, eds. S. L. Sutton, T. C. Whitmore & A. C. Chadwick, pp. 327-45. Blackwell, Oxford.

Jansen, A. E., Dighton, J. & Bresser, A. H. M. (eds.) (1988). *Ectomycorrhiza and Acid Rain*. CEC Air Pollution Research Report, **12**. Bilthoven, The Netherlands.

Jenson, V. & Holm, E. (1975). Associative growth of nitrogen fixing bacteria with other micro-organisms. In *Nitrogen Fixation by Free-living Micro-organisms*, ed. W. D. P. Stewart, pp. 101-119. Cambridge University Press, Cambridge, UK.

Keunen, J. G. & Robertson, L. A. (1988). Ecology of nitrification and denitrification. In *The Nitrogen and Sulphur Cycles*, eds. J. A. Cole & S. J. Ferguson, pp. 161-218. Cambridge University Press, Cambridge, UK.

Killham, K. (1986). Heterotrophic nitrification. In *Nitrification*, ed. J. I. Prosser, pp. 117-126. IRL Press,

Kumpfer, W. & Heyser, W. (1986). Effects of stemflow on the mycorrhiza of beech (*Fagus sylvaticus* L.). In *Mycorrhizae: Physiology and Genetics*, eds. V. Gianinazzi-Pearson & S. Gianinazzi, pp. 743-750. INRA, Paris.

Levi, M. P., Merrill, W. & Cowling, E. B. (1968). Role of nitrogen in wood deterioration. VI. Mycelial fractions and model nitrogen compounds as substrates for growth of *Polyporus versicolor* and other wood-destroying and wood inhabiting fungi. *Phytopathology*, **58**, 626-634.

Laheurte, F. & Berthelin, J. (1986). Interactions between endomycorrhizas and phosphate solubilizing bacteria: effects on nutrition and growth of maize. In *Mycorrhizae: Physiology and Genetics*, eds. V. Gianinazzi-Pearson & S. Gianinazzi, pp. 339-343. INRA, Paris.

Lang, E. & Jagnow, G. (1986). Fungi of a forest soil nitrifying at low pH values. *FEMS Microbiology Ecology*, **38**, 257-265.

Lindeberg, G. (1981). Role of litter-decomposing and ectomycorrhizal fungi in nitrogen cycling in the Scandinavian coniferous forest ecosystem. In *The Fungal Community: its Organisation and Role in the Ecosystem*, eds. D. T. Wicklow & G. C. Carroll, pp. 653-664. Marcel Dekker, New York.

Linkins, A. E. & Antibus, R. K. (1981). Mycorrhizae of *Salix rotundifolia* in coastal arctic tundra. In *Arctic and Alpine Mycology*, eds. G. A. Laursen & J. F. Ammirati, pp. 509-531. University of Washington Press, Washington.

Melillo, J. M., Aber, J. D. & Muratore, J. F. (1982). Nitrogen and lignin control of hardwood leaf litter decomposition dynamics. *Ecology*, **63**, 621-626.

Merrill, W. & Cowling, E. B. (1966). The role of nitrogen in wood deterioration: amount and distribution of nitrogen in fungi. *Phytopathology*, **56**, 1085-1090.

Millbank, J. W. (1969). Nitrogen fixation in moulds and yeasts - a reappraisal. *Archiv für Mikrobiologie*, **68**, 32-39.

Miller, H. G. (1979). The nutrient budgets of even-aged forests, In *The Ecology of Even-Aged Forest Plantations*, eds. E. D. Ford, D. C. Malcolm & J. Atterson, pp. 221-256. Institute of Terrestrial Ecology, Cambridge.

Miller, H. G. (1981). Forest fertilization: some guiding concepts. *Forestry*, **54**, 157-167.

Miller, H. G., Cooper, J. M., Miller, J. D. & Pauline, O. J. C. (1979). Nutrient cycles in pine and their adaptation to poor soils. *Canadian Journal of Forest Research*, **9**, 19-26.

Morgan, C. R. & Mitchell, M. J. (1987). The effects of feeding by *Oniscus asellus* on leaf litter sulfur constituents. *Biology and Fertility of Soils*, **3**, 107-111.

Newell, K. (1984a). Interaction between two decomposer basidiomycetes and a collembolan under Sitka spruce: distribution, abundance and selective grazing. *Soil Biology and Biochemistry*, **16**, 227-233.

Newell, K. (1984b). Interactions between two decomposer basidiomycetes and a collembolan under Sitka spruce: grazing and its potential effects on fungal distribution and litter decomposition. *Soil Biology and Biochemistry*, **16**, 235-239.

Oelbe-Farivar, M. (1985). Physiologische Reaktionen von Mykorrhizapilzen auf simulierte saure Bodenkbedingungen. Ph.D. Thesis, University of Göttingen.

Ovington, J. D. (1962). Quantitative ecology and the woodland ecosystem concept. *Advances in Ecological Research*, **1**, 103-192.

Park, D. (1976). Carbon and nitrogen levels as factors influencing fungal decomposers, In *The Role of Terrestrial and Aquatic Organisms in Decomposition Processes*, eds. J. M. Anderson & A. Macfadyen, pp. 41-59, Blackwell Scientific Publications, Oxford.

Parkinson, D., Visser, S. & Whittaker, J. B. (1979). Effects of collembolan grazing on fungal colonization of leaf litter. *Soil Biology and Biochemistry*, **11**, 529-555.

Paustian, K. & Schnürer, J. (1987). Fungal growth response to carbon to and nitrogen limitation: a theoretical model. *Soil Biology and Biochemistry*, **19**, 613-620.

Rayner, A. D. M. & Boddy, L. (1988). *Fungal Decomposition of Wood*. John Wiley, Chichester.

Read, D. J. (1986). Non-nutritional effects of mycorrhizal infection. In *Mycorrhizae: Physiology and Genetics*, eds. V. Gianinazzi-Pearson & S. Gianinazzi, pp. 189-176. INRA, Paris.

Reich, P. B., Schoettle, A. W., Stroo, H. F., Troiano, J. G Amundson, R. G. (1985). Effects of O_3, SO_2 and acidic rain on mycorrhizal infection in northern red oak seedlings. *Canadian Journal of Botany*, **63**, 2049-2055.

Reichle, D. E. (1977). The role of invertebrates in nutrient cycling. In *Soil Dynamics as Components of Ecosystems*, Ecological Bulletin 25, eds. U. Lohm & T. Persson, pp. 145-154. Swedish Natural Science Research Council, Stockholm.

Seastedt, T. R. & Crossley, D. A. Jnr. (1980). Effects of microarthropods on seasonal dynamics of nutrients in forest litter. *Soil Biology and Biochemistry*, **12**, 237-342.

Stark, N. (1972). Nutrient cycling pathways and litter fungi. *Bioscience*, **22**, 355-360.

Stroo, H. F. & Alexander, M. (1985). Effect of simulated acid rain on mycorrhizal infection of *Pinus strobus* L. *Water, Air, Soil Pollution*, **25**, 107-114.

Swank, W. T. & Waide, J. B. (1980). Interpretation of nutrient cycling research in a management context: evaluating potential effects of alternative management strategies on site productivity. In *Forests: Fresh Perspectives from Ecosystem Analysis*, ed. R. Wareing, pp. 137-158. Oregon State University, Corvallis.

Swift, M.J. (1977). The role of fungi and animals in the immobilisation and release of nutrient elements from decomposing branch-wood, In *Soil Organisms as Components of Ecosystems*, Ecological Bulletin 25, eds. U. Lohm & T. Persson, pp. 193-202. Swedish Natural Science Research Council, Stockholm.

Swift, M. J. & Boddy, L. (1984). Animal-microbial interactions during wood decomposition. In *Invertebrate-Microbial Interactions*, eds. J. M. Anderson, A. D. M. Rayner & D. W. H. Walton, pp. 89-131. Cambridge University Press, Cambridge, UK.

Swift, M. J., Heal, O. W. & Anderson, J. M. (1979). *Decomposition in Terrestrial Ecosystems*. Blackwell Scientific, Oxford.

Thompson, G. W. & Medve, R. V. (1984). Effects of aluminium and manganese on the growth of ectomycorrhizal fungi. *Applied Environmental Microbiology*, **48**, 556-560.

Thompson, W. & Boddy, L. (1983). Decomposition of suppressed oak trees in even-aged plantations. II. Colonisation of tree roots by cord and rhizomorph producing basidiomycetes. *New Phytologist*, **93**, 277-291.

Thompson, W. & Rayner, A. D. M. (1982). Structure and development of mycelial cord systems of *Phanerochaete laevis* in soil. *Transactions of the British Mycological Society*, **78**, 193-200.

Ullrich, B., Mayer, R. & Khanna, P. H. (1979). Deposition von Luftverunreinigungen und ihre Auswirkungen in Waldöksystemen im Solling. *Schriften aus der Forstlichen Fakultät der Üniversität Göttingen*, **58**, J. D. Sauerlands Verlag, Frankfurt.

Visser, S., Whittaker, J. B. & Parkinson, D. (1981). Effects of collembolan grazing on nutrient release and respiration of a leaf litter inhabiting fungus. *Soil Biology and Biochemistry*, **13**, 215-218.

Vogt, K. A., Grier, C. C. & Vogt, D. J. (1986). Production, turnover, and nutrient dynamics of above- and belowground detritus of world forests. *Advances in Ecological Research*, **15**, 303-377.

Vogt, K. A., Moore, E. E., Vogt, D. J., Redlin, M. J. & Edmunds, R. L. (1983). Conifer fine root and mycorrhizal root biomass within the forest floors of Douglas fir stands of different ages and site productivities. *Canadian Journal of Forest Research*, **13**, 429-437.

Wainwright, M. (1988). Metabolic diversity of fungi in relation to growth and mineral cycling in soil - a review. *Transactions of the British Mycological Society*, **90**, 159-170.

Watkinson, S. C. (1971). Phosphorus translocation in stranded and unstranded mycelium of *Serpula lacrimans*. *Transactions of the British Mycological Society*, **57**, 535-539.

Went, F. W. & Stark, N., (1968). The biological and mechanical role of soil fungi. *Proceedings of the National Academy of Sciences, U.S.A.*, **60**, 497-504.

Whitford, W. G., Meentemeyer, V., Seastedt, T. R., Cromack, K. Jr., Crossley, D. A. Jr., Santos, P., Todd, R. L. & Waide, J. B. (1981). Exceptions to the AET model: deserts and clear-cut forest. *Ecology*, **62**, 275-277.

Whittaker, R. H., Likens, G. E., Bormann, F. H., Eaton, J. S. & Siccama, T. G. (1979). The Hubbard Brooke ecosystem study: forest nutrient cycling and element behaviour. *Ecology*, **60**, 203-220.

Index

A

E

F

N

O

R

S

X

Z